WENNER-GREN INTERNATIONAL SERIES
VOLUME 63

Light and Biological Rhythms in Man

Wenner-Gren International Series

Vol. 60 **Neuro-immunology of fever** *ed.* T. Bartfai *and* D. Ottoson
Vol. 61 **Functional organisation of the human visual cortex** *ed.* B. Gulyas, D. Ottoson *and* P. E. Roland

New in 1993

Light and biological rhythms in man *ed.* L. Wetterberg
Trophic regulation of the basal ganglia *ed.* K. Fuxe *et al.*

Light and Biological Rhythms in Man

Edited by

L. WETTERBERG
Karolinska Institute, Stockholm, Sweden

PERGAMON PRESS
OXFORD · NEW YORK · SEOUL · TOKYO

UK Pergamon Press Ltd, Headington Hill Hall,
Oxford OX3 0BW, England

USA Pergamon Press Inc., 660 White Plains Road,
Tarrytown, New York 10591-5163, USA

KOREA Pergamon Press Korea, KPO Box 315, Seoul 110-603, Korea

JAPAN Pergamon Press Japan, Tsunashima Building Annex,
3-20-12 Yushima, Bunkyo-ku, Tokyo 113, Japan

First edition 1993

Library of Congress Cataloging in Publication Data
New data to be supplied

ISBN 0-08-0422799

Printed in Great Britain by
Butler & Tanner Ltd, Frome and London

Preface

THE OBJECTIVE of this book is to summarize the knowledge of light as a regulator of biological rhythms in man in relation to health and disease.

The first scientific meeting on biological rhythms in Sweden was held in 1937 in the small spa of Ronneby in the south of Sweden. About twenty scientists who called themselves "Rhythms enthusiasts" gathered and established a Society for Biological Rhythm Research. The second meeting on biological rhythms in Sweden was held in 1955 in Stockholm, still with a small group of scientists.

The presentations published in this book update what has happened during the last four decades and focus in particular on how light is influencing biological rhythms. In 1959 Aaron Lerner and coworkers discovered the time keeping substance melatonin. In one chapter in this book Lerner recounts the history of his efforts for four years to find the melanin tonizing hormone (melatonin) which blanched frog skin, until they found the substance and it could be said "This is it!". From there on many speculations about the mediator of biological rhythm regulation have become scientifically testable hypotheses. The results based on Lerner's work have led to solid data on which much of modern biological rhythm research is based.

That Aaron Lerner's historical aspects are included in this volume, as well as all the other presentations will we hope benefit younger generations of students and scientists for whom the research frontiers of today have a strong bearing on the clinical practice of tomorrow and well beyond the year 2000.

In different ways research is confronted with the fact that many of the fundamental rhythm processes are not easily observable and must many times be reconstructed from preliminary data. An interesting model to explain the origin of biological rhythms is the one of Erik Odeblad from University of Umeå, Sweden, who has proposed that it is based on the presence of collective vibrations (phonons) in loop-shaped molecules or molecular aggregates. Vibrational waves along a closed loop will give rise to self-interference as the basis for rhythms. Further experiments to measure phonons in tissues and organs are called for to test this hypothesis.

As illustrated in this book, interfacing disciplines progress steadily through experimentation, observation and theoretical interpretation, each one helping the advances of the other. The biological rhythms covered in this book range from short (infradian) and about 24-hour rhythms (circadian) to longer rhythms (ultradian) present in all mammalian species.

During the last decade it has become clear that humans also are in many ways influenced by biological rhythms. Time structures are present on several levels from population, individuals, organs (e.g. brain and heart) to tissues, cells and subcellular organization. An introductory chapter identifies the genetic machinery, including the light influenced proteins, which governs the clocks that regulate us all.

The new findings about the effect of light on biological rhythms has already lead to therapeutic trials in different conditions. The use of light to regulate menstrual cyclicity may have clinical value for rhythm contraceptive methods and for treatment of infertility. The further studies of the effect of light on the menstrual cycle may lead to new views of reproductive endocrinology. In some chapters, specialists in light perception and light reception have presented data on light treatment as an antidepressant in seasonal affective disorder. It is clear that clinical testing is needed to determine the optimal intensities, the timing and duration of light treatment for producing beneficial effects.

In the report of placebo-controlled studies it is stated that it will require further studies to establish how light treatment compares with placebo. Such work is also important to help understand the mechanisms and etiology of depression and other conditions.

To understand the rhythm regulating effects of light it is necessary to examine the ocular mechanisms which mediate the photic effect through the eye. When light reaches the retina, there are individual differences in the sensitivity of the photoreceptivity and the ability of the melatonin rhythm generating system to integrate photic stimuli, temporally.

The overall aim in the present volume is to provide a basis, both for scientific as well as clinical perspectives to establish the specific modes and measures which mediate therapeutic and physiologically beneficial effects of light as a regulator of biological rhythms.

I am truly indebted to all of the contributors to this volume and very grateful for the support and help of David Ottoson and Johan Beck-Friis. The generous financial contribution of the Wenner-Gren Center Foundation and the Swedish Medical Research Council is acknowledged.

Stockholm, 1993

LENNART WETTERBERG

Contents

Section II

Anatomy and Physiology of Light Regulated Endogenous Oscillators

Section III

Light Influence on Circadian Rhythms

Special Lecture

1

Biological Rhythms: From Gene Expression to Behavior*

JOSEPH S. TAKAHASHI

NSF Center for Biological Timing, Department of Neurobiology and Physiology, Northwestern University, Evanston, IL 60208–3520, USA

Abstract

Circadian rhythms regulate the function of living systems at virtually every level of organization. In the last decade, our understanding of the cellular and molecular processes involved in the generation and regulation of circadian rhythms has advanced considerably. New model systems for studying circadian oscillators have been developed, a potential regulatory role for cellular immediate-early genes in circadian behavior has been discovered, and critical periods of macromolecular synthesis for progression of the circadian clock through its cycle have been defined. These findings are of particular interest because independent approaches suggest that an important role for macromolecular synthesis exists at all levels of the circadian system.

Introduction

Circadian rhythms regulate the behavior, physiology and biochemistry of most living systems: from cyanobacterium to man.[3,15,38,50] Among animals, much is known about the physiology of circadian rhythms, and in all cases, circadian control is exerted by structures that contain circadian pacemakers within the central nervous system.[4,24,29,51,55] On the basis of a wide variety of evidence, the mechanism of the circadian clock appears to be cell autonomous and to involve periodic gene expression.[51,52,54] In addition, signal transduction pathways into the clock mechanism are present for conveying environmental information for entrainment of the clock. Furthermore, the clock mechanism has diverse output pathways for exerting circadian control at all levels of organismal biology. A number of generalizations can be made about circadian biology. Circadian rhythms are a property of all eukaryotic and some prokaryotic organisms and are entrained primarily by environmental cycles of light and temperature. Circadian rhythms are genetically determined and single-gene clock

* This paper is dedicated to Dr Aaron B. Lerner for his seminal work on melatonin.

mutants have been found in six organisms including mammals. Circadian oscillations are remarkably precise and can have a variation in cycle length of less than one part in a thousand. The period length of the oscillation is temperature-compensated and usually varies less than 20% for each 10°C change in temperature (Q_{10}=0.8–1.2). Finally, the biochemica! machinery responsible for generating circadian rhythms can be expressed at the level of individual cells. Thus, circadian rhythmicity is a fundamental organizing feature of virtually all organisms. The properties of these rhythms are unique and widely conserved among living systems.

This chapter will be organized in two parts. First, a brief review of vertebrate circadian organization will be presented. Second, a summary of our own work on photic entrainment and regulation of immediate early genes in mammals will follow.

Components of circadian systems

All circadian systems contain at least three elements: (1) an input pathway or set of input pathways that convey environmental information to the circadian pacemaker for entrainment; (2) a circadian pacemaker that generates the oscillation; and (3) an output pathway or set of output pathways by which the pacemaker regulates its various output rhythms.[58] In all systems, a photic entrainment pathway is present as an input. It is already clear from the diversity of photopigments and phototransduction pathways that photic entrainment pathways differ markedly in different organisms.[38] At the receptor level, the diversity spans the spectrum from phytochrome in plants to members of the rhodopsin family in animals. There appears to be comparable diversity at the second messenger level in photic signal transduction pathways. At the output level, diversity is even more extreme. Again, this type of regulation spans the clock control of photosynthesis in plants to endocrine and behavioral rhythms in animals. As I have argued previously,[51] the input and output pathways of the circadian clock within each organism appear to be specific to each system. Despite these differences in the coupling pathways (inputs and outputs) of the circadian clock, the core mechanism of the pacemaker appears to be fundamentally similar in all organisms. Whether these similarities in pacemaker mechanisms will ultimately be found to be functionally analogous or phylogenetically homologous remains to be seen.

Physiological organization of vertebrate circadian systems

Circadian pacemakers, which control the behavior and physiology of animals, have been localized as discrete structures within the nervous system.[45,48,55] At the physiological level, both "oscillator" and "pacemaker" function have been defined. A "circadian oscillator" is a structure that expresses a self-sustained oscillation under constant conditions (the absolute minimum number of cycles is two). Circadian oscillators have been localized by isolating the tissue in question *in vitro* and then demonstrating the persistence of circadian oscillations in the isolated tissue under constant conditions. In vertebrates, three diencephalic structures have been shown to contain circadian oscillators, the pineal gland of birds, reptiles and fish,[55] the hypothalamic suprachiasmatic nucleus (SCN) of mammals,[24,29] and the retinas of amphibians and birds.[4,37] A "circadian pacemaker" is a circadian oscillator that has been shown to drive and therefore control some overt rhythmic process such as locomotor behavior. Pacemaker function has been experimentally demonstrated by transplanting the structure in question and then showing that the phase or period of the recipient rhythm is regulated by the transplant. This result has been demonstrated in vertebrates for the pineal gland of sparrows and the SCN of hamsters, and in invertebrates for the optic lobes of cockroaches and the brain of *Drosophila*.[43,45]

Although an ocular-SCN-pineal axis underlies vertebrate circadian organization, it is difficult to make simple generalizations at the physiological level. The clearest division is a mammalian versus non-mammalian dichotomy. In mammals, the organization is the simplest. The dominant circadian pacemaker is located in the SCN, the photoreceptors for entrainment are exclusively retinal, and output pathways such as the pineal rhythm of melatonin are well defined.[24] In non-mammalian vertebrates, the organization is more complex;[35,45] and, multiple oscillators and photoreceptors are involved. The pineal gland, the SCN and the eyes can each play a dominant role in circadian behavior of birds depending on the species. For example, in passerine birds, such as the house sparrow, the pineal gland plays a dominant pacemaker role; whereas, in gallinaceous birds, such as chickens and quail, the pineal gland is not essential. The avian SCN appears to be necessary in all species examined, but its role as a pacemaker has not been determined. Finally, in Japanese quail, the eyes play a dominant role in circadian organization.[60] In addition to the multiplicity of oscillatory centers, multiple photoreceptors for entrainment are located in the retina, the pineal gland and the brain.[61] In summary, non-mammalian vertebrates appear to have distributed circadian oscillators with local photoreceptive input; whereas, mammals appear to have retained only a subset of these components.

The mammalian suprachiasmatic nucleus

A wealth of experimental evidence strongly argues that the hypothalamic SCN is the site of a circadian pacemaker that drives overt rhythms in mammals.[24,29,45,48] Six lines of evidence support this conclusion. First, the SCN receives direct input from the retina through a specialized visual pathway, the retinohypothalamic tract, which is required for entrainment to light cycles. Second, a wide variety of circadian rhythms in rodents are disrupted by SCN lesions, ruling out the possibility of a highly specific, limited effect of lesions on a particular subsystem, for instance, locomotor behavior. Third, the SCN expresses circadian rhythms of multi-unit electrical activity that persist after neural isolation *in vivo*. Fourth, SCN explants continue to express circadian rhythms of single-unit electrical activity, vasopressin release and metabolic activity *in vitro*. Fifth, circadian rhythmicity can be restored to arrhythmic SCN-lesioned animals by transplantation of fetal tissue containing SCN cells. Sixth, transplantation of SCN tissue derived from *Tau* mutant hamsters, which express short period rhythms, demonstrates that the genotype of the donor SCN determines the period of the restored rhythm. Taken together, these results demonstrate that the SCN plays a dominant role in the generation of circadian rhythms in mammals.

Functional properties of photic entrainment in mammals

Photic entrainment in mammals is mediated by retinal photoreceptors that project to the SCN via the retinohypothalamic tract.[29] We have used light-induced phase shifts of the circadian rhythm of wheel-running activity to measure the photic sensitivity of the circadian system of the golden hamster. Previously we showed that the spectral sensitivity function for phase-shifting matches an opsin-based photopigment with a peak (λ_{max}) around 500 nm.[53] Although the peak sensitivity is similar to that of rhodopsin, two features of this photoreceptive system are unusual: the threshold of the response is high, especially for a rod-dominated retina like that of the hamster, and the reciprocal relationship between intensity and duration holds for extremely long durations.[33] Figure 1 shows the phase-shifting response to 300 sec light pulses at circadian time (CT) 19. The phase-shifting response increases with light intensity and can be fit with a four parameter logistic function. The sensitivity to stimulus *duration* was assessed by measuring the magnitude of phase-shift responses to photic stimuli of different irradiance and duration (Figure 2). The hamster circadian system is more sensitive to the irradiance of longer duration stimuli than to that of briefer stimuli. The system is maximally sensitive to the irradiance of stimuli of 300 sec and longer in duration (Figure 3A). As shown previously the threshold for photic stimulation of the hamster circadian pacemaker is high. The threshold irradiance (the maximum irradiance necessary to induce statistically significant responses) is approximately 10^{11} photons cm^{-2} sec^{-1} for optimal stimulus durations. This threshold is equivalent to a luminance at the cornea of about 0.1 $cd \cdot m^{-2}$. We also measured the sensitivity of this visual pathway to the

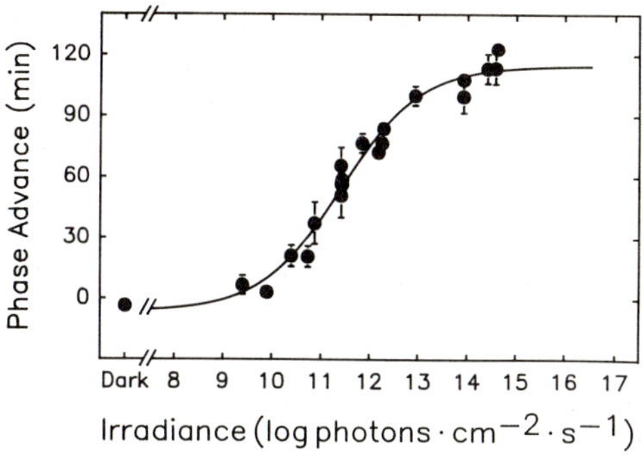

FIG. 1. Magnitude of phase shift of hamster activity rhythm in response to 300 sec 503 nm light pulses of different irradiance at circadian time 19 (CT19). The points represent the mean ± SEM. The continuous line is a modified Naka-Rushton function fitted to the data. (From Nelson and Takahashi,[33] copyright by *J. Physiol.* (*Lond.*).)

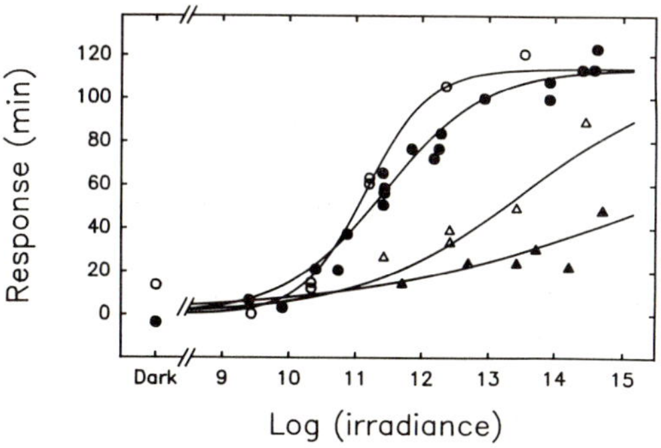

FIG. 2. Magnitude of phase shift of hamster activity rhythm in response to 503 nm light pulses of varying durations and irradiances at CT19. Open circles are 3,600 sec; closed circles are 300 sec; open triangles are 30 sec; and closed triangles are 3 sec stimuli. Curves are best fit modified Naka-Rushton functions. (From Nelson and Takahashi,[33] copyright by *J. Physiol.* (*Lond.*).)

total number of photons in a stimulus (Figure 3B). Surprisingly, the system is maximally sensitive to photic stimuli between 30 and 3,600 sec in duration. The maximum quantum efficiency of photic integration occurs in 300 sec stimuli.

To summarize, the photic threshold for entrainment, even under optimal conditions, is relatively insensitive and corresponds to thresholds reported for cone photoreceptors. In addition, the system can integrate light over very long durations. From a functional perspective, these characteristics are clearly adaptive. The threshold is just above the level of full moonlight and thus entrainment would not be disrupted by this inappropriate light source. Furthermore, the optimum for long duration pulses would also render the system insensitive to very intense but brief

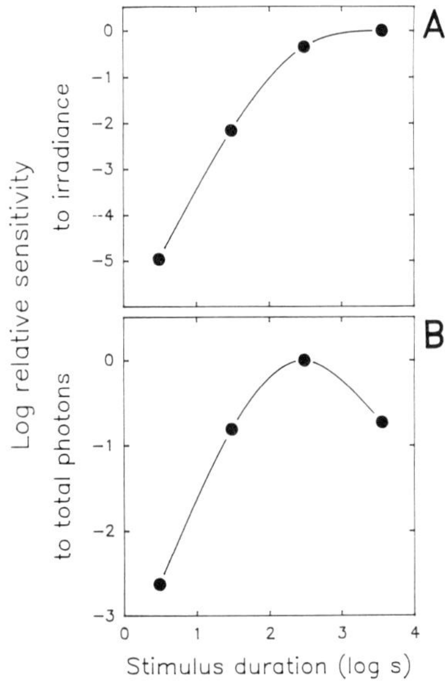

FIG. 3. Sensitivity curves to irradiance and total number of photons in a pulse measured at CT19. A, the relative sensitivity to irradiance is plotted as a function of stimulus duration. B, the relative sensitivity to total photons as a function of stimulus duration. (From Nelson and Takahashi,[33] copyright by *J. Physiol.* (*Lond.*).)

sources of light such as lightening. Thus, both the threshold and integration characteristics of the photic entrainment pathway in hamsters make this system optimally tuned to respond to the daily light–dark cycle, while at the same time remaining unresponsive to environmental "noise" that would interfere with stable entrainment. The nature of the photoreceptors mediating photic entrainment remain to be established; however, it is already clear that specialized retinal elements must exist. For example, experiments using the *rd* mutation in mice strongly argue that rod photoreceptors are not essential for the phase-shifting response.[10] In addition, recent experiments from our laboratory show that the spectral sensitivity for phase-shifting in another species, the Djungarian hamster, has a peak around 475 nm which is clearly different from that of rhodopsin.[62] Taken together these results suggest that the "circadian photoreceptor" is not a rod and is either a cone or some other unidentified photoreceptor class in the retina.

Role of macromolecular synthesis in mammalian circadian rhythms

It has been known for a number of years that inhibitors of protein synthesis on 80S ribosomes produce changes in the period length or phase of circadian rhythms of microorganisms and invertebrates, including *Euglena, Acetabularia, Gonyaulax, Neurospora* and *Aplysia*.[17,18,22,32,46,58,63] More recently, we have demonstrated that the acute administration of two protein synthesis inhibitors (anisomycin and cycloheximide), with two different mechanisms of action, are capable of inducing pronounced phase shifts in the circadian clock of hamsters, and that the site of action of these inhibitors appears to be on cells within the SCN region.[16,56,64] Interestingly, the phase response curves for anisomycin and cycloheximide in hamsters are similar to those measured for protein synthesis inhibitors in microorganisms and invertebrates, suggesting that the biochemical mechanisms generating circadian oscillations in mammals may share common features with those found in very distantly related phylogenetic groups.

In addition to causing phase shifts by themselves, protein synthesis inhibitors have also been shown to block the phase-shifting effects of light pulses in *Aplysia* and *Neurospora*.[20,42] We have also found that light-induced phase shifts can be blocked in hamsters suggesting that the signal transduction pathways for photic entrainment may involve macromolecular synthesis (Takahashi *et al.*, unpublished results). Indeed, as described below, light does induce the expression of a set of genes in the SCN.

Although a requirement for new RNA synthesis has been inferred from genetic experiments,[6,12,44] this issue has only recently been addressed with the use of the reversible RNA synthesis inhibitor, 5,6-dichloro-1-β-D-ribofuranosylbenzimidazole (DRB), in the *Aplysia* eye.[41] Pulses of DRB caused phase-dependent delays at phases between circadian time (CT) 20 and 10, and had no effects from CT10 to CT20. Continuous treatment with DRB caused a dose-dependent lengthening of the circadian period. Similar effects of DRB on phase and period have been found in chick pineal cells.[36] The DRB experiments suggest that a critical period for transcription of specific genes involved in the generation of circadian rhythms occurs from CT20 to CT10 in *Aplysia*, and from CT18 to CT19 in chick pineal cells.

While the inhibitor pulse experiments suggest that a critical period for macromolecular synthesis exists within the circadian cycle, these experiments do not provide evidence that such treatments can arrest or stop the motion of the circadian pacemaker. Khalsa *et al.*[23] have addressed this issue by applying very long (5–44 hours) inhibitor pulses of cycloheximide to *Bulla* eyes *in vitro*. Shorter pulses that do not extend past CTO have minimal effects on the rhythm. However, longer pulses that extend beyond CTO delay the phase of the subsequent rhythm precisely by the duration that the inhibitor pulse extends past CTO. The results are consistent with the interpretation that the motion of the circadian pacemaker is arrested during these long inhibitor pulses. By extrapolating the phases of the rhythms “backwards” towards the end of the inhibitor pulses, it appears that the critical period for protein synthesis in *Bulla* begins in the late subjective night (near CTO). By analogy to work in the cell cycle, the "START" of the circadian cycle therefore appears to be near CTO. Thus, both the inhibitor phase shift and the circadian cycle progression experiments are concordant and suggest that the circadian clock requires macromolecular synthesis during a critical period for progression through the cycle. In addition to inhibitor experiments, there is now direct evidence that the *Drosophila* period gene products express circadian oscillations which appear critical for the expression of behavioral rhythms.[13,14]

Photic regulation of immediate-early genes in the SCN

A number of groups have reported that the products of the proto-oncogene, c-*fos*, are stimulated by light in the SCN of rodents.[25] In the hamster there is a profound induction of Fos immunoreactivity[47] and c-*fos* mRNA[26] in the SCN following light exposure during the "subjective" night. To begin to address whether Fos mediates transcriptional events involved in phase shifting of the circadian pacemaker, we have further characterized the photic induction of c-*fos* mRNA utilizing *in situ* hybridization techniques. By using defined light stimuli for which the effects on the hamster's circadian rhythm of activity have been measured, we have compared the photic induction of c-*fos* mRNA and the behavioral phase shifts produced by light[26]. Specifically, we examined whether the photic induction of c-*fos* was quantitatively correlated with phase shifting of locomotor activity. Because Fos must act in concert with a member of the Jun family of proto-oncogenes to form the transcription factor, AP-1[30], we have also examined whether Jun and AP-1 are induced by light.[27]

*Photic induction of c-*fos* and* jun-B *mRNA in the SCN*

Light exposure of hamsters during the night, at CT19, causes a dramatic induction of c-*fos* mRNA levels in the SCN (Figure 4). The increase in c-*fos* mRNA levels illustrated here was induced by a 5 minute light pulse, after which the animal was returned to darkness for 25 minutes before sacrifice. The dark-field photomicrograph shows that hybridization to complementary RNA probes for c-*fos* is localized mainly in the ventrolateral portion of each SCN, and extends dorsally from the SCN toward the periventricular region. Similar results are seen with *jun-B* mRNA levels.[27] In contrast to the dramatic enhancement of *jun-B* mRNA expression, light exposure at CT19 causes only a modest increase of c-*jun* mRNA hybridization in the SCN. No other effect of light on c-*fos* or *jun-B* mRNA was observed in

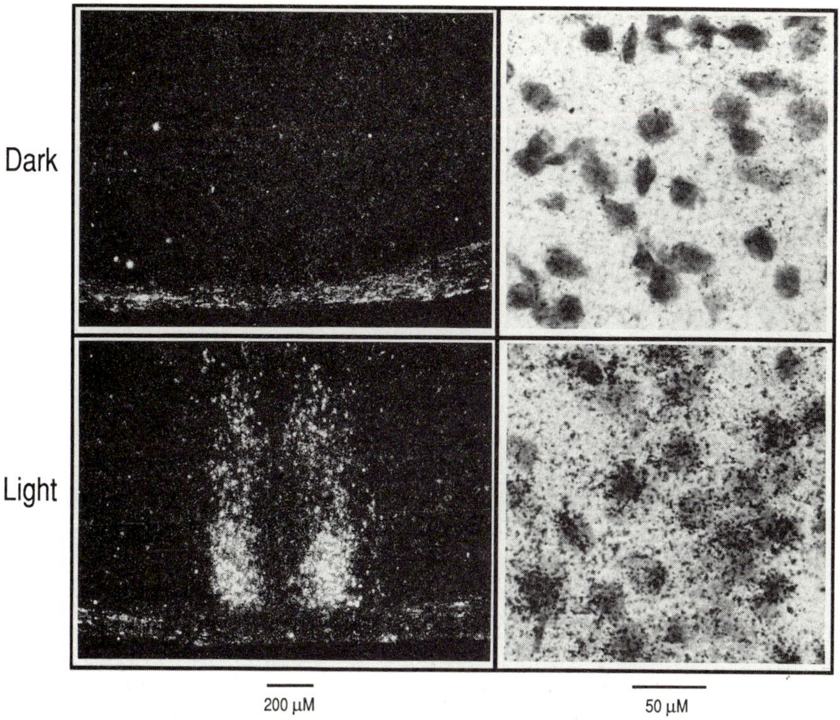

FIG. 4. *In situ* hybridization of c-*fos* mRNA in the SCN of hamsters 30 minutes after the onset of a 300 sec light pulse at CT19. Top panels show the SCN region from a dark control animal. Left side is a dark field photomicrograph of emulsion autoradiography. Right side is a higher magnification bright field photomicrograph. Lower panels show the SCN region from a hamster receiving light at CT19. (From Kornhauser *et al.*,[26] copyright by *Cell Press*.)

other areas of the brain, including retinorecipient regions such as the intergeniculate leaflet (IGL) (although we have detected a modest photic induction in the IGL with Fos immunocytochemistry). The elevation of c-*fos* mRNA levels following light stimulation is rapid and transient, with maximal levels occurring about 30 minutes after the onset of light.[26] Similar to c-*fos* mRNA, the highest levels of *jun-B* mRNA are detected 30 minutes after the onset of the light pulse, and have returned nearly to basal levels by 120 minutes.[27] Clearly, this temporal correlation of the expression of c-*fos* and of *jun-B*, suggests that Fos and *jun-B* proteins are available to dimerize with one another to form the transcription factor, AP-1 (see below).

Photic threshold of c-fos mRNA induction

As described above, the response of the hamster circadian system to varying levels of illumination has been characterized. The magnitude of the phase shift in locomotor activity exhibits a monotonic, saturable dependence on the irradiance of the light stimulus. Figure 5 illustrates the amount of behavioral phase shift produced by 5 minute light pulses (503 nm wavelength) of varying irradiance occurring at CT19.[26] To compare the sensitivity to light of the c-*fos* mRNA induction and of this photic response of the circadian system, we performed *in situ* hybridization following 5 minute light stimuli of different illumination levels, and quantified the amount of specific c-*fos* hybridization within the SCN. The threshold for induction of c-*fos* is indistinguishable from the threshold for light-induced phase shifts, consistent with the idea c-*fos* is involved in the entrainment mechanism of the pacemaker.[26]

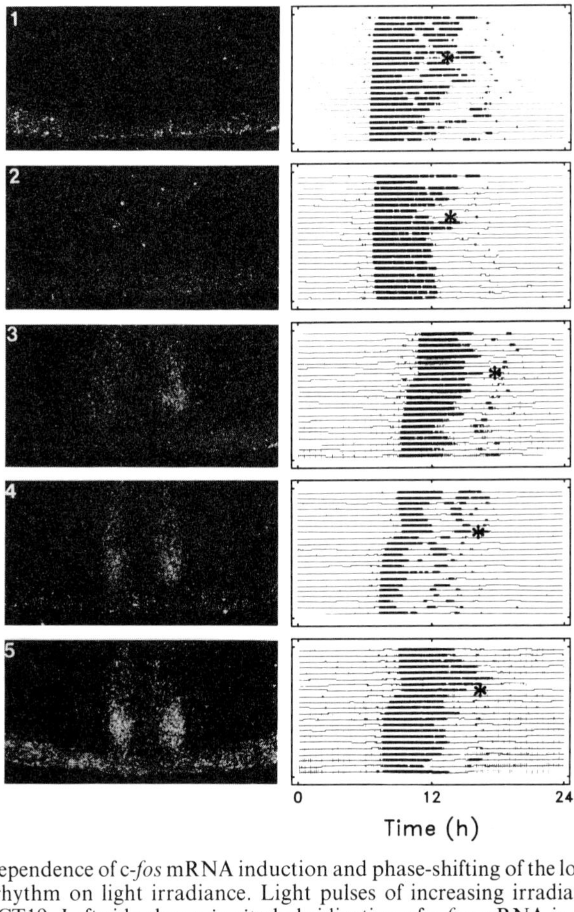

FIG. 5. Dependence of c-*fos* mRNA induction and phase-shifting of the locomotor activity rhythm on light irradiance. Light pulses of increasing irradiance were given at CT19. Left side shows *in situ* hybridization of c-*fos* mRNA in the SCN region. Panel 1 shows no light (dark control); panels 2–5 show increasing light irradiance. Right side shows the locomotor activity rhythms of hamsters exposed to light pulses at CT19 at the same irradiance. (From Kornhauser *et al.*,[26] copyright by *Cell Press*.)

*Circadian gating of c-*fos *and* jun-B *induction by light*

Light exposure during the subjective night of the hamster's circadian cycle causes a phase shift in the activity rhythm of the animal. Light exposure during the subjective day, by contrast, produces no effect on the phase of the rhythm of locomotor activity (Figure 6A). The photic induction of c-*fos* gene expression displays a similar phase-dependence: the ability of

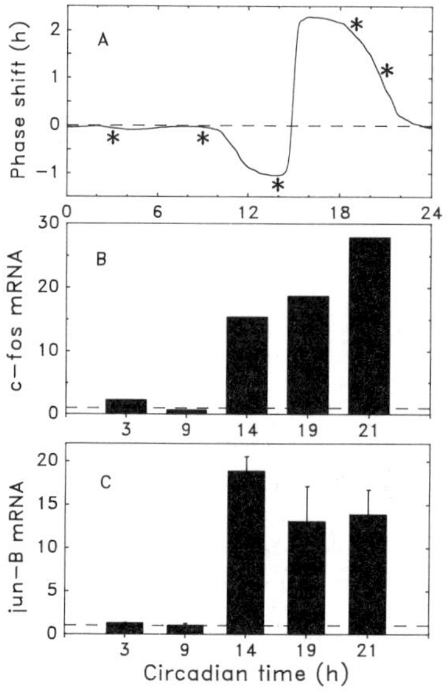

FIG. 6. Circadian phase-dependence of light-induced behavioral phase shifts (A), c-*fos* mRNA induction in the SCN (B), and *jun-B* mRNA induction in the SCN of hamsters. Photic induction of c-*fos* and *jun-B* were measured at five phases of the cycle. Photic induction was seen only during the subjective night and was correlated with light-induced phase shifts of the activity rhythm. (From Kornhauser *et al.*,[27] copyright by *Science*.)

light to stimulate c-*fos* is restricted to the same times of the circadian cycle when light causes a behavioral phase shift (Figure 6B). *In situ* hybridization studies demonstrate that the induction of *jun-B* mRNA following a 5 minute light pulse are likewise dependent on circadian phase (Figure 6C). Light pulses at CT14, which produce a phase-delay in the hamster's activity rhythm, dramatically elevate *jun-B* mRNA levels; light at CT19 and at CT21 causes a phase-advance and also induces *jun-B* mRNA. During the subjective day, at CT3 and CT9, light neither produces phase shifts nor induces the expression of *jun-B*. The expression of c-*fos* and *jun-B* mRNA, then, are both gated by the circadian pacemaker, and are both induced under the same temporal conditions.[27]

Photic regulation of AP-1 DNA binding activity in the SCN

The co-induction of *jun-B* and c-*fos* mRNA by light suggests that levels of the heterodimeric AP-1 complex in the SCN are increased by light stimulation. To determine whether light indeed alters amounts of AP-1 factor capable of specifically binding to a consensus DNA recognition site, we performed DNA-binding gel mobility shift assays.[27] Whole-cell extracts were prepared from microdissections of hamster brain tissue containing the SCN. These SCN extracts were incubated with a ^{32}P-labeled oligonucleotide containing consensus AP-l sites and then electrophoresed. A protein complex of retarded mobility was detected using extract prepared from hamsters 60 or 120 minutes after the onset of a 5 minute light pulse, presented at CT19 SCN extracts from hamsters receiving the same handling, but no light pulse, contained low levels of AP-1 binding activity measured by this *in vitro* assay. The time course of AP-1 induction was elevated at 1 hour and peaked at 2 hours after light stimulation. As in the case of c-*fos* and *jun-B* mRNA, the induction of AP-1 by light was also phase-dependent. Light induced AP-1 at CT14 and CT19, but did not induce AP-1 at CT6. Thus, the circadian clock also gates the photic induction of AP-1 activity. These results demonstrate that light increases levels of AP-1 protein complex implying that photic stimulation is coupled to AP-1 regulated changes in gene transcription in the SCN. Taken together, these experiments suggest that photic entrainment in mammals may involve transcriptionally regulated signal transduction processes, and that AP-1 may play a role in conveying photic information within the circadian system.

Light induces phosphorylation of the transcription factor, CREB, in the SCN

In cell culture systems, the signaling pathways that regulate the expression of c-*fos* have been extensively studied.[49] In PC12 cells, the cAMP response element binding protein, CREB, mediates the transcriptional activation of c-*fos* in response to stimuli that elevate the second messenger cAMP or Ca^{2+} signaling pathways. These pathways converge at the level of CREB phosphorylation on Ser^{133} and this phosphorylation event is necessary for the transcriptional activation of CREB-regulated target genes.[49] Because Ca^{2+} is a likely second messenger in neurotransmitter regulated pathways such as the retinohypothalamic input to the SCN, we asked whether CREB may play a role in the photic regulation of c-*fos*. In a very fruitful collaboration with David Ginty, Michael Greenberg and colleagues,[11] we have examined the phosphorylation of CREB using a novel antibody that specifically recognizes the Ser^{133} site of CREB. Light exposure of hamsters at CT19 caused the rapid phosphorylation of CREB in the SCN without any detectable changes in the amount of "total CREB". DNA mobility shift assays for CREB using the Ca^{2+}/cAMP response element (Ca-CRE, sequence TGACGTTT) of the c-*fos* promoter as a probe, suggests that the major Ca-CRE binding factor in the SCN is CREB rather than another member of the CREB family such as ATF-1 or CREM.[11]

To determine whether the light-induced phosphorylation of CREB is phase dependent, we examined hamsters during the subjective daytime at CT6.[11] At this time, light fails to cause both behavioral phase shifts and immediate-early gene induction.[27] Light also failed to induce CREB phosphorylation in the SCN at CT6. The absence of CREB phosphorylation was not due to a decrease in CREB because no changes in the level of total CREB could be detected in the SCN at different phases of the circadian cycle. Thus, light-induced CREB phosphorylation and c-*fos* induction are both gated by the circadian clock. These experiments strongly suggest that the circadian gating mechanism regulating immediate-early genes acts upstream of the phosphorylation of CREB. It will be of great interest to examine kinases responsibility for CREB phosphorylation, as well as other regulatory sites in the c-*fos* promoter such as the serum response element and the factors that interact with it.

Unresolved issues concerning Fos/AP-1

Thus far the relationship between behavioral phase shifting and Fos/AP-1 induction is correlative. Are these two responses causally related, with Fos/AP-1 acting as a step in the photic entrainment pathway? Or, alternatively, are these two responses coincidental? At the present time it has not been possible to obtain a definitive answer to this issue. The current state is as follows. The photic induction of Fos is anatomically specific and restricted to the SCN and the IGL. The photic induction of Fos in the SCN is tightly correlated with the behavioral phase-shifting effects of light under a number of conditions. These include phase dependence and photic threshold. The Fos response is modality specific and is associated with photically induced but not non-photically induced phase shifts.[19,28] This indicates that Fos is related to the photic input rather than to a phase-shifting mechanism *per se*. The requirement for Fos induction has tested by pharmacological blockade by NMDA antagonists,[1,8,61] by non-NMDA antagonists[2,5] and by mecamylamine.[67] In all three cases, agents that blocked behavioral phase shifting to light also attenuated Fos induction. These experiments are all consistent with the hypothesis that Fos is a component of the photic entrainment pathway in rodents. However, experiments that selectively remove Fos by other means such as antisense oligonucleotides or gene knock out[21] have not yet been completed. Clearly, experiments that directly and specifically test the requirement for Fos/AP-1 in photic entrainment are important goals for future work.

Conclusions

Although we are far from a concrete understanding of the clock mechanism in mammals, substantial progress on a number of fronts using diverse approaches suggests the following five generalizations.

1. Vertebrate circadian systems

There is a substantial diversity in the organization of circadian systems at the physiological level, but only a few discrete loci within the nervous systems of animals act as circadian pacemakers. The mammalian SCN, the avian pineal gland and the amphibian retina define three central elements of the circadian system in vertebrates in which oscillator and pacemaker functions have been demonstrated.[4,24,29,45,55,59]

2. A cellular circadian clock

In spite of the diversity at the physiological level, the circadian clock appears to be a cellular entity even in multicellular organisms. The strongest case for this is seen in chick pineal cells in which a circadian oscillator system with a photic entrainment pathway can be studied in cell culture.[55] Photoreception, clock function and melatonin biosynthesis all appear to be cellular properties of a single cell type, the pinealocyte. Recently the SCN and the retina have also been shown to be capable of expressing circadian rhythms in cell culture.[31,37]

3. Coupling pathways

The coupling pathways (inputs and outputs) of the circadian clock within the organism are specific to each system, and these specific differences in coupling can explain many apparent differences in "clock mechanism" reported in various systems. Among vertebrates, the differences between the coupling pathways in the mammalian SCN and the chick pineal gland are self-evident. Entrainment pathways of SCN cells are neurally coupled,[29] whereas those in chick pineal cells are not.[55] These differences in input pathways can explain, for example, why neurally coupled circadian oscillators such as the SCN are phase-shifted by neurotransmitters and second messengers,[39,40] while non-neurally coupled circadian oscillators such as the chick pineal gland are not.[34,65]

4. A unified clock mechanism?

The "core mechanism" of the circadian clock, when stripped of its inputs and outputs, will be fundamentally the same in all organisms. This assertion is based more on faith than on fact, but as in the case of the cell cycle, I believe a unified mechanism is likely. At the molecular genetic level, this argument is still tenuous because only two clock genes in *Drosophila* and *Neurospora* have been well characterized.[6,7,12,44] Interspecific homologies in clock genes have been found, however.[6,12] In addition, an argument can be made at the pharmacological level because inhibitors of macromolecular synthesis have strikingly similar effects in all organisms studied.[54]

5. Gene expression

The mechanism of the circadian clock requires gene expression on an ongoing (daily) basis for the expression of periodic clock proteins that are essential for the generation of circadian oscillations. There is now direct evidence for this idea in *Drosophila* with the *per* gene.[9,13,14,66] The convergence upon this theme from independent approaches in *Drosophila*, molluscan eyes and vertebrates suggests that this idea has some merit.[52] Clearly, the isolation and molecular genetic analysis of circadian clock genes in vertebrate systems are critical goals for the future.

Acknowledgments

I thank Drs. Lennart Wetterberg, David Ottoson and Johan Beck-Friis for organizing an outstanding symposium. I also thank Dr. Wetterberg for his patience in the preparation of this manuscript. Research was supported by grants from the NIMH, NEI and NSF to J.S.T.

References

1. Abe H., Rusak B. and Robertson H.A. (1991) Photic induction of Fos protein in the suprachiasmatic nucleus is inhibited by the NMDA receptor antagonist MK-801. *Neurosci. Lett.* **127**, 9–12.
2. Abe H., Rusak B. and Robertson H.A. (1992) NMDA and non-NMDA receptor antagonists inhibit photic induction of Fos protein in the hamster suprachiasmatic. *J. Neuroendocrinol.* **28**, 831–835.
3. Aschoff J. (1981) *Handbook of Behavioral Neurobiology, Biological Rhythms*, 4 pp. Plenum Press, New York.
4. Cahill G.M., Grace M.S. and Besharse J.C. (1991) Rhythmic regulation of retinal melatonin: Metabolic pathways, neurochemical mechanisms, and the ocular circadian clock. *Cell. Mol. Neurobiol.* **11**, 529–560.
5. Colwell C.S. and Menaker M. (1992) NMDA as well as non-NMDA receptor Antagonists can prevent the phase-shiftingeffects of light on the circadian system. *J. Biol. Rhythms* **7**, 125–136.
6. Dunlap J.C. (1990) Closely watched clocks: Molecular analysis of circadian rhythms in *Neurospora* and *Drosophila*. *Trends Genet.* **6**, 159–165.

7. Dunlap J.C. (1993) Genetic analysis of circadian clocks. *Annu. Rev. Physiol.* **55**, 683–729.
8. Ebling F.J.P., Maywood E.S., Staley K., Humby T., Hancock D.C., Waters C.M., Evan G.I. and Hastings M.H. (1991) The role of N-methyl-D-aspartate-type glutamatergic neurotransmission in the photic induction of immediate-early gene expression in the suprachiasmatic nuclei of the Syrian hamster. *J. Neuroendocrinol.* **3**, 641–652.
9. Ewer J., Hamblin-Coyne M., Rosbash M. and Hall J.C. (1990) Requirement for period gene expression in the adult and not during development for locomotor activity rhythms of imaginal *Drosophila melanogaster*. *J. Neurogenet.* **7**, 31–73.
10. Foster R.G., Provencio I., Hudson D., Fiske S., De Grip W. and Menaker M. (1991) Circadian photoreception in the retinally degenerate mouse (*rd/rd*). *J. Comp. Physiol. A.* **169**, 39–50.
11. Ginty D.D., Kornhauser J.M., Thompson M.A., Bading H., Mayo K.E., Takahashi J.S. and Greenberg M.E. (1993) Regulation of CREB phosphorylation in the suprachiasmatic nucleus by light and a circadian clock. *Science* **260**, 238–241.
12. Hall J.C. (1990) Genetics of circadian rhythms. *Annu. Rev. Genet.* **24**, 659–697.
13. Hardin P.E., Hall J.C. and Rosbash M. (1990) Feedback of the *Drosophila period* gene product on circadian cycling of its messenger RNA levels. *Nature* **343**, 536–540.
14. Hardin P.E., Hall J.C. and Rosbash M. (1992) Circadian oscillations in period gene mRNA levels are transcriptionally regulated. *Proc. Natl. Acad. Sci. USA* **89**, 11711–11715.
15. Hastings J.W., Rusak B. and Boulos Z. (1991) Circadian rhythms: The physiology of biological timing. In *Neural and Integrative Animal Physiology* (ed. Prosser C.L.), pp. 435–546. Wiley-Liss, Inc., New York.
16. Inouye S.T., Takahashi J.S., Wollnik F. and Turek F.W. (1988) Inhibitor of protein synthesis phase shifts a circadian pacemaker in mammalian SCN. *Am. J. Physiol.* **255**, R1055–R1058.
17. Jacklet J.W. (1977) Neuronal circadian rhythms: phase shifting by a protein synthesis inhibitor. *Science* **198**, 69–71.
18. Jacklet J.W. (1980) Protein synthesis requirement of aplysia circasdian clock tested by active and inactive derivative of the inhibitor anisomycin. *J. Exp. Biol.* **85**, 33–42.
19. Janik D. and Mrosovsky N. (1992) Gene expression in the geniculate induced by a nonphotic circadian phase stimulus. *Neuroreport* **3**, 575–578.
20. Johnson C.H. and Nakashima H. (1990) Cycloheximide inhibits light-induced phase shifting of the circadian clock in *Neurospora*. *J. Biol. Rhythms* **5**, 159–167.
21. Johnson R.S., Spiegelman B.M. and Papaioannou V. (1992) Pleiotropic effects of a null mutation in the c-*fos* proto-oncogene. *Cell* **71**, 577–586.
22. Karakashian M.W. and Schweiger H.G. (1986) Evidence for a cycloheximide-sensitive component in the biological clock of Acetabularia. *Am. J. Physiol.* **98**, 303–312.
23. Khalsa S.B.S., Whitmore D. and Block G.D. (1992) Stopping the circadian pacemaker with inhibitors of protein synthesis. *Proc. Natl. Acad. Sci. USA* **89**, 10862–10866.
24. Klein D.C., Moore R.Y. and Reppert S.M. (1991) *Suprachiasmatic Nucleus: The Mind's Clock*. pp. Oxford University Press, New York.
25. Kornhauser J.M., Mayo K.E. and Takahashi J.S. (1992) Immediate-early gene expression in a mammalian circadian pacemaker: The suprachiasmatic nucleus. In *Molecular Genetics of Biological Rhythms* (ed. Young M.W.), 2, pp. 271–307. Marcel Dekker, New York.
26. Kornhauser J.M., Nelson D.E., Mayo K.E. and Takahashi J.S. (1990) Photic and circadian regulation of c-*fos* gene expression in the hamster suprachiasmatic nucleus. *Neuron* **248**, 127–134.
27. Kornhauser J.M., Nelson D.E., Mayo K.E. and Takahashi J.S. (1992) Regulation of jun-B messenger RNA and AP-1 activity by light and a circadian clock. *Science* **255**, 1581–1584.
28. Mead S., Ebling F.J.P., Maywood E.S., Humby T., Herbert J. and Hastings M.H. (1992) A nonphotic stimulus causes instantaneous phase advances of the light-entrainable circadian oscillator of the Syrian hamster but does not induce the expression of c-*fos* in the suprachiasmatic nuclei. *J. Neurosci.* **12**, 2516–2522.

29. Meijer J.H. and Rietveld W.J. (1989) Neurophysiology of the suprachiasmatic circadian pacemaker in rodents. *Physiol. Rev.* **69**, 671–707.
30. Morgan J.L. and Curran T. (1991) Stimulus-transcription coupling in the nervous system: involvement of the inducible proto-oncogenes *fos* and *jun*. *Annu. Rev. Neurosci.* **14**, 421–451.
31. Murakami M., Takamure M., Takahashi K., Utunomiya K., Kuroda H. and Etoh T. (1991) Long-term cultured neurons from rat suprachiasmatic nucleus retain the capacity for circadian oscillation of vasopressin release. *Brain Res.* **545**, 347–350.
32. Nakashima H., Perlman J. and Feldman J.F. (1981) Cycloheximide-induced phase shifting of circadian clock of neurospora. *Am. J. Physiol.* **241**, R31–R35.
33. Nelson D.E. and Takahashi J.S. (1991) Sensitivity and integration in a visual pathway for circadian entrainment in the hamster, *Mesocricetus auratus*. *J. Physiol.* (*Lond*) **439**, 115–145.
34. Nikaido S.S. and Takahashi J.S. (1989) Twenty-four hour oscillation of cAMP in chick pineal cells: Role of cAMP in the acute and circadian regulation of melatonin production. *Neuron* **3**, 609–619.
35. Norgen R.B. (1990) Neural basis of avian circadian rhythms. *Bird Behav.* **8**, 57–66.
36. Ohi K. and Takahashi J.S. (1991) Effects of the transcription inhibitor , 5, 5-dichloro-1-b-D-ribofuranosylbenzimidazole (DRB), on the circadian melatonin rhythm of cultured chick pineal cells. *Soc. Neurosci. Abst.* **17**, 675.
37. Pierce M.E., Sheshberadaran H., Zhang Z., Fox L.E., Applebury M.L. and Takahashi J.S. (1993) Circadian regulation of iodopsin gene expression in embryonic photoreceptors in retinal cell culture. *Neuron* **10**, 579–584.
38. Pittendrigh C.S. (1981) Circadian systems: General Perspective. In *Handbook of Behavioral Neurobiology, Biological Rhythms* (ed. Aschoff J.), Vol. 4, pp. 57–80. Plenum Press, New York.
39. Prosser R.A. and Gillette M.U. (1989) The mammalian circadian clock in the suprachiasmatic nuclei is reset *in vitro* by cAMP. *J. Neurosci.* **9**, 1073–1081.
40. Prosser R.A., Miller J.D. and Heller H.C. (1990) A serotonin agonist phase-shifts the circadian clock in the suprachiasmatic nuclei *in vitro*. *Brain Res.* **534**, 336–339.
41. Raju U., Koumenis C., Nunez-Regueiro M. and Eskin A. (1991) Alteration of the phase and period of a circadian oscillator by a reversible transcription inhibitor. *Science* **253**, 673–675.
42. Raju U., Yeung S.J. and Eskin A. (1990) Involvement of proteins in light resetting ocular circadian oscillators of *Aplysia*. *Am. J. Physiol.* **258**, R256–R262.
43. Ralph M.R., Foster R.G., Davis F.C. and Menaker M. (1990) Transplanted suprachiasmatic nucleus determines circadian period. *Science* **247**, 975–978.
44. Rosbash M. and Hall J.C. (1989) The molecular biology of circadian rhythms. *Neuron* **3**, 387–398.
45. Rosenwasser A.M. (1988) Behavioral neurobiology of circadian pacemakers: A comparative perspective. *Prog. Psychobiol. Physiol. Psychol.* **13**, 155–226.
46. Rothman B.S. and Strumwasser F. (1979) Phase shifting the circadian rhythm of neural activity in the isolated Aplysia eye with puromycin and cycloheximide: Electrophysiological and biochemical studies. *J. Gen. Physiol.* **68**, 359–384.
47. Rusak B., Robertson H.A., Wisden W. and Hunt S.P. (1990) Light pulses that shift rhythms induce gene expression in the suprachiasmatic nucleus. *Science* **248**, 1237–1240.
48. Rusak B. and Zucker I. (1979) Neural regulation of circadian rhythms. *Physiol. Rev.* **59**, 449–526.
49. Sheng M. and Greenberg M.E. (1990) The regulation and function of c-*fos* and other immediate early genes in the nervous system. *Neuron* **4**, 477–485.
50. Sweeney B.M. and Borgese M.B. (1989) A circadian rhythm in cell division in a prokaryote, the cyanobacterium *Synechocccus* WH7803. *J. Phycol.* **25**, 183–186.
51. Takahashi J.S. (1991) Circadian rhythms: From gene expression to behavior. *Cur. Opin. Neurobiol.* **1**, 556–561.
52. Takahashi J.S. (1993) Circadian clock regulation of gene expression. *Cur. Opin. Genet. Devel.* **3**, 301–309.
53. Takahashi J.S., DeCoursey P.J., Bauman L. and Menaker M. (1984) Spectral

sensitivity of a novel photoreceptive system mediating entrainment of mammalian circadian rhythms. *Nature* **308**, 186–188.

54. Takahashi J.S., Kornhauser J.M., Koumenis C. and Eskin A. (1993) Molecular approaches to understanding circadian oscillations. *Annu. Rev. Physiol.* **55**, 729–753.
55. Takahashi J.S., Murakami N., Nikaido S.S., Pratt B.L. and Robertson L.M. (1989) The avian pineal, a vertebrate model system of the circadian oscillator: Cellular regulation of circadian rhythms by light, second messengers, and macromolecular synthesis. *Rec. Prog. Hormone Res.* **45**, 279–352.
56. Takahashi J.S. and Turek F.W. (1987) Anisomycin, an inhibitor of protein synthesis, perturbs the phase of a mammalian circadian pacemaker. *Brain Res.* **405**, 199–203.
57. Takahashi J.S. and Zatz M. (1982) Regulation of circadian rhythmicity. *Science* **217**, 1104–1111.
58. Taylor W.R., Dunlap J.C. and Hastings J.W. (1982) Inhibitors of protein synthesis on 80s ribosome sphase shift the gonyalaux clock. *J. Exp. Biol.* **97**, 121–136.
59. Underwood H., Barrett R.K. and Siopes T. (1990) The quail's eye: A biological clock. *J. Biol. Rhythms* **5**, 257–265.
60. Underwood H. and Groos G.A. (1982) Vertebrate circadian rhythms: retinal and extraretinal photoreception. *Experientia* **38**, 1113–1121.
61. Vindlacheruvu R.R., Ebling F.J.P., Maywood E.S. and Hastings M.H. (1992) Blockade of glutamatergic neurotransmission in the suprachiasmatic nucleus prevents cellular and behavioral responses of the circadian system to light. *Eur. J. Neurosci.* **4**, 673–679.
62. Vitaterna M.H., Milette J.J., Turek F.W. and Takahashi J.S. (1993) Spectral sensitivity of a photoreceptive system regulating reproductive responses to light in a mammal. (Submitted).
63. Walz B. and Sweeney B.M. (1979) Kinetics of the cycloheximide-induced phase changes in the biological clock in Gonyalaux. *Prox. Natl. Acad. Sci. USA* **76**, 6443–6447.
64. Wollnik F., Turek F.W., Majewski P. and Takahashi J.S. (1989) Phase shifting the circadian clock with cycloheximide: Response of hamsters with an intact or split rhythm of locomotor activity. *Brain Res* **496**, 82–88.
65. Zatz M. and Mullen D.A. (1988) Norepinephrine, acting via adenylate cyclase, inhibits melatonin but does not phase-shift the pacemaker in cultured chick pineal cells. *Brain Res.* **450**, 137–143.
66. Zerr D.M., Hall J.C., Rosbash M. and Siwicki K.K. (1990) Circadian fluctuations of period protein immunoreactivity in the CNS and the visual system of *Drosophila. J. Neurosci.* **10**, 2749–2762.
67. Zhang Y., Zee P.C., Kirby J.D., Takahashi J.S. and Turek F.W. (1993) A cholinergic antagonist, mecamylamine, blocks light-induced Fos immunoreactivity in specific regions of the hamster suprachiasmatic nucleus. *Brain Res.* **615**, 107–112.

SECTION I

Light and Light Perception

2

Light—Definitions and Measurements

ROGER WIBOM

NIOH, National Institute of Occupational and Health, 17184 Solna, Sweden

THE SUBJECT in this presentation is the modern technology involved with light. Much of our information about the external world is gained through the visual sense. Hence adequate lighting is of major importance in everyday life.

We also know that light has non-visual effects. I think that most of you visiting this conference are even more interested in these effects.

The first part of this presentation begins with a short survey of some of the physical aspects of light and definitions of quantitative measures. The second part concerns measurement of light and a few demonstrations.

Light

Light is a form of energy that can pass from one material body to another without the need for any material substance in the intervening space. Such energy transfer has come to be called radiation, a term which implies that the energy flows out in straight lines in all directions from the source, although in fact straight-line flow does not always occur, particularly when material substance is traversed.

Some forms of radiation are best described as particles, for example those which are emitted by radioactive materials, and light was at one time thought to consist of a shower of particles.

Experiments later showed that the behaviour of light rays could be better described in terms of waves, the ray directions being the direction in which the waves are travelling. About one hundred years ago it became clear that light waves are electromagnetic in character and occupy only a very small part of a huge range of wavelengths that constitute the electromagnetic spectrum (Figure 1).

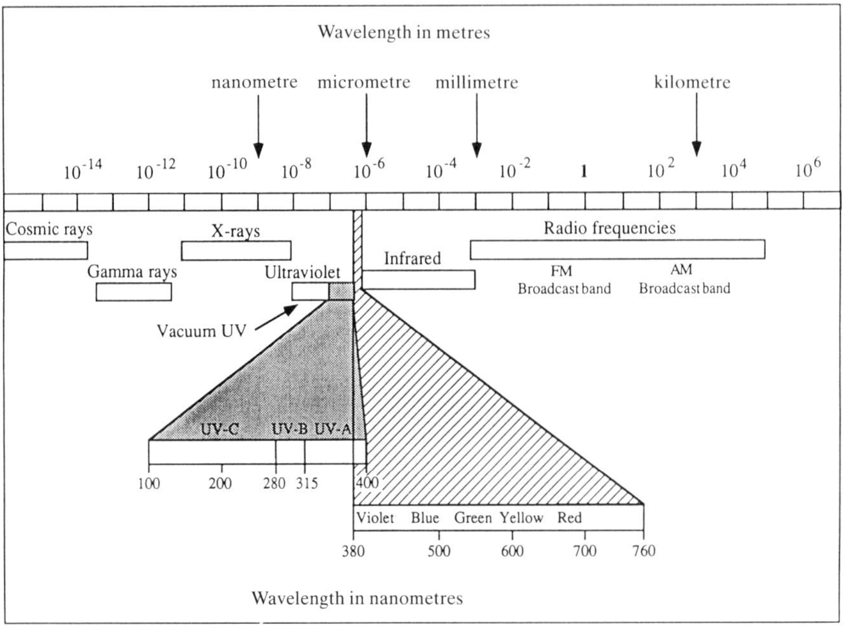

FIG. 1. The radiant energy (electromagnetic) spectrum.

At the long-wave end of this spectrum there are electromagnetic waves used for radio communications, with wavelengths ranging from tens of kilometres down to a few millimetres. At the other end of the spectrum there are X-rays and gamma-rays.

The wavelength of the electromagnetic spectrum ranges from 10^{-15} to 10^4 metre.

The visible spectrum

The visible portion of the spectrum covers the wavelength range from 380 to 760 nm and the eye discriminates between different wavelengths within this range by the sensation of colour. Violet and blue correspond to the short wavelengths, yellow and green to the middle and red to the long visible range of wavelengths.

Radiation reaching the earth's surface from the sun covers a range of wavelengths from about 290 nm to 1700 nm, which is considerably wider than the visible spectrum. Solar wavelengths shorter than 290 nm are absorbed by ozone in the upper level of the earth's atmosphere and wavelengths beyond 1700 nm are strongly absorbed by water vapour and carbon dioxide in the lower atmosphere.

Ultraviolet radiation

Wavelengths just beyond the violet end of the visible spectrum are known as ultraviolet radiation (UVR) and cover the region from 1 nm to 400 nm. The region from 1 nm to 100 nm is vacuum UV, 100 nm to 280 nm is UVC, 280 nm to 315 nm is UVB and the wavelengths between 315 and 400 nm is UVA (Figure 1).

Photometric quantities

The fundamental quantity used in photometry is luminous flux. This is a measure of the rate of flow of radiant energy modified according to the human spectral sensitivity. In other words, luminous flux is radiant flux multiplied by the spectral sensitivity of the visual system.

Radiant flux is measured in watts and luminous flux in lumens.

Luminous flux is an important quantity in determining the total light output of light sources and luminaires, but it is not the only useful quantity. The light distribution from luminaire is described by the luminous intensity. This intensity is the flux emitted/unit solid angle in a specified direction. The unit quantity is called *candela.*

Luminous flux and luminous intensity are both associated with area measures. The luminous flux falling on the unit area of a surface is called the *illuminance*. The luminous intensity projected area of a surface in a given direction is the *luminance*. The unit of illuminance is lumen/metre^2, usually called *lux* and the unit of luminance is *candela/metre*^{2} (Figure 2).

Luminance is the physical correlate to the psychological sensation of brightness.

The sensitivity of human eye is, however, not uniform over the visible spectrum, but varies with wavelength as shown in Figure 3. The right-hand

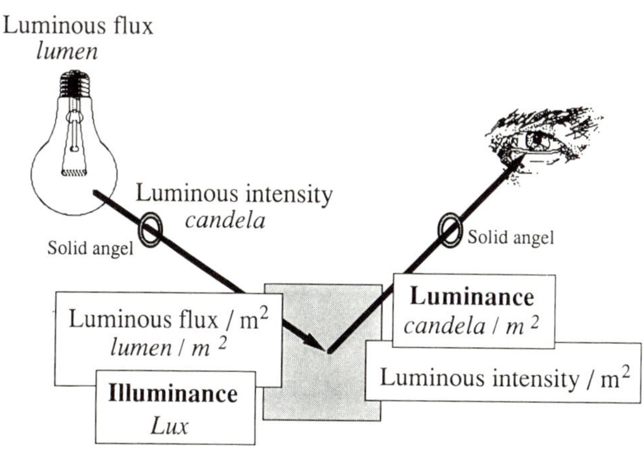

FIG. 2. Photometric quantities and units.

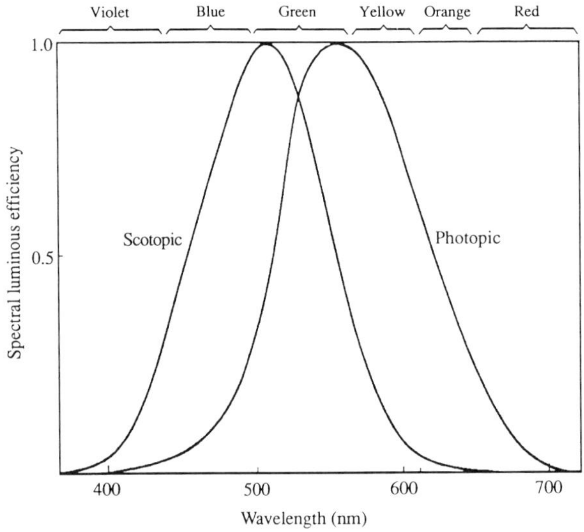

FIG. 3. Relative spectral sensitivity of human eye. The spectral luminous for the CIE* standard observer.

curve is valid for bright viewing conditions when the cones in retina are operating. These are starting to work at 3 cd/m^2 and are the receptors used in photopic vision.

The left-hand curve is the sensitive of the dark adapted eye when the rods are operating. These are sensitive to intensities below 0.1 cd/m^2 and cause the scotopic vision. For visual field luminances between 0.1 and 3 cd/m^2 an intermediate state exists, called mesopic, in which both cones and rods operate. Lighting technology is concerned with relatively high brightness and hence photopic vision attracts greatest attention.

Measurement

The illuminance is invisible to the eye. What the eye can record is the luminance.

The illuminance decreases with the square of the distance from the light source. Hence, a measurement of illumination needs to be accompanied by data on the distance from the light source and on the position at which the illumination has been measured.

In contrast, the luminance is dependent of the distance between the observer and the object. Thus, it is the unit to use when one relates to the visual sensation.

When light treatment is discussed, the quantity illuminance is almost exclusively used. Levels between 1,000 to 10,000 lux are often reported.

* CIE = *Commission Internationale de l'Eclairage* (*International Commission on Illumination*).

The question then is, where have the measurements been made. On the floor, half a meter above the floor, in the middle of the room, etc.? We are not informed about the exposure of the retina.

The illuminance does not become visible until it has been reflected on a surface, so we can say nothing about the retinal light exposure unless we know the reflectance of the surface.

To demonstrate the difference between illuminance and luminance, I have brought these two boxes.

It is easy to vary the record illuminance if the detector is moved. In principle you can get any illuminance you wish by moving the detector from high to low. What value do you want to have?

The type and number of fluorescent tubes is the same in both boxes, and thus the illuminance is the same. Nevertheless, the eye sees the intensity as very different. The reason is that the surface of this box has a much lower reflectance.

When we report our experimental of light treatment, luminance is a far better quantity to use than illuminance.

We all agree that the light must enter the eye, to acheive the therapeutic effect. The question is how much light shall we give to obtain best results, and which duration and spectrum shall we use.

If we maintain to measure the illuminance, the measurement needs to be made of the eye (cornea), so we can estimate how many lux that hit the retina. By recording this, results from different laboratories can be compared.

You may ask why the illuminance is so often used in the context of SAD-research? The probable reason is that illuminance is the quantity most often used in designing lighting at workplaces. This quantity is useful to describe how many luminaires you need, but when a lighting designer wants to control the vision ergonomics at the workplace he should use the quantity luminance.

When planning a workplace the lighting designer now considers the demand set by our normal vision. But, in the future the non-visual effects of light should also be considered when a good work environment is designed. If so, it will be very useful if we talk the same "language".

Another thing is that illuminance is very easy to measure using cheap and simple instruments.

To measure luminance is, however, not much more complicated than to measure illuminance. The main difference is that the measuring instruments are more complicated and much more expensive.

In collaboration with Professor Lennart Wetterberg and Docent Bengt Kjellman, I have built a light treatment room at St Görans Hospital in Stockholm. The room is homogeneously illuminated, all the surface is white and the light from the luminieres is indirect. The boxes I show are a miniature copy of the room.

The result of studies using this room is presented in Chapter 19 of this volume.

One question I often get about this room is: How many lux are in the room? The answer is that the illuminance can be everything between 1,000 to 10,000 lux, depending on where I measure.

What we know exactly is how much light that enters the eye. The luminance at the surrounding surface in the room is always 350 cd/m^2, irrespective of where in the room the detector is placed.

References

Boyce P.R. (1981) *Human Factors in Lighting*. Applied Science Publishers, London.

Hopkinson R.G. (1970) *The Ergonomics of Lighting*. MacDonald Technical and Scientific, London.

IES Lighting Handbook (1981) Reference Volume, Illuminating Engineering Society of North America.

3

Mechanisms in the Eye that Mediate the Biological and Therapeutic Effects of Light in Humans

GEORGE C. BRAINARD, JAMES R. GADDY, FELIX M. BARKER, JOHN P. HANIFIN and MARK D. ROLLAG

Departments of Neurology and Psychiatry, Jefferson Medical College, Philadelphia, PA, Pennsylvania College of Optometry, Philadelphia, PA, Department of Anatomy, Uniformed Services University of Health Sciences, Bethesda, MD, USA

Abstract

Research on the therapeutic potency of "bright" light has outpaced the study of the fundamental mechanisms by which light produces its beneficial effects. Originally, it was thought that illuminances equal to or above 2,500 lux were necessary to induce biological change or therapeutic response. Data are presented which show that illuminances as low as 5 lux of monochromatic green light or 100 lux of broad band white light also can produce significant suppression of melatonin in normal human volunteers. To understand how such low levels of illumination can produce strong biological and therapeutic responses, it is necessary to examine the ocular mechanisms which mediate the photic effects. Specifically, in humans it is shown that: (1) gaze behavior relative to a light source can significantly alter the amount of light reaching the retina; (2) the age of the ocular lens can significantly modify the balance of wavelengths reaching the retina; (3) near-ultraviolet radiation traditionally considered outside of the visible spectrum can stimulate visual responses in young adults at least up to the age of 25 years; (4) pupillary dilation modulates the capacity of a light stimulus to suppress melatonin; and (5) the entire retinal field participates in melatonin regulation. Although the specific photoreceptors or photopigments which mediate the biological and therapeutic effects of light in humans presently are not known, basic data are presented on the effects of different wavelengths elucidating melatonin suppression and SAD treatment. In summary, it is useful from both a scientific as well as a clinical perspective to establish the specific mechanisms in the eye which mediate the nonvisual therapeutic and biological effects of light.

Introduction

In 1980, it was demonstrated that exposing the eyes of healthy humans to bright light at night causes a strong suppression of plasma melatonin, a hormone secreted by the pineal gland.[53] This finding opened the door to numerous studies on the biological and therapeutic effects of light in humans. Since 1980, investigators have tested bright light as a stimulus for regulating neuroendocrine and circadian physiology and producing therapeutic benefits in patients with depression,[47,51,52,80,81,94–96] sleep disorders,[31,82] menstrual difficulties,[46,69,70] as well as problems associated with jet lag and shift work.[12,27,28,30,35,62,63,87,98] In fact, the use of bright light stimuli as a therapeutic intervention has far outpaced research of the fundamental mechanisms of how light induces such biological and behavioral changes. The aim of this paper is to examine mechanisms related to the eye that mediate the biological and therapeutic effects of light in humans.

Decades before the discovery that light could influence human physiology, researchers had shown that animals exhibit strong responses to light stimuli. In most vertebrate species, it is known that light enters the eyes and stimulates the retina. Neural signals are sent from the retina to the visual centers of the brain and permit the sensory capacity of vision. In addition, neural signals are sent from the retina into the hypothalamus, a non-visual part of the brain. Most, but not all, of the fibers projecting from the retina into the hypothalamus terminate specifically in the suprachiasmatic nuclei (SCN).[44,61,71] This part of the brain is considered to be an endogenous oscillator—a fundamental component of the "biological clock", or circadian system, which regulates the body's physiological and behavioral rhythms. The circadian system is responsible for controlling daily rhythms such as sleep and wakefulness, body temperature and hormonal secretion. It is now clear that light is the primary stimulus for regulating the circadian system, although other external stimuli such as sound, temperature and social cues may also influence the body's timing functions.[5,27,52,60,107]

The SCN relay retinal information to many of the major control centers in the nervous system.[61] One nerve pathway that carries non-visual photic information extends from the SCN to the pineal gland via a multisynaptic pathway with connections being made sequentially in the paraventricular hypothalamus, the upper thoracic intermediolateral cell column, and the superior cervical ganglion.[44,61] Cycles of light and darkness relayed by the retina entrain SCN neural activity which, in turn, entrain the rhythmic production and secretion of melatonin from the pineal gland. In all vertebrate species studied to date, including humans, high levels of melatonin are secreted during the night and low levels are secreted during the day.[43,54,74,75,101,102,105]

Light-induced suppression of melatonin in humans

In addition to entraining melatonin secretion from the pineal gland, light can have an acute suppressive effect on melatonin. Specifically, exposure of the eyes to light during the night can cause a rapid decrease in the high nocturnal synthesis and secretion of melatonin. This acute light-induced suppression of nocturnal melatonin synthesis was first observed[45] in 1972 and has been used in numerous animal studies to help determine the neural and biochemical mechanisms of melatonin regulation.[37,38,44,79] In addition, this response has been used as a model for testing the capacity of different light parameters for their relative influence on neuroendocrine and/or circadian physiology.[8,14–17,20,21,65,72,100] Complete dose-response curves show that exposure to 1.2 lux and 0.06 $\mu W/cm^2$ (0.2 lux) of white light during the night is sufficient to suppress melatonin in rats and hamsters, respectively.[16,59] Further studies in hamsters demonstrate that even lower illuminances at 0.022 $\mu W/cm^2$ (0.05 lux) of monochromatic (500 nm) green light can suppress pineal melatonin.[65,72] Other studies testing relatively few illuminances of white light have shown a wide range of sensitivity for melatonin suppression among different species.[73] Although it has not been firmly established, this diverse range of sensitivities for melatonin suppression may be related to a species' activity pattern.

Early human studies failed to demonstrate the acute suppressive influence of light on melatonin that previously had been demonstrated in other mammalian species.[2,43,54,101,102,105] In 1980, however, Lewy and colleagues[53] demonstrated that exposure of the eyes of normal volunteers to 2,500 lux of white light during the night induced an 80% decrease in circulating melatonin within 1 hour. In contrast, volunteers exposed to 500 lux of white light exhibited no significant melatonin suppression.[53] The earlier attempts to suppress melatonin in humans failed when investigators used typical indoor light levels reported to be between 100 to 800 lux.[3,43,54,101,102,105] Whereas such typical room light would be sufficient for suppressing melatonin in many animal species,[16,65,72,73,79] and would be adequate for human vision, it was not enough to suppress melatonin in those experiments. Simply put, it takes much more light to suppress melatonin in humans than is required for human vision or for regulating the circadian and neuroendocrine systems of some other animal species. The discovery that much brighter light is needed to suppress melatonin in humans provided the groundwork for numerous studies on biological responses of humans to bright artificial light including the confirmation that bright light ($\geq$2,500 lux) entering the eyes of humans is a potent stimulus for controlling circadian rhythms.[27,28,30,52,60,107] However, the notion that only "bright" light can suppress melatonin or regulate circadian rhythms in humans is not *entirely* accurate.

Several years after the landmark study of Lewy and colleagues,[53] a study was done to determine more precisely the dosages of light needed to suppress melatonin in normal volunteers.[14] In that study, six normal males were exposed to carefully controlled intensities of monochromatic green light at 509 nm (10 nm half-peak bandwidth) for one hour during the night after a three hour period of dark-adaptation. Specifically, the volunteers were continuously exposed to the experimental light between 2:00 a.m. and 3:00 a.m. with their pupils fully dilated by a mydriatic agent, their heads held steady relative to the light source by an ophthalmic head holder, and with translucent white integrating hemispheres covering both eyes. This procedure produced a constant and uniform illumination of the whole retina during the entire light exposure. The data from this experiment (Figure 1 and Table 1) demonstrated that light affects a human hormone in a dose-response fashion: i.e. the brighter the photic stimulus the greater the suppression of melatonin.[14]

It is interesting to note that all of the stimuli used in this study activated the visual system—i.e. both the volunteers and the experimenters saw each of the different light intensities and accurately reported them to be green. The lower light intensities, however, did not change hormone levels whereas the higher intensities induced a 60–80% decrease in this hormone. Thus, light sufficient to activate vision does not necessarily cause neuroendocrine change. It appears to be generally true in both animals and humans that much more light is needed for biological effects than for vision.[2,14,16,27,28,30,43,52–54,60,65,66,72,73,79,92,101,102,105,107]

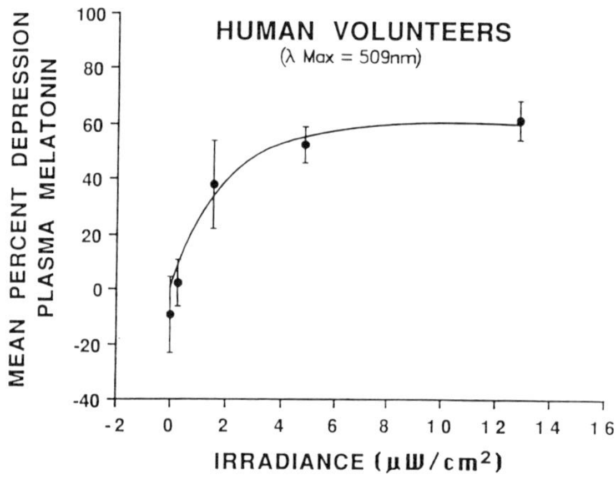

FIG. 1. The dose-response relationship between green monochromatic light exposure of normal volunteers eyes and suppression of the hormone melatonin. Data points indicate mean + SEM.[14]

The demonstration of the dose-response function for light suppression of melatonin in humans produced an unexpected result: *very bright light is not necessarily needed for melatonin suppression*. As demonstrated by Table 1, the mean threshold illuminance for producing a statistically significant melatonin suppression was between 5 and 17 lux in normal volunteers—a level of illuminance equal to civil twilight and well below typical indoor light. This means that under the proper conditions, 25 to 100 times less light can suppress melatonin than originally believed.[14]

TABLE 1.
Radiometric and Photometric Stimuli Used in the Melatonin Dose-Response Curve for 509 nm[14]

Irradiance (μW/cm^2)	Quanta (photons/cm^2)	Illuminance (photopic lux)	(scotopic lux)	% Melatonin suppression
0.01	9.2×10^{13}	0.03	0.17	−9.67
0.30	2.8×10^{15}	1.03	5.25	1.83
1.60	1.5×10^{16}	5.50	27.98	37.33
5.00	4.6×10^{16}	17.18	85.90	51.67
13.00	1.2×10^{17}	44.66	227.37	60.67

It is reasonable to ask why ambient room light at levels much higher than 17 lux did not suppress melatonin in the earlier attempts to suppress melatonin in humans.[2,43,54,101,102,105] The answer to this can be found by examining the ocular mechanisms involved in light stimulation of the circadian system. Specifically, it is important to consider the nature of the light stimulus and its relation to the eyes, gaze direction, status of the ocular media, pupillary dilation, retinal field exposure and photoreceptors. Each of these factors contribute to determining if a photic stimulus will or will not be effective in producing a biological or therapeutic response to light.

Influence of gaze direction on light effects

In the early human studies testing for melatonin suppression,[2,43,54,101,102,105] neither the exposure conditions nor the light stimuli were optimized. Often the experimental light stimulus consisted of turning on the overhead light provided with the experimental room. Furthermore, measurement of the experimental illuminance often consisted of placing a light meter at desk height and aimed directly at the overhead lights. Such a measure may over- or under-estimate the actual corneal illuminance experienced by the subjects. Furthermore, in almost any given room, it is possible to vary the light illuminance entering the eyes by a factor of 10 simply by changing the direction of gaze. Thus, in a room characterized as having a "typical" illumination level of 500 lux, the occupants may be able to see up to 500 lux if they look directly towards the light fixtures, but if they look at the floor or walls, the light reaching their eyes may drop to 50 lux or lower. Light entering the eyes can be further reduced if the volunteers close their eyes, squint or gaze into shadowy areas.

Experimental circumstances in which subjects are free to alter gaze and distance from a single light source are subject to loss of 80% to 99% of the putative corneal irradiance when subjects distance themselves and avert their gaze from the source.[29] The loss of corneal illuminance is compounded by the eye's optics which focus an image of the light source on the retina rather than diffusing the light across it. In a recent study which examined the influence of gaze direction relative to specific light sources on corneal illuminance, subjects were positioned 91 cm (3 feet) in front of a 61×122 cm (24×48 inch) light source which was entirely visible with a straight-on gaze. A 45° rotation of the subjects from the straight-on position resulted in a 60% reduction of the light source luminant surface image averaged over both eyes. A 90° rotation produced an 85% loss. The degree of loss was more extreme for a 16×24 inch light source (45°: 82%; 90°: 96%).[36] Thus, gaze aversion not only reduces general corneal illuminance, it also reduces the total area of the retinal image produced by a discrete light source. Beyond the motion of the eye relative to the environment and light sources, there are mechanisms within the eye itself that determine effectiveness of photic stimuli for inducing biological change.

Influence of ocular media transmission on light effects

An important ocular variable that determines both the quantity and quality of light reaching the retina is the transmittance characteristic of the ocular media. It is well documented that the cornea, aqueous humor, and vitreous humor of the healthy human eye are clear tissues which transmit nearly 100% of visible light and ultraviolet wavelengths down to 300 nm.[10,49,85] In the healthy human, there is little change in the transmission characteristics of these tissues over age. In contrast, the crystalline lens yellows with age and acts as a filter that greatly attenuates radiant energy of the shorter wavelength portion of the spectrum (i.e. less than 1% transmission between 300 nm and 380 nm after age 30; reduced by 50% at 470 nm at age 75).[6]

To study the lenticular transmission of ultraviolet, visible and infrared wavelengths, postmortem lenses ($n = 288$) were collected from human donors ranging from 0 to 91 years of age.[6] The spectral transmittance characteristics of these lenses were measured between 200 nm and 2500 nm with a Beckman UV-5240 spectrophotometer. Figure 2 illustrates the lenticular transmission of lenses collected from two different donors.

The data in Figure 2 show how lenticular transmission varies greatly over the human life span. Note the progressive loss of ultraviolet and short wavelength visible transmission with age. The overall results of this study indicated that longer, infrared wavelength transmission is not substantially different between age groups, while significant, gradually occurring differences do exist between age groups in the shorter wavelength and visible ranges. The averaged percent transmittance data from the 20–29 year and the 50–59 year age groups are shown in Figure 3 opposite.[6]

A portion of the averaged percent transmittance data from the 20–29

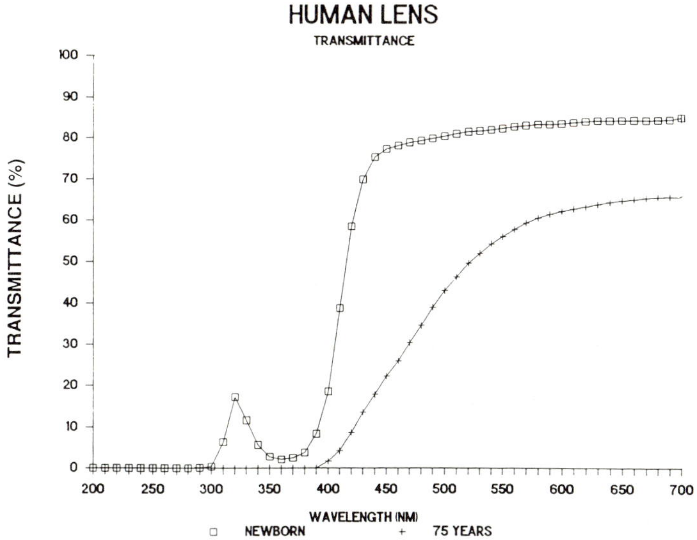

FIG. 2. Lenticular transmission of single human lenses collected from a newborn and 75 year old donor.[6]

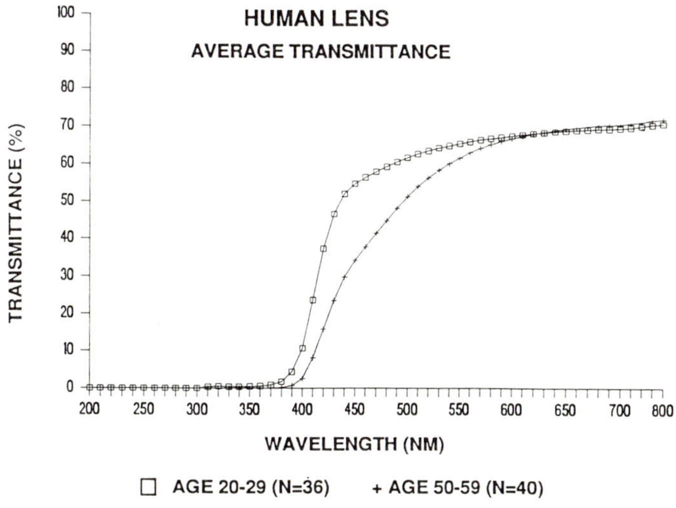

FIG. 3. The mean % transmittance of visible and ultraviolet wavelengths by postmortem human lenses from 36 individuals aged 20 to 29, and from 40 individuals aged 50 to 59.[6]

year and the 50–59 year age groups are shown in the Table 2. These data were analyzed using a 2-tailed "t" test and probability values are not corrected for multiple comparisons.

These data show that the age of the human lens significantly modulates the balance of wavelengths reaching the retina. As the eye ages, the increased yellowing of the lens diminishes the transmission of shorter wavelength energy. Hence, illumination measured at the level of the cornea is not necessarily indicative of retinal illumination among different age groups. In terms of using light as a biological or therapeutic stimulus, for any given broad spectrum light source, there will be a progressive decrease in energy reaching the retina in the short wavelength visible and ultraviolet spectrum as patients age.

In examining the data in Table 2, it is obvious that there are statistically significant differences in lenticular transmission between different age groups—particularly in the shorter wavelength ranges. It is reasonable to ask if the very small percentage transmittance of near-ultraviolet radiation (an average of less than 1%) in young adults allows an effective photic stimulus to reach the retina. International standards define the visible spectrum as wavelengths of electromagnetic radiation between 380 nm and 780 nm.[25] There are, however, data demonstrating that the spectrum of radiation visible to the human eye actually extends further below 380 nm into the near-ultraviolet range. Recent studies have demonstrated that the visual system of young humans is responsive to ultraviolet radiation at least as low as 340 nm.[13,84] In one of those studies, a dose-response relationship between varying irradiances of monochromatic UV-A (340 nm) and visual evoked potentials (VEP) in young subjects was established. The VEP is a gross electrical signal generated by the visual cortex in response to repetitive photic stimulation. Its waveform, peak latency, and amplitude vary according to experimental conditions.[4,7,88]

To probe for potential sensitivity to ultraviolet radiation at 340 nm, healthy, young adults 20 to 25 years-of-age with normal color vision ($n = 10$) were tested.[84] The VEP recording techniques and parameters are described elsewhere.[3,13] The 340 nm stimulus was administered by a lamp suspended 15 cm above the right eye and the left eye was covered with a patch. Seven different flash irradiances plus an auditory control were presented to each subject under an ambient photopic condition of overhead white fluorescent light at 550 lux. Data from this study were analyzed by ANOVA with statistical significance determined by the Student Newman Keuls multiple range test.

The data shown in Figure 4 illustrate a dose-dependent response between irradiance and VEP N_2-P_3 amplitude for young adults.[84] This illustrates that, as irradiance decreases, the N_2-P_3 amplitude also decreases. At an irradiance of 0.37 $\mu W/cm^2$, only two of the ten subjects in

TABLE 2
Lenticular transmittance in two age groups (mean and SE)[6]

AGE		340 nm	440 nm	540 nm	640 nm	750 nm
20–29	X	0.394	51.99	64.63	68.29	69.74
($n=36$)	SE	0.077	1.68	1.68	1.56	1.49
50–59	X	0	29.62	59.92	68.64	70.94
($n=40$)	SE	0	1.38	1.30	1.20	1.15
p (2 tail)		0.0001	0.0001	0.028	0.857	0.521

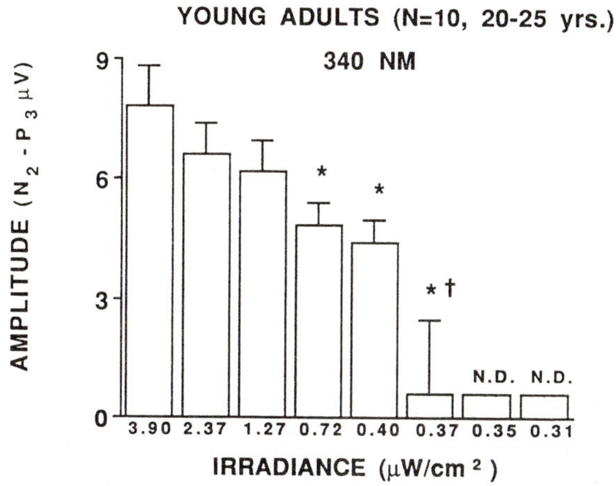

FIG. 4. This graph illustrates the visual evoked response amplitude in young adults exposed to different irradiances of near-ultraviolet radiation at 340 nm.[84] Asterisks (*) indicate statistical significance of at least $p<0.05$ compared to the maximum N_2-P_3 amplitude. The dagger (†) indicates that only two of the ten subjects were able to detect radiation at this irradiance. N.D. = not detected by any subject.

this age group could detect this radiation and at the two lowest irradiances, none of the subjects could detect this radiation.

It is known that the lens will fluoresce when illuminated with ultraviolet wavelengths. In this study, for all but the lowest irradiances, all subjects were able to clearly discriminate the shape of the filaments suspended within the flash lamp during stimulation. Therefore, if only secondary lenticular fluorescence accounted for all the energy reaching the retina during stimulation, the subjects would not have been able to focus the image. Tan provides an elegant argument against such fluorescence-based UV vision.[93] Thus, this study demonstrates that monochromatic near-UV radiation can elicit visual evoked potentials in a dose-dependent manner in the young human eye.[84] The study also confirms that the very low percentage lenticular transmission of energy at 340 nm as shown in Table 2 is meaningful in that it can induce visual responses.

Given that some ultraviolet energy is transmitted through the lenses of young humans, and that ultraviolet radiation can stimulate the visual system, it is useful to question the potential role of ultraviolet radiation in biological and therapeutic responses to light. In several rodent species, ultraviolet radiation down to 300 nm is transmitted through the lenses and can participate in the regulation of the circadian and neuroendocrine systems.[8,13,15,19,72,90,91] Indeed, very recent findings indicate that the rat retina may contain a specific photopigment that has its peak sensitivity in the near-ultraviolet spectrum.[42] Thus, mammals may have ultraviolet photoreceptors similar to those which have been identified in various insect, fish, reptile and bird species.

It is currently unclear whether or not the human retina contains a specific ultraviolet photoreceptor or if ultraviolet radiation can be an effective circadian stimulus for humans. Further work is required to determine if detection of ultraviolet radiation (UV) by the young human eye serves some purpose other than vision. A practical issue debated among researchers concerns the role of UV in light therapy. Most of the early studies on seasonal affective disorder (SAD) therapy successfully utilized fluorescent lamps that emitted white light containing a portion of UV wavelengths.[80,96] Those early results erroneously led to the suggestion that UV wavelengths are necessary for successful therapy. The literature, however, shows clearly that SAD symptoms can be reduced by lamps which emit little or no UV.[9,48,50,52,64,68,108] Hence, UV wavelengths do not appear to be necessary for eliciting positive therapeutic results. Whether or not UV wavelengths can contribute to the optimum balance of wavelengths for SAD therapy remains an open question. It also must be recalled that overexposure to ultraviolet radiation—particularly wavelengths below 300 nm—can be a hazard to both skin and eye tissues.[40,49,85,99,103] Similarly, overexposure to visible light and infrared radiation can also be damaging to these tissues.[49,85,103] Thus, it is becoming increasingly important to discern between wavelengths and doses of light that are effective therapeutically yet are safe in terms of potential tissue damage.

Clearly, the transmittance characteristics of the clear ocular media, and particularly the crystalline lens, influence the quality and quantity of light reaching the retina. Thus, these tissues are involved in determining the biological impact of a photic stimulus in terms of neuroendocrine and circadian regulation. Another potential ocular mechanism for modifying a photic stimulus before it reaches the retina involves the iris and the pupil.

Pupillary diameter mediates biological effects of light

The iris of the eye is a heavily pigmented layer of muscle tissue which adjusts the pupil of the eye. Pupil dilation leads to increased amounts of light reaching the retina while pupil constriction results in decreased illumination reaching the retina. The human pupil ranges in size from 2 mm to 9 mm depending on the brightness of light being viewed. Extensive studies have been done on the iris as it relates to enhancing vision, providing visual comfort and protecting the retina from damage.[76,85] Little has been done, however, to assess the role of the iris and pupil in moderating the effects of light stimuli on the therapeutic or biological effects of light. To test the capacity of pupil dilation to affect light-induced melatonin suppression,[38] a group of eight healthy male and female subjects were studied (mean age + SEM = 24.1 + 0.74 years). All subjects were in good general health and passed the Ishihara test for color blindness. Each subject was tested on three separate nights with at least one week between each night of study.

On two of the three experimental nights, subjects stared at a uniform visual field delivering a corneal illuminance of 100 lux of white light. On one of these exposure nights, pupils were free to constrict while on the other, they were pharmacologically dilated. The third night was a control condition in which the subject remained blindfolded throughout the 90 minute period. The order of treatment varied systematically between subjects. Each study began at midnight of a given night in a laboratory which was dimly lit to permit approximately a 10 lux exposure while the volunteers talked with each other, listened to a personal stereo, read, or slept. At 01:00 to 01:30 hours, 1–2 drops of 0.5% cyclopentolate (an anticholinergic pupillary dilator) or placebo (artificial tears) were placed in the subjects' eyes. At this time, all subjects were blindfolded. Cyclopentolate was used on both the night of dilated-pupil light exposure and on the night of no light exposure. Placebo drops were used on the night of 100 lux exposure when the pupils were free to constrict.

In order to provide a spatially-uniform light stimulus that filled the visual field, three Goldmann perimeters were modified by aiming a fiber optic light guide into the top of the perimeter. A 150 W quartz-halogen lamp provided uniformly reflected white light from the perimeter's interior. Spot sampling of the perimeter's surface with a luminance meter showed a homogeneous luminance across the visual field. Blood samples were taken just prior to the beginning of light exposure at 02:00 hours and again just after light exposure was completed at 03:30 hours. Plasma samples were analyzed for melatonin using a modification[37] of the radioimmunoassay method described by Rollag and Niswender.[78]

Treatment with cyclopentolate produced a marked dilation of the subjects' pupils (mean = 7.3, SEM = 0.3). In contrast, the pupils treated with placebo drops on the night of light exposure were constricted (mean = 3.3, SEM = 0.1). As shown in Figure 5A, in both the dilated and free pupil conditions, 100 lux of corneal illuminance significantly suppressed plasma melatonin compared to the control night

(F(2,14) = 10.6; $p<0.002$). The data in Figure 5B demonstrate that pupil dilation produced stronger light-induced melatonin suppression at 100 lux corneal illuminance as compared to the free pupil condition ($t=2.78$; $df=7$, $p=0.027$).

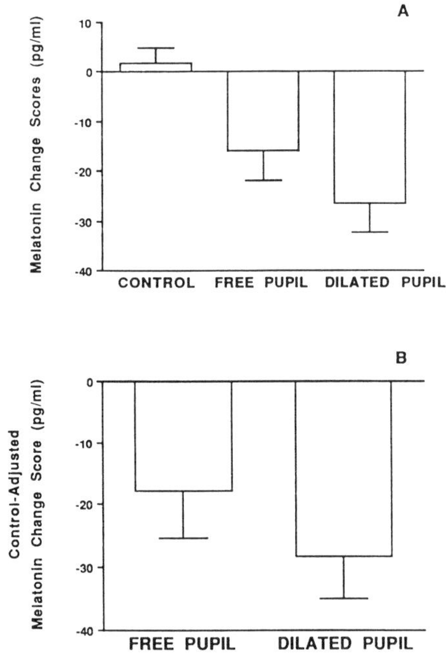

FIG. 5. (A) These data illustrate mean (+SEM) change scores from baseline plasma melatonin levels (02:00 hours) to final melatonin levels (03:30 hours) in eight subjects studied on three separate nights.[38] (B) Mean (+SEM) control-adjusted change scores for plasma melatonin at 100 lux of corneal illuminance in both free-pupil and dilated pupil conditions. Suppression is expressed as the change from baseline on the testing night less the change from baseline on the control (no light) night.[38]

This result suggests that the threshold for white light-induced melatonin suppression in both the free- and dilated-pupil conditions is somewhere below 100 lux of corneal illuminance for young, healthy adults. This result also confirms that the pupil is important in controlling light-induced melatonin suppression. When the pupil was dilated, melatonin suppression increased significantly. This finding parallels the classic visual psychophysical principal that, as pupil diameter increases, visual stimulation increases.[76] These similarities suggest that the regulation of nonvisual effects of light occurring via the retina are accessible to classical psychophysical methods and that application of these methods should lead to increased understanding of basic mechanisms. In addition, these results suggest that pupil diameter may be a factor in the effectiveness of light stimuli used to shift circadian rhythms or to treat seasonal depression or sleep disorders.

To reiterate, the original notion that only relatively bright white light above 2,500 lux can suppress melatonin[53] is not entirely valid. Other studies have shown that as little as 200 to 400 lux of white light can suppress plasma melatonin levels in humans[11,37,56] and the data illustrated above show that as little as 100 lux of white light can suppress melatonin significantly when gaze direction and exposure conditions are carefully controlled.[38] These data not only emphasize the importance of behavioral movement of the eyes and uniformity of light presentation but also suggest that the spatial location of retinal photoreceptors has a role in mediating the therapeutic and biological effects of light. Specifically, when the entire visual field is steadily exposed, relatively low levels of illumination (5 lux of monochromatic green light or 100 lux of white light) can suppress melatonin.[14,38]

The finding that relatively low levels of light can suppress melatonin when gaze behavior is rigorously controlled and the entire retinal field is evenly illuminated suggests that light stimulation of the entire retinal field leads to a stronger biological response than when the retinal field is only partially stimulated. A recent study on normal humans demonstrated that when illuminance measured at the cornea is equal, stronger melatonin suppression is observed when the whole retinal field is illuminated as opposed to illumination of approximately half of the retinal field.[37] It appears that there may be a mechanism for spatial summation of photic stimuli at the level of the retina—the more retinal area exposed to light, the stronger the resulting biological response. Neuroanatomical studies help support this concept. Morphological evidence from rodent and non-human primates indicates that retinal ganglion cells projecting to the suprachiasmatic nuclei are distributed across the entire retina.[57,71] This suggests that photoreceptors from the entire retinal field contribute to stimulation of the SCN. Further support for this concept comes from a study which examined melatonin suppression in humans with 1,000 lux of white light on two occasions: one illuminating the central visual field 5° from the center of gaze, and one directed to the peripheral visual field 60° lateral to the direction of gaze. That study showed that melatonin suppression did not differ between central and peripheral illumination.[1] At this time, many questions remain unanswered about the how the retina processes photic stimuli for melatonin regulation. Do certain areas of the retina send stronger signals to the SCN than others? Does the superior retinal field have a greater role than the inferior retinal field? A relatively homogenous spread of retinal ganglion cells in the retina does not guarantee a homogeneity of physiological function. Further studies are needed to determine more precisely how the physical location of photoreceptors in the retina mediates the therapeutic or biological effects of light stimuli.

Wavelength sensitivity for light effects

Currently, there are no definitive data that determine what photoreceptors and photopigments transduce light stimuli for the non-visual biological and therapeutic effects in humans. In examining animal studies from many laboratories, some data have indicated that the peak sensitivity of the circadian and neuroendocrine system is around 500 nm.[8,17,21,22,65,66,72,92,97] This supports the hypothesis that rhodopsin or a rhodopsin-based molecule is the primary receptor for circadian and neuroendocrine regulation. In contrast, other data have suggested that one or more cone photopigments may be involved in these regulatory effects.[8,15,17,41,58,72] In considering the potential photoreceptor cell type(s) that transduce photic information to the SCN, it should be kept in mind that the rodent retina contains both cone and rod photoreceptors.[23,24,106] The retinally degenerate mouse (rd/rd), however, shows normal circadian responses to 515 nm light pulses despite nearly total loss of classical visual photoreceptors.[34] Further, in a pilot study on a single 18 year-old congenitally blind man, melatonin suppression was still achieved with bright light over 6,000 lux. Thus, the neuroendocrine sensitivity to light remained despite a loss of pupillary reflex, outer retinal functioning as determined by electroretinographic testing or conscious perception of the light stimulus.[55] Such data suggest that light detection for the circadian and neuroendocrine systems may not rely on the rod or cones used for vision.[34,55]

It is important to note that while most data from rodents show that the highest sensitivity is in the blue–green range, other wavelengths may participate in circadian and neuroendocrine regulation. For example, short wavelengths in the ultraviolet region of the spectrum[8,13,15,19,72,90,91] and longer wavelengths in the red portion of the spectrum[20,67,100] are quite capable of suppressing melatonin, entraining circadian rhythms and influencing reproductive responses in rodents given a sufficiently high intensity. There is considerable diversity in the cellular structure and function of the retina across mammalian species.[77] Thus, further studies are required to conclusively identify which specific photoreceptors and photopigments are involved in regulating the circadian and neuroendocrine systems among different species. Whereas data from non-human species may be helpful in guiding research at the human level, caution should be exercised in generalizing results from rodents to humans.

To date, only one study has been done to examine the relative impact of different wavelengths on melatonin synthesis in humans.[14] After completing the dose-response work illustrated in Figure 1, a *preliminary* test was done to compare the efficacy of different monochromatic wavelengths at equal photon densities for suppressing nocturnal plasma melatonin. The same exposure method described above was employed and two or three of the same subjects were tested at six different wavelengths. These preliminary results are shown in Table 3.

TABLE 3
The effects of different monochromatic wavelengths on melatonin suppression in normal volunteers[14]

Wavelength (nm)	Subjects (*n*)	% Melatonin Suppression	SEM
448	3	−12.00	22.00
474	2	26.00	8.00
509	3	63.67	6.94
542	3	19.67	13.35
576	3	16.33	14.38
604	2	−11.50	8.50

These data suggest that the peak sensitivity for melatonin suppression may be in the blue–green range as seems to be the case in some lower mammals. It is premature, however, to put much emphasis on such a conclusion since relatively few subjects were tested in this pilot study and analysis of variance shows no statistical differences between melatonin suppression and the different wavelength stimuli. Only a more complete action spectrum determined with a larger subject sample is useful for determining which photoreceptors or photopigments are involved in mediating melatonin regulation in humans.

Theoretically, it should also be possible to determine an action spectrum for the therapeutic effects of light. Current evidence supports the hypothesis that light therapy for SAD works by way of light shining into the eyes as opposed to light on the skin.[104] To begin characterizing which ocular photoreceptors or photopigments mediate the therapeutic benefits of light in winter depression, three consecutive studies have been done to specifically compare different portions of the spectrum for clinical efficacy in treating SAD.[18,68,89] In the first study, 18 patients were treated with an equal photon dose of white, blue or red light for a period of one week.[18] The photon density of 2.3×10^{15} photons/cm^2/sec was selected because this particular photon density of broad spectrum white light has been shown in many previous studies to be clinically effective in one week of therapy.[80,96] Patients' clinical status before and after light therapy was evaluated by means of the 21-item Hamilton Depression Rating Scale (HDRS), a standard scale for measuring symptoms associated with depression.[39] The results of this study are illustrated in Figure 6.

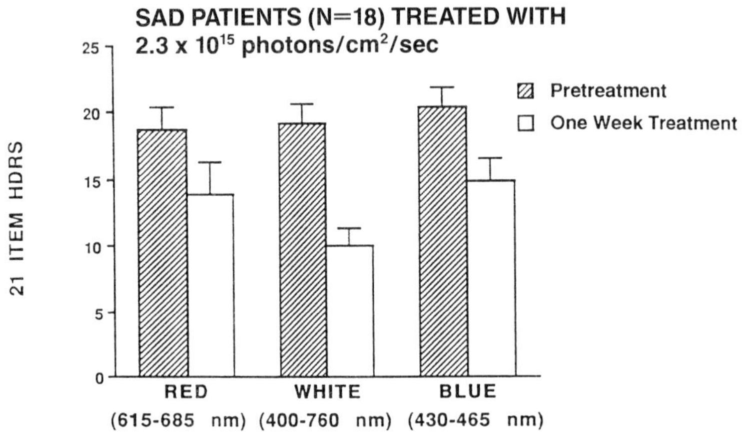

FIG. 6. The bars in this graph indicate mean +SEM HDRS values for patients before treatment (hatched bars) and after 1 week of treatment with equal photon densities of different light spectra (open bars). Numbers in parentheses indicate the half-peak bandwidth of the light source.[18]

This study was the first step towards defining an action spectrum of light therapy for winter depression. As shown in Figure 6, one week of light therapy with each of the three light sources produced an improvement in depression symptoms among the groups of patients tested. Specifically, the percent drop in mean HDRS scores were 26%, 47% and 27% for the red, white and blue light sources, respectively. Thus, the photon density emitted from the white light source elicited a significantly stronger clinical response compared to the results obtained from an equal photon density from the blue and red light sources.[18] This suggests that broad spectrum white light at this particular photon density is superior to restricted bandwidths of light in the red and blue portions of the visible spectrum. That result implies that light sources for SAD light therapy could not be improved by substituting narrower bandwidths of blue or red light in place of broadband white light.

A second study was done comparing green light to red light at 2.3×10^{15} photons/cm^2/sec for treating SAD.[68] As in the first study, patients' clinical status before and after one week of light therapy was followed by means of the 21-item HDRS. The results of this study are shown in Figure 7.

As illustrated in Figure 7, one week of light therapy with either green or red light sources produced an improvement in depression symptoms in the groups of patients tested. The percent reduction in mean HDRS scores was 51% and 30% for the green and red light sources, respectively. Hence, at this photon density, green light was significantly stronger than the red light for treating winter depression.[68] Considered alongside the results from the study comparing red, white and blue light therapy at the same photon density (Figure 6), the results of this study (Figure 7) suggest that broad

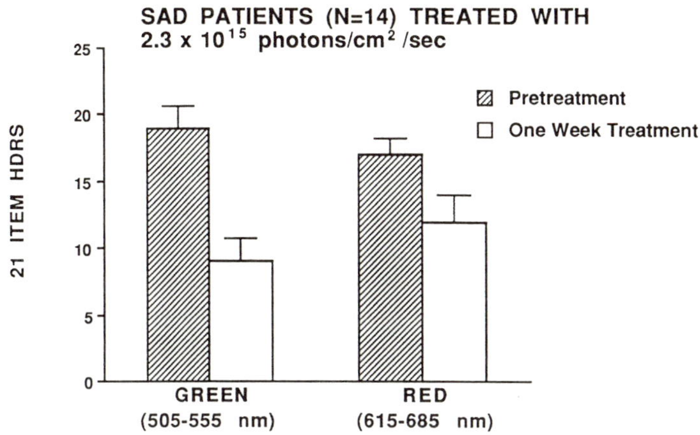

FIG. 7. The bars in this graph indicate mean +SEM HDRS values for patients before treatment (hatched bars) and after 1 week of treatment with equal photon densities of green or red light (open bars). Numbers in parentheses indicate the half-peak bandwidth of the light source.[68]

spectrum white light and narrower band green light are equivalent in their capacity to reduce symptoms of SAD. Between the two studies, white and green light treatments were associated with a 48% and 53% reduction in HDRS scores, respectively. Comparisons of group responses between different studies, however, are not conclusive.

Are white and green light really equivalent in their phototherapeutic strength? To answer this question, 12 patients were given 1 week of light therapy for SAD with either green or white light at an equal photon density.[89] Since therapy with white and green light appeared to cause roughly equivalent HDRS reductions across the first two studies, the experimental photon density was lowered to 1.23×10^{15} photons/cm^2/sec in the third study. Again, patients' clinical status before and after 1 week of light therapy was followed using the 21-item HDRS. The results of this study are illustrated in Figure 8.

As shown in Figure 8, 1 week of therapy using each of the light sources produced an improvement in depression symptoms. Specifically, the percent drop in mean HDRS scores was 22% for the green light and 46% for the white light sources. At this lower photon density, white light was superior to the green light in treating SAD.[89]

Together, these three studies form the ground work for determining the action spectrum for SAD light therapy.[18,68,89] The traditional approach to defining a complete action spectrum, however, requires testing with narrower bandwidth light stimuli with more tightly controlled light exposures.[26,86] A thoroughly defined action spectrum can guide the development of light treatment devices that emit the optimum balance of wavelengths for treating SAD. Furthermore, an action spectrum will yield

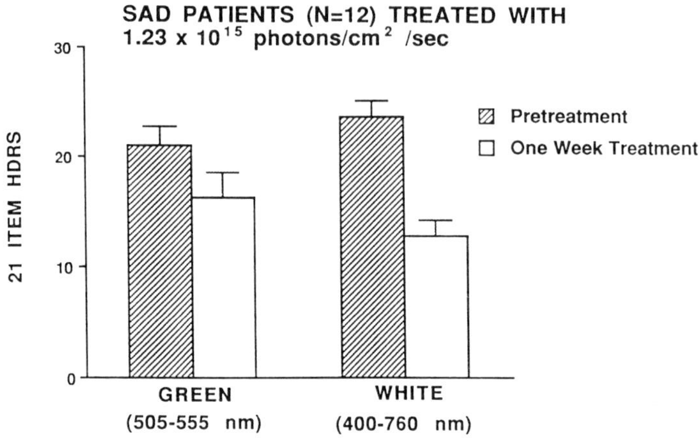

FIG. 8. The bars in this graph indicate mean +SEM HDRS values for patients before treatment (hatched bars) and after 1 week of treatment with equal photon densities of white or green light (open bars). Numbers in parentheses indicate the half-peak bandwidth of the light source.[89]

important information about the photosensory mechanism(s) responsible for the beneficial effects of light therapy. Currently, it is premature to predict what photopigment(s) or photoreceptor(s) mediate the antidepressant effects of light.

Several methodological problems will have to be overcome before further progress can be made in defining an action spectrum for SAD light therapy. One complication for nearly all studies on SAD involves the fact that they are done on an outpatient basis. Hence, patient compliance on treatment timing, frequency and duration cannot be closely controlled even with the most cooperative subjects. Furthermore, as discussed earlier, very small changes in gaze direction and patient position relative to the light source can cause great variability in the amount of light transmitted to the patients' eyes.[29,36] Did patients have different gaze behaviors or different patterns of light usage with the different wavelength light sources? The optimum method of comparing different wavelengths—or any other photic parameter—for SAD therapy is to work with more carefully controlled exposures.

Across the wavelength studies outlined above, each light treatment produced some therapeutic improvements. Does this indicate that each light was at least partially effective in treating SAD symptoms, or are some of the therapeutic benefits of light therapy due to a non-specific or placebo response? Since patient expectations of treatment outcome are thought to contribute significantly to the placebo effect, evaluation of expectations before treatment is one strategy for approaching this question. Prior to any light treatment, subject expectations were systematically probed in each of the three wavelength studies. In general, all subjects had positive expectations about the success of light therapy but there were no differences between the expectations for the different light spectra in these studies.[18,68,89] This evidence supports the idea that some of the therapeutic benefit of the different light spectra may have been due to a placebo response but that the differential therapeutic responses to the different light spectra were not merely an extension of the patients' preconceived beliefs.

It has been well documented that patients with a wide range of disorders—depression, schizophrenia and anxiety as well as cancer, diabetes and ulcers—can successfully respond to inactive or placebo treatments.[32,33,83] Hence, it would be remarkable if SAD patients did not show some level of placebo response to light therapy. In fact, therapeutic improvements are almost always observed with light treatments regardless of light intensity, wavelength or duration.[80,96] Although it is obvious that light therapy indeed will reduce patients' depression symptoms, the critical question is how much of the patients' response to light therapy is due to a non-specific placebo response versus a genuine clinical response? This remains an open question in the SAD field and has been discussed most insightfully by Eastman.[32,33] The inability to accurately separate placebo responses from genuine clinical antidepressant responses causes an element of "noise" in phototherapy data which seriously hinders the accurate discrimination of differential wavelength effects in light therapy. Unfortunately, until this question is resolved, a more conclusive action spectrum for SAD phototherapy may not be possible.

Conclusion

Over the past decade, considerable progress has been made in exploring the potential of bright light therapy for affective disorders and disturbances of the circadian system in humans. In fact, the research on the therapeutic potency of light has outpaced the work on fundamental mechanisms by which light produces its beneficial effects. It is useful from both a scientific as well as a clinical perspective to establish the specific mechanisms in the eye which mediate the non-visual therapeutic and biological effects of light. Studies with animal models most certainly can guide the way for an improved understanding of how light may produce its beneficial effects in man. Animal studies alone, however, are not sufficient and must be matched by direct human studies to determine the relative impact of gaze behavior, status of the ocular media, pupillary response, retinal field and photoreceptor sensitivity on the biological and therapeutic effects of light in humans.

Acknowledgments

The authors would like to thank F. L. Ruberg and D. L. Brainard for their thorough editorial assistance and manuscript preparation, and J. Georgiou for library and reference support. This work was supported by the NIH Grant #RO3-MH-44890-01A1, the FDA Grant #785346, the National Electrical Manufacturer's Assoc. (#LRI 89:DR:1), the Lighting Research Inst. (#LRI 88:SP:LREF:6), the Dean's Overage Research Program of Jefferson Medical College, USUHS Grant C07049, and the Philadelphia Chapter of the Illuminating Engineering Society.

References

1. Adler J.S., Kripke D.F., Loving R.T. and Berga S.L. (1992) Peripheral vision suppression of melatonin. *J. Pineal Res.* **12**, 49–52.
2. Akerstedt, T., Froberg, J.E., Friberg, Y. and Wetterberg, L. (1979) Melatonin excretion, body temperature and subjective arousal during 24 hours of sleep deprivation. *Psychoneuroendocrinology* **4**, 219–225.
3. American Electroencephalographic Society. (1984) Guidelines for clinical evoked potential studies. *J. Clin. Neurophysiol.* **1**, 17–18.
4. Apkarian P. and Spekreijse H. (1990) In *Visual Evoked Potentials* (ed. Desmedt J.), pp. 169–223. Elsevier, Amsterdam.
5. Aschoff J. (ed) (1981) *Handbook of Behavioral Neurobiology*, pp. 1–563, Plenum Press, New York.
6. Barker F.M. and Brainard G.C. (1993) The direct spectral transmittance of the excised human lens as a function of age. *Vision Res.* (in press).
7. Barnet A.B., Freedman S.J., Weiss I.P., Ohlrich F.S., Shanks B. and Lodge A. (1980) VEP development in infancy and childhood: A longitudinal study. *Electroenceph. Clin. Neurophysiol.* **49**, 579–589.
8. Benshoff H.M., Brainard G.C., Rollag M.D. and Lynch G.R. (1987) Suppression of pineal melatonin in Peromyscus leucopus by different monochromatic wavelengths of visible and near-ultraviolet light (UV-A). *Brain Res.* **420**, 397–402.
9. Bielski, R.J., Mayor J. and Rice J. (1992) Phototherapy with broad spectrum white fluorescent light: A comparative study. *Psychiatry Res.* **43**, 167–175.
10. Boettner E.W. and Wolter J.R. (1962) Transmission of the ocular media. *Invest. Opthalmol.* **1**, 776–783.
11. Bojkowski C.J., Aldhous M.E., English J., Franey, C., Poulton, A., Skene, D. and Arendt, J. (1987) Suppression of nocturnal plasma melatonin and 6-sulphatoxymelatonin by bright and dim light in man. *Horm. Metab. Res.* **19**, 437–440.
12. Brainard G.C. (1992) Biological Effects of Light in Humans: The Regulation of Physiology, Mood and Behavior. In *Biologic Effects of Light* (eds. Holick M.F. and Kligman A.M.), pp. 133–154. Walter de Gruyter, Berlin.
13. Brainard G.C., Beacham S., Hanifin J.P., Sliney D. and Streletz L. (1992) Ultraviolet regulation of neuroendocrine and circadian physiology in rodents and the visual evolved response in children. In *Physiological Effects of UV-A Radiation* (ed. Urbach F.), pp. 261–272. Valdenmar, Overland Park, Kansas.
14. Brainard G.C., Lewy A.J., Menaker M., Miller L.S., Fredrickson R.H., Weleber R.G., Cassone V. and Hudson D. (1988) Dose-response relationship between light irradiance and the suppression of melationin in human volunteers. *Brain Res.* **454**, 212–218.
15. Brainard G.C., Podolin P.L., Leivy S.W., Rollag M.D., Cole C. and Barker F.M. (1987) Near ultraviolet radiation (UV-A) suppresses pineal melatonin content. *Endocrinology* **119**, 2201–2205.
16. Brainard, G.C., Richardson B.A., King T.S., Matthews S.A., and Reiter R.J. (1983) The suppression of pineal melatonin content and N-acetyltransferase activity by different

light irradiances in the Syrian hamsters: A dose-response relationship. *Endocrinology* **113**, 293–296.

17. Brainard G.C., Richardson B.A., King T.S. and Reiter R.J. (1984) The influence of different light spectra on the suppression of pineal melatonin content in the Syrian hamster. *Brain Res.* **294**, 333–339.
18. Brainard G., Rosenthal N., Sherry D., Skwerer R., Waxler M. and Kelly D. (1990) Effects of different wavelengths in seasonal affective disorder. *J. Affect. Disord.* **20**, 209–216.
19. Brainard G.C., Vaughan M.K. and Reiter R.J. (1986) Effect of light irradiance and wavelength on the Syrian hamster reproductive system. *Endocrinology* **119**, 648–654.
20. Broker, B.J., Hanifin J.P., Rollag M.D., Thornton W.A. and Brainard G.C. (1990) Suppression of pineal melatonin content in Long Evans Hooded rats: Dose-response curve at 640nm. Abstract: Society for Neuroscience, Phoenix, Arizona.
21. Bronstein D.M., Jacobs G.H., Haak K.A., Neitz J. and Lytle L.D. (1987) Action spectrum of the retinal mechanism mediating nocturnal light-induced suppression of rat pineal gland N-acetyltransferase. *Brain Res.* **406**, 352–356.
22. Cardinali D.P., Larin F. and Wurtman R. (1972) Action spectra for effects of light on hudroxyindole-o-methyltransferases in rat pineal, retina and Harderian gland. *Endocrinology* **91**, 877–886.
23. Carter-Dawson L.D. and LaVail M.M. (1979) Rods and cones in the mouse retina. *J. Comp. Neurol.* **188**, 245–262.
24. Cicerone C.M. (1976) Cones survive rods in the light damaged eye of the albino rat. *Science* **194**, 1183–1185.
25. CIE International Lighting Vocabulary. (1987) CIE Publication No. 17.4, Vienna.
26. Coohill T.P. (1991) Action spectra again? *Photochem. Photobiol.* **6**, 276–296.
27. Czeisler C.A., Allan J.S., Strogatz S.H., Ronda J.M., Sanchez R., Rios C.D., Freitag W.O., Richardson G.S. and Kronauer R.E. (1986) Bright light resets the human circadian pacemaker independent of the timing of the sleep–wake cycle. *Science* **233**, 667–671.
28. Czeisler C.A., Johnson M.P., Duffy J.F., Brown E.N., Ronda J.M. and Kronauer R.E. (1990) Exposure to bright light and darkness to treat physiologic maladaptation to night work. *New Eng. J. Med.* **322**, 1253–1259.
29. Dawson D. and Campbell S.S. (1990) Bright light treatment: Are we keeping our subjects in the dark. *Sleep* **13**, 267–271.
30. Eastman C.I. (1990) Circadian rhythms and bright light: Recommendations for shift work. *Work and Stress.* **4**, 245–260.
31. Eastman C.I. (1991) Squashing versus nudging circadian rhythms with artificial bright light: solutions for shift work? *Persp. Biol. Med.* **34**, 181–195.
32. Eastman C. (1990) What the placebo literature can tell us about SAD. *Psychopharm. Bull.* **26**, 495–504.
33. Eastman C.I., Young M.A. and Fogg L.F. (1993) A comparison of two different placebo-controlled SAD light treatment studies. In *Light and Biological Rhythms in Man.* (ed. Wetterberg L.), Pergamon Press, Stockholm, in press.
34. Foster R.G., Provencio D., Hudson S., Fiske S., DeGrip W. and Menaker M. (1991) Circadian photoreception in the retinally degenerate mouse (rd/rd). *J. Comp. Physiol.* **169**, 39–50.
35. French J., Hannon P.R. and Brainard G.C. (1990) Effects of bright illuminance on body temperature and human performance. *Ann. Rev. Chronopharmacol.* **7**, 37–40.
36. Gaddy J.R. (1990) Sources of variability in phototherapy. *Sleep Res.* **19**, 394.
37. Gaddy J.R., Edleson M., Stewart K., Brainard G.C. and Rollag M.D. (1992) Possible retinal spatial summation in melatonin suppresion. In *Biological Effects of Light* (eds. Holick M.F. and Kligman A.M.), pp. 196–204. Walter de Gruyter & Co., Berlin.
38. Gaddy, J.R., Rollag M.D. and Brainard G.C. Pupil size regulation of threshold of light-induced melatonin suppression. *J. Clin. Endocrin. Metab.* (in press).
39. Hamilton M. (1967) Development of a rating scale for primary depressive illness. *Brit. J. Soc. Clin. Psychol.* **6**, 276–296.
40. Holick M.F. and Kligman A.M. (eds.) (1992) *Biological Effects of Light.* Walter de Gruyter, Berlin.

41. Holtz M.M., Milette J.J., Takahashi J. and Turek W.F. (1990) Spectral sensitivity of the circadian clock's response to light in Djungarian hamsters. In *Second Meeting, Society for Research on Biological Rhythms* , Jacksonville, Florida.
42. Jacobs G.H., Neitz J. and Deegan III J.F. (1991) Retinal receptors in rodents maximally sensitive to ultraviolet light. *Nature* **353**, 655–656.
43. Jimerson D.C., Lynch H.J., Post R.M., Wurtman R.J. and Bunney W.E. (1977) Urinary melatonin rhythms during sleep deprivation in depressed patients and normals. *Life Sci.* **20**, 1501–1508.
44. Klein D.C., Smoot R., Weller J.L., Higa S., Markey S.P. and Creed G.J. (1983) Lesions of the paraventricular nucleus area of the hypothalamus disrupt the suprachiasmatic spinal cord circuit in the melatonin rhythm generating system. *Brain Res. Bull.* **10**, 647–652.
45. Klein D.C. and Weller J.L. (1972) A rapid light-induced decrease in pineal serotonin N-acetyltransferase activity. *Science* **177**, 532–533.
46. Kripke D.F. (1993) Light regulation of the menstrual cycle. In *Light and Biological Rhythms in Man.* (ed. Wetterberg L.), Pergamon Press, Stockholm (in press).
47. Kripke D.F., Mullaney D.J., Savides T.J. and Gillin J.C. (1989) Phototherapy for nonseasonal major depressive disorders. In *Seasonal Affective Disorders and Phototherapy* (eds. Rosenthal N.E. and Blehar M.C.), pp. 342–356. Guilford, New York.
48. Lam R.W., Buchanan A., Clark C., Remick R. (1991) Ultraviolet vs. nonultraviolet light therapy for SAD. *J. Clin. Psych.* **52**, 213–216.
49. Lerman S. (1980) *Radiant Energy and the Eye*, pp. 1–321. MacMillan, New York.
50. Levitt A.J., Joffee R.T., Martin L.S., Moul D.E., Lam R.W., Teicher M.H., Lebeue B., Oren D.A., Buchanan A., Glod C.A., Murry M.G. and Brown J. (1991) Light Visor Treatment of SAD. *Abstr. Ann. Mtg. Soc. Light Treatment Biol. Rhythms*, Toronto, June 14.
51. Lewy A.J., Herbert K.A., Rosenthal N.E. and Wehr T.A. (1982) Bright artificial light treatment of a manic depressive patient with a seasonal mood cycle. *Am. J. Psych.* **139**, 1496–1498.
52. Lewy A.J., Sack R.L., Miller L.S. and Hoban T.M. (1987) Antidepressant and circadian phase-shifting effects of light. *Science* **235**, 352–354.
53. Lewy A.J., Wehr T.A., Goodwin F.K., Newsome D.A. and Markey S.P. (1980) Light suppresses melatonin secretion in humans. *Science* **210**, 1267–1269.
54. Lynch H.J., Jimmerson D.C., Ozaki Y., Post R.M., Bunney Jr. W.E. and Wurtman R.J. (1978) Entrainment of rhythmic melatonin secretion in man to a 12-hour phase shift in the light/dark cycle. *Life Sci.* **23**, 1557–1564.
55. Martens H., Klein T., Rizzo III J.F., Shanahan T.L. and Czeisler C.A. (1992) Light-induced melatonin suppression in a blind man. In *Programs and Abstracts: Society for Research on Biological Rhythms*, p. 52. Jacksonville, FL.
56. McIntyre I.M., Norman T.R., Burrows G.D. and Armstrong S.M. (1989) Human melatonin suppression by light is intensity dependent. *J. Pineal. Res.* **6**, 149–156.
57. Mick G., Cooper H. and Magnin M. The retinal projection to the suprachiasmatic nucleus in mammals: Comparative and quantitative analysis of bilateral distribution. *J. Comp. Neur.* (in press).
58. Millette J.J., Holtz M.H., Takahashi J.S. and Turek F.W. (1987) Characterization of wavelength of light nessasary for the initiation of neuroendocrine gonadal activity in the male Djungarian hamster. In *Abstract 20th Annual Meeting Society for Study of Reproduction*, p. 110. University of Illinois, Urbana, IL.
59. Minneman K.P., Lynch H.J. and Wurtman R.J. (1974) Relationship between environmental light intensity and retina-mediated suppression of rat pineal serotonin-N-acetyltransferase. *Life Science* **15**, 1791–1796.
60. Minors D.S., Waterhouse J.M. and Wirz-Justice A. (1991) A human phase-response curve to light. *Neurosci. Lett.* **133**, 36–40.
61. Moore R.Y. (1983) Organization and function of a central nervous system circadian oscillator: The suprachiasmatic nucleus. *Fed. Proc.* **42**, 2783–2789.
62. Moore-Ede M.C., Czeisler C.A. and Richardson G.S. (1983) Circadian timekeeping in health and disease. *N. Eng. J. Med.* **309**, 530.

63. Moore-Ede M.C., Sulzman F.M. and Fuller C.A. (1982) *The Clocks that Time Us.*, pp. 1–408. Harvard University Press, Cambridge, Massachusetts.
64. Moul D.E., Rosenthal, N.E., Helleckson C.J., Oren D.A., Frank A., Brainard G.C., Murray M.G. and Wehr T.A. (1993) A multinuclear study of the light visor for seasonal affective disorder: no difference in efficacy between two different intensities. *Neuropsychopharm.* **8**, 151–160.
65. Nelson D.E. and Takahashi J.S. (1991) Comparison of visual sensitivity of pineal melatonin and circadian phase-shifting in the golden hamster. *Brain Res.* **554**, 272–277.
66. Nelson D.E. and Takahashi J.S. (1991) Sensitivity and integration in a visual pathway for circadian entrainment in the hamster. *J. Physiol.* **439**, 115–145.
67. Nguyen D.C., Hanifin J., Rollag M., Stetson M. and Brainard G.C. (1990) The influence of different photon densities of 620 nm light on pineal melatonin in Syrian hamsters. *Anat. Rec.* **226**, 72A.
68. Oren D.A., Brainard G.C., Joseph-Vanderpool J.R., Johnston S.H., Sorek E. and Rosenthal N.E. (1991) Treatment of seasonal affective disorder with green light versus red light. *Am. J. Psych.* **148**, 509–511.
69. Parry B., Rosenthal N., Tamarkin L. and Wehr T. (1987) Treatment of a patient with premenstrual syndrome. *Am. J. Psych.* **144**, 762–766.
70. Parry B., Berga S., Mostofi N., Sependa P., Kripke D. and Gillin J. (1989) Morning versus evening bright light treatment of late luteal phase dysphoric disorder. *Am. J. Psych.* **146**, 1215–1217.
71. Pickard G.E. and Silverman A. (1981) Direct retinal projections to the hypothalamus, periform cortex, and accessory optic nuclei in the Golden hamster as demonstrated by a sensitive anterograde horseradish peroxidase technique. *J. Comp. Neurol.* **196**, 155–172.
72. Podolin P.C., Rollag M.D. and Brainard G.C. (1987) The suppression of nocturnal pineal melatonin in the Syrian hamster: Dose-response curves at 500 nm and 360 nm. *Endocrinology* **121**, 266–270.
73. Reiter R.J. (1993) The mammalian pineal gland as an end organ of the visual system. In *Light and Biological Rhythms in Man* (ed. Wetterberg L.), Pergamon Press, Stockholm (in press).
74. Reiter R. J. (1981) The mammalian pineal gland: structure and function. *Am. J. Anat.* **162**, 287–313.
75. Reiter R. J. (1991) Pineal Gland: Interface between the photoperiodic environment and the endocrine system. *Trends Endocrinol. Metab.* **2**, 13–19.
76. Riggs L.A. (1965) Light as a stimulus for vision. In *Vision and Visual Perception* (ed. Graham C., et al.), pp. 1–38. John Wiley & Sons, New York.
77. Rodieck R.W. (1973) *The Vertebrate Retina*, pp. 1–1044. W.H. Freeman and Company, San Franscisco.
78. Rollag M.D. and Niswender G.D. (1976) Radioimmunoassay of serum concentrations of melatonin in sheep exposed to different lighting regimens. *Endocrinology* **98**, 482–489.
79. Rollag M.D., Panke E.S., Trakulrungski W., Trakulrungski C. and Reiter R.J. (1980) Qualification of daily melatonin synthesis in the hamster pineal gland. *Endocrinology* **106**, 231–236.
80. Rosenthal N.E., Sack D.A., Skwerer R.G., Jacobsen F.M. and Wehr T.A. (1988) Phototherapy for seasonal affective disorder. *J. Biol. Rhyth.* **3**, 101–120.
81. Rosenthal N.E., Sack D.A., Gillin J.C., Lewy A.J., Goodwin F.K., Davenport Y., Mueller P.S., Newsome D.A. and Wehr T.A. (1984) Seasonal affective disoder: A description of the syndrome and preliminary findings with light therapy. *Arch. Gen. Psych.* **41**, 72–80.
82. Rosenthal N.E., Joseph-Vanderpool J.R., Levendosky A.A., Johnston S.H., Allen R., Kelly K.A.,Souetre E., Schultz P.M. and Starz K.E. (1990) Phase-shifting effects of bright morning light as treatment for delayed sleep phase syndrome. *Sleep* **13**, 354–361.
83. Ross M. and Olson J.M. (1981) An expectancy-attribution model of the effects of placebos. *Psychol. Rev.* **88**, 408–437.
84. Sanford B., Brainard G., Beacham S., Hanifin J., Markoff J., Streletz L., Hannon P. and Sliney D. Dose dependent effects of UV-A on visual evoked potential in humans. In

Biologic Effects of Light (eds. Holick M. and Kligman A.), pp. 133–154. Walter de Gruyter, New York..

85. Sliney D. and Wolbarsht M. (1980) *Safety with Lasers and Other Optical Sources*, pp. 1–1035. Plenum Press, New York.
86. Smith K. C. (ed.) (1989) *The Science of Photobiology*, pp. 1–426. Plenum Press, New York.
87. Society for Light Treatment and Biological Rhythms (1991) Consensus statements on the safety and effectiveness of light therapy of depression and disorders of biological rhythms. *Light Therap. Biol. Rhythms*, 45–50.
88. Sokol S. (1976) Visually evoked potentials: Theory, techniques and clinical applications. *Survey Opthal.* **21**, 18–44.
89. Stewart K.T., Gaddy J.R., Byrne B., Miller S. and Brainard G.C. (1991) Effects of green or white light for treatment of seasonal depression. *Psych. Res.* **38**, 261–270.
90. Stewart K.T., Hanifin J.P., Rollag M.D., Stetson M.H. and Brainard G.C. (1990) Effects of near-ultraviolet radiation on melatonin rhythms and reproductive development in Siberian hamsters. In *Society for Research on Biological Rhythms*. Amelia Island, FL.
91. Stewart K.T., Rollag M.D., Stetson M.H. and Brainard G.C. (1990) Effects of long or short day cycles of near-ultraviolet radiation on melatonin rhythms of Siberian hamsters. *Anat. Rec.* **226**, 98A.
92. Takahashi J.S., Decoursey P.J., Bauman L. and Menaker M. (1984) Spectral sensitivity of a novel photoreceptive system mediating entrainment of mammalian circadian rhythms. *Nature* **308**, 186–188.
93. Tan K.E.W.P. (1971) Vision in the Ultraviolet. Thesis. University of Utrecht, pp. 1–92. Ultrecht, The Netherlands.
94. Terman J.S., Terman M., Schlager D., Rafferty B., Rosofsky M., Link M.J. and Quitkin F.M. (1990) Efficacy of brief intense light exposure for treatment of winter depression. *Psychopharm. Bull.* **26**, 3–11.
95. Terman M. and Terman J.S. (1992) Light therapy for winter depression In *Biologic Effects of Light* (eds. Holick M.F. and Kligman A.M.), pp. 133–154. Walter de Gruyter, Berlin.
96. Terman M., Terman J.S. and Quitkin F.M. (1989) Light therapy for seasonal affective disorder: a review of efficacy. *Neuropsychopharmacology*. **2**, 1–21.
97. Thiele G. and Meissl H. (1987) Action spectra of the lateral eyes recorded fron mammalian pineal glands. *Brain Res.* **424**, 10–16.
98. U.S. Congress, Office of Technology Assessment (1991) *Biological Rhythms: Implications for the Worker*. OTA-BA-463, U.S. Government Printing Office, 1–249.
99. Urbach F. (ed.) (1992) *Biological Responses to Ultraviolet A Radiation*, pp. 1–443. Valdenmar, Overland Park, Kansas.
100. Vanecek J. and Illnerova H. (1982) Night pineal N-acetyltransferase activity in rats exposed to white or red light pulses of various intensities and duration. *Experientia* **38**, 1318–1320.
101. Vaughan G., Pelham R., Pang S., Loughlin K., Wilson M., Sandock K., Vaughan M., Koslow S. and Reiter R. (1976) Nocturnal elevation of plasma melatonin and urinary 5-hydroxyindoleacetic acid in young men: Attempts at modification by brief changes in environmental lighting and sleep by automatic drugs. *J. Clin. Endocrin. Met.* **42**, 752–764.
102. Vaughan G.M., Bell R. and De La Pena A. (1979) Nocturnal plasma melatonin in humans: episodic pattern and influence of light. *Neurosci. Lett.* **14**, 81–84.
103. Waxler M. and Hitchens V.M. (1986) *Optical Radiation and Visual Health*, pp. 1–215. CRC Press, Boca Raton, FL.
104. Wehr T.A., Skwerer R.G., Jacobsen F.M., Sack D.A. and Rosenthal N.E. (1987) Eye versus skin phototherapy of seasonal affective disorder. *Am. J. Psych.* **144**, 753–757.
105. Wetterberg L. (1978) Melatonin in humans: Physiological and Clinical studies. *J. Neural Transm.* **13**, 289–310.
106. Wietstruck L. (1949) Histologische untersuchung der goldhamster-retina. PhD Thesis, Humboldt University, East Berlin, East Germany.

107. Wurtman R.J., Baum M.J. and Potts J.T. (eds.) (1985) *The Medical and Biological Effects of Light*. Ann. N.Y. Acad. Sci., New York.
108. Yerevanian B.I., Anderson J.L., Grota L.J. and Bray M. (1986) Effects of bright incandescent light on seasonal and nonseasonal major depressive disorder. *Psych. Res.* **18**, 355–364.

4

The Mammalian Melatonin Rhythm Generating System

DAVID C. KLEIN

Section on Neuroendocrinology, Laboratory of Developmental Neurobiology, National Institute of Child Health and Human Development, National Institutes of Health, Bethesda, MD 20892, USA

A DAY/NIGHT rhythm in circulating melatonin occurs in all vertebrates, with high values always coinciding with the dark period of the night. An example of this rhythm, as detected in cerebral spinal fluid of a non-human primate[60,60a], is presented in Figure 1. The nocturnal increase in circulating melatonin results from the increased production of melatonin by the serotonin→*N*-acetylserotonin→melatonin pathway, as outlined in Figure 2. This rhythm is driven by an endogenous ~24-hour clock, which is synchronized with the 24-hour day/night cycle; the synchronizing signal is light.

In non-mammalian species the clock and the photodetection system are typically located within the pineal gland itself. However, in mammals these elements are located outside the pineal gland. The clock is located in the suprachiasmatic nucleus of the hypothalamus (SCN)[32] and light is detected by the retina. In constant darkness the rhythm persists because the SCN continue to drive the pineal gland (Figure 1); however, the rhythm becomes slightly longer or shorter than 24 hours. The term "melatonin rhythm generating system" was coined to emphasize that the normal rhythm in melatonin in mammals reflects the integrated functioning of the pineal gland, SCN and eyes.[34] Lesions at any point in the system dramatically disrupt the rhythm, resulting in either no detectable rhythm or a rhythm which is not synchronized to the environmental lighting cycle, a circumstance described as free-running (see Figure 1).

The components of the melatonin rhythm generating system are given in Figure 3. In addition to the critical functional components of the system—the pineal gland, SCN and eye—other elements of the system function as wiring and relay points.

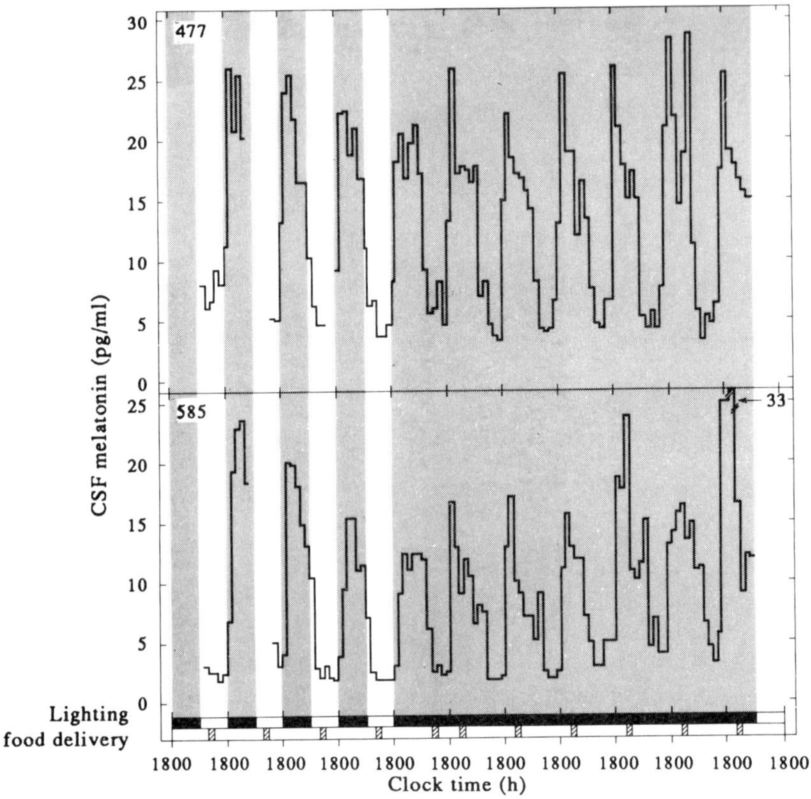

FIG. 1. Daily rhythm in melatonin in the CSF of a non-human primate. Night is represented by darkness. For 3 days animals were maintained in a light/dark cycle. After that animals were maintained in constant darkness, during which time the rhythm in melatonin persisted. CSF was continually withdrawn; melatonin was measured by radioimmunoassay (from reference 60a).

Historical perspective

As is true of many aspects of pineal biochemistry, physiology and pharmacology, the concept of a multicomponent system which regulates the melatonin rhythm can be traced to the pioneering work of McCord and Allen in 1917,[46] who first demonstrated that small amounts of extracts of the pineal gland cause frog skin to lighten. This was followed up by Aaron Lerner and his coworkers at Yale University in the late 1950s.[43,44] They extracted large amounts of cow pineal glands and were able to isolate a skin lightening compound and determine its structure. The compound was called melatonin.

The next major step was to demonstrate that melatonin could be synthesized within the pineal gland. This occurred soon after the structure of melatonin became known, as a result of the work of Julie Axelrod and his coworkers.[6,91,92] They obtained bovine pineal tissue preparations from

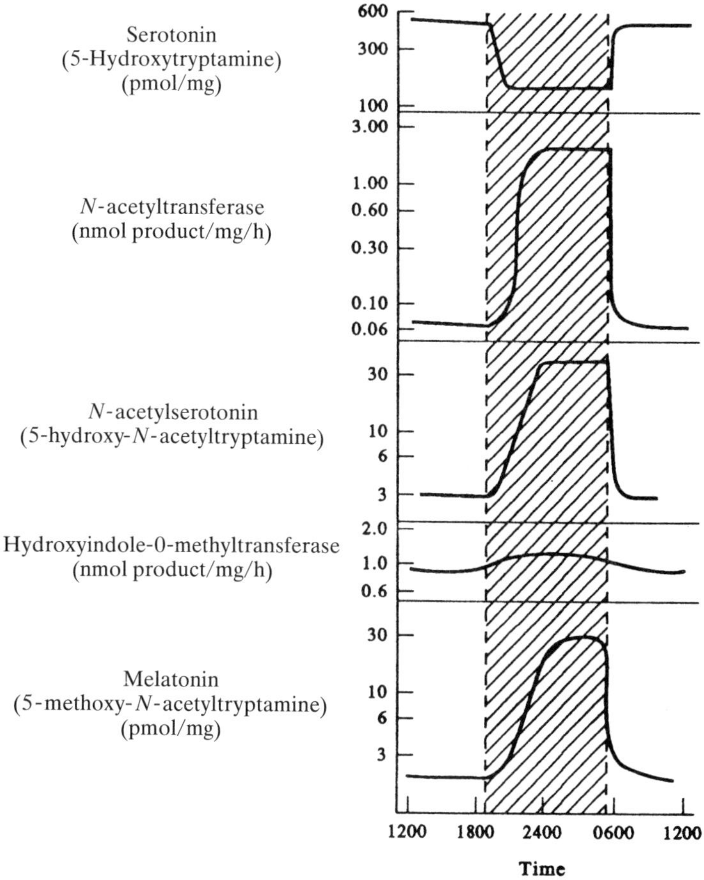

FIG. 2. Daily rhythms in the serotonin→melatonin pathway in the pineal gland. Rhythms in the serotonin→*N*-acetylserotonin→melatonin pathway are controlled by large changes in the activity of serotonin *N*-acetyltransferase activity (after reference 25).

Lerner and using these quickly found that the pineal gland could convert serotonin to *N*-acetylserotonin and *N*-acetylserotonin to melatonin. It had already been known that the pineal gland contained an abundant amount of serotonin.[18] Taken together, these advances established that the pineal gland could make melatonin.

The fact that there were large day/night changes in the amounts of indole compounds in the pineal gland was recognized by several workers,[17,57–59,69] who found that the amounts of serotonin and melatonin in the pineal gland exhibited day/night differences.

The next critical finding was our discovery of a very large (~100-fold) rhythm in the enzyme which converts serotonin to *N*-acetylserotonin, serotonin *N*-acetyltransferase (arylalkylamine *N*-acetyltransferase E.C.

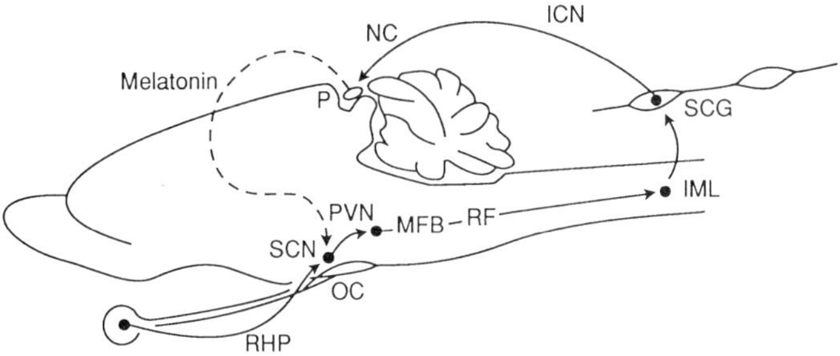

FIG. 3. Melatonin rhythm generating system: The rhythm is generated in the suprachiasmatic nuclei (SCN) and stimulatory signals are transmitted to the pineal gland (P) by the indicated pathway. Light acts through retinohypothalamic projections (RHP) to adjust the clock by resetting in and by adjusting the duration of the period during which the SCN will stimulate the pineal gland. Light also acts to block transmission of signals from the SCN to the pineal gland. The dotted line represents the melatonin→feedback loop. The following abbreviations are used: OC, optic chiasm; SCN, suprachiasmatic nuclei; PVN, paraventricular nucleus of the hypothalamus; MFP, medial forebrain bundle; RF, reticular formation; IML, intermediolateral cell column of spinal cord; SCG, superior cervical ganglia; ICN, internal carotid nerve; NC, nervi conarii (from reference 36).

2.3.1.87).[3,50,89] Based on this we proposed that the daily rhythm in melatonin was generated by the rhythm in serotonin *N*-acetyltransferase activity.[25,39] This rhythm produced a large nocturnal increase in *N*-acetylserotonin, which in turn generated a large nocturnal increase in melatonin production through a mass action effect mediated by hydroxyindole-O-methyltransferase; large day/night changes in the activity of this enzyme do not appear to occur.

The discovery of the rhythm in *N*-acetyltransferase activity was not only of significant intellectual interest to scientists interested in understanding the molecular basis of the melatonin rhythm. It also was of great practical importance because it provided investigators with a sensitive means of monitoring pineal function, at a time before methods to measure melatonin were available.

These early findings stimulated research on many aspects of the melatonin rhythm: What anatomical structures are involved in generating the rhythm? How are they connected? What transmitters control melatonin production? What is the biochemical basis of the increased synthesis of melatonin?

Functional anatomy

The elements comprising the mammalian melatonin rhythm generating system appear in Figure 3.[27]

The eye

The importance of the eye in the melatonin rhythm generating system was demonstrated in studies which extended the discovery that the nocturnal increase in *N*-acetyltransferase activity is suppressed by light.[38] Experiments determined that light does not act when the eyes were removed or when the optic nerves were severed. This indicated that the eyes mediated the effects of light. It was also found that light exposure at night causes a very rapid decrease in *N*-acetyltransferase activity, and that this effect is also mediated by the eyes.[39] These findings were subsequently confirmed in humans.[66]

Peripheral structures

The pineal gland is innervated by postganglionic projections from the superior cervical ganglia (SCG), which course with the inferior carotid nerve and then branch off to form the nervi conarii. The latter travel over the cerebellum to the pineal gland. The importance of these nerves in regulating melatonin synthesis came from studies which demonstrated that removal of the SCG blocks the large nocturnal increase in *N*-acetyltransferase activity and in melatonin production.[36,42]

The rhythm in *N*-acetyltransferase activity is not only dependent upon the presence of the SCG, but also upon central innervation of the ganglia. This was demonstrated by the finding that the rhythms in *N*-acetyltransferase activity and melatonin production are absent following decentralization.[36,42] These findings indicated that the clock regulating the pineal gland is not located in the pineal gland or SCG, but elsewhere.

Subsequent investigations have indicated that the pineal-controlling neuronal circuit coursing through the SCG is insulated from general sympathetic traffic.[42b] Stress, for example, has relatively little effect on pineal function.[54]

The role of the SCG in controlling the melatonin rhythm in man is evident from the observation that there is no melatonin rhythm in individuals with spinal cord injury blocking central innervation of the ganglia.[42a]

Central structures

With the evidence that central innervation regulated pineal function, it was proposed that the endogenous clock controlling the pineal gland was located in the brain. The available experimental evidence pointed to the hypothalamus as a likely site of the clock, according to studies by Richter, who had found that massive lesions in that region disrupted locomotor activity rhythms.[63] Our interest focused specifically on the SCN, because of new evidence from amino acid tracing experiments which established that a retino-hypothalamic projection terminated there.[47a]

The role of the SCN in the melatonin rhythm generating system was investigated in rats and a non-human primate using lesioning experiments.[31,47,61] In both cases it was found that lesioning the SCN profoundly altered the rhythms in *N*-acetyltransferase activity and melatonin. These observations and related findings established that the timing mechanisms regulating the pineal gland are located in the SCN.

The functional importance of the retino-hypothalamic project in the melatonin rhythm generating system was also established by lesioning experiments.[31,47] These demonstrated that cutting all other visual inputs from the retina did not block photic regulation of *N*-acetyltransferase activity, but that lesions which destroyed the retino-hypothalamic projection block all effects of light on pineal *N*-acetyltransferase activity. These effects include entraining the clock, modulating the duration of the period the clock can stimulate the pineal gland, and suppressing SCN stimulation of the pineal gland. (See Chapter 11.)

The neural elements in the SCN→pineal circuit which linked the SCN to the SCG appear to include projections from the SCN to the paraventricular nucleus of the hypothalamus (PVN) and monosynaptic projections from the PVN to the SCG. This was first indicated by tracing experiments which demonstrated that SCN→PVN and PVN→SCG projections exist.[8,70] These observations stimulated experiments which demonstrated that destruction of the PVN blocks the rhythm in *N*-acetyltransferase activity and melatonin production;[36] and, that electrical stimulation of the PVN stimulates these biochemical end points.[95]

Pineal signal transduction

Norepinephrine is contained in the sympathetic nerves which innervate the pineal gland and is released at night.[10a,19,94] Using organ culture experiments it was found that treatment with norepinephrine produces the same changes in pineal biochemistry which occur at night, including an increase in *N*-acetyltransferase activity and melatonin production.[5,40] Based on these findings, it appears that norepinephrine is the transmitter which controls large changes in melatonin production in the mammalian pineal gland. Organ culture studies together with whole animal studies established that the large day/night changes in melatonin production are due to large changes in the activity of serotonin *N*-acetyltransferase activity. Accordingly, interest in the molecular basis of how melatonin production is regulated usually is focused on *N*-acetyltransferase and the mechanisms through which receptors control the second messengers which regulate the activity of this enzyme (Figure 4), namely cyclic AMP and Ca^{++}.[28,29,96]

Norepinephrine acts on the pineal gland through at least three adrenergic receptors, which control interacting processes to regulate melatonin production. These interactions are mediated by electronic-like biochemical "AND" gates. Activation of such gates by any one receptor

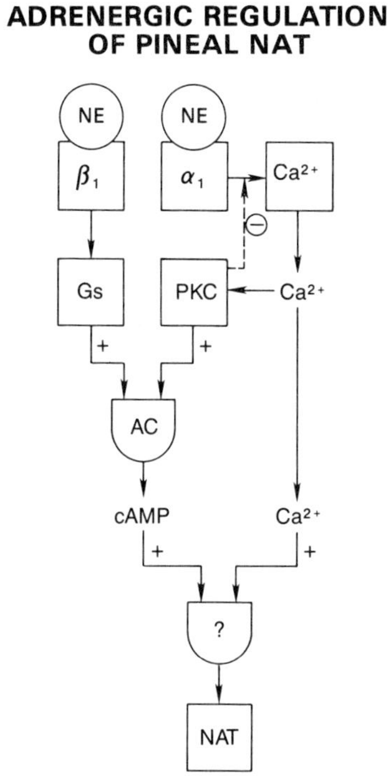

FIG. 4. "AND" Gate regulation of pineal *N*-acetyltransferase activity. Cyclic AMP is controlled by a dual receptor mechanism. Full stimulation of the enzyme appears to require activation of all three receptors. β-adrenergic stimulation activates adenylylate cyclase(AC) through the GTP binding stimulatory protein (Gsα). α_1-Adrenergic stimulation elevates $[Ca^{++}]_i$, which results in maximal activation of AC through a Ca^{++}, phospholipid dependent protein kinase (PKC) mechanism. Cyclic AMP and Ca^{++} act in concert to stimulate *N*-acetyltransferase activity.

has little influence. However, simultaneous activation of these gates by two receptors produces dramatic responses (Figure 4).

The adrenergic receptors involved belong to the β_1- and α_1-subgroups. The β_1-adrenergic receptor stimulates adenylate cyclase *via* a GTP binding regulatory protein.[7,55] This results in a partial increase in the intracellular concentration of cyclic AMP. Activation of this receptor is essential for any stimulation of melatonin production to occur. The second receptor involved is the α_1-adrenergic receptor.[72,75] Agonist occupancy of this receptor has several effects. The most rapid is an increase in $[Ca^{++}]_i$.[78] Another is α_1-adrenergic stimulation of phosphatidyl inositol turnover, which generates diacyglycerol.[21,23] Together, the increase in $[Ca^{++}]_i$ and diacylglycerol activates Ca^{++}, phospholipid dependent protein kinase (PKC).[22,74] This activation potently enhances β- adrenergic stimulation of cyclic AMP production and *N*-acetyltransferase activity.[77,79,86] Together, these elements form a PKC/G protein "AND" gate to control adenylate cyclase. Ca^{++} also acts downstream with cyclic AMP to stimulate the production of *N*-acetyltransferase, resulting in an increase in melatonin production. Most recently, α_{2D}-adrenergic receptors have been identified in the mammalian pineal gland.[67,68] The function of these receptors remains unclear.[68a]

The large changes in *N*-acetyltransferase are initiated by the action of Ca^{++} and cyclic AMP; Ca^{++} strongly potentiates the effects of cyclic AMP.[96] One effect of adrenergic stimulation is to increase gene expression, which occurs during a period of about 90 minutes after the start of adrenergic stimulation; it is possible that both Ca^{++} and cyclic AMP are involved in this, although this has not been proven. Thereafter, protein synthesis is required for the enzyme to reach full stimulation.[28,40] In addition, the stimulation of *N*-acetyltransferase activity requires ongoing cyclic AMP-dependent activation of the enzyme.[30] Accordingly, a rapid decrease in norepinephrine causes a rapid decrease in cyclic AMP, which results in inactivation of *N*-acetyltransferase and cessation of melatonin synthesis.

In summary, at least three elements are required for stimulation of enzyme activity, transcription, translation and activation. The second messengers involved appear to be cyclic AMP and Ca^{++}, which act through a cyclic AMP/Ca^{++} "AND" gate to stimulate *N*-acetyltransferase activity.[96] The simplest scheme is that cyclic AMP and Ca^{++} enhance generation of mRNA encoding serotonin *N*-acetyltransferase, which then directs synthesis of the protein; and, that the protein undergoes cyclic AMP, and/or Ca^{++}-dependent activation. However, this may not be the case because the possibility exists that *N*-acetyltransferase activity is regulated by an activator molecule which is controlled by the scheme outlined above.

In addition to the stimulatory actions detailed above, inhibitory-negative feedback-actions of norepinephrine also exist,[71] involving protein kinase C (Figure 4). These mechanisms may provide a rapid means of down-regulating signal transduction at the membrane level.

The rapid inactivation of *N*-acetyltransferase may involve protein thiol:disulfide exchange.[33,41,51] Cystamine and similar disulfides act through this mechanism to inactivate the enzyme in the intact cell and in broken cell preparations.[8,52,53] It is possible that cyclic AMP controls protein thiol:disulfide exchange by controlling an enzyme which converts a reservoir of free thiols to potent inactivating disulfides in the cell, thereby triggering inactivation of the enzyme.

It should be noted that cyclic AMP and Ca^{++} have been found to control expression of genes in other systems; this is one possible site of interaction of the cyclic AMP/Ca^{++} "AND" gate. Other possibilities include the mediation of cyclic AMP dependent and of Ca^{++} dependent kinases, or of the mediation of a cyclic AMP/Ca^{++} activated kinase.

Another interesting possible mechanism through which *N*-acetyltransferase is regulated is membrane potential (46). An intriguing observation is that the stimulation of *N*-acetyltransferase activity requires membrane hyperpolarization,[55] and that this hyperpolarization is produced by the cyclic AMP/Ca^{++} activation of a K^{+} channel.[13] The biochemical significance of hyperpolarization is not yet clear. It might reflect changes in a transmembrane spanning kinase or reductase, whose structure and activity are influenced by transmembrane potential. For example, hyperpolarization might activate an enzyme which keeps *N*-acetyltransferase in a reduced stated.

Although it is obvious that the rhythmic production of melatonin is regulated by the rhythm in *N*-acetyltransferase activity, the importance of other steps in the tryptophan→melatonin pathway should not be underestimated.[73] Changes in any one of the enzymes and cofactors required for this conversion could alter the amount of melatonin produced when the activity of *N*-acetyltransferase is elevated.

The melatonin→SCN feedback loop

Several lines of evidence indicate that melatonin feeds back into the melatonin rhythm generating system and that the primary melatonin target in the system is the SCN.[12] High affinity melatonin binding sites have been identified in the SCN of most mammals, using [^{125}I]2-iodomelatonin. (See Chapter 6.)[14] The concentration of melatonin binding sites in the rat SCN is more than 10-fold higher than in essentially all other tissues in mammals, except for the neonatal pituitary gland and the pars tuberalis.

Melatonin acts on cells in the SCN, as indicated by electrophysiological analysis of rat tissue.[19] Based on studies using other cells with melatonin binding sites, including melanocytes, cells in the pars tuberalis, and neonatal gonadotrophs, it appears that melatonin acts through the same type of high affinity binding site found in the SCN.[48,49,80,81,85,90,93] This establishes these sites as bona fide receptors.

Melatonin receptors are linked to GTP-binding regulatory proteins (G-proteins) which are characterized by sensitivity to pertussis toxin.[11,48,49,64] Studies on the neonatal rat pituitary cell indicate that a primary action of melatonin is hyperpolarization of the plasma membrane. This can alter cell physiology through a variety of voltage-sensitive channels. For example, melatonin-induced hyperpolarization blocks the GnRH-induced increase in $[Ca^{++}]_i$ in a subpopulation of melatonin-sensitive gonadotrophs by inhibiting Ca^{++} influx through voltage-sensitive Ca^{++} channels.[84] It seems possible that hyperpolarization in other cells could have similar effects on agonist-induced increase in $[Ca^{++}]_i$. These features provide reason to place melatonin receptors in a group of receptors whose actions are generally similar, including both hyperpolarization and inhibition by pertussis toxin.[3]

Hyperpolarization, however, is not the only effect of melatonin. Melatonin has also been found to block forskolin-induced increases in cAMP production in several systems, probably through an effect on adenylate cyclase mediated by an inhibitory G-protein.[48,49] Effects on other second messenger systems have also been reported.[87,88] Accordingly, it would seem that melatonin could act on SCN cells through a variety of mechanisms.

The most consistent physiological effect of administration of melatonin on biological rhythms is to modify entrainment.[3,45,66] This has been found in both experimental animals and in humans, as reviewed in detail in Chapter 12. The work on humans has been done in part with blind subjects and clearly demonstrates that melatonin administration can serve as a synchronizing cue which prevents free running circadian rhythms. It appears likely that melatonin is acting directly on the SCN because administration of melatonin entrains several circadian rhythms, which suggests that a common site of action is involved.[4] In addition, melatonin administration produces a phase response curve—advancing the clock at certain times and retarding it at others—most likely reflecting SCN involvement.[45,66]

The finding that melatonin acts on entrainment is consistent with the view that melatonin is useful in preventing jet lag.[1,2] In addition, it is thought that melatonin might be used successfully to minimize undesirable effects of phase shifting in circumstances where humans are subject to frequent changes in working schedules, such as in east–west intercontinental travel and shift work; and, it may be of practical value in entraining blind subjects to the environmental lighting schedule.[2a]

The melatonin→SCN feedback loop may also be involved in photic-induced changes in physiology which occur on a seasonal basis.[16,24,97] These changes can be mimicked by administration of melatonin. Melatonin only acts during a narrow window of time, late in the afternoon.[20] The precise details of this have not been worked out. However, it seems probable that the SCN is involved.

To summarize, a melatonin→SCN feed back loop appears to exist in many mammalian species and probably plays a physiological role in the melatonin rhythm generating system, perhaps fine tuning the process of entrainment. The existence of melatonin sensitivity in the SCN provides investigators and clinicians with the opportunity of accessing and altering circadian function through prudent time-dependent use of melatonin.

Development

The development of the melatonin rhythm generating system reflects the ontogeny of each element.[26,34] One of the earliest to develop is clock function in the SCN.[62] A circadian rhythm in *N*-acetyltransferase is first detectable in the rat pineal gland at the end of the first week of life.[15] One of the last elements to develop is the enzymatic capacity of the pineal gland to convert *N*-acetylserotonin to melatonin.[34] This topic has been covered in several books.[26,59] Of special developmental interest is the observation that melatonin mediates striking seasonal control over reproductive development seen in some mammals (see Chapter 6).[65]

Evolution

The evolution of the mammalian eye→SCN→pineal melatonin rhythm generating system is beginning to be understood. As indicated in the Introduction, the system seems to have evolved late in vertebrate evolution, because it is absent or redundant in non-mammalian vertebrates, which appear to have a melatonin rhythm generating system contained within their pineal gland. The evolution of the mammalian melatonin rhythm generating system reflects a distinct improvement—one which centralizes circadian functions—an optimization which uses the SCN as the single master circadian clock driving all rhythms.

In considering evolutionary aspects, it is important to note that the pineal gland and retina share a number of striking biochemical and anatomical features not found elsewhere.[54] These include phototransduction molecules and melatonin synthesizing enzymes and photoreceptors. From lower to higher vertebrates there is a gradient of cross-over in function and expression of these features, with both melatonin production and phototransduction evident in both the pineal and retina of lower forms; in mammals, however, melatonin production is generally restricted to the pineal gland and photoreception is limited to the retina. The ability of the pineal gland and retina of lower forms to detect light and make melatonin probably reflects the evolution of both tissues from a common primitive photochemical transduction system which may function in ancestral vertebrate forms.

The association of melatonin with darkness may have its origins very early in the history of life on earth. This is based on the finding that the last enzyme in melatonin synthesis, hydroxyindole-O-methyltransferase (HIOMT), is homologous to an enzyme found in *Rhodobacter capsulata*, hydroxyneurosporene-O-methyltransferase (HNOMT).[35] The latter O-methylates carotenoids, which function in *R. capsulata* to absorb light and transfer energy to the photosynthetic apparatus; they also protect against photo-oxidative damage. Within a 180 amino acid-long region of

HIOMT, 55 (31%) of the residues are identical and an additional 78 (43%) represent conservative substitutions. The similarity of the primary structures of HIOMT and HNOMT indicates both enzymes evolved from a common ancestral protein.

R. capsulata is thought to be closely related to a primitive endosymbiont, existing 2 to 3 billion years ago, which evolved within the primitive eukaryotic cell to become the mitochondrion.

Relevant to the melatonin—darkness link is the observation that the activity of each enzyme decreases as light intensity increases from low to high levels and vice versa. In the case of *R. capsulata* the entire photosynthetic apparatus, including HNOMT, appears to increase in dim light. This is an adaptive survival strategy which increases the efficiency of light energy capture when light intensity is low; such an adaptation is an obvious advantage to a photosynthetic organism, especially several billion years ago when light was the energy source and the atmosphere was devoid of oxygen.

Based on these observations, it would appear that photic regulation of O-methylation originated as part of a general adaptive strategy to increase photosynthetic efficiency, and was recruited to support melatonin production. Although there is no evidence that *R. capsulata* can synthesize melatonin, it is still fascinating that traces of the melatonin rhythm generating system are found at the earliest times in the evolution of life closely associated with the photic environment.

References

1. Arendt J. (1991) Melatonin in humans jet-lag and after. In *Advances in Pineal Research* (eds. Arendt J. and Pevet P.) Vol. 5, pp. 299–302. J. Libbey, London.
2. Arendt J., Aldhous M. and Marks V. (1988) Further evaluation of the treatment of jet lag by melatonin a double blind crossover study. *Ann. Rev. Chronopharmacol.* **5**, 53–55.

2a. Arendt J., Aldhous M. and Wright J. (1988) Synchronisation of a disturbed sleep-wake cycle in a blind man by melatonin treatment. *Lancet* **April**, 772–773.

3. Armstrong D.L. and White R.E. (1992) An enzymatic mechanism for potassium channel stimulation through pertussis-toxin-sensitive G proteins. *Trends Neurosci.* **15**, 403–408.
4. Armstrong S.M. and Redman J. (1985) Melatonin administration effects on rodent circadian rhythms. In *Photoperiodism, Melatonin and the Pineal* (eds. Evered D. and Clark S.) pp. 188–207. Ciba Foundation Symposium 117. Pitman, London.
5. Axelrod J., Shein H.M. and Wurtman R.J. (1969) Stimulation of C^{14}-tryptophan by norepinephrine in rat pineal in organ culture. *Proc. Natl. Acad. Sci. USA* **62**, 544–549.
6. Axelrod J. and Weissbach H. (1960) Enzymatic *O*-methylation of *N*-acetylserotonin to melatonin. *Science* **131**, 1312.
7. Axelrod J. and Zatz M. (1977) The β-adrenergic receptor and the regulation of circadian rhythms in the pineal gland. In *Biochemical Actions of Hormones* (ed. Litwack G.), Vol. 4, pp. 249–268. Academic Press, New York.
8. Berk M.L. and Finklestein J.A. (1981) An autoradiographic determination of the afferent projections of the suprachiasmatic nucleus of the hypothalamus. *Brain Res.* **226**, 1–13.
9. Binkley S., Klein D.C. and Weller J.L. (1976) Pineal serotonin *N*-acetyltransferase activity: Protection of stimulated activity by acetyl-CoA and related compounds. *J. Neurochem.* **26**, 51–55.

10. Bondareff W. and Gordon B. (1966) Submicroscopic localization of norepinephrine in sympathetic nerves of rat pineal. *J. Pharmacol. Exp. Ther.* **153**, 42–47.
10a. Brownstein M. and Axelrod J. (1974) Pineal gland 24 hour rhythm in norepinephrine turnover. *Science* **184**, 163–165.
11. Carlson L.L., Weaver D.R. and Reppert S.M. (1989) Melatonin signal transduction in hamster brain: Inhibition of adenylyl cyclase by a pertussis toxin-sensitive protein. *Endocrinology* **125**, 2670–2676.
12. Cassone V.M. (1991) Melatonin and suprachiasmatic nucleus function. In *Suprachiasmatic Nucleus* (ed. Klein D.C., Moore R.Y. and Reppert S.M.), pp. 309–323. Oxford Press, New York.
13. Ceña V., Halperin J., Yeandle S. and Klein D.C. (1991) Norepinephrine stimulates potassium efflux from pinealocytes: Evidence for involvement of biochemical "AND" gate operated by calcium and adenosine 3′,5′-monophosphate. *Endocrinology* **128**, 559–569.
14. Dubocovich M.L. and Takahashi J.S. (1987) Use of 2-[^{125}I]-iodomelatonin to characterize melatonin binding sites in chicken retina. *Proc. Natl. Acad. Sci. USA* **84**, 3916–1920.
15. Ellison N., Weller J.L. and Klein D.C. (1972) Development of a circadian rhythm in the activity of pineal serotonin *N*-acetyltransferase. *J. Neurochem.* **19**, 1335–1341.
16. Evered, D and Clark. C. (eds) (1985) *Photoperiodism, Melatonin and the Pineal.* Ciba Foundation Symposium 117, Pittman Press, London.
17. Fiske V.M. (1964) Serotonin rhythm in the pineal organ: control by the sympathetic nervous system. *Science* **146**, 253–254.
18. Gairman N.J. and Day M. (1959) Presence of biogenic amines in the bovine pineal body. *Biochem. Pharmacol.* **1**, 235.
19. Gillette M.U. (1991) SCN electrophysiology *in vitro*: Rhythmic activity and endogenous clock properties. In *Suprachiasmatic Nucleus* (eds. Klein D.C., Moore R.Y. and Reppert S.M.), pp. 125–143. Oxford Press, New York.
20. Goldman B., Hall V., Hollister C., Roychoudhury P., Tamarkin L. and Westrom W. (1979) Effects of melatonin on the reproductive system in intact and pinealectomized male hamsters maintained under various photoperiods. *Endocrinology* **104**, 82–88.
21. Ho A.K., Chik C.L. and Klein D.C. (1988) Permissive role of calcium in a_1-adrenergic stimulation of pineal phosphatidylinositol phosphodiesterase (Phospholipase C) Activity. *J. Pineal Res.* **5**, 553–564.
22. Ho A.K., Thomas T.P., Anderson W., Chik C.L. and Klein D.C. (1988) Protein kinase C: Subcellular redistribution by increased Ca^{++} influx. *J. Biol. Chem.* **263**, 9292–9297.
23. Ho A.K. and Klein D.C. (1987) Phosphatidylinositol phosphodiesterase (Phospholipase C) activity in the pineal gland: characterization and photoneural regulation. *J. Neurochem.* **48**, 1033–1038.
24. Karsch F.J., Woodfill C.J.I., Malpaux B., Robinson J.E. and Wayne N.L. (1991) Melatonin and mammalian photoperiodism: Synchronization of annual reproductive cycles. In *Suprachiasmatic Nucleus* (eds. Klein D.C., Moore R.Y. and Reppert S.M.), pp. 217–232. Oxford Press, New York.
25. Klein D.C. (1974) Circadian rhythms in indole metabolism in the rat pineal gland. In *The Neurosciences; Third Study Program* (eds. Schmitt F.O. and Worden F.G.), pp. 509–516. MIT Press, Cambridge.
26. Klein D.C. (ed.) (1982) *The Melatonin Rhythm Generating System: Developmental Aspects.* Kargar Press, Basal.
27. Klein D.C. (1985) Photoregulation of the mammalian pineal gland. In *Photoperiodism, Melatonin and the Pineal.* (eds. Evered D. and Clark S.) pp. 38–56. CIBA Foundation symposium. Pitman, London.
28. Klein D.C., Berg G.R. and Weller J.L. (1970) Melatonin synthesis: Adenosine 3′,5′-monophosphate and norepinephrine stimulate *N*-acetyltransferase. *Science* **168**, 979–980.
29. Klein D.C., Berg G.R., Weller J.L. and Glinsmann W. (1970) Pineal gland: Dibutyryl cyclic adenosine monophosphate stimulation of labelled melatonin production. *Science* **167**, 1738–1740.
30. Klein D.C., Buda M.J., Kapoor C.L. and Krishna G. (1978) Pineal serotonin *N*-

acetyltransferase activity: Abrupt decrease in adenosine 3′-5′-monophosphate may be the signal for "turn-off". *Science* **199**, 309–311.

31. Klein D.C. and Moore R.Y. (1979) Pineal *N*-acetyltransferase and hydroxyindole-*O*-methyltransferase: Control by the retinohypothalamic tract and the suprachiasmatic nucleus. *Brain Res.* **174**, 245–262.
32. Klein D.C., Moore R.Y. and Reppert S.M. (eds.) (1991) *Suprachiasmatic Nucleus: The Mind's Clock*. Oxford Press, New York.
33. Klein D.C. and Namboodiri M.A.A. (1982) Control of the circadian rhythm in pineal serotonin N-acetyltransferase activity: possible role of protein thiol:disulfide exchange. *Trends Biochem. Sci.* **7**, 98–102.
34. Klein D.C., Namboodiri M.A.A. and Auerbach, D.A. (1981) The melatonin rhythm generating system: Developmental aspects. *Life Sci.* **28**, 1975–1986.
35. Klein D.C., Roseboom P.H., Donohue S.J. and Marrs B.L. (1992) Evolution of melatonin as a night signal: Contribution from a primitive photosynthetic organism. *Mol. Cell. Neurosci.* **3**, 181–183.
36. Klein D.C., Smoot R., Weller J.L., Higa S., Markey S.P., Creed G.J. and Jacobowitz D.M. (1983) Lesions of the paraventricular nucleus area of the hypothalamus disrupt the suprachiasmatic spinal cord circuit in the melatonin rhythm generating system. *Brain Res. Bulletin* **10**, 647–652.
37. Klein D.C., Sugden D. and Weller J.L. (1983) Postsynaptic α-adrenergic receptors potentiate the β-adrenergic stimulation of pineal serotonin N-acetyltransferase. *Proc. Natl. Acad. Sci. USA* **80**, 599–603.
38. Klein D.C. and Weller J.L. (1970) Indole metabolism in the pineal gland: A circadian rhythm in *N*-acetyltransferase. *Science* **169**, 1093–1095.
39. Klein D.C. and Weller J.L. (1972) Rapid light-induced decrease in pineal serotonin *N*-acetyltransferase activity. *Science* **177**, 532–533.
40. Klein D.C. and Weller J.L. (1973) Adrenergic-adenosine 3′,5′-monophosphate regulation of serotonin *N*-acetyltransferase activity and the temporal relationship of serotonin N-acetyltransferase activity to synthesis of ^{3}H-melatonin in the cultured rat pineal gland. *J. Pharmacol. Exp. Ther.* **186**, 516–527.
41. Klein D.C., Weller J.L., Auerbach D.A. and Namboodiri M.A.A. (1980) Regulation of pineal N-acetyltransferase activity: Focus on "turn-off". In: *Enzymes and Neurotransmitters in Mental Disease* (eds. Usdin E., Sourkes T.L. and Youdim M.B.H.), pp. 603–627. John Wiley, London.
42. Klein D.C., Weller J.L. and Moore R.Y. (1971) Melatonin metabolism: Neural regulation of pineal serotonin:acetyl coenzyme A *N*-acetyltransferase activity. *Proc. Natl. Acad. Sci. USA* **68**, 3107–3110.
43. Lerner A.B., Case J.D., Takahashi Y., Lee T.H. and Mori W. (1958) Isolation of melatonin, the pineal gland factor that lightens melanocytes. *J. Am. Chem. Soc.* **80**, 2587.
44. Lerner A.B., Case J.D. and Heinzelman R.V. (1959) Structure of melatonin. *J. Am. Chem. Soc.* **81**, 6084–6085.
45. Lewy A.J., Ahmed S., Jackson J.M.L. and Sack R.L. (1992) Melatonin shifts human circadian rhythms according to a phase-response curve. *Chronobiol. Int.* **9**, 380–392.

45a. Lingappa J.R. and Zigmond R.E. (1984) A comparison of the regulation of serotonin *N*-acetyltransferase]activity in normal pineal glands and pineal glands transplanted in *oculo*. *J. Steroid Biochem.* **20**, 1457.

45b. Lingappa J.R. and Zigmond R.E. (1987) Pineal transplants in-oculo limitations on the ability of collateral]sprouts of foreign neurons to establish normal function. *J. Neurosci.* **7**, 3525–3528.

46. McCord C.P. and Allen F.P. (1917) Evidence associating pineal gland function with alterations in pigmentation. *J. Exp. Zool.* **23**, 207–224.
47. Moore R.Y. and Klein D.C. (1974) Visual pathways and the central neural control of a circadian rhythm in pineal serotonin N-acetyltransferase activity. *Brain Res.* **71**, 17–33.

47a. Moore R.Y. and Lenn N.J. (1972) A retino hypothalamic projection in the rat. *J. Comp. Neurol.* **146**, 1–14.

48. Morgan P.J., Lawson W., Davidson G. and Howell H.E. (1989) Melatonin inhibits

cyclic AMP production in cultured ovine pars tuberalis cell. *J. Mol. Endocrinol.* **3**, R5–R8.

49. Morgan P.J., Lawson W., Davidson G. and Howell H.E. (1989) Guanine nucleotides regulate the affinity of melatonin receptors on the ovine pars tuberalis. *Neuroendocrinology* **50**, 359–362.
50. Namboodiri M.A.A., Dubbels R. and Klein D.C. (1986) Arylalkylamine *N*-acetyltransferase (mammalian pineal). *Methods Enzymol.* **142**, 583–596.
51. Namboodiri M.A.A., Favilla J.T. and Klein D.C. (1981) Pineal *N*-acetyltransferase is inactivated by disulfide-containing peptides: Insulin is the most potent. *Science* **213**, 571–573.
52. Namboodiri M.A.A., Nakai C. and Klein D.C. (1979) Effects of selected treatments on stability and activity of pineal serotonin *N*-acetyltransferase. *J. Neurochem.* **33**, 807–810.
53. Namboodiri M.A.A., Weller J.L. and Klein D.C. (1980) Evidence of inactivation of pineal indoleamine N-acetyltransferase by protein thiol: disulfide exchange. *J. Biol. Chem.* **255**, 6032–6035.
54. O'Brien P.J. and Klein D.C. (eds.) (1986) *Pineal and Retinal Relationships*. Academic Press, Orlando.

54a. Parfitt A.G. and Klein D.C. (1976) Sympathetic nerve endings in the pineal gland protect against acute stress-induced increase in *N*-acetyltransferase (E.C. 2.3.1.5) activity. *Endocrinology* **99**, 840–851.

55. Parfitt A., Weller J.L. and Klein D.C. (1976) Beta adrenergic blockers decrease adrenergically stimulated N-acetyltransferase activity in pineal glands in organ culture. *Neuropharmacology* **15**, 353–358.
56. Parfitt A., Weller J.L., Sakai K.K., Marks B.H. and Klein D.C. (1975) Blockade by ouabain or elevated potassium ion concentration of the adrenergic and adenosine cyclic 3′,5′-monophosphate-induced stimulation of pineal serotonin *N*-acetyltransferase activity. *Mol. Pharmacol.* **11**, 241–255.
57. Quay W.B. (1963) Circadian rhythm in rat pineal serotonin and its modification of estrous cycle and photoperiod. *Gen. Compl. Endocrinol.* **3**, 473–479.
58. Quay W.B. (1964) Circadian and estrous rhythms in pineal melatonin and 5-hydroxyindole-3-indole acetic acid. *Proc. Soc. Exp. Biol. Med.* **115**, 710–714.
59. Quay W.B. and Halevy A. (1962) Experimental modification of the rat pineal's content of serotonin and related indole amines. *Physiol. Zool.* **35**, 1–7.

59a. Reppert S.M. (ed) (1989) *Development of circadian rhythmicity and photoperiodism in mammals*. Perinatology Press, New York.

60. Reppert S.M., Perlow M.J., Tamarkin L. and Klein D.C. (1979) A diurnal melatonin rhythm in primate cerebrospinal fluid. *Endocrinology* **104**, 295–301.

60a. Reppert, S.M., Perlow, M.J., Tamarkin, L., Orloff, D. and Klein, D.C. (1981) The effects of environmental lighting on the daily melatonin rhythm in primate cerebrospinal fluid. *Brain Research* **223**, 313–323.

61. Reppert S.M., Perlow M.J., Ungerleider L., Mishkin M., Tamarkin L., Orloff D.G., Hoffman H. and Klein D.C. (1981) Effects of damage to the suprachiasmatic area of the anterior hypothalamus on the daily melatonin and cortisol rhythms in the Rhesus monkey. *J. Neurosci.* **1**, 1414–1425.
62. Reppert S.M. and Weaver D.R. (1991) A biological clock is oscillating in the fetal suprachiasmatic nuclei. In *Suprachiasmatic Nucleus* (eds. Klein D.C., Moore R.Y. and Reppert S.M.), pp. 405–418. Oxford Press, New York.
63. Richter C.P. (1967) Sleep and activity: their relation to the 24-hour clock. *Res. Publ. Assoc. Res. Nerv. Ment. Dis.* **45**, 8–29.
64. Rivkees S.A., Carlson L.L. and Reppert S.M. (1989) Guanine nucleotide-binding protein regulation of melatonin receptors in lizard brain. *Proc. Natl. Acad. Sci. USA* **86**, 3882–3886.
65. Rivkees S.A. and Reppert S.M. (1991) Development of SCN structure and function in an opossum. In *Suprachiasmatic Nucleus* (eds. Klein D.C., Moore R.Y. and Reppert S.M.), pp. 419–428. Oxford Press, New York.
66. Sack R.L., Lewy A.J., Blood M.L., Keith L.D. and Nakagawa H. (1992) Circadian

rhythm abnormalities in totally blind people incidence and clinical significance. *J. Clin. Endocrinol. Metab.* **75**, 127–134.
67. Schaad N.C. and Klein D.C. (1992) Characterization of α_2-adrenergic receptors of rat pinealocytes. *Endocrinology* **130**, 2804–2810.
68. Simonneaux V., Ebadi M. and Bylund D.B. (1991) Identification and characterization of α_{2D}-adrenergic receptors in bovine pineal gland. *Mol. Pharmacol.* **40**, 235–241.
69. Snyder S.H., Zweig M., Axelrod J. and Fischer J.E. (1965) Control of the circadian rhythm in serotonin content of the rat pineal gland. *Proc. Natl. Acad. Sci. USA* **53**, 301–305.
70. Stephen F.K., Berkley K.J. and Moss R.L. (1981) Efferent projections of the rat suprachiasmatic nucleus. *Neuroscience* **6**, 2625–2641.
71. Sugden D., Ho A.K., Sugden A.L. and Klein D.C. (1988) Negative feedback mechanisms: evidence that desensitization of pineal a_1-adrenergic responses involves protein kinase C. *Endocrinology* **123**, 1425–1432.
72. Sugden D. and Klein D.C. (1983) Rat pineal alpha$_1$-adrenoceptors: Identification and characterization using [^{125}I]iodo-2-[β-(4-hydroxyphenyl)-ethylaminomethyl]-tetralone ([^{125}I]HEAT). *Endocrinology* **114**, 435–440.
73. Sugden D. and Klein D.C. (1983) Adrenergic stimulation of rat pineal hydroxyindole-*O*-methyltransferase. *Brain Res.* **265**, 348–351.
74. Sugden D. and Klein D.C. (1988) Activators of protein kinase C act at a post-receptor site to amplify cAMP production in rat pinealocytes. *J. Neurochem.* **32**: 149–155.
75. Sugden D., Namboodiri M.A.A., Klein D.C., Pierce J.E., Grady R.K., Jr. and Mefford I. (1985) Ovine pineal α_1-adrenoreceptors: Characterization and evidence for a functional role in the regulation of serum melatonin. *Endocrinology* **116**, 1960–1967.
76. Sugden D., Vanecek J., Klein D.C., Thomas T.P. and Anderson W.B. (1985) Activation of protein kinase C potentiates isoprenaline-induced cyclic AMP accumulation in rat pinealocytes. *Nature* **314**, 359–361.
77. Sugden D., Vanecek J., Klein D.C., Thomas T.P. and Anderson W. (1985) Activation of protein kinase C potentiates isoprenaline-induced cyclic AMP accumulation in rat pinealocytes. *Nature* **314**, 359–361.
78. Sugden L., Sugden D. and Klein D.C. (1987) α_1-Adrenoceptor activation elevates cytosolic calcium in rat pinealocytes by increasing net influx. *J. Biol. Chem.* **262**, 741–745.
79. Sugden L., Sugden D. and Klein D.C. (1987) Essential role of calcium influx in the adrenergic regulation of cAMP and cGMP in rat pinealocytes. *J. Biol. Chem.* **261**, 11608–11612.
80. Vanecek J. (1988) Melatonin binding sites. *J. Neurochem.* **51**, 1436–1440.
81. Vanecek J. (1988) The melatonin receptors in rat ontogenesis. *Neuroendocrinology* **48**, 201–203.
82. Vanecek J. and Klein D.C. (1992) Melatonin inhibits GnRH stimulation of $[Ca^{++}]_i$ in neonatal rat pituitary cells. *Endocrinology* **130**, 701–707.
83. Vanecek J. and Klein D.C. (1992) Sodium-dependent effects of melatonin on membrane potential of neonatal rat pituitary cells. *Endocrinology* **131**, 939–946.
84. Vanecek J. and Klein D.C. A subpopulation of neonatal gonadotrophs is sensitive to melatonin. *Endocrinology* (submitted).
85. Vanecek J., Pavlik A. and Illnerova H. (1987) Hypothalamic melatonin receptor sites revealed by autoradiography. *Brain Res.* **435**, 359–362.
86. Vanecek J., Sugden D., Weller J.L. and Klein D.C. (1985) Atypical synergistic α_1- and β-adrenergic regulation of adenosine 3′,5′-monophosphate and guanosine 3′,5′-monophosphate in rat pinealocytes. *Endocrinology* **116**, 2167–2173.
87. Vanecek J. and Vollrath L. (1989) Melatonin inhibits cyclic AMP and cyclic GMP accumulation in the rat pituitary. *Brain Res.* **505**, 157–159.
88. Vanecek J. and Vollrath L. (1990) Melatonin modulates diacylglycerol and arachidonic acid metabolism in the rat anterior pituitary. *Neurosci. Lett.* **110**, 199–203.
89. Voisin P., Namoodiri M.A.A. and Klein D.C. (1984) Arylamine N-acetyltransferase and arylalkylamine *N*-acetyltransferase in the mammalian pineal gland. *J. Biol. Chem.* **259**, 10913–10918.

90. Weaver D.R., Rivkees S.A., Carlson L.L. and Reppert S.M. (1991) Localization of melatonin receptors in mammalian brain. In *Suprachiasmatic Nucleus* (eds. Klein D.C., Moore R.Y. and Reppert S.M.), pp. 289–308. Oxford Press, New York.
91. Weissbach H., Redfield B.G. and Axelrod J. (1960) Biosynthesis of melatonin; Enzymic conversion of serotonin to *N*-acetylserotonin. *Biochim. Biophys. Acta* **43**, 352–353.
92. Weissbach H., Redfield B.G. and Axelrod J. (1961) The enzymatic acetylation of serotonin and other naturally occurring amines. *Biochim. Biophys. Acta* **54**, 190–192.
93. White B.H., Sekura R.D. and Rollag M.D. (1987) Pertussis toxin blocks melatonin-induced pigment aggregation in Xenopus dermal melanophores. *J. Comp. Physiol.* B **157**, 153–159.
94. Wolfe D., Potter L., Richardson K. and Axelrod J. (1962) Localizing tritiated norepinephrine in sympathetic axons by electron microscopic autoradiography. *Science* **138**, 440–442.
95. Yanovski J., Witcher J., Adler N., Markey S. and Klein, D.C. (1987) Stimulation of the paraventricular nucleus area of the hypothalamus elevates urinary 6-hydroxymelatonin during daytime. *Brain Res. Bulletin* **19**, 129–133.
96. Yu L., Schaad N. and Klein D.C. (1991) Calcium potentiates cyclic AMP stimulation of pineal N-acetyltransferase. *J. Neurochem.* (in press).
97. Zucker I., Lee T.M. and Dark J. (1991) The suprachiasmatic nucleus and annual rhythms of mammals. In *Suprachiasmatic Nucleus* (eds. Klein D.C., Moore R.Y. and Reppert S.M.), pp. 246–259. Oxford Press, New York.

5

Circadian Photoreception in Mammals and Other Vertebrates

RUSSELL G. FOSTER and MICHAEL MENAKER

Department of Biology and NSF Center for Biological Timing, University of Virginia, Charlottesville, VA 22903, USA

CIRCADIAN RHYTHMS are driven by endogenous pacemakers that have periods close to, but rarely exactly, 24 hours. In order to function adaptively in the real world, circadian *systems*, which include pacemakers and the many rhythms that they control must be synchronized (entrained) with the astronomical day—their periods must be constrained to become exactly 24 hours (at least on average). Entrainment regulates not only period but more importantly, phase. When entrained, the pacemaker of the circadian system must, of necessity, adopt a determinate phase relationship with the astronomical day while individual circadian rhythms may each adopt particular and different phase relationships depending on the specifics of their coupling with the pacemaker. It is phase relationships with the environment and with other rhythms that are adaptive and have therefore been subject to natural selection. Indeed, it is the sole function of the circadian system to regulate the times at which biological events occur; times either in relation to specific features of the daily environmental cycle such as dawn or dusk or in relation to other cyclic events within the organism. Viewed from a circadian perspective, an organism is an ensemble of adaptive phase relationships among multiple rhythms at all levels of organization from the molecular to the behavioral.

Thus the process of entrainment is central to circadian function and must be understood in detail if we are to understand circadian systems. This fact has been recognized for some time, and as a result a good deal of effort has been expended on experiments which have addressed various aspects of entrainment. Central to almost all of these experiments has been the recognition that the dominant environmental entraining agent is the daily alternation of light and darkness. Although other environmental cycles (principally temperature) can entrain circadian systems, it is the light cycle which, because of its precision and reliability, has been used by natural selection as the major external temporal reference. That is not at all surprising, however, many of the most "obvious" assumptions about the mechanisms that underlie photic entrainment of circadian systems have turned out on experimental examination to be unwarranted. The first of these is that, for vertebrates, photic entrainment is mediated by the eyes; it is this assumption that we examine in detail below.

Although it had been known for some time that some vertebrates possessed extraretinal, non-visual photoreceptors,[18,48] the first report that such receptors mediated circadian entrainment was the work of Menaker[35] who showed that house sparrows (*Passer domesticus*) would entrain to laboratory light cycles even though they had been blinded by bilateral enucleation. In a series of papers that followed, Menaker and his colleagues demonstrated that:

(1) The intensity of light passing through the skull and reaching the brain of blinded birds could be manipulated by the removal of feathers and the subdural injection of India ink, and that whether or not the birds entrained to a particular light cycle depended on how much light reached the brain.[36]

(2) Not only entrainment to light cycles, but also the effect of constant light on circadian period length and the effect of bright constant light in making birds arrhythmic, as well as the reproductive effects of different artificial day lengths (photoperiodism) were partially mediated by extraretinal photoreceptors (see Figure 1).[37,38]

(3) The eyes were involved in three of the above four responses to light but were not involved at all in photoperiodic photoreception.[31] Furthermore, in mediating entrainment and period length, the effects of light perceived by the eyes were additive with those of light perceived by extraretinal photoreceptors, while for the effects of bright constant light in producing arrhythmicity there was an absolute requirement for light perceived by the eyes.[32,33,34]

(4) The presence of the pineal gland was not required for blind birds to entrain to light cycles or to respond reproductively to photoperiod (involvement of the pineal in the effects of constant light could not be ascertained, as its removal renders sparrows arrhythmic in either constant light or constant darkness).[20,39]

(5) Many of the same results were obtained when similar experiments were performed on several species of lizards.[57]

These results were extended by several laboratories to other birds and

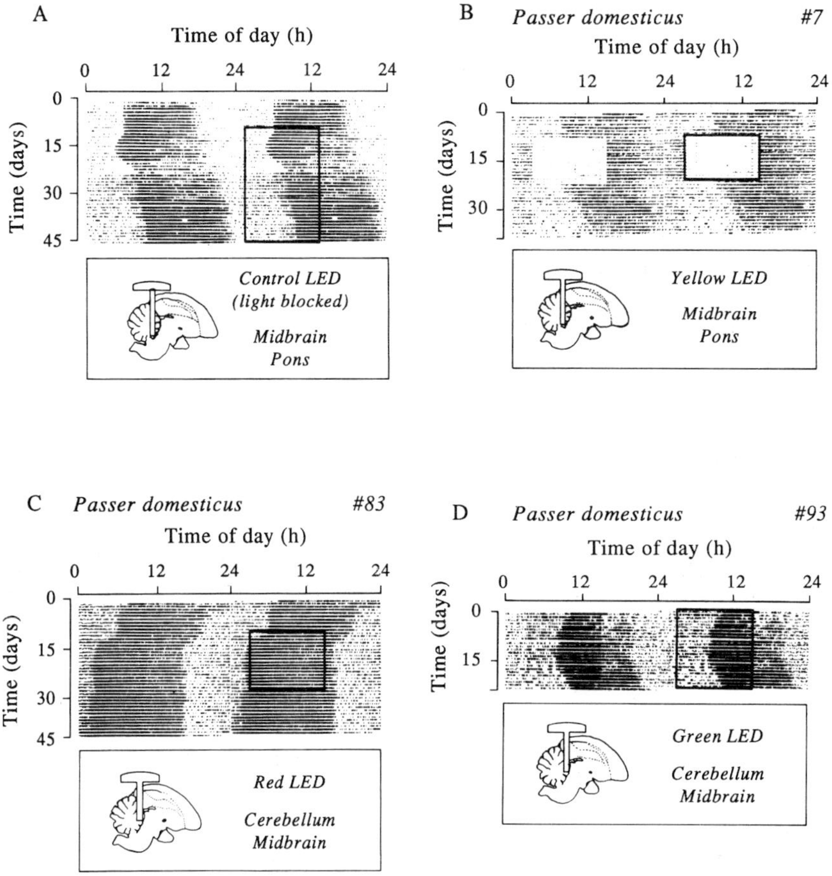

FIG. 1. Activity records (double-plotted) from four house sparrows in which low amplitude light cycles from light-emitting diodes (LED) of different colors were delivered directly to the midbrain. The LEDs were imbedded in black plastic in a shallow stainless steel cup mounted on the skull, but were optically coupled to a light guide within a stainless steel tube which was implanted stereotactically in the brain. The LEDs were connected by fine wires through a commutator to a clock-controlled power supply and the rotation of the commutator as the bird moved was sensed by an infrared detector directed at stripes painted on its surface. Thus "activity" reflects small motions of the head (as little as a turn of 3°) as well as gross movements, rather than the more usually recorded "perch hopping." Bold rectangles on the right halves of each panel indicate the times during which current was passed. At other times the birds were in constant darkness. **A** is a control in which current was passed through the LED, but no light was permitted to escape; **C** and **D** show, respectively, weak entrainment and weak masking by the lights-off transition; **B** is the most interesting record, for it shows clearly that although the bird's circadian system was not entrained by the light cycle, the light had a strong inhibitory effect on movement; indeed, the bird hardly turned its head during the entire 12 hours of lights-on each day. This one record suggests that there may well yet be unappreciated effects of extraretinally perceived light on the central nervous system (Menaker M. and Vaaler G., unpublished).

reptiles as well as to fish and amphibians. Representative species from all vertebrate classes, except mammals, have extraretinal photoreceptors. For example, in the eel (*Anguilla anguilla*) the synchronization of circadian locomotor rhythms requires extraocular and extrapineal photoreceptors.[58] Similar findings have been reported in other fish species.[53] In the frog (*Rana esculenta*) electrophysiological responses to light have been recorded from the basal brain in blinded and pinealectomized animals,[5,6] although no studies have so far attempted to link these photoreceptors with circadian physiology.

In addition to their role in the circadian system, extraretinal photoreceptors have also been shown to play an important part in the regulation of seasonal reproductive cycles in non-mammalian vertebrates. The primary factor controlling seasonal breeding in temperate-zone birds is the annual variation in daylength, but birds do not require their eyes to perceive this change. Blinded ducks exposed to long photoperiods grow large testes.[2,3] More recent studies, involving enucleation, sectioning of the optic nerve and shielding light from entering the brain confirmed this original finding in a range of bird species.[11] Many of these studies showed that, unlike circadian responses to light, gonadal growth occurred at maximal rates in the absence of the eyes, suggesting that ocular photoreceptors did not even contribute to the detection of the daylength signal. Initially, the pineal was considered as the prime candidate for this extraocular photoreceptor, but several studies demonstrated that the pineal appeared to play no significant role in avian photoperiodic responses. In quail, for example, illuminating the pineal with radioluminous paint does not stimulate gonadal growth[26] and pinealectomy of either blinded or intact quail leaves photoperiodic induction unaffected.[50,51] In contrast, illumination of the basal brain with luminous paint or fiber optics in blinded and pinealectomized animals[42,44,61] causes gonadal growth at maximal rates. These experiments provided strong additional evidence for the existence of a "deep brain photoreceptor" located within the basal brain of birds.

Taken together, these studies have led to the current, generally held view that all non-mammalian vertebrates have several kinds of non-visual, extraretinal, extraocular photoreceptors and use them for circadian entrainment as well as a variety of other photically regulated activities. Vertebrate photoreceptors include deep brain photoreceptors, the parietal eye (when present), the pineal gland (the photoreceptive capacity of which has been clearly and repeatedly demonstrated and is probably required for entrainment of its own internal circadian oscillator) in addition to the eyes themselves. The widespread occurrence of extraretinal photoreceptors in non-mammalian vertebrates (Figure 2) is in direct contrast to the situation in mammals which use retinal photoreceptors exclusively (but see Torres and Lytle[55]).

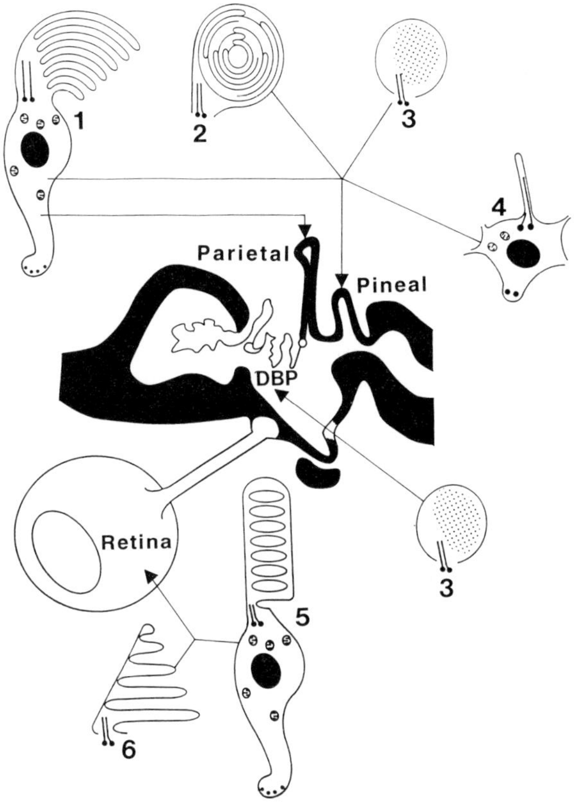

FIG. 2. Schematic drawing of the vertebrate fore-brain showing the diversity of opsin-containing cells within the retina, pineal complex (pineal and parietal organs) and basal brain. Opsin-immunoreactive cells can be broadly classified as follows: **1** Cell with an outer segment resembling a retinal cone; **2** Cell with an outer segment with "disorganized" membrane whorls; **3** Cell with an outer segment containing few or no membrane structures; **4** Cell with no outer segment; **5** Retinal rod cell; **6** Retinal cone cell. These opsin-containing cells show the following taxonomic distribution: In the parietal organ or parapineal organ of cyclostomes (lampreys) and lacertilians (lizards), or frontal organ of anurans (frogs), the dominant cell type resembles **1**. The pineal organ is present in some form in almost all vertebrates. In the fish and amphibia = cell type **1** dominates; reptiles = cell types **1** and **2**; birds = cell types **2** and **3**; mammals- = cell type **4**. Note that in mammals, opsin has been identified within pineal cells, but these cells are not thought to be photoreceptive, lacking photopigment chromophore.[15] Opsin immunoreactive CSF-contacting cells have only recently been identified within the basal brain of lamprey larvae,[19] lizards[17] and birds.[49] We postulate that these cells are the deep brain photoreceptors of the non-mammalian vertebrates. On the basis of earlier ultrastructural analysis of CSF-contacting cells, these cells most closely resemble cell type **3**. The retina of all vertebrates contains either rod or cone photoreceptors or more usually some mixture of both (cell types **5** and **6**; see Oksche and Hartwig[43] for further details).

Identification of circadian photoreceptors in non-mammalian vertebrates

While it has been relatively easy to demonstrate that particular tissues are photosensitive by surgical removal, directed illumination or isolation *in vitro*, the identification of the photoreceptive cells within these tissues has proved more difficult because to do so it is necessary to demonstrate photopigments localized within particular cells. In general, photopigments can be characterized on the basis of the physiological responses they mediate and their characteristic biochemistry. All vertebrate photopigments have been found to be remarkably similar, consisting of a type of opsin protein coupled to a chromophore derived from an 11-*cis* form of vitamin A retinaldehyde.[60] The characterization of circadian photoreceptors has involved the identification and localization of these photopigment characteristics (opsin and chromophore) within the cells and tissues suspected of circadian photoreception.

Action spectrum techniques were particularly helpful in the initial characterization of pineal and deep brain photoreceptors. All opsin and 11-*cis* retinaldehyde based photopigments have a characteristic shape to their absorbance and action spectra (even though these photopigments may have a very different wavelength of maximum sensitivity). In the isolated chick pineal, an action spectrum for suppression of the activity of the enzyme *N*-acetyltransferase by light demonstrated that the response was mediated by a photopigment maximally sensitive to 500 nm, and with the characteristic shape of an opsin:11 *cis*-retinal based photopigment.[8] Almost identical results have been demonstrated in the pineals of several fish, reptiles and amphibians using a range of different assays of sensitivity, including melatonin release and electrical activity.[21] An action spectrum for the deep brain photoreceptors mediating photoperiodic responses in quail was strikingly similar to the action spectra of pineal responses although the receptors involved were clearly not in the pineal.[11,12] These action spectra suggested the presence of opsin and 11-*cis* retinaldehyde within the pineal and basal brain; such photopigment components have now been directly demonstrated.

In the pineal, antibodies against vertebrate visual pigment opsins have immunolabeled pinealocytes, which in many respects resemble the photoreceptors of the developing retina. These cells are distinguished as having a short outer segment comprising lamellar whorls or membrane stacks. In addition to opsin proteins, many pinealocytes contain other proteins associated with phototransduction.[13] A functional photopigment requires a chromophore, and in the limited number of studies undertaken, it has always been possible to identify chromophore (11-*cis* retinoid) within the pineal of non-mammalian vertebrates.[45] Anatomical and biochemical studies have also provided evidence that pineals of various vertebrates contain multiple types of photoreceptors.[14] To date, this surprising complexity has either been ignored or has stimulated only limited speculation.

In contrast to the pineal, the identification of candidate photoreceptors within the basal brain has been difficult. Numerous studies over the past 25 years have failed to provide an exact anatomical localization.[17] However, three recent studies have suggested a location for these cryptic photoreceptors. Silver *et al.*[49] demonstrated that cerebrospinal fluid (CSF)-contacting neurons within the septal and tuberal areas of the brain in the ring dove, quail and duck could be labeled with an anti-opsin antibody. Conclusive opsin identification was not possible in their studies because immunoblot analysis of the putative photoreceptor-containing brain regions failed to identify opsin-like proteins. The second, and most detailed study to date, has been on the lizard *Anolis carolinensis*.[17] In this species anti-opsin antiserum was shown to bind to CSF-contacting neurons in the septal area of the brain (Figure 3). Immunoblot analysis

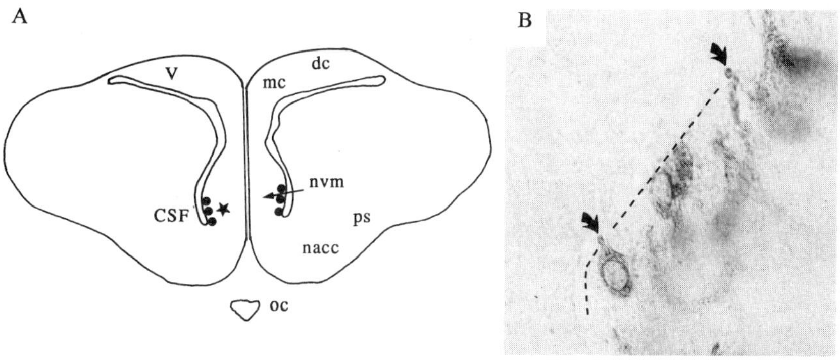

FIG. 3. Opsin immunoreactivity in the basal brain of the lizard *Anolis carolinensis*. **A** Schematic drawing of a frontal section of the brain. Opsin-immunoreactive periventricular CSF-contacting neurons (*dots*) were identified within the nucleus ventromedialis of the septum. **B** Photomicrograph of CSF-contacting neurons stained with anti-opsin antibodies, showing the presence of apical processes (*arrows*) protruding into the lateral ventricle (*dotted line* = surface of the ventricle). The location of these cells within the brain has been indicated by a *star* in **A**. Western blot analysis indicates that our anti-opsin antibodies recognize a single 40 kD protein in anterior brain extracts. In addition this area of the brain contains retinoids associated with phototransduction (11-*cis* and all-*trans* retinoid). These data suggest that CSF-contacting neurons are photosensitive, and that these cells could be responsible for circadian responses to light in the absence of retinal and pineal photoreceptors. *Abbreviations*: *CSF* cerebrospinal fluid-contacting neurons; *dc* dorsal cortex; *mc* medial cortex; *nacc* nucleus accumbens; *nvm* nucleus ventromedialis of the septum; *oc* optic chiasm; *ps* paleostriatum; *V* lateral ventricle (from Foster *et al.*[17]).

showed that the anti-opsin antibodies recognized a single 40 kD protein in ocular, pineal and septal brain areas, but not in other areas of the brain. In addition, the anterior brain of *Anolis* was shown to contain specific retinoids associated with phototransduction (11-*cis* and all-*trans*-retinoid). A third study examined a "primitive" vertebrate, the larval (Ammocoetes) lamprey.[19] In this species, anti-opsin antisera labeled a population of CSF-contacting neurons within the basal brain (postoptic commissural nucleus and ventral hypothalamic nucleus), and in addition, antibodies raised against the α-subunit of retinal G protein (α-transducin) labeled the same cells. Collectively these data suggest that a subpopulation of CSF-contacting neurons are photosensory and that these cells are likely to be the "deep brain" photoreceptors that mediate extraocular and extrapineal photoreception in non-mammalian vertebrates. While this interpretation requires direct experimental confirmation, at least we now have candidate cells upon which to focus our efforts.

Retinal, pineal and parapineal photoreceptors, and CSF-contacting neurons are strikingly similar in that they all have specialized ciliated dendrites of a sensory type (9+0) which protrude into a cavity derived from the ventricles.[59] CSF-contacting neurons are different from retinal and pineal photoreceptors in that they seem to lack photoreceptor membrane lamellae or vesiculations. However, laminated membranes are probably not obligatory for photoreception.[23,43] In our studies of lamprey and *Anolis*, the ciliated dendrite of CSF-contacting neurons was often more strongly labeled than the cell body with anti-opsin and α-transducin antibodies. If CSF-contacting cells are indeed photosensitive, then the most obvious organelle responsible for phototransduction would be the ciliated dendrite (intraventricular process), which may be homologous with a photoreceptor outer segment.

It is not clear why previous attempts to localize opsin proteins within the brain have failed.[13,59] The use of newly obtained cone-specific antibodies may partially explain the more recent results; however, the selection of particular species may be equally important. Using methodologies and antibodies identical to those used to study *Anolis*, we have failed to identify any opsin immunopositive cells within the brains of either chicken, quail or sparrows in spite of the fact that they are almost certainly present.[17]

Identification of circadian photoreceptors in mammals

The identification of extraretinal photoreceptors in non-mammalian vertebrates stimulated a search for similar photoreceptors in mammals. Early studies showed that the mammalian photoperiodic response needed an intact retina/eye[9,24,25,54] and that the suppression of pineal melatonin synthesis by light was abolished if the sympathetic input to the pineal was severed, isolating the gland from the retina and suprachiasmatic nuclei (SCN).[56] Early studies also showed that circadian rhythms could not be entrained by light cycles in the absence of the eyes.[22] A more recent study by Nelson and Zucker[41] on both diurnal and nocturnal rodents housed outdoors and exposed to natural day-night cycles, demonstrated that blinded animals failed to show any sign of circadian entrainment.

Although the physiological evidence suggests that mammals use their eyes for all forms of light detection, the mammalian pineal does contain a range of so-called "photoreceptor proteins", including rhodopsin kinase,[52] the 48 kD protein S-antigen,[27,40] interphotoreceptor retinoid-binding protein, cellular retinal-binding protein,[4] and opsin.[15,28] The identification of these photoreceptor proteins within the mammalian pineal led, not unreasonably, to speculation that phototransduction may take place within this tissue and mediate some unrecognized aspect of the animal's physiology. To address this issue Foster[15] investigated whether 11-*cis* and all-*trans* retinaldehyde were present within the hamster pineal, and whether, if present, photoisomerization of the 11-*cis* chromophore to the all-*trans* form could be demonstrated. Some 400 pineals were collected and assayed for chromophore content, but none was found (cf. avian or lizard pineal). These data suggest that it is unlikely that the opsin within the mammalian pineal is coupled to a chromophore and reinforce the view that the mammalian pineal is incapable of phototransduction.

Although there is no evidence that adult mammals have extraretinal photoreceptors, there is a strong suggestion of extraretinal photoreception in neonatal mammals. In carefully controlled experiments Torres and Lytle[55] have shown that light-induced suppression of pineal *N*-acetyltransferase occurs in blinded rat pups 4 days old. In animals older than 6 days this response is lost. Their report extends and confirms earlier studies and suggests that further investigation of this phenomenon will be fruitful.

Although it is clear that circadian photoreception in adult mammals is exclusively retinal, we do not yet know which cells within the retina are responsible. We have used mice with hereditary retinal disorders and transgenically modified retinas in attempts to identify the photoreceptors within the mammalian retina that mediate circadian responses to light. In mice homozygous for the autosomal recessive allele *rd* (retinally degenerate) all rod cells degenerate by 60 days of age, and between 90 and 150 days of age electrophysiological and behavioral visual responses to bright light disappear. In the normal mouse retina, approximately 97–98% of all photoreceptors are rods and although all rods degenerate in the *rd*/*rd* retina, a few cone cells (2–5% of the original population) remain in animals over one year of age. It should be stressed that the remaining cone cells in these aged *rd*/*rd* mice lack outer segments, and with increasing age cone cell loss continues to completion. In view of this massive photoreceptor damage we expected circadian responses to light to decline in parallel with the loss of visual function, but this did not occur.[16]

We assayed circadian photosensitivity by measuring the size of the phase shift in the freerunning locomotor rhythm produced by a single 15 minute pulse of light (515 nm) (Figure 4). We varied the light intensity (irradiance) of the 15 minute pulse to compare the dynamic range of circadian responses to light in three genotypes of mice (*rd/rd*, *rd*/+, +/+) from the same C57BL strain (Figure 5). Despite the loss of visual photoreceptors in *rd/rd* mice, these animals show circadian responses to light that are indistinguishable from those of mice with phenotypically normal retinas (*rd*/+, +/+). The irradiance required to produce both saturating and half-saturating responses are the same for all groups. This result demonstrates not only that some photosensitivity remains in mice with degenerate retinas, but that the circadian photosensitivity shown by these animals is not different from the sensitivity of animals with normal retinas. In addition, the site of circadian photoreception must reside within the eye because bilateral enucleation of *rd/rd* animals abolishes all circadian responses to light.[16]

The progression of photoreceptor degeneration in *rd/rd* mice commences early in post natal development with the loss of rods, followed by a more protracted loss of cone cell bodies. If the surviving cone cell bodies mediated circadian responses to light, then one would expect the sensitivity of the circadian system to light in *rd/rd* mice to decline with age in parallel with the loss of cones from the retina. Circadian responses to light of *rd/rd* and +/+ mice up to 800 days of age were examined. The results indicate that the phase shifts of *rd/rd* and +/+ mice remain indistinguishable. Circadian responses to light do not parallel either rod or cone cell loss.[46]

The light input pathway to the SCN was examined to determine whether the anatomy or physiology of the retinohypothalamic tract is affected by the *rd/rd* mutation. Several laboratories have shown that activation of the immediate early gene c-*fos* within the SCN is a useful marker of neural activation of the light input pathway to the circadian system.[29,47] If the rods and/or cones of the retina contribute to light information reaching the SCN, one would expect a reduced expression of c-*fos* in the SCN of *rd/rd* mice in response to light. On the other hand, if circadian photoreceptors are not affected by this mutation, then light activated c-*fos* expression should be the same in the SCN of *rd/rd* and +/+ mice. Using immunocytochemical techniques, we have been unable to detect any differences between the two genotypes in the distribution, number and density of *c-fos*-labeled SCN cells.[7]

While all of our studies have shown that there is no effect of the *rd/rd* mutation on circadian responses to light, a study by Ebihara and Tsuji[10] found that *rd/rd* mice required more light to entrain to a light:dark cycle

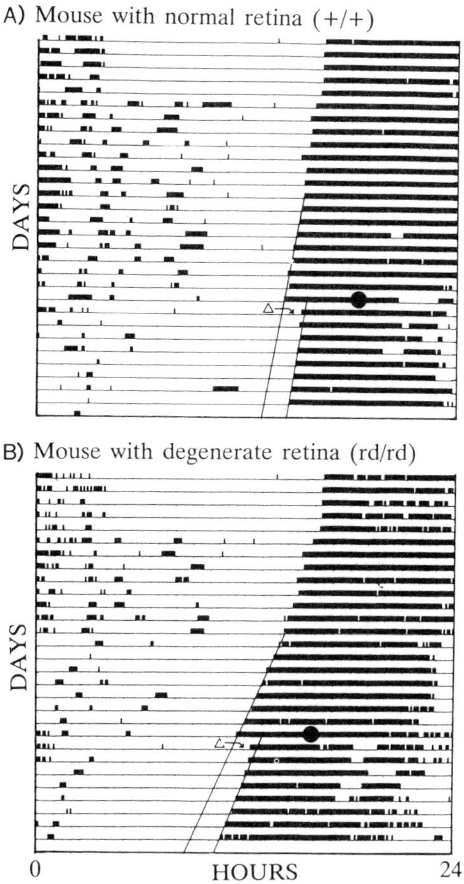

FIG. 4. Locomotor activity (wheel running behavior) of C57BL mice with either: **A** normal retinas and visual responses (+/+), or **B** retinally degenerate mice homozygous for the *rd* gene (*rd/rd*), with massive loss of visual function. Wheel running activity is shown for 29 days (1 line = 1 day); animals were housed within a light-tight box. For the first 5 days, each mouse was exposed to a light:dark cycle. Note that the animal started activity at the same time each day. The mouse was then transferred to constant darkness, and the free-running rhythm was observed. The period of the free-running rhythm for both animals was shorter than 24 hours. As a result the activity starts a little earlier on each subsequent day. On the 16th day in constant darkness, the mouse was removed from the light tight-box, placed in a standard chamber and exposed to a 15 minute light pulse at circadian time 16 (CT 16), shown as *filled circles* on the record. By convention, activity onset under free-running conditions is designated circadian time 12 (CT 12). As a result, CT 16 will be 4 hours after activity onset. The duration, wavelength and amount of light delivered to the animal was quantitatively determined. After a further 8 days the effect of the light pulse (i.e. the size of the phase shift) was determined. The magnitude of the phase shift produced by the light pulse was measured as the difference between the steady state phase of the free-running rhythm before and after the pulse. The phase shift is shown by a △ in both records. In both animals the light pulse caused a phase delay (animal started activity later) of about 90 minutes (from Foster *et al.*[16]).

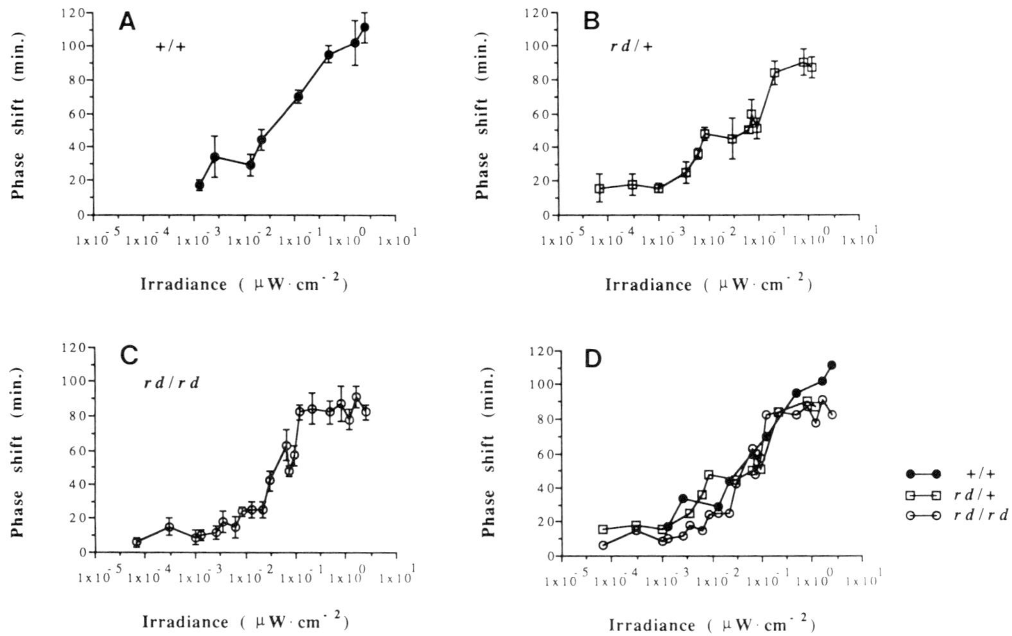

FIG. 5. The magnitude of phase shifts produced by a 15 minute light pulse as a function of light intensity in: **A** mice with normal retinas (+/+); **B** mice with normal retinas (*rd*/+); **C** mice with degenerate retinas (*rd*/*rd*). For comparison the three irradiance response curves (**A**, **B**, **C**) have been superimposed in **D**. Despite the massive damage to the visual photoreceptors in *rd*/*rd* mice, these animals show circadian responses to light that are indistinguishable from those of mice with normal retinas. Error bars ± SEM (from Foster *et al.*[16]).

compared to +/+ controls. While the authors concluded that this difference was due to the loss of rods in *rd/rd* animals their results are open to several interpretations. The study compared two strains of mice, C3H *rd/rd* with wild-type C57BL. It is therefore possible that the differences observed are the result of mouse strain and have nothing directly to do with retinal photoreceptors. We have recently compared the amount of light required to entrain C3H *rd/rd* and +/+ mice with C57BL *rd/rd* and +/+ mice and found no differences in sensitivity between these groups. This would suggest that strain cannot account for the differences in sensitivity that Ebihara and Tsuji observed. We are currently collaborating with the Ebihara laboratory to resolve these questions. It is possible that the Japanese mice have a different or additional mutation that may modify circadian responses to light. If true, this would be a particularly useful tool in identifying the retinal components that regulate circadian responses.

Another retinal mutation, *rds* (retinally degenerate slow) in the C3H mouse strain, has provided a second, somewhat different approach to the same question. In *rds/rds* mice, the retina undergoes normal development and differentiation of cells until the first postnatal week, but at this time both rod and cone photoreceptors fail to develop outer segments and gradually degenerate. This is in contrast to *rd/rd* animals which develop outer segments that later degenerate. In *rds/rds* mice approximately half of all photoreceptor cell bodies have degenerated by 3 months, and virtually all are gone by 1 year of age. Data from these animals, including data from aged animals, suggest that circadian light detection is identical in *rds/rds*, *rd/rd* and +/+ genotypes.[1]

We have also examined circadian responses in transgenic mice that incorporated a genetic construct consisting of the gene coding for diphtheria toxin A-chain, controlled by the rhodopsin promotor (mice generously donated by Maureen McCall, University of Wisconsin-Madison and Toshimichi Shinohara, NIH). This construct eliminates all rod photoreceptors but has no obvious effect on neurons in the outer retinal layers or more central visual structures. Cone cells survive for several weeks, although they lack outer segments, and gradually disappear from the retina. In preliminary studies, the circadian responses of these animals are indistinguishable from those of normal control mice.

Collectively, these data suggest that circadian photoreception can be maintained either by a small number of cone cells without outer segments, or alternatively by an as yet unrecognized class of retinal photoreceptive cell that is unaffected by the *rd/rd* mutation, the *rds/rds* mutation, or the destruction of rod photoreceptors by transgenic modification of the retina. We are attempting to distinguish between these two alternatives by identifying the circadian photopigment in reduced retinas, and then localizing this photopigment within the retina.

Preliminary action spectrum data for phase-shifting the circadian rhythm in *rd/rd* and +/+ mice suggest that the wavelength of maximum sensitivity is near 500 nm and that the shape of the curve is typical of animal photopigments based upon an opsin and 11-*cis*-retinaldehyde chromophore (Provencio I. and Foster R.G. unpublished). Using High Performance Liquid Chromatography we have identified significant amounts of 11-*cis* retinaldehyde within the dark adapted eyes of *rd/rd* mice (1–2% of the levels found in +/+ individuals). If this chromophore is associated with the rods or cones of the degenerating *rd/rd* retina, then it should decline in parallel with their loss and ultimately disappear. Alternatively, if the chromophore is associated with some other circadian photoreceptor which does not decline with age, then its level should remain low but constant. Our early results indicate that after an initial rise and decline during the first 14 days of life, chromophore levels remain at constant low levels in the *rd/rd* eye.[46]

The genes and cDNAs for the rod opsin, blue cone opsin and green/red cone opsin in the +/+ mouse have been successfully characterized by Meredith Applebury's laboratory at the University of Chicago (personal communication). Because circadian responses to light as well as chromophore levels remain stable in aged *rd/rd* mice we anticipate that any opsin mediating circadian responses to light would also remain at a constant low level. Northern blot analysis using opsin probes made from Applebury's cDNAs has shown that rod opsin mRNA rapidly disappears from the eye, followed by the slower decline in levels of blue cone opsin and green/red cone opsin mRNA. These data are open to two interpretations, either (1) one or more of the known opsins mediates circadian responses to light and occurs in the degenerate retina at very low levels, or (2) none of the known opsins mediates circadian responses to light, and there is a unique "circadian" opsin in the retina.[1]

The loss of rods and cones in mutant mice produces a parallel decline in visual function, yet circadian responses to light remain unaffected in these animals. Full interpretation is not yet possible; however, the facts are consistent with the hypothesis that retinal rods and cones do not normally play a significant part in the regulation of the mammalian circadian system by light, and that some as yet unidentified retinal photoreceptor performs this role.

Implications

Early in their evolutionary history mammals may well have gone through what could be called a "nocturnal bottleneck". Extraretinal photoreceptors either in the pineal or the deep brain that may have served well for diurnal animals exposed to bright light for hours at a time, were simply not sensitive enough to define dawn and dusk for primitive mammals that spent the day hiding in dark caves and holes. As a consequence, brain photoreceptors may have been "moved out" into the retina where they had better access to the dim light which illuminates the world of strongly nocturnal creatures. Alternatively and perhaps more reasonably, both deep brain and specialized retinal circadian photoreceptors were already present in the diurnal mammal-like reptiles and the extraretinal receptors became non-functional and were therefore lost when primitive mammals entered the nocturnal niche. In either case, those mammalian species which subsequently and secondarily became diurnal have apparently failed to re-evolve extraretinal photoreception. The above speculation will of course be difficult to substantiate experimentally, and, it should be noted, does not address the interesting related question of the adaptive significance of separating the perception and processing of visual from that of circadian photoreception.

Correct or not, the reasoning outlined above has led us to search for retinal cell types that resemble and are perhaps homologous to the CSF-contacting neurons that are our best candidates for deep brain photoreceptors. The cell type known as Landolt's clubs[59] are modified bipolar cells of unknown function, widely scattered throughout the retina. Landolt's clubs have a ciliated dendritic process which extends through the outer nuclear layer and terminates in the space between the pigmented epithelium and inner and outer segments of the photoreceptors. Vigh and Vigh-Teichmann[59] classify these cells as CSF-contacting neurons and have noted their resemblance to pinealocytes. They would seem good candidates for further study.

It is worth examining the implications for human beings of what we know about circadian photoreception in vertebrates in general. There is nothing to suggest that we are different from other mammals in the retinal location of our circadian photoreceptors. Like them we probably have specialized retinal photoreceptors that convey information about the environmental light conditions more or less directly to the circadian oscillators in the SCN and not to the higher visual centers. Thus it is entirely possible that our circadian oscillators are kept informed of relevant changes in external light conditions without this information ever reaching consciousness. If this description is correct, it will have at least two consequences: (1) As a genetic defect, some individuals may lack the specialized circadian photoreceptor cell types without being directly aware of their absence. By analogy with the color blindness that results when individuals lack a class of cone cell photoreceptors, such individuals would be "time blind"—unable to synchronize their circadian oscillators with the outside world and as a result, drifting with their innate circadian periodicity across the strictly 24-hour day which the rest of us inhabit. Less severe forms of this defect might result in a reduced number of circadian photoreceptors with consequent increase in the intensity of light required

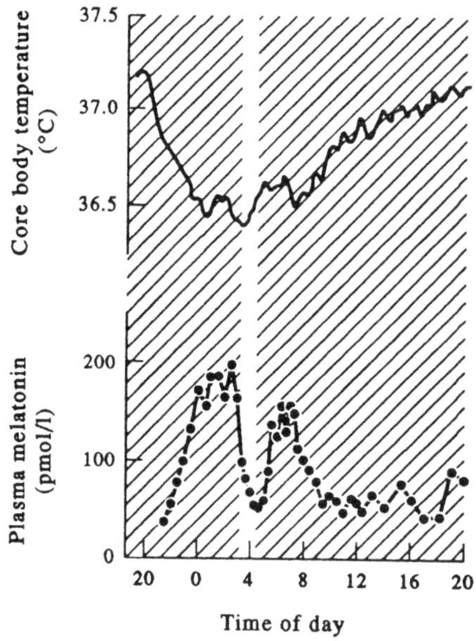

FIG. 6. Melatonin suppression by light in a human subject with no conscious light perception. Exposure to 1.5 hours of bright light (open vertical bar, 6,300 to 8,400 lux in the direction of gaze) centered around the minimum of the endogenous core body temperature rhythm (upper panel) during the nighttime induced an immediate and complete suppression of circulating plasma melatonin levels, which rebounded to nocturnal values immediately upon return to darkness (stippled area, light intensity 0.02 lux) (from Martens *et al.*[30]).

for entrainment and perhaps abnormal phase relationships with the day–night cycle. (2) Some individuals who are visually blind might retain circadian photoreception without being aware of it. These people would entrain to the day–night cycle, have normal sleep patterns and be temporally well integrated in society although they might attribute this to social signals rather than light. It goes without saying that every effort should be made to protect and enhance circadian photoreceptive function in such cases. In this connection, a recent report from Czeisler's laboratory is of great interest.[30] Figure 6, which is taken from that report, shows clearly that in a human subject with no conscious light perception, plasma melatonin levels can be suppressed by light. Since it is well known that at least some of the effects of light on melatonin synthesis are mediated by retinal input to the SCN which then suppresses the pineal gland via a multi-synaptic pathway, it is hard to escape the conclusion that at least some of the retinal receptors that supply light information to the SCN have been spared in this completely "blind" subject and that as a consequence, it might be possible for him to entrain to light cycles.

References

1. Argamaso S.M., Boggs B.R. and Foster R.G. (1992) Effects of retinal degeneracy on locomotor rhythms in mice: Circadian and molecular analysis of *rd/rd* and *rds/rds* mutants. *Soc. Res. Biol. Rhythms Abs.* **3**, 59.
2. Benoit J. (1935a) Le rôle des yeux dans l'action stimulante de la lumière sur le développement testiculaire chez le canard. *C.R. Soc. Biol.* (*Paris*) **118**, 669–671.
3. Benoit J. (1935b) Stimulation par la lumière artificielle du développement testiculaire chez des canards aveugles par section du nerf optique. *C.R. Soc. Biol.* (*Paris*) **120**, 133–136.
4. Bridges C.D.B., Foster R.G., Landers R.A., Liou G.I. and Fong S.-L. (1987) Interstitial retinol-binding protein and cellular retinal-binding protein in the mammalian pineal. *Vision Res.* **27**, 2049–2060.
5. Cadusseau J. and Galand G. (1980) Electrophysiological evidence for white light sensitivity of the encephalon in eyeless and pinealectomized frogs. *Exp. Brain Res.* **40**, 339–341.
6. Cadusseau J. and G. Galand (1981) Electrophysiological recordings of an extraocular and extrapineal photoreceptor in the frog encephalon. *Brain Res.* **219**, 439–444.
7. Colwell C.S. and Foster R.G. (1992) Photic regulation of Fos-like immunoreactivity in the suprachiasmatic nucleus of the mouse. *J. Comp. Neurol.* **324**, 135–142.
8. Deguchi T. (1981) Rhodopsin-like photosensitivity of isolated chicken pineal gland. *Nature* **290**, 702–704.
9. Dixit V.P., Sharma O.P. and Agrawl M. (1977) The effects of light deprivations/blindings on testicular function of gerbil (*Meione horrianae*). *Endocrinology* **70**, 13–18.
10. Ebihara S. and Tsuji K. (1980) Entrainment of the circadian activity rhythm to the light cycle: Effective light intensity for a Zeitgeber in the retinal degenerate C3H mouse and the normal C57BL mouse. *Physiol. Behav.* **24**, 523–527.
11. Foster R.G. and Follett B.K. (1985) The involvement of a rhodopsin-like photopigment in the photoperiodic response of the Japanese quail. *J. Comp. Physiol. A* **157**, 519–528.
12. Foster R.G., Follett B.K. and Lythgoe J.N. (1985) Rhodopsin-like sensitivity of extra-retinal photoreceptors mediating the photoperiodic response in quail. *Nature* **313**, 50–52.
13. Foster R.G., Korf H.G. and Schalken J.J. (1987) Immunocytochemical markers revealing retinal and pineal but not hypothalamic photoreceptor systems in the Japanese quail. *Cell. Tissue Res.* **248**, 161–167.
14. Foster R.G., Schalken J.J., Timmers A.M. and De Grip W.J. (1989a) A comparison of some photoreceptor characteristics in the pineal and retina: I. The Japanese quail (*Coturnix coturnix*). *J. Comp. Physiol. A.* **165**, 553–563.
15. Foster R.G., Timmers A.M., Schalken J.J. and De Grip W.J. (1989b) A comparison of some photoreceptor characteristics in the pineal and retina: II. The Djungarian hamster (*Phodopus sungorus*). *J. Comp. Physiol. A.* **165**, 565–572.
16. Foster R.G., Provencio I., Hudson D., Fiske S., De Grip W. and Menaker M. (1991) Circadian photoreception in the retinally degenerate mouse (*rd/rd*). *J. Comp. Physiol. A.* **169**, 39–50.
17. Foster R.G., Garcia-Fernandez J.M., Provencio I. and De Grip W.J. (1993) Opsin localization and chromophore retinoids identified within the basal brain of the lizard *Anolis carolinensis*. *J. Comp. Physiol. A.* **172**, 33–45.
18. Frisch K. von (1911) Beiträge zur Physiologie der Pigmentzellen in der Fischhaut. *Pflüger's Arch.* **138**, 319–387.
19. Garcia-Fernandez J.M. and Foster R.G. (1993) Immunocytochemical identification of photoreceptor proteins in hypothalamic CSF-contacting neurons of the larval lamprey (*Petromyzon marinus*). *Cell. Tissue Res.* (in press).
20. Gaston S. and Menaker M. (1968) Pineal function: The biological clock in the sparrow? *Science* **160**, 1125–1127.
21. Groos G. (1982) The comparative physiology of extraocular photoreception. *Experientia* **38**, 989–1128.
22. Halberg F., Visscher M.B. and Bittner J.J. (1954) Relation of visual factors to eosinophil rhythms in mice. *Am. J. Physiol.* **179**, 229–235.

23. Hartwig H.-G. (1982) Comparative aspects of retinal and extraretinal photosensory input channels of entraining endogenous rhythms. In *Vertebrate Circadian Systems* (eds. Aschoff J., Daan S. and Groos G.), pp. 25–30. Springer, Berlin, Heidelberg, New York.
24. Herbert J., Stacey P.M. and Thorpe P.H. (1978) Recurrent breeding seasons in pinealectomized or optic-nerve sectioned ferrets. *J. Endocrinol.* **78**, 389–397.
25. Hoffman R.A. and Reiter R. (1965) Pineal gland: Influence on gonads of male hamsters. *Science* **148**, 1609–1611.
26. Homma K., Ohta M. and Sakakibara Y. (1980) Surface and deep photoreceptors in photoperiodism in birds. In *Biological Rhythms in Birds* (eds. Tanabe Y., Tanabe K. and Ookawa T.), pp. 149–156. Springer, Berlin, Heidelberg, New York.
27. Kalsow C.M. and Wacker W.B. (1977) Pineal reactivity of anti-retina sera. *Invest. Ophthalmol. Visual Sci.* **16**, 181–186.
28. Korf H.-W., Foster R.G., Ekström P. and Schalken J.J. (1985) Opsin-like immunoreaction in the retinae and pineal organs of four mammalian species. *Cell. Tissue Res.* **242**, 645–648.
29. Kornhauser J.M., Nelson D.E., Mayo K.E. and Takahashi J.S. (1990) Circadian regulation of c-fos gene expression in the hamster suprachiasmatic nucleus. *Neuron* **5**, 127–134.
30. Martens H., Klein T., Rizzo J.F. III, Shanahan T.L. and Czeisler C.A. (1992) Light-induced melatonin suppression in a blind man. *Soc. Res. Biol. Rhythms Abs.* **3**, 52.
31. McMillan J.P., Underwood H., Elliott J.A., Stetson M.H. and Menaker M. (1975) Extraretinal light perception in the sparrow. IV: Further evidence that the eyes do not participate in photoperiodic photoreception. *J. Comp. Physiol.* **97**, 205–213.
32. McMillan J.P., Keatts H.C. and Menaker M. (1975) On the role of eyes and brain photoreceptors in the sparrow: Entrainment to light cycles. *J. Comp. Physiol.* **97**, 251–256.
33. McMillan J.P., Elliott J.A. and Menaker M. (1975) On the role of eyes and brain photoreceptors in the sparrow: Aschoff's Rule. *J. Comp. Physiol.* **102**, 257–262.
34. McMillan J.P., Elliott J.A. and Menaker M. (1975) On the role of eyes and brain photoreceptors in the sparrow: Arrhythmicity in constant light. *J. Comp. Physiol.* **102**, 263–268.
35. Menaker M. (1968) Extraretinal light perception in the sparrow I: Entrainment of the biological clock. *Proc. Natl. Acad. Sci. USA* **59**, 414–421.
36. Menaker M. (1968) Light perception by extraretinal receptors in the brain of the sparrow. *Proc. 76th Ann. Conv. Am. Psychol. Assoc.*, 299–300.
37. Menaker M. and Keatts (1968) Extraretinal light perception in the sparrow II: Photoperiodic stimulation of testis growth. *Proc. Natl. Acad. Sci. USA* **60**, 146–151.
38. Menaker M., Roberts R., Elliott J.A. and Underwood H. (1970) Extraretinal light perception in the sparrow III: The eyes do not participate in photoperiodic photoreception. *Proc. Natl. Acad. Sci. USA* **67**, 320–325.
39. Menaker M. and Underwood H. (1976) Extraretinal photoreception in birds. *Photochem. Photobiol.* **23**, 299–306.
40. Mirshahi M., Faure J.P., Brisson P., Falcon J. and Guerlotte J.P. (1984) S-antigen immunoreactivity in retinal rods and cones and pineal photosensitive cells. *Biol. Cell.* **52**, 195–198.
41. Nelson R.J. and Zucker I. (1981) Absence of extraocular photoreception in diurnal and nocturnal rodents exposed to direct sunlight. *Comp. Biochem. Physiol.* **69A**, 145–148.
42. Oishi T. and Kato M. (1968) Pineal organ as a possible photoreceptor; photoperiodic testicular responses in Japanese quail. *Mem. Fac. Sci. Kyoto Univ.* **2**, 12–18.
43. Oksche A. and Hartwig H.G. (1975) Photoneuroendocrine systems and the third ventricle. In *Brain-Endocrine Interactions. II. The Ventricular System, 2nd Int. Symp Shizuoka*, 1974, (eds. Knigge K.M., Scott D.E., Kobayashi H. and Ishii S.), pp. 40–53. Karger, Basel.
44. Oliver J., Jallageas M. and Baylé J.D. (1979) Plasma testosterone and LH levels in male quail bearing hypothalamic lesions or radioluminous implants. *Neuroendocrinology* **28**, 114–122.

45. Provencio I. and Foster R.G. (1993) Vitamin A_2-based photopigments within the pineal gland of a fully terrestrial vertebrate. *Neurosci. Lett.* **155**, 223–226.
46. Provencio I., Tennant G.S., Card J.P. and Foster R.G. (1992) Circadian photoreception in aged retinally degenerate (*rd/rd*) mice. *Soc. Res. Biol. Rhythms Abs.* **3**, 59.
47. Rusak B, Robertson H.A., Wisden W., Hunt S.P. (1990) Light pulses that shift rhythms induce gene expression in the suprachiasmatic nucleus. *Science* **248**, 12376–1240.
48. Scharrer, E. (1928) Die Lichtempfindlichkeit blinder Elritzen. I. Untersuchungen über das Zwischenhirn der Fische. *Z. Vergl. Physiol.* **7**, 1–38.
49. Silver R., Witkovsky P., Horvath P., Alones V., Barnstable C.J. and Lehman M.N. (1988) Coexpression of opsin- and VIP-like-immunoreactivity in CSF-contacting neurons of the avian brain. *Cell Tissue Res.* **253**, 189–198.
50. Simpson S.M., Urbanski H.F. and Robinson J.E. (1983) The pineal gland and the photoperiodic control of luteinizing hormone secretion in intact and castrated Japanese quail. *J. Endocrinol.* **99**, 281–287.
51. Siopes T.D. and Wilson W.O. (1974) Extraocular modification of photoreception in intact and pinealectomised *Coturnix*. *Poultry Sci.* **53**, 2035–2041.
52. Somers R.L. and Klein D.C. (1985) Rhodopsin kinase activity in the mammalian pineal and other tissues. *Science* **226**, 182–184.
53. Tabata M., Minh-Nyo M. and Oguri M. (1989) Thresholds of retinal and extraretinal photoreceptors measured by photobehavioral response in catfish, *Silurus asotus*. *J. Comp. Physiol. A.* **164**, 797–803.
54. Thomson A.P.D. (1954) The onset of oestrus in normal and blinded females. *Proc. Royal Soc. Lond. B.* **142**, 126–135.
55. Torres G. and Lytle L.D. (1989) Extraretinal mechanisms mediate light-induced changes in neonatal rat pineal gland N-acetyltransferase activity. *J. Pineal Res.* **7**, 211–220.
56. Underwood H. and Groos G. (1982) Vertebrate circadian rhythms: Retinal and extraretinal photoreception. *Experientia* **38**, 1013–1021.
57. Underwood H. and Menaker M. (1976) Extraretinal photoreception in lizards. *J. Comp. Physiol.* **83**, 187–222.
58. Veen Th. van, Hartwig H.-G. and Müller K. (1976) Light dependent motor activity and photonegative behavior in the eel (*Anguilla anguilla L.*). *J Comp. Physiol.* **111**, 209–219.
59. Vigh B. and Vigh-Teichmann I. (1988) Comparative neurohistology and immunocytochemistry of the pineal complex with special reference to CSF-contacting neuronal structures. *Pineal Res. Rev.* **6**, 1–65, 1988.
60. Wald, G. (1968) Molecular basis of visual excitation. *Science* **162**, 230–239.
61. Yokoyama, K., Oksche A., Darden T.R. and Farner D.S. (1978) The sites of encephalic photoreception in the photoperiodic induction of growth of the testes in the white crowned sparrow, *Zonotrichia leucophrys gambelii*. *Cell. Tissue Res.* **189**, 441–467.

6

Maternal Entrainment of a Fetal Biological Clock

STEVEN M. REPPERT and DAVID R. WEAVER

Laboratory of Developmental Chronobiology, Children's Service, Massachusetts General Hospital and Harvard Medical School, Boston, MA 02114, USA

Abstract

A biological clock oscillates in the mammalian fetus, and the fetal clock is entrained by redundant circadian signals from the mother. Recent evidence suggests that a dopamine system within the fetal hypothalamus is a final common pathway through which maternal circadian signals entrain the fetus. An entrainable biological clock during fetal life helps the developing mammal more readily prepare for life in the outside world.

Introduction

DURING FETAL life, studies in several mammalian species show that the biological clock in the suprachiasmatic nuclei (SCN) is oscillating in time (phase) with the environmental light–dark cycle. Entrainment of the fetal SCN involves maternal communication of circadian signals to the fetus. This chapter reviews these studies in rodents and discusses recent results which suggest a cellular mechanism through which the fetal SCN senses maternal entraining signals.

A biological clock is oscillating in the fetus

In mammals, circadian rhythms in physiology and behavior are not overtly expressed until postnatal life.[34] Deguchi[7] was the first to suggest that before rhythm expression a biological clock might be oscillating in the mammalian fetus. He determined in rats the phase of a rhythm monitored under constant conditions during the postnatal period to estimate the phase of the biological clock at earlier developmental stages. Because circadian rhythms persist (free run) under constant conditions with a cycle length close to 24 hours, the phase of a rhythm in the postnatal period can be used to infer the phase of the biological clock at earlier developmental stages.[4] Deguchi's findings suggested that a circadian clock oscillates at or before birth and that its phase is coordinated with the dam.

Postnatal paradigms cannot conclusively show that a biological clock actually functions *in utero*. It is possible that some rhythmic aspect of the birth process itself could start or set the timing of the developing clock. Demonstrating prenatal function of the biological clock requires a method that can measure an intrinsic, functionally relevant property of the clock itself. A method proven useful for monitoring the oscillatory activity of the SCN in adult rats is ^{14}C-labeled 2-deoxyglucose (2-DG) autoradiography which provides a biochemical means of visualizing the metabolic (functional) activity of discrete structures in the central nervous system *in vivo*.[44]

The 2-DG method was used by Reppert and Schwartz[35] to delineate a circadian rhythm of metabolic activity in the SCN of fetal rats. This study showed that the fetal SCN exhibit a circadian variation of metabolic activity that is "in time" (coordinated) with the rhythm in the dam and with the external lighting cycle (Figure 1). Remarkably, this fetal rhythm can be detected as early as day 19 of gestation (2–3 days before birth; all gestational ages have been standardized to day 0 = day of sperm positivity).[36] Recently, a rhythm of 2-DG in the fetal SCN has also been demonstrated *in vitro*.[46] This finding suggests that the metabolic activity rhythm in SCN is endogenously generated in the fetus and not passively driven by the mother.

Neurogenesis of the rat SCN occurs between days 13 and 16 of gestation.[1,16] A few synapses first appear in the SCN on day 19 of gestation, with the vast majority of synapses appearing postnatally.[25,28] A clear day-night rhythm of SCN action potentials is first apparent in fetal hypothalamic slices on day 21 of gestation[45] at a time when the metabolic activity rhythm is quite prominent. Thus, the SCN begin displaying circadian oscillations in metabolic and electrical activity when the neurons are virtually devoid of neural connections. This early circadian function in the absence of neural connections has led to the notion that individual SCN neurons may function as individual circadian pacemaking units.[28]

Vasopressin mRNA levels can also be used as an intrinsic marker of the oscillatory activity in the SCN during fetal life.[39] A day–night oscillation in vasopressin mRNA levels was found in the fetal SCN beginning on day 21 of gestation, the earliest time that vasopressin mRNA levels are detectable in SCN. The fetal rhythm was in phase with the circadian time of the dam.[39]

The maternal SCN transduces circadian information for the fetus

Entrainment of the fetus is due to maternal communication of circadian signal(s) to the fetus (Figure 2). Blind pregnant rats were used to show that environmental lighting acts through the maternal circadian system to entrain the rhythm of fetal SCN metabolic activity;[35] the fetal rhythm was

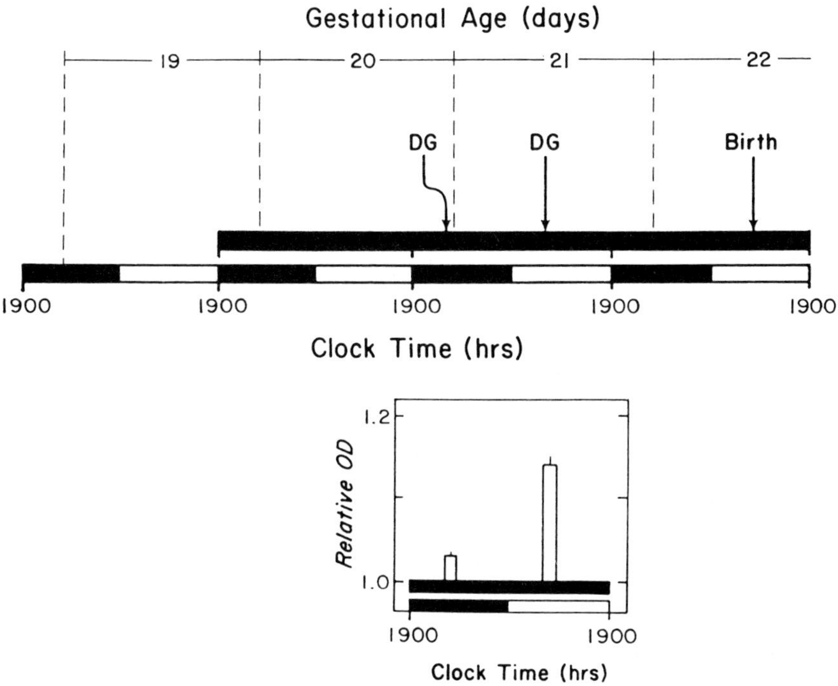

FIG. 1. Deoxyglucose experiment showing that the fetal SCN manifest a day–night rhythm in metabolic activity. Four pregnant Sprague-Dawley rats were housed with lights on from 07.00 to 19.00 hours until gestational day 19, when the animals were placed in constant darkness in preparation for the deoxyglucose injection (experimental paradigm depicted in upper panel). Two animals were injected i.v. with 2-deoxy[^{14}C]glucose during the period when the lights would have normally been off (subjective night) on gestational day 20, and the other two were injected during the period when the lights would have normally been on (subjective day) on gestational day 21. At 45 minutes after injection, the animals were killed and four fetal brains from each pregnant animal were randomly chosen, sectioned and processed for autoradiography. The optical density (OD) of each fetal suprachiasmatic nucleus was measured and the OD of adjacent hypothalamus was used as an internal reference standard for each fetal brain. The data are thus expressed as relative OD (OD of SCN/OD of adjacent hypothalamus). Each vertical bar in the lower panel gives mean relative OD ($\pm$SEM) for the SCN of eight fetal brains. The fetal SCN exhibit a clear day-night rhythm of metabolic activity ($p > 0.01$); the nuclei are metabolically active during the mother's subjective day and inactive during the mother's subjective night. (Modified from reference 35.)

synchronous with the circadian time of the blind mothers and not affected directly by ambient lighting. Investigators have used a variety of pre- and postnatal rhythms to confirm the prenatal coordination of maternal and fetal rhythms in several species.[4,42]

The involvement of the maternal SCN in the generation of entraining signals for the fetus has been demonstrated in rats and hamsters by destroying the maternal SCN early in gestation.[5,13,14,38,53] These experi-

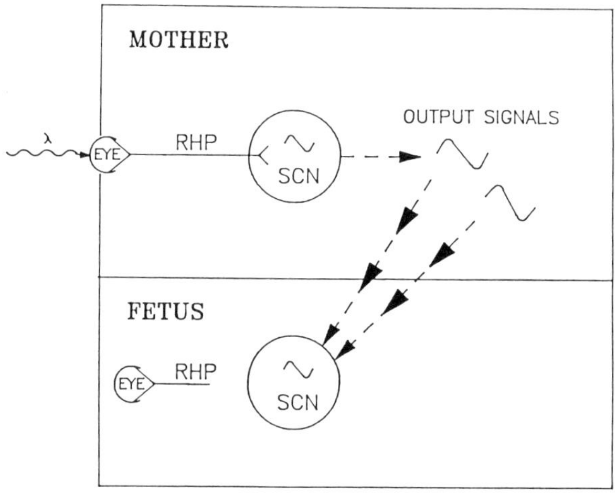

FIG. 2. Conceptual model of maternal–fetal communication of circadian phase. Light-induced neural signals are conveyed to the SCN by her retinohypothalamic pathway (RHP), entraining her circadian rhythms. Maternal output signals then entrain the fetal clock at a time when the innervation of the fetal SCN by the RHP is incomplete.

ments indicate that the entraining signals depend on the integrity of the maternal SCN (Figure 2); maternal SCN lesions cause desynchronization of the individual fetal clocks within a litter, with circadian rhythms in each animal persisting.

Our laboratory showed that during the postnatal period, rat pups born to SCN-lesioned dams respond appropriately (entrain) to light–dark cycles despite the absence of prenatal maternal entrainment (Weaver and Reppert, unpublished data). Furthermore, pups from sham-operated and SCN-lesioned dams express free-running rhythms with similar cycle lengths.[38] Thus, normal circadian function (e.g. light-dark entrainment and free-running cycle length) in adulthood does not appear to require prenatal maternal entrainment.

What are the maternal entraining signal(s)

Because the *in utero* environment provides a rich source of rhythmic hormones from the mother (e.g. prolactin, corticosterone and melatonin), studies have focused on the possibility that the maternal signal entraining the fetus is a hormone. Melatonin, the principal hormone of the pineal gland, was a prime candidate for the maternal signal communicating phase information to the fetus.[33] There is a robust rhythm of melatonin in the maternal circulation, and because melatonin can cross the placenta, a melatonin rhythm is reflected in the fetal circulation.[17,27,40,57–59]

However, maternal pinealectomy (which eliminates measurable levels of melatonin[26]) does not abolish maternal coordination of fetal circadian phase in rats,[37] hamsters (F.C. Davis, personal communication), or spiny mice (Weaver and Reppert, unpublished results). Furthermore, removal of the maternal adrenals, thyroid-parathyroids, pituitary or ovaries (in separate experiments) does not abolish the clear day–night rhythm of metabolic activity in the fetal rat SCN when performed on or before day 7 of gestation (the time when maternal SCN lesions are disruptive).[37] The maternal eyes, a potential source of both neural and endocrine signals, are also not necessary for the communication, since dams enucleated on day 2 of gestation synchronize the circadian clocks of their fetuses.[35]

Another approach to determine which aspect of maternal rhythmicity entrains the fetus is to artificially restore rhythmicity in an SCN-lesioned dam (who has no endogenous rhythmicity). Normally, there is a circadian rhythm in food consumption; this rhythm is disrupted by SCN lesions.[29] Restricted access to food in SCN-lesioned rats artificially produces a rhythm in food consumption and can also induce rhythms in locomotor activity and temperature.[22,48,53] Using a food access restriction paradigm (food cue) in SCN-lesioned pregnant rats, we and others have shown that the rhythmic ingestion of food can entrain fetuses of SCN-lesioned dams.[12,53]

Rhythmic food ingestion clearly would cause rhythmic fluctuations in nutrient levels in the blood, and this could in turn induce physiological responses to the presence of nutrients. In SCN-lesioned dams, food restriction may cause generation of a nutrient-related signal which entrains the fetuses in a manner parallel to that which occurs in intact dams. Alternatively, restricted feeding of SCN-lesioned dams can lead to the generation of other rhythms (e.g. in temperature or activity) which may be involved in setting the phase of the fetus.

Because the fetus is exposed to a multitude of maternal rhythms (behavioral rhythms, as well as hormonal), it is also possible that multiple maternal rhythms act in concert to entrain the fetal biological clock (Figure 2). Thus, eliminating any one of these rhythms would not be sufficient to disrupt maternal entrainment. This redundancy may explain some recent, seemingly contradictory data.

As mentioned above, the maternal pineal gland is not necessary for maternal entrainment of the fetus. However, Davis and Mannion[6] have convincingly shown that timed injections of pharmacological doses of melatonin into SCN-lesioned Syrian hamsters restore fetal synchrony. This finding supports the redundancy hypothesis: while melatonin is not necessary for fetal synchronization, it is one of several circadian signals capable of synchronizing the fetal biological clock.

Another line of evidence implicating melatonin as an important mediator of time-of-day information from mother to fetus is the recent discovery of putative melatonin receptors in the fetal SCN of several species.[41,54,56] Thus, an anatomical substrate exists for a direct action of maternal melatonin on the fetal biological clock.

Discovery of a functional dopamine system within the fetal SCN

We have begun to address the issue of how diverse maternal signals might act at a cellular level to set the time of the fetal SCN. As a first approach to address this issue, we have mapped neurotransmitter receptor gene expression in the fetal SCN. We discovered that a high level of D_1-dopamine receptor mRNA is expressed in the fetal SCN.[55] Furthermore, we showed that activation of these D_1-dopamine receptors induces a high level of *c-fos* gene expression.[55]

We examined whether D_1-dopamine receptors in the fetal SCN can activate *c-fos* gene expression by administering cocaine to pregnant rats. Cocaine was used because it acts on monoamine release and reuptake at presynaptic terminals[19,23] so that only receptors receiving an endogenous monoamine input should be affected by cocaine. Maternally administered cocaine (10–30 mg/kg) activated *c-fos* gene expression in the fetal SCN in a dose-dependent manner. The induction of *c-fos* mRNA in the fetal SCN appeared to be the result of a direct action of cocaine within the fetus, because cocaine administered directly to rat fetuses delivered by cesarean section on day 20 of gestation also induced *c-fos* expression in SCN. Interestingly, the ability of cocaine to induce *c-fos* expression in the fetal biological clock was not dependent on the time of day of administration, unlike the effects of light on SCN *c-fos* expression in adult rodents.[3,10,20,21,31,43]

Cocaine induction of *c-fos* expression in the fetal SCN was shown to occur in part through D_1-dopamine receptor activation.[55] Pretreatment with the D_1-receptor antagonist SCH 23390 (0.2 mg/kg) blocked ca. 50% of the cocaine-induced induction of *c-fos* mRNA in the fetal SCN. Furthermore, the D_1-dopamine selective agonist SKF 38393 (10 mg/kg) induced high levels of *c-fos* gene expression in the fetal SCN. An SKF 38393-induced increase in *c-fos* expression has been recently shown in the SCN of fetal mice and Syrian hamsters, as well as in fetal rats (Figure 3). Thus, D_1-dopamine receptor-mediated *c-fos* induction occurs in three rodent species in which maternal-fetal entrainment has been demonstrated.

The *c-fos* response to D_1-dopamine receptor stimulation in the fetal rat SCN is apparent on GD 18 when the nuclei are immature and devoid of synapses, as discussed above. Nonetheless, neurotransmission could occur in the absence of conventional synapses, since monoamines can be released from growth cones.[24] Tyrosine hydroxylase-positive neurons and fibers have been reported in the SCN region of the the fetal rat brain and could

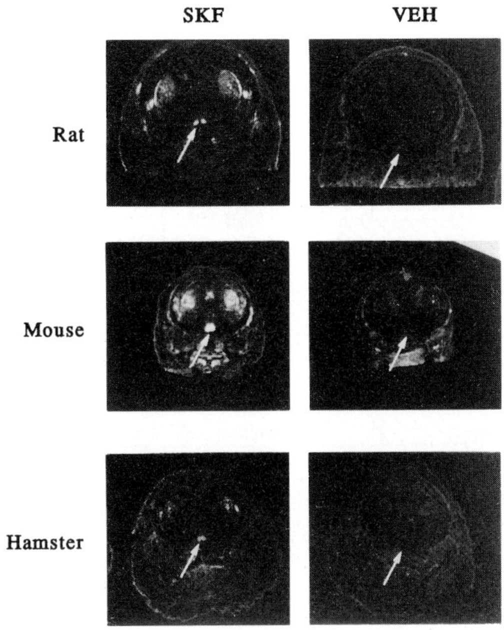

FIG. 3. D_1-dopamine receptor activation induces *c-fos* expression in the fetal SCN of three rodent species. Depicted are dark-field photomicrographs of film autoradiograms of coronal brain sections through the fetal SCN (arrows). Dams were injected ip with the specific D_1-dopamine agonist SKF 38393 (10 mg/kg) or vehicle. Fetal brains were collected 40 minutes after injection. Weaver and Reppert, unpublished data.

provide a source of dopamine for D_1-receptor activation within the fetal SCN.[32,51,52]

The functional consequences of D_1-dopamine-receptor-mediated *c-fos* induction in fetal SCN may be substantial. In the adult SCN, light exposure at night induces expression of *c-fos* mRNA and Fos protein[3,10,20,21,31,43] and causes phase shifts in circadian rhythms.[18] Furthermore, the quantitative features of light-induced *c-fos* expression and light-induced phase shifts are very similar suggesting that Fos may be a molecular component of the photic entrainment pathway for mammals.[20,21] We recently showed by Western blot analysis that Fos protein is expressed in the fetal hypothalamus (unpublished data). Thus, an activatable fetal dopamine system may represent an important unexpected mechanism for maternal signals to entrain the fetal SCN (Figure 4). Indeed, a dopamine system in the fetal hypothalamus may represent a final common pathway by which diverse entraining stimuli

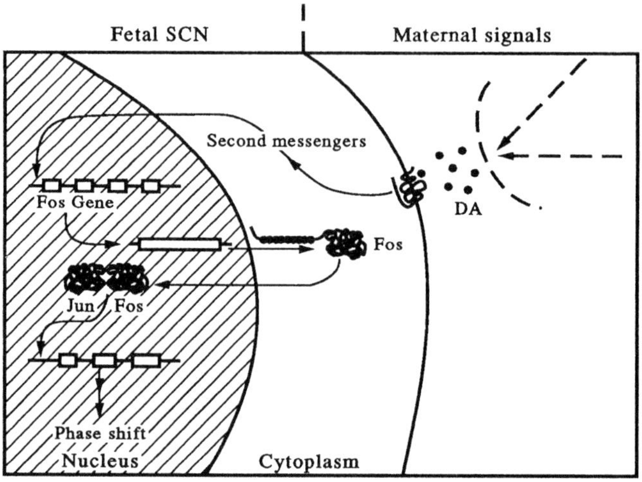

FIG. 4. Hypothetical model of a fetal dopamine system serving as a final common pathway for the action of maternal entraining signals on fetal SCN. Maternal signals act upon a dopamine system within the fetal SCN to regulate local dopamine release. Dopamine binds to its D1-receptor in SCN and via second messenger systems activates *c-fos* and other immediate early genes. Fos and Jun proteins are translocated to the nucleus, dimerize and interact with AP-1 binding sites on target genes ultimately leading to a phase shift.

converge upon the fetal SCN. As previously mentioned, the two identified entraining signals are rhythmic food ingestion and timed injections of melatonin.[6,53] The rhythmic ingestion of food by the mother could provide a rhythmic precursor pool of tyrosine, driving rhythm fetal brain dopamine production;[2,11] maternally derived melatonin could directly modulate the release of dopamine in the fetal SCN as demonstrated in other systems.[8,60] Studies to determine whether D_1-dopamine-receptor-stimulants can actually entrain the fetal SCN are currently underway.

Postnatal maternal entrainment

During the postnatal period, the maternal circadian system of altricial rodents continues to coordinate the timing of the developing circadian system.[49] It appears that maternal influences during the postnatal period serve to maintain or reinforce the coordination of phase that has been established during the prenatal period. In rats, the postnatal maternal influence persists until the pup develops the potential for direct light–dark entrainment through its own eyes (via the retinohypothalamic tract) at the end of the first week of life.[9,47,50]

Potential functions of an entrained fetal clock

The existence of prenatal communication of circadian phase in mammals suggests that this phenomenon is of adaptive value. Maternal entrainment would presumably coordinate the pups to the dam and to the environment, so that when physiological and behavioral rhythmicity later develops, the rhythms are expressed in proper relationship to one another and to the 24-hour day. This allows the young animal to more easily assume its temporal niche.

If the pups of altricial rodents were not coordinated to the environment by the mother, they might exhibit inappropriate timing of emergence from the burrow or disadvantageous behavioral patterns when they emerged.[30] Furthermore, if each pup were to develop its own rhythmicity, there would be a period of disorganization of the litter that could be detrimental to some or all members of the litter. For example, coordination of a dam's willingness to nurse and of all pups in the litter to suckle would clearly be of benefit. Circadian disorganization at the level of the individual or litter could thus threaten survival of individual pups.[15]

References

1. Altman J. and Bayer S.A. (1978) Development of the diencephalon in the rat. I. Autoradiographic study of the time of origin and settling patterns of neurons of the hypothalamus. *J. Comp. Neurol.* **182**, 945–947.
2. Arevalo R., Castro R., Palarea M.D. and Rodriguez M. (1987) Tyrosine administration to pregnant rats induce behavioral persistent modifications in the male off-spring. *Physiol. Behav.* **39**, 447–451.
3. Aronin N., Sagar S. M., Sharp F.R. and Schwartz W.J. (1990) Light regulates expression of a Fos-related protein in rat suprachiasmatic nuclei. *Proc. Natl. Acad. Sci. USA* **87**, 5959–5962.
4. Davis F.C. (1989) Use of post-natal behavioral rhythms to monitor pre-natal circadian function. In *Development of Circadian Rhythmicity and Photoperiodism in Mammals.* (ed. Reppert S.M.), pp. 45–66. Perinatology Press, Ithaca, NY.
5. Davis F.C. and Gorski R.A. (1988) Development of hamster circadian rhythms III. Role of the maternal suprachiasmatic nucleus. *J. Comp. Physiol.* **162**, 601–610.
6. Davis F.C. and Mannion J. (1988) Entrainment of hamster pup circadian rhythms by prenatal melatonin injections. *Am. J. Physiol.* **255**, R439-R448.
7. Deguchi T. (1975) Ontogenesis of a biological clock for serotonin: acetyl coenzymeA N-acetyltransferase in the pineal gland of rat. *Proc. Natl. Acad. Sci. USA* **72**, 2914–2920.
8. Dubocovich M.L. (1988) Pharmacology and function of melatonin receptors. *FASEB J.* **2**, 2765–2773.
9. Duncan M.J., Banister M.J. and Reppert S.M. (1986) Developmental appearance of light-dark entrainment in the rat. *Brain Res.* **369**, 326–330.
10. Earnest D.J., Iadarola M., Yeh H.H. and Olschowka J.A. (1990) Photic regulation of *c-fos* expression in neural components governing the entrainment of circadian rhythms. *Exp. Neurol.* **109**, 353–361.
11. Garabal M.V., Arevalo R.M., Diaz-Palarea, Castro R. and Rodriguez M. (1988) Tyrosine availability and brain noradrenaline synthesis in the fetus: control by maternal tyrosine ingestion. *Brain Res.* **457**, 330–337.
12. Honma K. and Honma S. (1978) In *Fifth SNCC Symposium.* (ed. Segawa M.), pp. 113–129. Sanposha, Tokyo.
13. Honma S., Honma K., Shirakawa T. and Hiroshige T. (1984) Effects of elimination of

maternal circadian rhythms during pregnancy on the postnatal development of circadian corticosterone rhythm in blinded infantile rats. *Endocrinology* **114**, 44–50.
14. Honma S., Honma K., Shirakawa T. and Hiroshige T. (1984) Maternal phase setting of fetal circadian oscillation underlying the plasma corticosterone rhythm in rats. *Endocrinology* **114**, 1791–1796.
15. Hudson R. and Distel H. (1989) Temporal pattern of suckling in rabbit pups: a model of circadian synchrony between mother and young. In *Development of Circadian Rhythmicity and Photoperiodism in Mammals.* (ed. Reppert S.M.), pp. 83–103. Perinatology Press, Ithaca, NY.
16. Ifft J.D. (1972) An autoradiographic study of the time of final division of neurons in rat hypothalamic nuclei. *J. Comp. Neurol.* **144**, 193–204.
17. Klein D.C. (1972) Evidence for the placental transfer of ^{3}H-acetyl melatonin. *Nature* **237**, 117–118.
18. Klein D.C., Moore R.Y. and Reppert S.M. (eds.) (1991) *Suprachiasmatic Nucleus: The Mind's Clock*. Oxford University Press, New York.
19. Koe B.K. (1976) Molecular geometry of inhibitors of the uptake of catecholamines and serotonin in synaptosomal preparations of rat brain. *J. Pharmacol. Exp. Ther.* **199**, 649–661.
20. Kornhauser J.M., Nelson D.E., Mayo K.E. and Takahashi J.S. (1990) Photic and circadian regulation of *c-fos* gene expression in the hamster suprachiasmatic nucleus. *Neuron* **5**, 127–134.
21. Kornhauser J.M., Nelson D.E., Mayo K.E. and Takahashi J.S. (1992) Regulation of *jun*-B messenger RNA and AP-1 activity by light and a circadian clock. *Science* **255**, 1581–1583.
22. Kreiger D.T., Hause L. and Krey L.C. (1977) Suprachiasmatic nuclear lesions do not abolish food-shifted circadian adrenal and temperature rhythmicity. *Science* **197**, 398–399.
23. Langer S.Z. and Arbilla S. (1984) *Trends Pharmacol. Sci.* **4**, 387–390.
24. Lauder J.M. (1988) Neurotransmitters as morphogens. In *Progress in Brain Research* (eds. Boer G.J., Feenstra M.G.P., Mirmiran M., Swaab D.F. and Van Haaaren F.), Vol. 73, pp. 365–387. Elsevier, Amsterdam.
25. Lenn N.J., Beebe B. and Moore R.Y. (1977) Post-natal development of the suprachiasmatic nucleus of the rat. *Cell Tissue Res.* **178**, 463–475.
26. Lewy A.J., Tetsuo M., Markey S.P., Goodwin F.K. and Kopin I.J. (1980) Pinealectomy abolishes plasma melatonin in the rat. *J. Clin. Endocrinol. Metab.* **50**, 204–207.
27. McMillen I.C. and Nowak R. (1989) Maternal pinealectomy abolishes the diurnal rhythm in plasma melatonin concentrations in the fetal sheep and pregnant ewe during late gestation. *J. Endocrinol.* **120**, 459–464.
28. Moore R.Y. and Bernstein M.E. (1989) Synaptogenesis in the rat suprachiasmatic nucleus demonstrated by electron microscopy and synapsin I immunoreactivity. *J. Neurosci.* **9**, 2151–2162.
29. Nagai K., Nishino T., Nakagawa S. and Fukuda Y. (1978) Effects of bilateral lesions of the suprachiasmatic nuclei on the circadian rhythm of food intake. *Brain Res.* **142**, 384–389.
30. Pratt B.L. and Goldman B.D. (1986) Maternal influence on activity rhythms and reproductive development in Djungarian hamster pups. *Biol Reprod.* **34**, 655–663.
31. Rae M.A. (1989) Light increases Fos-related protein immunoreactivity in the rat suprachiasmatic nuclei. *Brain Res. Bull.* **23**, 577–581.
32. Reisert I., Schuster R., Zienecker R. and Pilgrim C. (1990) *Dev. Brain Res.* **53**, 222–229.
33. Reppert S.M. (1982) Maternal melatonin: a source of melatonin for the immature mammal. In *Melatonin Rhythm Generating System* (ed. Klein D.C.), pp. 182–191. Karger, Basel.
34. Reppert S.M. (1987) Circadian rhythms: basic aspects and pediatric implications. In *Current Concepts in Pediatric Endocrinology* (ed. Styne D.M.), pp. 91–125. Elsevier, Amsterdam.
35. Reppert S.M. and Schwartz W.J. (1983) Maternal coordination of the fetal biological clock in utero. *Science* **220**, 969–971.
36. Reppert S.M. and Schwartz W.J. (1984) The suprachiasmatic nuclei of the fetal rat:

characterization of a functional circadian clock using 14C-labeled deoxyglucose. *J. Neurosci.* **4**, 1677–1682.

37. Reppert S.M. and Schwartz W.J. (1986) Maternal endocrine extirpations do not abolish maternal coordination of the fetal circadian clock. *Endocrinology* **119**, 1763–1767.
38. Reppert S.M. and Schwartz W.J. (1986) The maternal suprachiasmatic nuclei are necessary for maternal coordination of the developing circadian system. *J. Neurosci.* **6**, 2724–2729.
39. Reppert S.M. and Uhl G.R. (1987) Vasopressin messenger ribonucleic acid in the supraoptic and suprachiasmatic nuclei: appearance and circadian regulation during development. *Endocrinology* **102**, 2483–2487.
40. Reppert S.M., Chez R.A., Anderson A. and Klein D.C. (1979) Maternal-fetal transfer of melatonin in a non-human primate. *Pediatr. Res.* **13**, 788–791.
41. Reppert S.M., Weaver D.R., Rivkees S.A. and Stopa E.G. (1988) Putative melatonin receptors in a human biological clock. *Science* **242**, 78–81.
42. Reppert S.M., Weaver D.R. and Rivkees S.A. (1989) Pernatal function and entrainment of a circadian clock. In *Development of Circadian Rhythmicity and Photoperiodism in Mammals* (ed. Reppert S.M.), pp. 25–44. Perinatology Press, Ithaca, NY.
43. Rusak B., Robertson H.A., Wisden W. and Hunt S.P. (1990) Light pulses that shift rhythms induce gene expression in the suprachiasmatic nucleus. *Science* **248**, 1237–1240.
44. Schwartz W.J. (1991) SCN metabolic activity in vivo. In *Suprachiasmatic Nucleus: The Mind's Clock* (eds. Klein D.C., Reppert S.M. and Moore R.Y.), pp. 144–156. Oxford University Press, New York.
45. Shibata S. and Moore R.Y. (1987) Development of neuronal activity in the rat suprachiasmatic nucleus. *Dev. Brain Res.* **34**, 311–315.
46. Shibata S. and Moore R.Y. (1988) Development of fetal circadian rhythm after disruption of the maternal circadian system. *Dev. Brain Res.* **41**, 313–317.
47. Stanfield B. and Cowan W.N. (1976) Evidence for a change in the retinohypothalalmic projection in the rat following early removal of one eye. *Brain Res.* **104**, 129–133.
48. Stephan F.K. (1981) Limits of entrainment of periodic feeding in rats with suprachiasmatic lesions. *J. Comp. Physiol.* **143**, 401–410.
49. Takahashi K., Ohi K. Shimoda K. Yamada N. and Hayashi S. (1989) Postnatal maternal entrainment of circadian rhythms. In *Development of Circadian Rhythmicity and Photoperiodism in Mammals* (ed. Reppert S.M.), pp. 67–82. Perinatology Press, Ithaca, NY.
50. Terubayashi H., Fujisawa H., Itoi M. and Ibata Y. (1985) HRP-Histochemical detection of retinal projections to the hypothalamus in neonatal rats. *Acta Histochem. Cytochem.* **18**, 433–438.
51. Umgrumov M.V., Taxi J., Tixier-Vidal A., Thibault J. and Mitskevich M.S. (1989) Ontogenesis of tyrosine hydroxylase-immunopositive structures in the rat hypothalamus. An atlas of neuronal cell bodies. *Neuroscience* **29**, 135–156.
52. Umgrumov M.V., Tixier-Vidal A., Taxi J., Thibault J. and Mitskevich M.S. (1989) Ontogenesis of tyrosine hydroxylase-immunopositive structures in the rat hypothalamus. Fiber pathways and terminal fields. *Neuroscience* **29**, 157–166.
53. Weaver D.R. and Reppert S.M. (1989) Periodic feeding of SCN-lesioned pregnant rats entrains the fetal biological clock. *Dev. Brain Res.* **46**, 291–296.
54. Weaver D.R., Namboodiri M.A.A. and Reppert S.M. (1989) Iodinated melatonin mimics melatonin action and reveals discrete binding sites in fetal brain. *FEBS Lett.* **228**, 123–127.
55. Weaver D.R., Rivkees S.A. and Reppert S.R. (1992) D_1-dopamine receptors activate *c-fos* expression in the fetal suprachiasmatic nuclei. *Proc. Natl. Acad. Sci. USA* **89**, 9201–9204.
56. Williams L.M., Martinoli M.G., Titchener L.T. and Pelletier G. (1991) The ontogeny of central melatonin binding sites in the rat. *Endocrinology* **128**, 2083–2090.
57. Yellon S.M. and Longo L.D. (1987) Melatonin rhythms in fetal and maternal circulation during pregnancy in sheep. *Am. J. Physiol.* **252**, E799-E802.
58. Yellon S.M. and Longo L.D. (1988) Effect of maternal pinealectomy and reverse

photoperiod on the circadian melatonin rhythm in the sheep and fetus during the last trimester of pregnancy. *Biol. Reprod.* **39**, 1093–1099.

59. Zemdegs I.Z., McMillen I.C., Walker D.W., Thorburn G.D. and Nowak R. (1988) Diurnal rhythms in plasma melatonin concentrations in the fetal sheep and pregnant ewe during late gestation. *Endocrinology* **123**, 284–289.
60. Zisapel N. and Laudon M. (1982) Dopamine release induced by electrical field stimulation of rat hypothalamus in vitro: inhibition by melatonin. *Biochem. Biophys. Res. Commun.* **104**, 1610–1616.

SECTION II

Anatomy and Physiology of Light Regulated Endogenous Oscillators

7

Daily Melatonin Infusion Entrains Free-Running Activity in Syrian and Siberian Hamsters

R. KIRSCH, S. BELGNAOUI, S. GOURMELEN and P. PÉVET

CNRS-URA 1332 "Neurobiologie des fonctions rythmiques et saisonnières", Université L. Pasteur, Strasbourg, France

Introduction

IN RECENT years it has become clear that, in mammals, the pineal gland is essential for the regulation of photoperiodic responses and that melatonin, one of the hormones produced by this gland, is directly involved in such phenomena (for review see Reiter[26] and Pévet[21]). Melatonin synthesis, indeed, occurs during the dark period of the light/dark cycle and the duration of the secretion is directly dependent upon the duration of the dark phase.[10,16,17,21] Moreover, subcutaneous administration of melatonin in pinealectomized animals by a method designed to produce a daily peak corresponding to either a long or a short duration can reproduce the effect of short or long photoperiod on e.g. their sexual activities.[6,7,9,11,20,23] These data clearly demonstrate that it is the duration of the nocturnal peak of melatonin which is the critical parameter in the transmission of the photoperiodic information to the central nervous system.

Another aspect of melatonin secretion, which has not to be forgotten is that its rhythm is a circadian one which, like many other circadian rhythms, is directly generated in the suprachiasmatic nuclei of the hypothalamus. This means that it persists in the absence of light/dark cycle (period slightly different from 24 hours) and is entrained to a 24 hour period by the light/dark cycle. Beside its role as the transducing signal for daylength, the periodic secretion of melatonin might thus be used as a circadian zeitgeber, directly or indirectly involved in the control of circadian rhythmicity.

Considerable speculation has surrounded this possible role of melatonin as a circadian zeitgeber, especially in humans for the treatment of pathologically or socially induced disturbances of biological rhythms (disturbed sleep in shift-workers or blind peoples, jet lag, etc. . . . for review see Arendt[1]).

Although till now there is still little experimental evidences in mammals, all these informations as well as the clear data obtained in non-mammalian vertebrates (for review see Cassone[12]) suggest a role of melatonin in a direct control of circadian rhythmicity and, at least, ask for an expansion of our knowledge on this phenomenon.

In general, to date, as animal studies are concerned, there is only evidence for minor effects of the pineal on circadian timing in mammals (review in Cassone[13]). Injections of melatonin at the same time everyday, however, entrain the circadian locomotory pattern of male and female rats free running in constant conditions.[3,4,5,25] Whether this observation is the consequence of a pharmacological effect rather than a physiological one is still a debate, but it clearly established the useful properties of melatonin.

Daily melatonin injection induces, however, daily manipulation of the animals and this might interfere in the synchronisatory process studied. To avoid this problem, and with the purpose to analyze the properties of melatonin, we have decided to administer the hormone by infusion. This technique, indeed, permit to mimic the daily nocturnal peak of plasmatic melatonin concentration.

The present paper reports the results already obtained in two rodents species, the Syrian and Siberian hamsters.

Material and methods

Young adult male Syrian hamster (*Mesocricetus auratus*) were purchased from a commercial supplier (Centre de Production animale, Olivet, France) where they had been raised under a photoperiod of 14hL/10hD). Prior to experimentation the animals were adapted to the laboratory conditions for at least 2 weeks (photoperiod 14 hL/10hD, lights off at 18 hours, temperature 20 ± 1°C).

Male Siberian hamster (*Phodopus sungorus*) were raised in the laboratory (bred from animals kindly provided by Dr Steinlechner in 1985). Since birth, they were raised, under constant artificial photoperiod (16 hL/8hD) with lights on at 4 hours, temperature 20 ± 1°C, until they were used for the present experiment (2 months old).

During the experiment, animals were housed individually in infusion cages, each of these cages being placed inside a light-tight and sound isolated ventilated wooden-box. Each box was photoregulated at a given condition (see below). In each infusion cage, wheel-running activity was permanently monitored.

Pinealectomy was performed on animals anesthetized with either equithesin or pentobarbital. A hole centered on the lambda was made in the skull. The pineal was removed from the sinus surrounding it with fine forceps. Animals were given 1 week for recovery after the operation before being attached to the infusion system.

Cannulation of the animals for infusion was performed using a polyethylene tube. At 2 cm from the proximal end of the cannula a polyethylene collar was fixed and the cannula was thermoformed into a right angle. The animals were then anesthetized with halothane gas and the collar and the end of the cannula were introduced under the skin through a 3 cm incision on the neck. The collar prevented the cannula from being chewed and the extension of the cannula ensured a good efflux of the hormone solution. The distal end of the cannula was placed on a swivel fixed to a bar hanging on the top of the cage. The swivel was connected via polyethylene tube to a syringe. Infusions were performed using infusion pumps controlled by electronic timers. The syringes placed outside the wooden boxes were refilled every 2 or 3 days with freshly prepared solution. The cannula system was built in such a way that it did not prevent the animal to run within the wheel. The entire system enabled the animals to be infused subcutaneously during several months (e.g. 7 months in Figure 4). Evidently to obtain such data during a very long period of time is very difficult. Often after few weeks the cannulae were broken or obstructed, the animals were sick, refused to go within the wheel or even died. In all these situations, the partial data obtained, although often confirming our complete observations were systematically rejected. The complete data presented in these five illustrations were obtained from three Syrian and seven Siberian hamsters.

At the beginning of the experiment the Siberian hamsters were placed in their infusion cages and submitted to constant dim light (0.1–5 lux) (such constant conditions permit the regular cleaning of the cages). At the same time infusion of Ringer (6 or 8 hours/day depending of the animals, 33 μl/hours) were started. After a few weeks (3 or 4) when the free running conditions were well-established, melatonin infusion started (melatonin disolved in Ringer; 6 or 8 hours/day depending of the animals, 33 μl/hours; 16.5 μg melatonin/hour). After a few weeks (4 to 10) Ringer infusion was started again, a period which in few animals was followed by a new period of melatonin infusion.

For Syrian hamster, in order to test a possible effect of Ringer infusion, the animals were prepared and placed in the system a few weeks before starting the Ringer infusion. The lighting and infusion conditions were similar to those described above.

As the locomotor activity is concerned, the number of revolutions of the running wheel were sampled every 10 minutes by computer via a real time data acquisition system (built in our lab). Activity was double plotted over a 48 hour scale.

During all the experiment the animals received tap water and food pellets *ad libitum*.

Results

All animals used in this study, showed in constant dim-light conditions a classical free-run in their circadian locomotory pattern. Most of the animals, either Siberian or Syrian hamsters, free-ran with a period slightly higher than 24 hours, which is classical in nocturnal species. Few animals, however, free-ran with period inferior to 24 hours and results obtained with one of such animals are presented in Figure 5.

Infusion of Ringer, which in the Siberian hamster started immediately, did not interfere with the circadian locomotory pattern, at least in the present experimental conditions, and a clear free run was observed (Figures 1, 2 and 3). This observation was confirmed in the Syrian hamsters, where the Ringer infusion started 3 weeks after the beginning of locomotor activity registration (Figures 4 and 5).

In such experimental conditions, in both species tested, daily infusions of melatonin (8 hours, Figure 1, 6 hours Figures 2, 3, 4 and 5) were able to entrain the circadian locomotor pattern to a period of 24 hours.

Some animals were entrained as soon as the infusions started (Figure 2), others only after few days (Figure 1). Sometimes considerable longer time, up to 59 days (Figure 4) was needed to get entrainment. This was due to the fact that melatonin did not influence the activity rhythm until the beginning of the infusion reached a certain moment in the temporal organization of the locomotor activity. Interestingly, a clear species difference was observed in this phenomenon.

In the Siberian hamster, melatonin had no effect unless the beginning of the infusion period did coincide with the animal's activity onset (Figures 1, 2 and 3). In most of the cases the onset of activity was immediately following the beginning of melatonin infusion (Figures 2–3). In few situations (Figure 1), however, the onset of activity immediately preceeded the infusion period.

In the Syrian hamster, to the contrary, melatonin was able to entrain the activity rhythm only when the beginning of the infusion period coincided with the second part of the activity period. In this species, the onset of activity preceded always the beginning of melatonin infusion from 3 to 5 hours (Figures 4 and 5).

In both species, when the beginning of infusion coincided with the adequate period of the activity rhythm, melatonin was able to entrain this rhythm to a 24 hour period and that, even if the endogenous period of the animal was shorter than 24 hours (Figure 5).

Some animals, but not all of them (Figure 3) remained entrained as long as melatonin was administered (until 36 days in the experiments represented in Figures 1 and 2). When melatonin administration was stopped, the activity rhythm of some Siberian hamsters free ran almost immediately (Figure 1), while in others it took 4 to 14 days (Figures 2 and

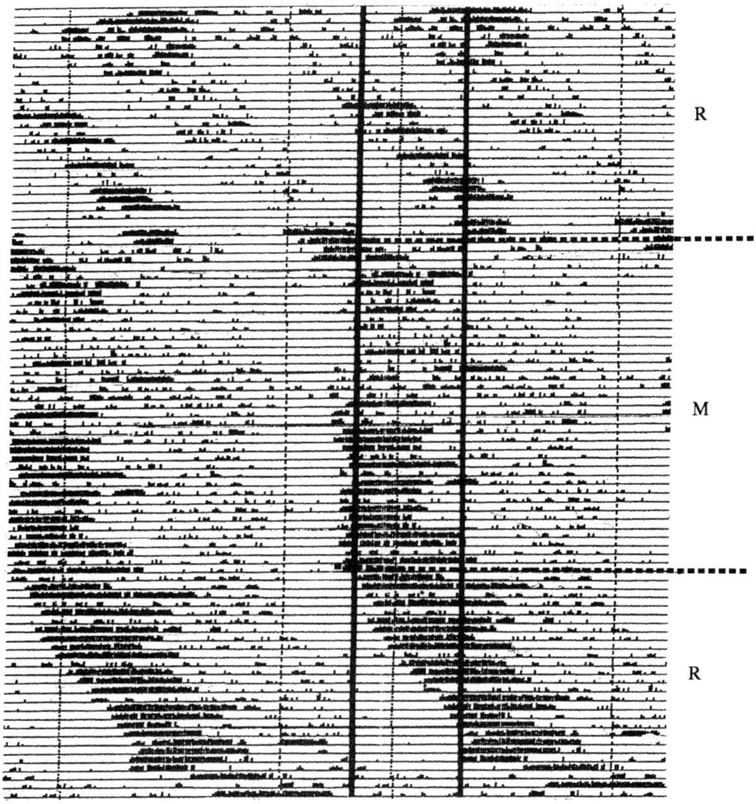

FIG. 1. Double plotted running wheel activity records of a Siberian hamster. Free-running activity is unaffected by Ringer (R) infusions. After a few days of melatonin (M) treatment (8 hours), an entrainment of the activity rhythm is observed when the beginning of the infusion coincided with the onset of activity. This entrainment ceased immediately when melatonin was replaced by Ringer (R). In this animal the onset of activity immediately preceeded the melatonin infusion period. The infusion period is indicated by the vertical black lines. The two vertical dotted lines trace onset and termination of the subjective night respectively.

3). In the two Syrian hamsters where the registration has been long enough to study this phenomenon, considerable time (Figure 4) elapsed from the termination of melatonin administration to the re-establishement of the original free-running period.

In very few animals—all Siberian hamsters—it has been possible to study the effect of a second period of melatonin infusion. Again, melatonin was able to entrain the free-running activity rhythm to a period of 24 hours and again this was possible only when the beginning of the infusion coincided with the onset of activity (Figure 2).

In Siberian hamsters, a different effect of melatonin infusion has also

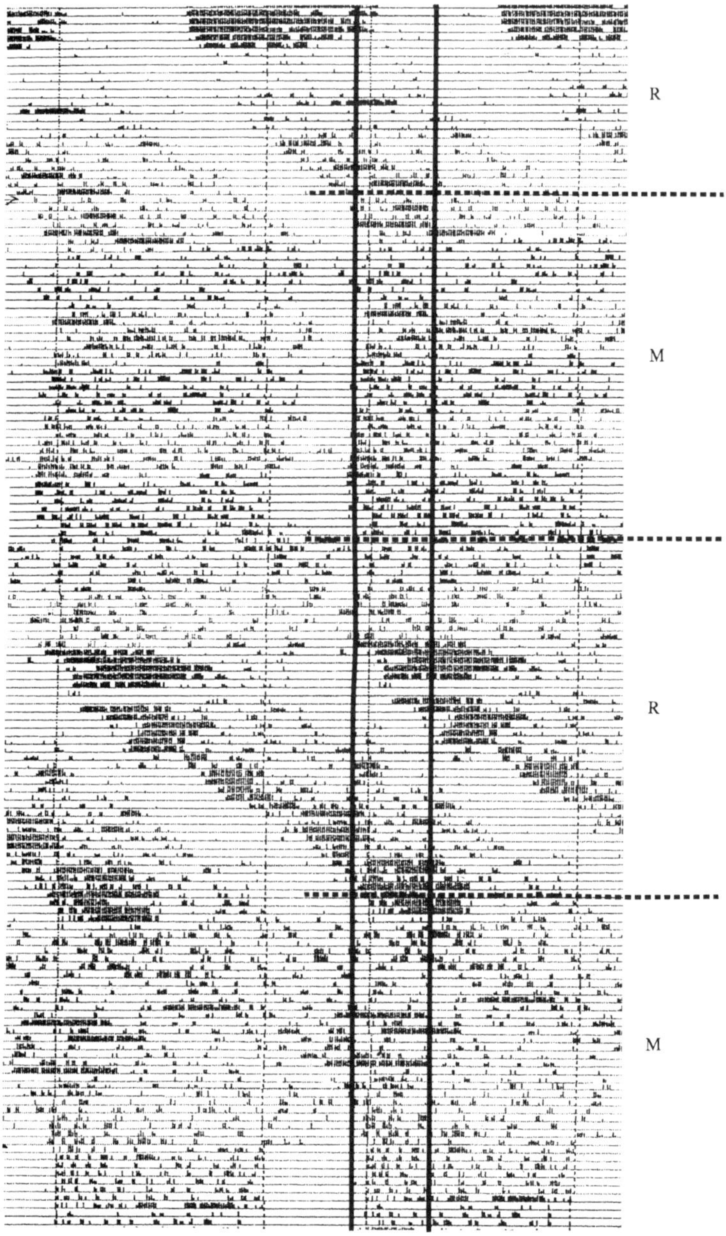

FIG. 2. Siberian hamster activity double plot showing the clear effect of melatonin infusions (6 hours). In this animal the onset of activity is immediately following the beginning of infusion period. When melatonin treatment was terminated a disorganization of the activity pattern is observed before the full re-establishment of the free-run. When a new sequence of melatonin infusions started a similar disorganization of the activity pattern is noted before to obtain entrainment.

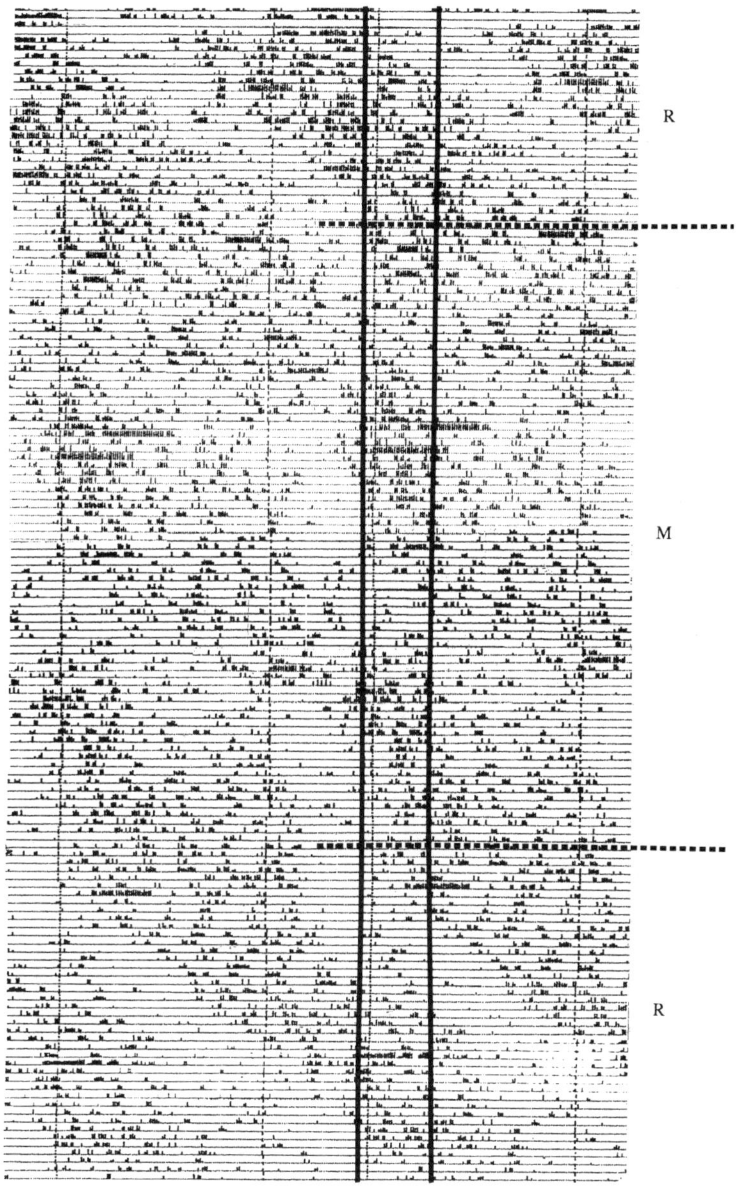

FIG. 3. In this Siberian hamster, infusion of melatonin (6 hours) immediately entrains the free-running activity rhythm (coincidence between onset of activity and beginning of the infusion period). After 1 week, however, a disorganization of the activity pattern is observed. This period is itself followed by new period (approximately 2 weeks) of entrainment and then again appears a perturbated pattern itself followed by a new synchronization. When melatonin treatment is terminated, after a few days, the free-running rhythm reappeared.

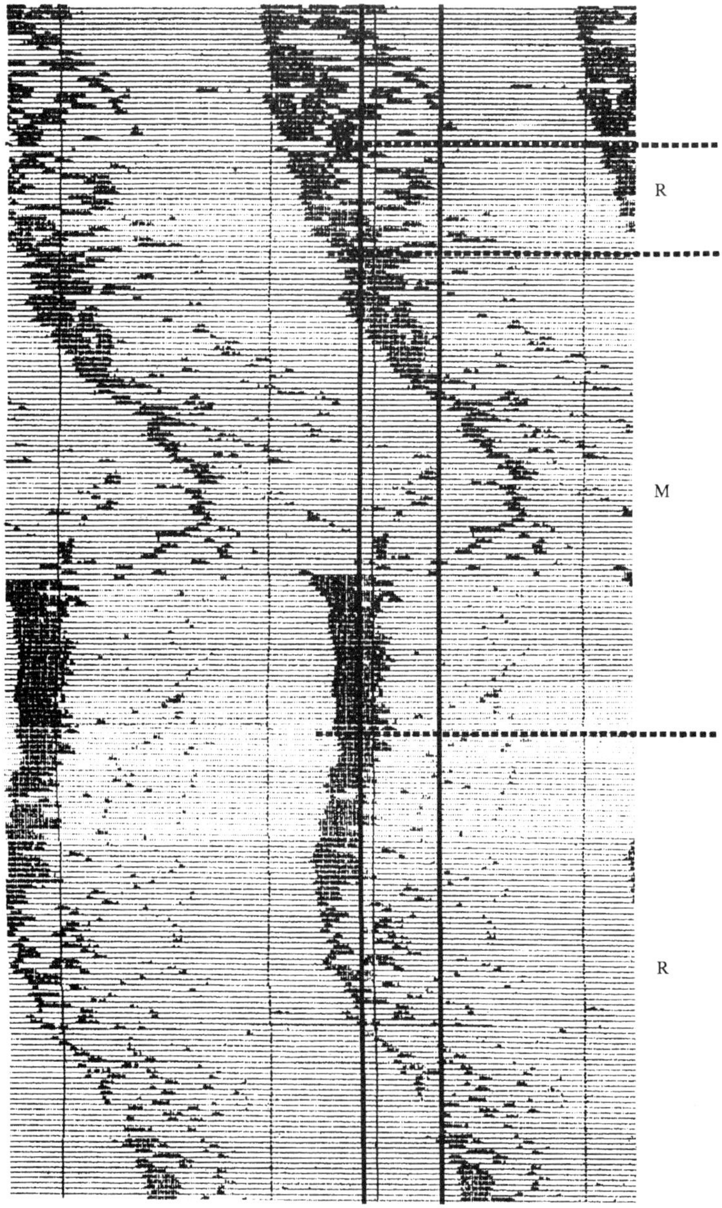

FIG. 4. Double plotted running wheel activity records of a Syrian hamster. Free-running activity is unaffected by Ringer infusion. An effect of melatonin (6 hours infusion) is here observed only after a very long period of time (59 days) and then the animal remained entrained as long as melatonin was administered. The entrainment is obtained when the beginning of melatonin infusion coincided with the second part of the activity. In this animal also, when melatonin infusion was terminated, the original free run took longer time to be re-established.

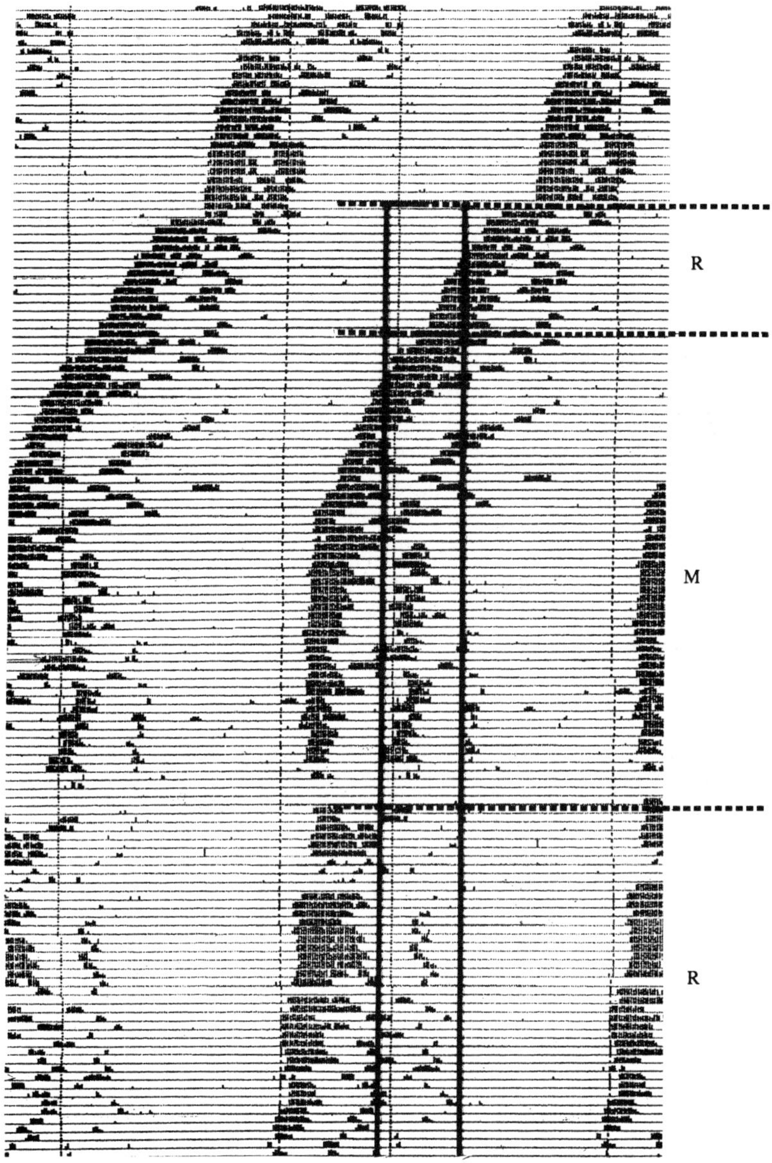

FIG. 5. Double plotted running wheel activity record of a Syrian hamster free-running with a period smaller than 24 hours. Again an entrainment of the activity was observed when the beginning of melatonin infusion coincided with the second part of the activity period. This effect ceased when melatonin was replaced by Ringer.

been observed. As shown in Figure 2 in the second period of melatonin infusion, after a few days, melatonin clearly affected the system by disturbing the circadian pattern of activity. When that happened, it was only after a few weeks that entrainment to a period of 24 hours was obtained. Such phenomenon of disorganization of the circadian pattern of activity was frequently observed. For example, looking at Figure 3 it appears that melatonin entrains the rhythm initially for a few days and that afterwards this is followed by a disorganized period which itself is followed by a new entrainment period and so on.

Discussion

Syrian hamsters, contrary to rats, are not entrained by daily injections of melatonin (for details see review of Cassone;[13] Armstrong[3]). The present results, however, clearly demonstrate that melatonin is able to entrain the free running rhythm of locomotor activity of pinealectomized Syrian and Siberian hamsters when the hormone is administered in a way which duplicates its endogenous plasmatic nocturnal peak. The observed effect corresponds to a true entrainment, since termination of melatonin infusions reestablished the free-running rhythm.

The data confirm the hypothesis that in mammals[3,25] like in non-mammalian vertebrates,[12] melatonin is involved in the control or the expression of circadian rhythms.

The question of major interest concerning this effect of melatonin is its precise site and mechanism of action.

The SCN, the circadian clock underlying rhythmicity in mammals, contains melatonin receptors[15,30,33,34] and its metabolism is known to be directly affected by melatonin (review in Cassone[12]). Moreover, in rats kept in constant dark, daily injections of melatonin entrain free-running circadian locomotor rhythm, a result not observed in SCN lesioned animals.[14] Entrainment takes place when the onset of activity has just passed injection time so that the active phase follows injection while the quiescent one precedes it. In these conditions melatonin induces a daily phase-advance. All these results strongly suggest that the circadian system is directly affected by melatonin and, based on the phase-advancing properties of exogenous melatonin, some authors have postulated that melatonin is working on the clock mechanism itself.

Apparently our results on the Syrian and the Siberian hamster confort such a hypothesis. A detailed analysis of our data, however, led us to think carefully with such an interpretation.

When melatonin infusion started, the entrainment was not observed immediately; a period from a few days to several weeks was needed. We have to assume that, like in rat, melatonin's entraining action was limited to a critical period of the circadian rest-activity cycle. In constant DD conditions the period of the free-run is slightly different from 24 hours and consequently the activity period shifted a few minutes every day. Depending on the starting time of the infusion, a few days to several weeks are needed for the infusion period to reach the critical period of entrainment ability.

In Siberian hamsters, melatonin was able to entrain the free-running activity rhythm to a period of 24 hours only when the beginning of the infusion period coincided with the onset of activity. Again that agrees with the work done in the rat. The difference, however, is that in individual animals the onset of activity followed (Figure 2) or preceeded (Figure 1) the apparition of melatonin in the circulation. This observation clearly indicates that if melatonin is able to entrain the circadian activity rhythm its role in this system is not as it has been suggested (Thomas and Armstrong, 1988) to switch from behavioral quiescence to activity.

The result obtained in the Syrian hamster strongly support this conclusion. Indeed, in this species, melatonin was again able to entrain the rhythm but only when the beginning of the infusion coincided with the second part of the activity period. In these conditions the onset of activity preceeded melatonin for 3 to 5 hours. In fact the experimental paradigm used is mimicking the natural occurring covariation between endogenous melatonin release and behavioral activity. Indeed, in the Siberian hamster, as in the rat, the locomotor activity starts at the beginning of the dark phase as does the melatonin synthesis. To the contrary, in the Syrian hamster, although the onset of activity is observed at the beginning of the dark phase, melatonin synthesis is always observed 3 to 5 hours later (Skene *et al.*, 1987). This difference in temporal organization implies a different mode of action of melatonin in these two species.

In nocturnal animal, kept under constant condition, the free-run period is slightly larger than 24 hours. Consequently the onset of activity is delayed for a few minutes every day. In order to entrain this rhythm to a 24 hour period, melatonin thus needs to induce every day an advance in the onset of activity. Based on these results, is it possible to speak of a phase advance effect of melatonin, a terminology which implies a direct effect of the hormone on the clock? Few of the animals used in this study presented a free-running period shorter than 24 hours. In these conditions, perfusion of melatonin, in the golden hamster (Figure 5) was also able to entrain the rhythm to a 24 hour period. To do this, however, melatonin has to induce a daily phase delay. Interestingly, this effect is also observed only when melatonin perfusion started 3 to 5 hours after the onset of activity, this means when the situation observed normally in Syrian hamster is mimicked. If we think in terms of a direct effect of melatonin on the clock, we have thus to accept that melatonin infused at the same circadian time is able to induce, depending of the animals, either a phase advance or a phase delay. That is difficult to conceive. However, an effect of melatonin on the

VIP-neurons of the SCN, neurons which are known to be the "gate" of the "clock" for synchronizing information, might explain our results.

Another possible mechanism of action of melatonin has to be considered. "Trigger circuits" initiating specific patterns of hormone secretion or behaviors (e.g. locomotory activity) that possess a circadian component do not appear to be directly innervated by SCN neurons. Main target areas of SCN efferents are the paraventricular nuclei of the hypothalamus (PVN) and of the thalamus (PVT).[18,31,32] As far as the present work is concerned, this last structure is of interest. The most remarkable feature of the PVT is its prominent projection to the nucleus accumbens, olfactory tubercule, central amygdala and prefrontal cortex,[8,22] thus providing the SCN with an almost direct input to the mesolimbo-cortical complex, traditionally considered as the main center in the central nervous system for the initiation and maintenance of (loco)motor behavior. Melatonin binding sites have been observed in the PVT of the Siberian as well as of the Syrian hamster.[33,34] An effect of melatonin at the level of the PVT might also explain the presently obtained data.

Are these results of physiological importance?

In our experimental conditions, the hamsters received 16.5 μg/hour which corresponds to approximately a total amount of 100 μg of melatonin infused per 6 hours and 120 μg/8 hours. Lower doses have not yet been tested. Such dosage probably results in supraphysiological blood levels of the hormone, at least at the beginning of the experiment. What happens however after several months of infusion (7 months in Figure 4). Pitrosky *et al.*,[23] indeed, have observed no significant difference in plasma melatonin concentration when Syrian hamsters were infused for 2 months with doses of melatonin varying from 20 to 80 ng/hour and concluded that within 8 weeks, the metabolism of melatonin is stimulated when melatonin doses is increased. It has not been possible to measure melatonin plasmatic concentration in the Syrian hamster after our experiments. In the Siberian hamster, however, after 2 months, a melatonin concentration of 12.35 ± ng/ml has been measured in the middle of the infusion period. This value is supraphysiological but far from the high value found after an acute administration of melatonin (e.g. 250 ng/ml of plasma 15 minutes after injection of 100 μg melatonin in the rat; 100 ng/ml and 25 ng/ml, 15 minutes after injection of respectively 25 and 2.5 μg in the golden hamster).[19,24,27] Whether such results are pharmacological or physiological is thus still a debate but the use of the present technique associated with lower concentrations of melatonin should help answer this question.

In conclusion, it appears that, in the Syrian as well as in the Siberian hamster, exogenous melatonin exerts an effect on the circadian locomotor activity. Nevertheless, at the present time, it is impossible to integrate such results into any global function for the mammalian pineal/SCN axis. It remains to be determined if the sites of action for this entraining effect of melatonin is at the level of the SCN or at the level of structures which otherwise require the SCN (e.g. PVT). In addition, it is not yet clear whether these results are pharmacological or physiological ones.

Acknowledgements

The authors wish to thank Drs. M. Masson-Pévet and J. Stehle for their suggestions and critical remarks and Miss F. Murro for her secretarial aid.

References

1. Arendt J. (1989) Melatonin and the pineal gland. In *Biological Rhythms in Clinical Practice* (eds. Arendt J., Minors D.S. and Waterhouse J.M.), pp. 184–206. Wright, London.
2. Arendt J. (1991) Melatonin in humans: jet lag and after. In *Advances in Pineal Research*, (eds. Arendt, J. and Pévet, P.), Vol. 5, pp. 299–302. John Libbey, London.
3. Armstrong S.M. (1989) Melatonin and circadian control in mammals. *Experientia* **45**, 932–938.
4. Armstrong S.M., Cassone V.M., Chesworth M.J., Redman J.R. and Short R.V. (1986) Synchronization of mammalian circadian rhythms by melatonin. In *Melatonin in humans* (eds. Wurtman R.J. and Waldhauser F.), pp. 375–394. Springer, New York.
5. Armstrong S.M. and Chessworth M.J. (1987) Melatonin phase-shifts a mammalian circadian clock. In *Fundamentals and Clinics in Pineal Research* (eds. Trentini G.P., De Gaetani C. and Pévet P.), pp. 195–198. Raven Press, New York.
6. Bartness T.J. and Goldman B.D. (1988a) Peak duration of serum melatonin and short-day responses in adult Siberian hamsters. *Am. J. Physiol.* **255**, R812-R822
7. Bartness T.J. and Goldman B.D. (1988b) Effects of melatonin on long-day responses in short-day housed adult Siberian hamsters. *Am. J. Physiol.* **255**, R823-R830.
8. Berendse H.W. and Groenewegen H.J. (1990) Organisation of the thalamostriatal projections in the rat, with special emphasis on the ventral striatum. *J. Comp. Neurol.* **299**, 187–228.
9. Bittman E.L. Dempsey R.J. and Karsch F.J. (1983) Pineal melatonin secretion drives the reproductive response to day-length in the ewe. *Endocrinology* **113**, 2276–2283.
10. Bittman E.L. and Karsch F.J. (1984) Nightly duration of pineal melatonin secretion determines the reproductive response to inhibitory day length in the ewe. *Biol. Reprod.* **30**, 585–593.
11. Bonnefond C., Martinet L. and Monnerie R. (1990) Effects of timed melatonin infusions and lesions of the suprachiasmatic nuclei on prolactin and progesterone secretions in pregnant and pseudopregnant mink (*Mustela vison*). *J. Neuroendocrinol.* **2**, 583–592.
12. Cassone V.M. (1990) Effects of melatonin on vertebrate circadian systems. *T.I.N.S.* **13**, 457–463.
13. Cassone V.M. (1991) Melatonin and suprachiasmatic nucleus function. In *Suprachiasmatic Nucleus: The Minds's Clock* (eds. Klein D.C., Moore R.Y. and Reppert R.M.), pp. 309–323. Oxford University Press.
14. Cassone V.M., Chesworth M.J. and Armstrong S. (1986) Entrainment of rat circadian rhythms by daily injection of melatonin depends upon the hypothalamic suprachiasmatic nucleus. *Phys. Behav.* **36**, 111–1121.
15. Gauer F., Masson-Pévet M. and Pévet P. (1992). Effect of constant light, pinealectomy and guanosine triphosphate gamma-s on the density of melatonin receptors in the rat suprachiasmatic nucleus: A possible implication of melatonin action. *J. Neuroendocrinol.* **4**, 455–459.
16. Goldman B.D. (1983) The physiology of melatonin in mammals. In *Pineal Research Review*, Vol. 1 (ed. Reiter R.J.), pp. 145–182. Alan R. Liss Inc. New York.

17. Hoffmann K. (1985) Interaction between photoperiod, pineal and seasonal adaptation in mammals. In *The Pineal Gland: Current State of Pineal Research* (eds. Mess B., Ruszas C., Tima L. and Pévet P.), pp. 211–227. Akademiai Kiado, Budapest.
18. Kalsbeek A., Buijs R.M., Van Heerikhuize J.J., Arts M. and Van der Woude T.P. (1992) Vasopressin-containing neurons of the suprachiasmatic nuclei inhibit corticosterone release. *Brain Res.* **580**, 62–67.
19. Mauviard F. (1992) Effet du 5-methoxypsoralene sur la concentration plasmatique de mélatonine : mécanisme d'action. Thèse, Université L. Pasteur, Strasbourg, France.
20. Maywood, E.S., Buttery R.C., Vance G.H.S., Herbert J. and Hasting M.H. (1990) Gonadal response of the male Syrian hamster to programmed infusions of melatonin are sensitive to signal duration and frequency but not to signal phase nor to lesions of the suprachiasmatic nuclei. *Biol. Reprod.* **43**, 174–182.
21. Pévet P. (1988) The role of the pineal gland in the photoperiodic control of reproduction in different hamster species. *Reprod. Nutr. Develop.* **28**, 443–458.
22. Phillipson O.T. and Griffiths A.C. (1985) The topographic order of inputs to nucleus accumbens in the rat. *Neurosciences* **16**, 275–296.
23. Pitrosky B., Masson-Pévet M., Kirsch R., Vivien-Roels B., Canguilhem B. (1991) Effect of different doses and durations of melatonin infusions on plasma melatonin concentrations in pinealectomized syrian hamsters: consequences in pinealectomized syrian hamsters : consequences at the level of sexual activity. *J. Pineal Res.* **11**, 149–155.
24. Raynaud F., Miguel J.L., Vivien-Roels B., Masson-Pévet M. and Pévet P. (1989). The effect of 5-methoxytryptamine on golden hamster gonads is no a consequence of its acetylation into melatonin. *J. Endocrinol.* **121**, 507–512.
25. Redman J.R., Armstrong S.M. and Ng K.T. (1983) Free running activity rhythms in the rat entrainment by melatonin. *Science* **219**, 1089–1091.
26. Reiter R.J. (1987) The melatonin message : duration versus coincidence hypotheses. *Life Sci.* **40**, 2119–2131.
27. Rollag M.D. and Stetson M.H. (1982) Melatonin injection into syrian hamsters. In *The Pineal and its Hormones* (ed. Reiter R.J.), pp. 143–152. Alan R. Liss, New York.
28. Skene D.J., Pévet P., Vivien-Roels B., Masson-Pévet M. and Arendt J. (1987). Effect of different photoperiods on concentrations of 5-methoxytryptophol and melatonin in the pineal gland of the syrian hamster. *J. Endocrinol.* **114**, 301–309.
29. Thomas E.M.W. and Armstrong S.M. (1988) Melatonin administration entrains female rat activity rhythms in constant darkness but not in constant light. *Am. J. Physiology* **255**, R237-R242.
30. Vanecek J., Pavlik A. and Illnerova H. (1987) Hypothalamic melatonin receptor sites revealed by autoradiography. *Brain Res.* **435**, 359–369.
31. Watts A.G. and Swanson L.N. (1987) Efferent projections of the suprachiasmatic nucleus II. Studies using retrograde transport of fluorescent dyes and simultaneous peptide immunohisto-chemistry in the rat. *J. Comp. Neurol.* **258**, 230–252.
32. Watts A.G., Swanson L.W. and Sanchez-Watts G. (1987) Efferent projections of the suprachiasmatic nucleus I. Studies using anterograde transport of *Phaseolus vulgaris* Leuco-agglutinin in the rat. *J. Comp. Neurol.* **258**, 204–229.
33. Weaver D.R., Rivkess S.A. and Reppert S.M. (1989) Localization and characterization of melatonin receptors in rodent brain by *in vitro* autoradiography. *J. Neuroscience* **9**, 2581–2590.
34. Williams L.M., Morgan P.J., Hastings M.H., Lawson W., Davidson G. and Howell H.E. (1989) Melatonin receptor sites in the syrian hamster brain and pituitary. Localization and characterization using ^{125}I-iodomelatonin. *J. Neuroendocrinology*, **1**, 315–320.

8

Distribution of the Melatonin Receptor in the Central Nervous System of Vertebrates. Kinetic Parameters and Signal Transduction Pathways

FRANCO FRASCHINI and BOJIDAR STANKOV

Chair of Chemotherapy, Department of Pharmacology, University of Milan, Milan, Italy

Abstract

Subtle species differences are observed in the distribution pattern of the high-affinity melatonin receptors in the central nervous system. Lower vertebrates' brains are enriched, while laboratory mammals display more restricted distribution. The pars tuberalis of the adenopituitary and the suprachiasmatic hypothalamic nuclei have been uncovered as the major melatonin targets in most of the species studied, while in others, neocortical areas also expressed significant receptor levels. The melatonin receptor is linked to a pertussis toxin-sensitive G-protein, regulating the adenylyl cyclase and possibly the Ca^{++} influx, but binding site(s) not linked to a G-protein have been described as well. Apparently melatonin interacts with lower affinity with the GABAa receptor, modulating the neuronal activity by potentiation of endogenous GABA effects.

Introduction

After the synthesis of 2-iodomelatonin in 1984, two valuable research tools for studying the melatonin binding sites became available: a radioactive ligand with high specific activity and a potent non-radioactive melatonin agonist.[27] Subsequently, a number of studies reported the distribution and the kinetic parameters of the melatonin receptor in the vertebrate brain; certain model systems were created, the signal transduction pathways and the mechanism of melatonin action were studied. Several melatonin analogs were synthesized and tested. This lecture will outline the recent achievements and will attempt to view in perspective the future trends in this rapidly developing field.

Distribution of the melatonin receptor

Most of the animal studies have been conducted on laboratory rodents, but in an increasing number of reports mapping and characterization of the receptor have been performed in the brains of phylogenetically close or distant vertebrates, such as fish, amphibians, reptiles, birds and various orders of mammals (for review see reference 22).

In lower vertebrates' brain, binding is widespread, and in fish, e.g. rainbow trout (*Salmo gairdneri*) and Atlantic salmon (*Salmo salar*), autoradiography discovered highest levels in the optic tectum, nucleus semicircularis, lobus inferioris and the molecular layer of the cerebellum.[1,11] The green anole lizard (*Anolis carolinensis*) expressed 2-[^{125}I]iodomelatonin binding with highest levels in the anterior cerebral cortex, anterior dorsal ventricular ridge, lateral geniculate nucleus, lentiform thalamic nucleus, nucleus rotundus, habenula, pretectum, tectum, pituitary and the suprachiasmatic nuclei.[1]

In the avian species studied (*Gallus domesticus*,[24] *Coturnix coturnix*, B. Cozzi *et al.*, unpublished), the visual suprachiasmatic nuclei, ovoid nuclei, accessory hyperstriatum and the neostriatum were intensely labeled. Areas, such as the tectum opticum, ectomammilary body, dorsolateral anterior thalamic nuclei, lateral geniculate nuclei, Edinger-Westphal nuclei, nucleus triangularis, nucleus rotundus and pars magnocellularis of the nucleus isthmi expressed high receptor concentrations. Lower levels were recorded throughout the brain. The visual suprachiasmatic nuclei are known to be the site of an endogenous pacemaker in birds and are thought to be the homologue to the mammalian suprachiasmatic nuclei[26]. The localization pattern observed in the low vertebrates suggests an involvement of melatonin mainly in the processing of visual information.

Melatonin binding sites in the CNS of a primitive mammalian species, the Bennet's wallaby (*Dendrolagus bennettianus*) were also widespread throughout the brain, comparable to most of the non-mammalian species.[17] The brain of the marsupials is relatively simple, however. It lacks corpus callosum, and the neopallium displays absence or very simple convolutions of the cerebral hemispheres.

In some of the higher mammals, however, intense 2-[^{125}I]iodomelatonin binding can also be found in many areas of the neocortex, apart from the paleocortex and the archicortex. In the brain of the Old World rabbit (*Oryctolagus cuniculus*) 2-[^{125}I]iodomelatonin binding was extensive, similarly to the lower vertebrates. Many of the phylogenetically old cortical areas, such as the hippocampus, indusium griseum and olfactory cortex were intensely labeled. In Lagomorphs indusium griseum is well developed and has not yet undergone the partial regression typical for the brain of the higher mammals and *Primates*. Binding, however, was strong in the greater part of the neocortex, as well[19,20] (Figure 1).

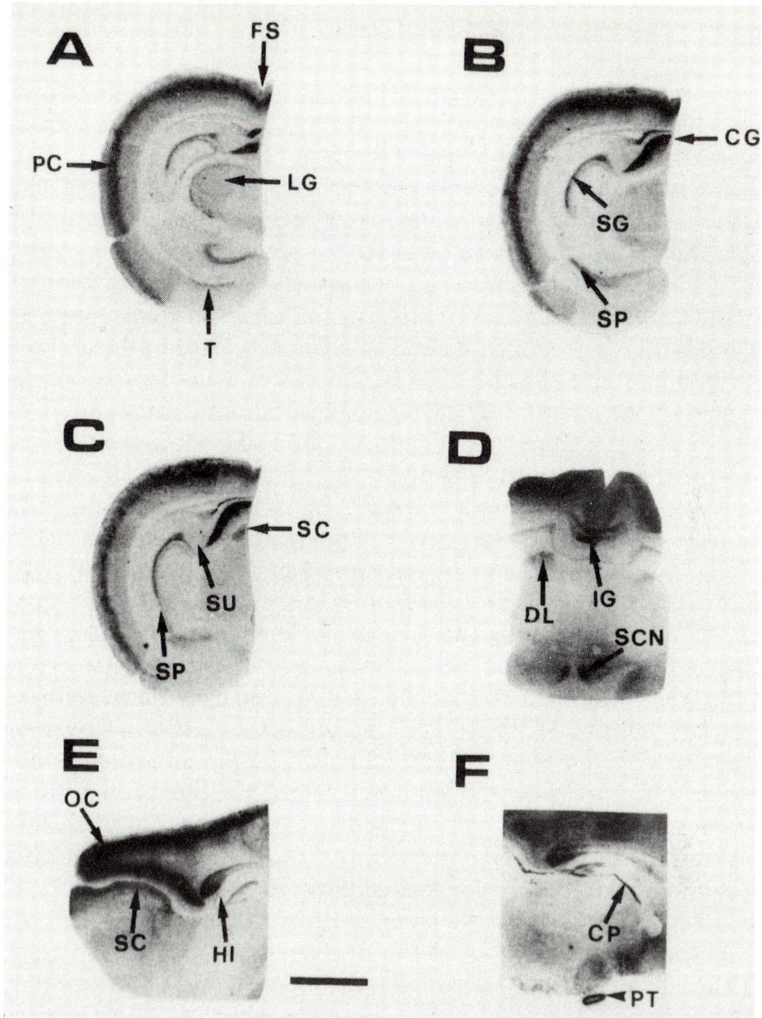

FIG. 1. Autoradiographs from coronal (A-D) and parasagittal (E, F) sections, showing the distribution of 2-[^{125}I]iodomelatonin binding in the rabbit brain. A, B and C are sections through the right hemisphere, 300 μm apart. D: coronal section through the central segment. CG, cingulate gyrus; CP, choroid plexus of the third ventricle; DL, dorsolateral thalamic nuclei; HI, hippocampus; IG, indusium griseum; LG, lateral geniculate ganglion; OC, occipital cortex; PC, parietal cortex; PT, pars tuberalis; SCN, suprachiasmatic nuclei; SC, superior colliculus; SG, stratum granulosum dentatum; SP, stratum piramidale; SU, subiculum; T, tapetum. Note the diffuse binding in the hypothalamus. Scale bar = 5 mm.

A common feature in laboratory rodent brains studied to date is the limited distribution of 2-[^{125}I]iodomelatonin binding. Three anatomical sites have been most frequently mentioned as potential melatonin targets: the pars tuberalis of the pituitary gland, the hypothalamic suprachiasmatic and the paraventricular thalamic nuclei.[5,25,28,30] Another putative site of melatonin action in the rat brain is the area postrema.[14,31] Binding in this area was not detected in any of the strains of laboratory mice,[24,30] Syrian (*Mesocricetus auratus*) and Djungarian (*Phodopus sungorus*) hamster brains.[31]

An apparent exception among rodents is a New World species, the white-footed mouse (*Peromyscus leucopus*), where the basal forebrain and the hypothalamus, the amygdala, thalamus and a number of neocortical areas were intensely labeled.[30] It is not clear why in the brain of this species melatonin receptor is expressed with distribution so different from the rest of the rodents examined until now. The difference could be due to the fact that the Connecticut and the Georgian stocks of white-footed mice have been bred in captivity for only few generations, compared to the laboratory albino rat, hamster and the common laboratory strains of mice. On the other hand, autoradiography may have overlooked some areas because of limits in resolution. Specific 2-[^{125}I]iodomelatonin binding was reported in other regions of the rat and hamster brain, such as cortex and hippocampus, detected by *in vitro* ligand-receptor binding.[10,33] Binding in the olfactory bulb was reported in the Syrian hamster[10] and the rabbit,[19,20] and autoradiography revealed melatonin receptors in the olfactory epithelium of the Siberian hamster (*Phodopus sungorus*).[5]

Of the large order of Artiodactyla, two species have been a subject of study so far: the domestic sheep (*Ovis aries*) and the goat (*Capra hircus*). In the sheep, 2-[^{125}I]iodomelatonin binding was reportedly confined to the pars tuberalis of the pituitary gland.[7] Later experiments demonstrated that melatonin receptor is more widespread in the brain of this species.[4,19] The suprachiasmatic nuclei were found to express low to undetectable levels of binding, diffuse labeling being observed in the immediate vicinity. The retrosplenial, temporal, occipital and entorhinal cortex, the subiculum, the molecular layers of the hippocampus adjacent to the dentate gyrus, the medial preoptic nucleus and the preoptic area expressed high to medium melatonin receptor levels. Low levels of binding were recorded as well in the anterior hypothalamus, medial basal hypothalamus, paraventricular thalamic and supramammilary nuclei. In the midbrain the binding was restricted to the ventral raphe complex and the inferior colliculus. In the caprine brain, melatonin receptor distribution was similar, with the apparent difference of higher binding levels recorded in the suprachiasmatic nuclei.[8]

Of the three recent families contained in the order *Perissodactyla*, only the horse (*Equus caballus*) has been a subject of investigation. In the brain areas examined by *in vitro* binding, melatonin receptor was detected with a limited distribution. Highest levels were recorded in the pars tuberalis/ median eminence region.[19]

In the brain of the only carnivore species examined, the ferret (*Mustela* sp.), binding was limited to the pars tuberalis and pars distalis of the pituitary gland.[32]

The first work on melatonin receptor distribution in the brain of Primates, described high affinity 2-[^{125}I]iodomelatonin binding in the suprachiasmatic nuclei of the human hypothalamus. Surprisingly, binding was not reported in the pars tuberalis of the pituitary gland.[18] In the baboon (*Papio ursinus*) and the vervet monkey (*Cercopithecus aethiops*), melatonin receptor was expressed with high density in the pars tuberalis of the pituitary gland (Figure 2), and the collar of pars distalis, present in Papio, and the suprachiasmatic nuclei were diffusely labeled. The neurohypophysis was devoid of binding (Stankov *et al.*, unpublished). Thus, of the mammals studied so far, the human is the only species where 2-[^{125}I]iodomelatonin binding was not reported in the pars tuberalis of the pituitary gland.

In general, there seems to exist a phylogenetic trend towards a restriction in the brain areas expressing melatonin receptors. Apparently, fewer areas possess high affinity melatonin binding sites in higher mammals. The number of comparative studies is still very limited and there are many species that have not been the subject of investigation to date.

Kinetic parameters and signal transduction pathways

The first results on the melatonin receptor kinetic parameters, obtained by using 2-[^{125}I]iodomelatonin, were controversial. The apparent affinity was reported either in the low or medium nanomolar range in the rat[33] and the Syrian hamster[10] brains. At the same time, picomolar affinity binding site was described in the rat median eminence and suprachiasmatic nuclei.[28] Considering that melatonin peripheral blood concentrations in all species studied to date are in the high picomolar or low nanomolar range, the picomolar Kd and Ki values seemed physiologically appropriate and suggested that 2-[^{125}I]iodomelatonin labeled a relevant receptor.

Later work again pointed out to the existence of two binding forms, a low and a high affinity. Thus, a hypothesis for the existence of two melatonin receptor types (ML-1 and ML-2) was forwarded.[9] Experiments on transduction of the melatonin signal at cellular level demonstrated that what has been described as high- and low affinity binding type, may be in fact the same receptor class, undergoing shifts in its apparent affinity according to the process of coupling and uncoupling to its G-protein;[5,20,20,30] a well documented phenomenon for several G-protein coupled receptors. Moreover, pharmacological identity of the high-affinity binding site was demonstrated in the avian and mammalian

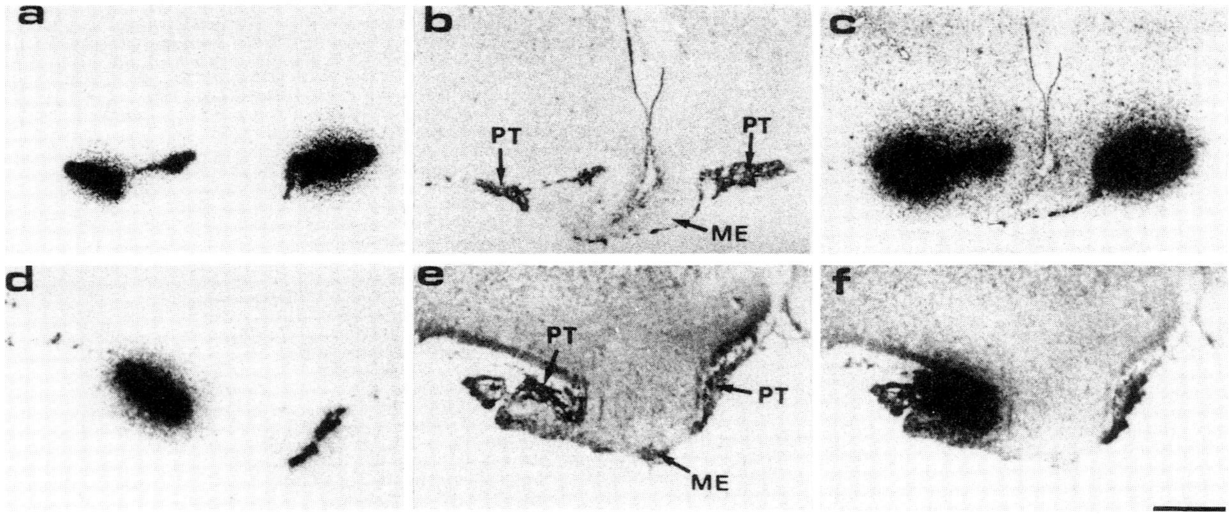

FIG. 2. High magnifications of the autoradiographs from the region of the median eminence of the vervet monkey (a) and baboon (d). The autoradiographic films (a, d) were apposed by micromanipulation over the generation microscopic slides (b, e, respectively) after staining with Cresyl violet, to produce the composite image of c and f, respectively. Note that the binding is confined to the layer of pars tuberalis cells (PT), and that the median eminence (ME) is devoid of binding even in its external zone, close to the PT. Scale bar = 1 mm.

brains.[23] Inhibition of forskolin-stimulated cAMP formation by physiological concentrations of melatonin was demonstrated in explants from hamster, rat, sheep and mouse pars tuberalis, and rabbit cortex (Figure 3, for review see reference 22). The inability of pertussis toxin to completely abolish the melatonin inhibitory effect on cAMP levels in some systems indicates partial sensitivity to pertussis toxin. The main effect of pertussis toxin-mediated rybosilation is to block the interaction between the receptor and its G-protein. Thus, the synergistic effect of both pertussis toxin and GTP in inhibiting high affinity melatonin binding in sheep pars tuberalis suggested that a single class of possibly two subtypes receptors is coupled to two G-proteins, pertussis toxin-sensitive and -insensitive, to regulate the adenylate cyclase.[15] However, melatonin binding sites that are not coupled to a G-protein have been observed in the neuronal retina[6] and the basal brain arteries (Stankov, unpublished data). Interestingly, melatonin receptors expressed in *Xenopus* oocytes following microinjection of poly(A)$^+$ RNA, extracted from ovine pars tuberalis, coupled to a signal transduction mechanism with characteristics similar of an inositol-phosphate/Ca$^+$ system.[13] In model systems that make use of pars tuberalis cells and/or explants it would be of extreme interest to investigate the molecular basis of melatonin action in the control of LH release. This exocytotic process has very complex control mechanisms, involving diverse second messenger systems and short feedback loops between LH and GnRH. Novel studies demonstrated that melatonin inhibited GnRH-stimulated Ca^{++} elevation in neonatal rat pituitary cells, presumably acting through a pertussis toxin-sensitive G-protein.[29]

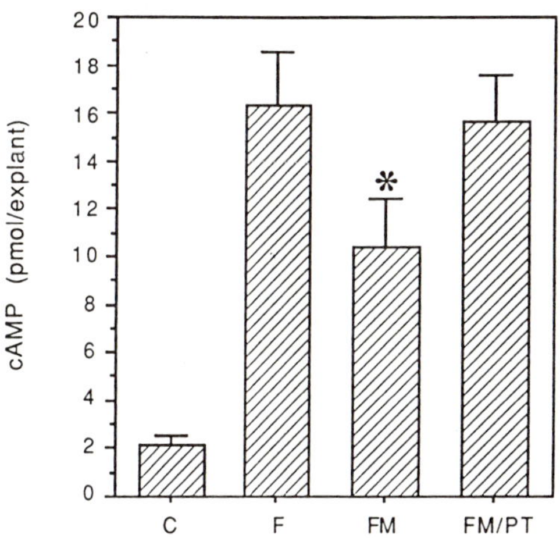

FIG. 3. Inhibitory effect of melatonin (10 nM) on forskolin-stimulated (10 μM) accumulation of cAMP in rabbit cortical explants and the blocking effect of pertussis toxin (1 μg/ml). C, control; F, forskolin; FM, forskolin plus melatonin; FM/PT, forskolin plus melatonin, preincubation with pertussis toxin. *, $p > 0.05$. From reference 21.

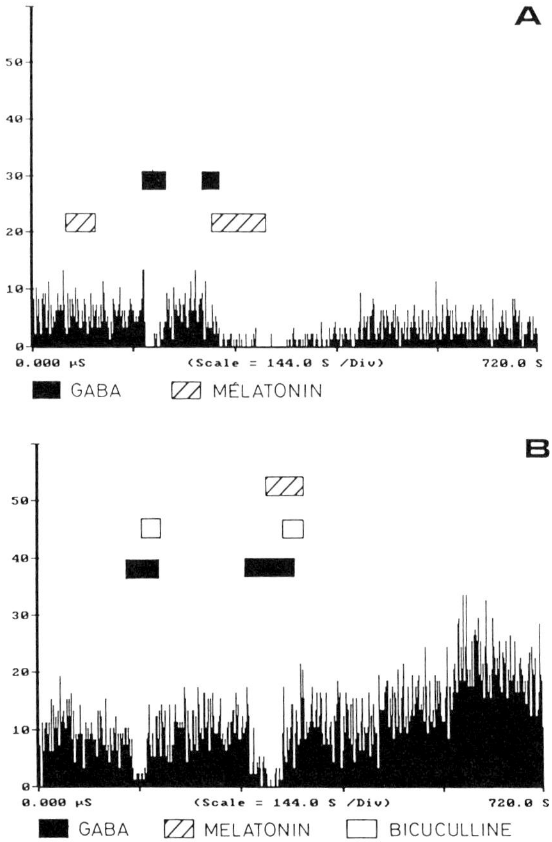

FIG. 4. Melatonin (100 nM concentration in the electrode, applied pressure 0.22 kg/cm^2) is ineffective *per se*, but strongly potentiates the effect of GABA in suppressing the spontaneous firing activity of the rat reticular thalamic nuclei neurons (A). The specific GABAa antagonists bicuculline counteracts the effects of either GABA or melatonin (B). From reference 2.

Recent experimental evidence also demonstrated that nanomolar concentrations of melatonin increased the inhibitory effect of GABA on the firing activity of single neurons in the rat reticular thalamic nuclei.[2] Furthermore, the potent melatonin agonists 2-iodomelatonin and 2-bromomelatonin expressed benzodiazepine-like effects, similar to what was described in the rabbit cortex.[20] The difference from the rabbit cortex was that melatonin *per se*, in low concentrations was without effect on the firing activity of the thalamic reticular nuclei neurons, but was able to increase significantly the inhibitory effect of GABA. This action could be counteracted by the GABAa receptor antagonist bicuculline[3] (Figure 4). High affinity melatonin binding sites have not been demonstrated in the rat reticular thalamic nuclei. Thus, it appears that melatonin can exert biological effects (benzodiazepine-like in this case), without necessarily binding to its own receptor, but utilizing other complex receptor systems, i.e. acting as a modulator. Recently allosteric modulation of *t*-[^{35}S]TBPS binding by melatonin was demonstrated in the rat brain.[16] Therefore, the action of melatonin and its potent agonists can apparently be exerted trough a dual mechanism: (i) interaction with its high affinity receptor in discrete brain areas, leading to an inhibition of the adenylyl cyclase; (ii) possibly through a non-GABA, non-benzodiazepine binding site melatonin is acting on the great GABA receptor complex, modulating the cellular chlorine influx and potentiating the effects of endogenous GABA. These interactions can explain the reported hypnotic properties of melatonin in humans and may serve as a basis for the development of clinical applications of the indole. Studies with healthy young human volunteers demonstrated that melatonin can successfully be used as supplement treatment to sleep-inducing and -maintenance benzodiazepine therapy, allowing for a reduction in the benzodiazepine doses with significant amelioration of the sleep, determined by the microstructural changes and the subjective assessment of the sleep quality.[12]

References

1. Aggelopoulos N. and Demaine C. (1990) Quantitative receptor autoradiography of melatonin receptors in three non-mammalian vertebrates: the green anole lizard (*Anolis carolinensis*), the edible frog (Rana esculenta) and the rainbow trout (*Salmo gairdneri*), In *Vth Colloquium of the European Pineal Study Group*, Guildford, U.K., abstr. 66.
2. Biella G., Panara C., Cozzi B. and Fraschini F. (1991) Melatonin enhances GABA-mediated effects when administered by micropressure ejection in single unit neuronal recordings. In *Role of Melatonin and Pineal Peptides in Neuroimmunomodulation* (eds. Fraschini F. and Reiter R.J.), pp. 191–200. Plenum Press, New York.
3. Biella G., Panara C., Stankov B., Ferini-Strambi L. and Fraschini F. (1992) Melatonin-induced modulation of GABAa synapses in the central nervous system. A model and a new theory. *Sleep* (in press).
4. Bittman E.L. and Weaver D.R. (1990) The distribution of melatonin binding sites in neuroendocrine tissues of the ewe. *Biol. Reprod.* **43**, 986–993.
5. Carlson L.L., Weaver D.R. and Reppert S.M. (1991) Melatonin receptors and signal transduction during development in Siberian hamsters (*Phodopus sungorus*). *Devel. Brain. Res.* **59**, 83–89.
6. Chong N.W.S. and Sugden D. (1991) Guanine nucleotides regulate 2-[^{125}I]iodo-melatonin binding sites in the chick retinal pigment epithelium, but not in in neuronal retina. *J. Neurochem.* **57**, 685–689 .
7. DeReviers M-M., Ravault J-P., Tillet Y. and Pelletier J. (1989) Melatonin binding sites in the sheep pars tuberalis. *Neurosci. Lett.* **100**, 89–93.
8. Deveson S., Howarth J., Arendt J. and Forsyth I.A. (1990) In vitro autoradiographical localization of melatonin (MT) binding sites in the caprine brain. In *Vth Colloquium of the European Pineal Study Group*, Guildford, UK, abstr. 67.
9. Dubocovich M.L. (1988) Pharmacology and function of melatonin receptors. *FASEB J.* **2**, 2765–2773.
10. Duncan M.J., Takahashi J.S. and Dubocovich M.L. (1988) 2-[^{125}I] Iodomelatonin binding sites in hamster brain membranes: pharmacological characteristics and regional distribution. *Endocrinology* **122**, 1825–1833.
11. Ekström P. and Vanecek J. (1992) Localization of 2-[^{125}I]iodomelatonin binding sites in the brain of the Atlantic salmon, *Salmo salar*. *Neuroendocrinology* **55**, 529–537.

12. Ferini-Strambi L., Zucconi M., Biella G., Stankov B., Fraschini F., Zamboni M.G. and Smirne S. (1991) Sleep laboratory study on melatonin effects in healthy subjects. In *Third Milano International Symposium on Sleep*, p. 31 (Abstr.). September 18–19.
13. Fraser S.P., Barrett P., Djamgoz M.B.A., and Morgan, P.J. (1991) Melatonin receptor mRNA expression in *Xenopus* oocytes: inhibition of G-protein-activated response *Neurosci. Lett.* **124**, 242–245.
14. Laitinen J.T., Flugge G. and Saavedra J.M. (1990) Characterization of melatonin receptors in the rat area postrema: modulation of affinity with cations and guanine nucleotides. *Neuroendocrinology* **51**, 619–624.
15. Morgan P.J., Davidson G., Lawson W. and Barrett P. (1990) Both pertussis toxin-sensitive and insensitive G-proteins link melatonin receptor to inhibit of adenylate cyclase in the ovine pars tuberalis. *J. Neuroendocrinol.* **2**, 773–778.
16. Niles L.P. and Peace C.H., (1990) Allosteric modulation of *t*-[^{35}S]butylbicyclophosphorothionate bindig in rat brain by melatonin. *Brain Res. Bull.* **24**, 635–637.
17. Paterson A.M., Chong N., Sugden D., Brinklow B.R. and Loudon A.S.I. (1990) Studies of the role of melatonin in determining the transition to seasonal reproductive quiescence and preliminary observations on melatonin binding sites in the brain of a marsupial, the Bennett's wallaby. In *Vth Colloquium of the European Pineal Study Group*, Guildford, UK, abstr. 60.
18. Reppert S.M., Weaver D.R., Rivkees S.A. and Stopa E.G.(1988) Putative melatonin receptors in human biological clock. *Science* **242**, 78–81.
19. Stankov B., Cozzi B., Lucini V., Fumagalli P., Scaglione F. and Fraschini F. (1991) Characterization and mapping of melatonin receptors in the brain of three mammalian species: rabbit, horse and sheep. *Neuroendocrinology* **53**, 214–221.
20. Stankov B., Cozzi B., Lucini V., Capsoni S., Fauteck J., Fumagalli P. and Fraschini F. (1991) Localization and characterization of melatonin receptors in the rabbit brain by autoradiography and in vitro ligand-receptor binding. *Neurosci. Lett.* **133**, 68–72.
21. Stankov B., Biella G., Panara C., Lucini V., Capsoni S., Fauteck J., Cozzi B. and Fraschini F. (1991) Melatonin signal transduction and mechanism of action in the Central Nervous System: Using the rabbit cortex as a model. *Endocrinology* **130**, 2152–2159.
22. Stankov B., Fraschini F. and Reiter R.J. (1993) The Melatonin Receptor: Distribution, Biochemistry and Pharmacology. In *Melatonin: Biosynthesis, Physiological Effects and Clinical Applications* (eds. Reiter R.J. and Yu H.S.), pp. 155–186. Chapter 7. CRC Press, Boca Raton, Florida.
23. Sugden D. and Chong N.W.S. (1991) Pharmacological identity of 2-[^{125}I]iodomelatonin binding sites in chicken brain and sheep pars tuberalis. *Brain Res.* **539**, 151–158.
24. Suiciak J.A., Krause D.N. and Dubocovich M.L. (1991) Quantitative pharmacological analysis of 2-[^{125}I]iodomelatonin binding sites in discrete areas of the chicken brain. *J. Neurosci.* **11**, 2855–2863.
25. Siuciak J. A., Fang J-M. and Dubocovich, M.L., (1990) Autoradiographic localization of 2-[^{125}I]-iodomelatonin binding sites in the brains of C3H/HeN and C57BL/6J strains of mice. *Eur. J. Pharmacol.* **180**, 387–390.
26. Underwood H. (1989) The pineal and melatonin: regulators of circadian function in lower vertebrates. *Experientia* **45**, 914–922.
27. Vakkuri O., Lamsa E., Rahkamaa E., Ruotsalainen H. and Leppaluoto J. (1984) Iodinated melatonin: preparation and characterization of the molecular structure by mass and 1H NMR spectroscopy. *Anal. Biochem.* **142**, 284–289.
28. Vanecek J., Pavlik A. and Illnerova H. (1987) Hypothalamic melatonin receptor sites revealed by autoradiography. *Brain Res.* **435**, 359–362.
29. Vanecek J. and Klein, D.C. (1992) Melatonin inhibits gonagotropin-releasing hormone-induced elevation of intracellular Ca^{++} in neonatal rat pituitary cells. *Endocrinology* **130**, 701–707.
30. Weaver D.R., Carlson L.L. and Reppert S.M. (1990) Melatonin receptors and signal transduction in melatonin-sensitive and melatonin-insensitive populations of white-footed mice (*Peromyscus Leucopus*). *Brain Res.* **506**, 353–357.
31. Weaver D.R., Rivkees S.A. and Reppert S.M. (1989) Localization and characterization

of melatonin receptors in rodent brain by *in vitro* autoradiography. *J. Neurosci.* **9**, 2581–2590.
32. Weaver D.R. and Reppert S.M. (1990) Melatonin receptors are present in the ferret pars tuberalis and pars distalis, but not in brain. *Endocrinology* **127**, 2607–2609.
33. Zisapel N., Nir I. and Laudon, M. (1988) Circadian variations in melatonin-binding sites in discrete areas of the male rat brain. *FEBS Lett.* **232**, 172–179.

Alphabetical list of the abbreviations used in the figures

CG, cingulate gyrus; CP, choroid plexus of the third ventricle; DL, dorsolateral thalamic nuclei; HI, hippocampus; IG, indusium griseum; LG, lateral geniculate ganglion; OC, occipital cortex; PC, parietal cortex; PT, pars tuberalis; SCN, suprachiasmatic nuclei; SC, superior colliculus; SG, stratum granulosum dentatum; SP, stratum piramidale; SU, subiculum; T, tapetum.

9

Neurotransmitters and Peptides in the Pineal Gland

A. YUWILER[1] and G. L. BRAMMER[2]

[1]*Neurobiochemistry Laboratory T-85, West Los Angeles Veterans Affairs Medical Center Brentwood Division, Los Angeles, CA 90073, USA*
[2]*Brain Research Institute and Department of Psychiatry and Biobehavioral Sciences, UCLA School of Medicine, University of California, Los Angeles, CA 90024, USA*

Abstract

While norepinephrine is regarded as the primary regulator of melatonin biosynthesis, an increasing number of "classical" transmitters and neuropeptides have been implicated as modulating pineal function. The anatomical and functional bases for these suggestions is briefly reviewed. In addition, some of the similarities and differences between the effects of β-adrenergic and VIP stimulation on induction of *N*-acetyltransferase activity are discussed.

Science moves in successive waves of accumulating data which eventually consolidate into general principles only to swell again as new data accumulates. The neurosciences are in the midst of such a wave. The growing number of identified neurally active compounds and neuropeptides, each with its assortment of receptors, appear to eventually activate a series of second messenger systems. Currently the number of such second messenger systems is much smaller than the product of the number of transmitters and their receptor subtypes. Since many cells receive multiple inputs from different transmitters using the "same" second messenger systems, the question arises over whether cells uniquely differentiate between different agonists stimulating the same second messenger system. If so, how? If not, what is the purpose of the redundant incoming code?

The pineal gland has been useful in explicating noradrenergic transmitter transduction, and the discovery of multiple receptor inputs into the pineal permits its use in defining transmitter interactions as well. New information brings new complexities, and even the very well established role of norepinephrine in melatonin synthesis in the rat[25,27,67] turns out to be at least one step more complex than originally believed. Not only does noradrenergic stimulation of the $\beta1$ receptor drive increased *N*-acetyltransferase activity but concomitant noradrenergic stimulation of postsynaptic $\alpha 1$ receptors, while having no effect of its own, augments β-stimulated *N*-acetyltransferase (NAT) synthesis. A growing list of other putative transmitters possibly implicated in regulating melatonin synthesis has also accumulated over the years. These include adenosine,[9,10,41,46,65] γ-aminobutyric acid (GABA),[51,52,53] substance P,[47] taurine[68,69,70] enkephalines,[2,3,37] serotonin,[61] and more recently the polypeptides vasoactive intestinal peptide (VIP),[22,72] peptide *N*-terminal histidine C-terminal isoleucine (PHI)[33,38,62] and neuropeptide Y (NPY).[36,42,47,64,71,75]

The actual case for these various substances as regulators of pineal function is generally incomplete, and species variations are great. There are four minimal requirements for a substance to be considered a regulator. First it must reach some active site. The route could be local production, neuronal release, or by body fluid transport. Second, it should be released in some physiologically relevant manner: if in nerves, it should be released by depolarization; if in blood or if produced locally, by some physiologically specific discharge. Third, it should produce some physiologically relevant effect. Last, it must have some mechanism for timely removal.

How well the various putative transmitters meet these requirements is summarized in Table 1. Clearly only norepinephrine currently satisfies all requirements, and even its exact role is species dependent. For example, while norepinephrine stimulates β-adrenergic receptors on pinealocytes to elevate NAT in the rat, in the chick norepinephrine activates a postsynaptic $\alpha 2$-receptor to oppose the stimulatory action of VIP on NAT activity.[45]

TABLE 1
Status of proposed modulators of pineal function

	Presence	Binding	Release	Action	Inactivation
Acetylcholine	17,44,50,57,59	14	—	15,30,56	—
Adenosine	*	55	—	9,10,40,41,46,65	—
Dopamine	—	13	—	—	—
GA	*	12	—	—	—
GABA	51	51	51,53	31,52	58
Norepinephrine	28	24	25	25,27	—
Serotonin	29	15	1,6	61	—
Taurine	8	—	68	68	—
Enkephalin	37	2,3	—	—	—
NPY	60,75	—	—	42,47,64	—
Substance P	48,60	11	—	—	—
PHI	33	62	—	38	—
VIP	34,35,63	19,32	—	45,67	—

Numbers refer to references in the Bibliography.
* signifies a general cell constituent.

The strength of the evidence for the other transmitters varies considerably. In one way or another all of the putative transmitters fulfill the first requirement to some degree. However, the small molecules GABA, adenosine, taurine and serotonin do so by being made in pinealocytes from which they are released during noradrenergic stimulation. Evidence for dopaminergic innervation is based on detection of its synthetic enzyme dopamine β-hydroxylase rather than dopamine itself in some nerves, and evidence for cholinergic innervation rests on immunochemical detection of choline acetyltransferase.[44] High affinity binding has been demonstrated for many of the candidate modulators, and for some, like glutamic acid. Substance P, and the enkephalines, support for modulatory activity is based solely on occurrence and binding. Functional release has only been demonstrated for norepinephrine unless noradrenergic effects on releasing GABA, adenosine, taurine and serotonin from pinealocytes presumably during noradrenergic stimulation is to be regarded as "functional". On the other hand, except for substance P, glutamic acid and the enkephalines, all of the compounds in Table 1 have been reported to have some effect on the pineal. The significance of some of these effects is unclear. For example, taurine weakly stimulates the β-receptor to increase NAT activity but taurine release is dependent upon norepinephrine which itself is a strong stimulator of both α and β receptors.[68] Some of the findings are controversial. This is the case for GABA, for example. One study found[54] GABA inhibition of norepinephrine-induced increases in NAT activity, while another[31] and our own results found GABA to have no effect either by itself or in inhibiting norepinephrine or isoproterenol induction on NAT. Similarly, NPY in one study inhibited norepinephrine-induced increases in NAT activity by pinealocytes,[42] yet another[64] reported NPY increased melatonin production and enhanced norepinephrine stimulation of melatonin, while a third[47] found that NPY given to intact rats increased NAT activity but nighttime activity in blinded rats was decreased. Our own experience with rat pineals in organ culture was that NPY had no effect by itself nor on agonist induction of NAT. Some of the findings are puzzling. For example, in the rat, adenosine analogues given to the intact animal increased pineal N-acetylserotonin and melatonin[10] while, *in vitro*, adenosine and its analogues increased cyclic nucleotides[41] but not NAT activity[40] or melatonin production.[65] Also, in the chick pineal adenosine inhibits or stimulates melatonin production depending on experimental conditions.[9] Some transmitters have functions other than those involved in melatonin production. This appears to be the case for acetylcholine, which has no effect on basal or induced activity,[16,43] but which increases synaptic ribbon membrane formation in the rat pinealocytes[16] and stimulates release of pineal arginine vasopressin.[56] Finally, reuptake systems exist to inactivate most of the small molecules which have been suggested to modulate pineal activity, but inactivation mechanisms have not yet been demonstrated for any of the peptides.

Perhaps the best studied and most consistent case for a modulator other than norepinephrine is that for vasoactive intestinal polypeptide. Although, as seen in Table 1, the two criteria of demonstrated functional release and efficient inactivation have not been met, there is now ample evidence and confirmation that in rat pinealocytes,[22,23,72] pineals in organ culture[72] and pineals *in situ*,[39,72] VIP increases cyclic neucleotides, NAT activity and melatonin production (Table 2). In these activities it resembles a weak β-agonist like isoproterenol but is an order of magnitude less potent. The resemblance is further strengthened in that just as concomitant stimulation with an α-adrenergic agonist like phenylephrine increases

TABLE 2
Effects of isoproterenol and VIP on pineal indole production

Compound	Location	Control	ISO 10 nM	VIP 1 μM
Melatonin	pineal	4 ± 0.2	13 ± 2	11 ± 1
	medium	27 ± 3	59 ± 11	61 ± 6
N-Ac5HT	pineal	0.1	18 ± 4	18 ± 1
	medium	3	61 ± 16	65 ± 6
5HIAA	pineal	10 ± 3	6 ± 1	7 ± 1
	medium	86 ± 17	73 ± 12	94 ± 7
5HIOH	pineal	2 ± 0.4	1 ± 0.2	1 ± 0.2
	medium	18 ± 3	17 ± 3	19 ± 3
5HTP	pineal	0.7 ± 0. 1	0.9 ± 0.1	0.6 ± 0.1
5HT	pineal	40 ± 8	16 ± 3	20 ± 2
NAT	pineal	0.2 ± 0.09	7.7 ± 1.1	6.8 ± 0.3

Pineals were cultured for 48 hours before challanged with agonist as indicated. Pineal indoles were assayed by HPLC using an Axxion C18 RP column and a mobil phase of 0.1 M NaAc pH 4.25 with 35% methanol (for melatonin and *N*-acetylmelatonin) or 12% in methanol (for other indoles). Indoles were quantitated by fluorescence at 345 nm upon activation at 285 nm. Values are MEAN ± SEM for 4–6 pineals or associated medium. Values for compounds are in ng/pineal. Units for NAT are nmoles/pineal/hour.

the effectiveness of isoproterenol,[66] α-agonists also increase the effectiveness of VIP.[74] Norepinephrine itself, of course, stimulates both α and β-receptors. The purpose of a second β-like agonist in the rat pineal is unclear. One possibility is that VIP may augment the effects of β-stimulation under special conditions. If so, the sum of maximal VIP and maximal isoproterenol stimulation should be additive. As seen in Table 3, however, concomitant VIP and isoproterenol, VIP and norepinephrine, or VIP and the weak beta agonist 6-methoxy-2-benzoxazolinone are additive when agonist concentrations were low but not when concentrations were high. This argues against augmentation and indicates that both receptors share the same limiting components of the NAT transduction system. Nor is temporal additivity the answer because VIP does not

TABLE 3
Effects of concomitant agonist stimulation on NAT activity

Agent	Conc (μM)	N-acetyltransferase No VIP	+VIP (0.1 μM)	(Expected)
None	—	0.4±0.1	2.7±0.3	2.3
ISO	0.0001	0.9±0.3	3.2±0.6	3.2
	0.001	2.5±0.4	5.1±0.3	4.8
	1	9.8±0.4	8.0±0.5	12.1
NE	1	5.8±0.6	8.3±0.6	8.1
6MBOA	1000	2.9±0.3	6.6±1.2	5.2

Pineals were cultured for 48 hours before agonist challange. *N*-acetyltransferase activity was determined by the method of Deguchi and Axelrod (1972). Values are MEAN±SEM in nmoles/gland/hour for NAT activity of 4–8 pineals. 6MBOA is 6-methoxy-2-benzoxazolinone.

appreciably extend the time course of NAT activity after beta stimulation nor is the time course of induction appreciably different.[72] Further, the VIP system, like the β-adrenergic system, shows circadian sensitivity to stimulation of cyclic AMP[20] and NAT[72] and to agonist content in the pineal.[20]

The adrenergic and VIP systems do show some differences, however, especially in response to chronic agonist challenge and to steroid exposure. Table 4 shows that chronic (48 hour) stimulation with high concentrations of VIP or isoproterenol diminishes the response of the pineal to

TABLE 4
Diminished response after chromic stimulation

48 Hour Pre-Treatment	Challenge Stimulation None	VIP (0.1 μM)	ISO (10 nM)
None	0.1±0.02	2.8±0.3	3.3±0.4
VIP 10 μM	0.2±0.04	1.1±0.2*	2.8±0.3
ISO 10 μM	0.7±0.1	1.0±0.1*	2.9±0.5

Values are MEAN±SEM in nmoles/gland/hour for NAT activity of 4–8 pineals.

* indicates a significant difference from No Pre-Treatment condition ($p<0.05$)

subsequent stimulation by VIP but less so to subsequent challenge with isoproterenol. The requirement for more than 24 hours of stimulation before a decrement in subsequent response can be detected suggests this phenomenon results from depletion of limiting components of the transduction sequence. The difference in relative sensitivities of the VIP and adrenergic systems to prior stimulation could be due to a tighter coupling between the beta receptor and adenylate cyclase and/or G protein and/or protein kinase A. A tighter coupling might also explain the

TABLE 5

Relative effect of corticosterone preincubation on pineal response to isoproterenol and vasoactive intestinal polypeptide

Hours	Untreated ISO	VIP	Corticosterone Treated ISO	VIP
0	—	—	9.1 ± 0.7	3.3 ± 0.4
48	—	—	4.2 ± 0.6*	2.7 ± 0.3
72	9.8 ± 0.8	3.8 ± 0.7	3.5 ± 0.4*	2.7 ± 0.1

Pineals were cultured in the presence or absence of corticosterone (50 μM) for 48 or 72 hours before challenge with 0.1 μM ISO or 1 μM VIP for 6 hours. Values are MEAN ± SEM in nmoles/gland/hour for NAT activity of 4-8 pineals.

* indicates a significant difference from zero-time controls ($p < 0.05$).

TABLE 6

Effect of corticosterone on the pineal response to various agonists

Agonist	N-acetyltransferase Control	Steroid treated
Isoproterenol 1 μM	(4) 7.4 ± 0.6	(4) 2.6 ± 0.04*
Cholera toxin 72 μg	(4) 2.7 ± 0.2	(4) 0.2 ± 0.04*
Forskolin 10 μM	(4) 6.9 ± 0.2	(4) 3.2 ± 0.3*
bCAMP 1 μM	(4) 8.0 ± 1.2	(4) 2.0 ± 0.4*
VIP 1 μM	(4) 1.4 ± 0.2	(4) 1.8 ± 0.2

Pineals were maintained for 48 hours as described previously in the presence or absence of 100 μM corticosterone before a 5 hour challenge with the agonists indicated. Values are (N) MEAN ± SEM in nmoles/gland/hour for NAT activity.

* indicates a significant difference from Control ($p < 0.05$).

TABLE 7

Response to agonists of pineals, cultured for 48 hours, upon acute ethanol exposure

EtOH	Agonist	ISO (1 nM)	VIP (100 nM)
0	0	(8) 0.12 ± 0.01	—
0	+	(22) 6.08 ± 0.58	(20) 2.15 ± 0.20
25 mM	+	(8) 7.62 ± 0.70	(12) 2.70 ± 0.19
50 mM	+	(21) 8.78 ± 0.34*	(17) 2.16 ± 0.19
75 mM	+	(8) 8.94 ± 0.72*	(12) 2.69 ± 0.27

Values are (N) MEAN ± SEM in nmoles/gland/hour for NAT activity.

* indicates a significant difference from No Ethanol ($p < 0.05$).

greater potency, on a molar basis, of isoproterenol in stimulating NAT activity and its somewhat longer duration of action. The effects of steroid pre-exposure are shown in Tables 5 and 6. In these cases VIP stimulation is relatively unaffected by steroid pretreatment while isoproterenol stimulation is markedly blocked. Ethanol also reveals differences between the VIP and β-adrenergic systems.[4] Concurrent α1- and β1-stimulation or α1- and VIP-stimulation increases both cAMP and cGMP. Ethanol exposure of pinealocytes does not affect the changes produced by combined α1-β1 adrenergic stimulation but does block the cyclic nucleotide increases produced by combined α1-VIP stimulation, apparently by interference with Ca dependent processes[5] likely linked to interactions between Kinase A and Kinase C. As seen in Table 7, ethanol at concentrations comparable to those used in the cyclic nucleotide studies does not affect the response of pineals in organ culture to either isoproterenol or VIP. At higher concentrations, response to isoproterenol is slightly increased. This effect may be due to changes in membrane fluidity and, since VIP response was not changed, may indicate that the strength of coupling between the transduction components differs in these systems.

The VIP and β-adrenergic systems respond differently to light shock. Both show increased agonist response[7,73] and increased receptor density[21,49] in light, when NAT activity is low, and decreases in these measures in the dark when NAT activity in high. Light exposure in the dark period produces a profound fall in NAT activity[18,26] and, as seen in Table 8, an unexpected fall in sensitivity to stimulation by isoproterenol as

TABLE 8
Response to light shock

Agonist	Group 4D	4DL
ISO	(14) 101 ± 23	(15) 44 ± 8
VIP	(16) 96 ± 12	(15) 98 ± 20

After 3 hours in the dark phase of a 14:10 L:D cycle, one group of animals (4DL) were exposed to 5 minutes of light and then returned to darkness while controls (4D) remained in dark. Pineals were obtained from both groups 1 hour later, incubated for 1 hour in organ culture, transferred to new media and stimulated with either ISO (10 nM) or VIP (100 nM) for 5 hours. Values are (N) Mean-%-of-Control ± SE for two pooled experiments. Actual mean values for controls (4D) in the two experiments were: Expt 1. ISO = 5.8 ± 1.6; VIP 2.8 ± 1.0; Expt 2 ISO = 7.0 ± 2.7; VIP = 3.0 ± 0.4 nmoles/gland/hour.

well. On the other hand sensitivity to stimulation by VIP is unaltered. This might be due to differences in receptor metabolism or, again, to coupling between receptor and transduction. Taken together, these results would seem to suggest that the VIP and β-adrenergic systems may differ in the coupling between component parts to their transduction systems.

The physiological significance of these differences between the VIP and β-adrenergic systems in the rat pineal gland is certainly still unclear. Whatever the differences may be, it seems more subtle than in the chick where VIP and norepinephrine act in opposition. The role, if any, of the other transmitter candidates is even more obscure. As illustrated above the pineal seems an excellent system to identify actions and interactions and, ultimately, functions. Until these subtleties and significance are resolved, the rat pineal poses the anomaly of an organ without parasympathetic balance of sympathetic actions, which seem only augmented by and not opposed by most putative secondary regulators.

References

1. Aloyo V.J. and Walker R.F. (1987) Noradrenergic stimulation of serotonin release from rat pineal glands *in vitro*. *J. Endocrinol.* **114**, 3–9.
2. Aloyo V.J., Lewis M.E. and Walker R.F. (1990) Opioid peptide mRNAs in the rat pineal. *Prog. Clin. Biol. Res.* **328**, 235–238.
3. Aloyo V.J. (1992) Identification and characterization of delta opioid binding sites in the bovine pineal. *J. Pharmacol. Exp. Ther.* **262**, 292–297.
4. Chik C.L., Ho A.K. and Klein D.C. (1987) Ethanol inhibits dual receptor stimulation of pineal cAMP and cGMP by Vasoactive Intestinal Peptide and Phenylephrine. *Biochem. and Biophys. Res. Comm.* **147**, 145–151.
5. Chik C.L. and Ho A.K. 1991 Inhibitory effects of ethanol on the calcium dependent potentiation of vasoactive intestinal peptide-stimulated cAMP and cGMP accumulation in rat pinealocytes. *Biochem. Pharmacol.* **42**, 1601–1608.
6. Chuluyan H.E., Rosenstein R.E. and Cardinali D.P.(1989) Serotonin release mechanisms in bovine pineal gland: stimulation by norepinephrine and dopamine. *Molec. Cell. Endocrinol.* **64**, 71–80.
7. Deguchi T. and Axelrod J. (1972) Control of circadian change of serotonin *N*-acetyltransferase activity in the pineal organ by the beta adrenergic receptor. *Proc. Natl. Acad. Sci. USA* **69**, 2547–2550.
8. Ebels I., Benson B. and Larsen B.R., (1980) Biosynthesis of taurine by rat pineals *in vitro*. *J. Neural. Trans.* **48**, 101–117.
9. Falcon J., Brun-Marmillon J., Claustrat B. and Collin J.P. (1988) Melatonin production in organ cultured chicken pineal: modulation by adenosine and its analogs. *Pflugers Archiv Eur. J. Physiol.* **413**, 93–95.
10. Gharib A., Reynaud D., Sarda N., Vivien-Roels B., Pevet P. and Pacheco H.A. (1989) Adenosine analogs elevate N-acetylserotonin and melatonin in rat pineal gland. *Neurosci. Lett.* **106**, 345–349.
11. Govitrapong P. and Ebadi M. (1986) Studies on high-affinity [^{3}H]substance P binding sites in bovine pineal gland. *Endocrine Res.* **12**, 293–304.
12. Govitrapong P., Ebadi M. and Murrin L.C. (1986) Identification of a Cl^-/Ca^{2+}-dependent glutamate (quisqualate) binding site in bovine pineal organ. *J. Pineal Res.* **3**, 223–234.
13. Govitrapong P., Hama Y., Pfeiffer R. and Ebadi M. (1989) Status of dopamine in bovine pineal glands and the stimulation of N-acetyltransferase activity by D2-dopaminergic receptor agonists in the rat pineal glands in culture. *J. Pineal Res.* **6**, 17–31.

14. Govitrapong P., Phansuwan-Pujito P. and Ebadi M. (1989) Studies on the properties of muscurinic cholinergic receptor sites on bovine pineal gland. *Comp. Biochem. Physiol.* **94C**, 159–164 .
15. Govitrapong P., Prapapanich V. and Ebadi M. (1991) Identification of serotonin 5HT2 receptors in bovine pineal gland. *J. Pineal Res.* **11**, 182–187.
16. Gupta, B.B.P., Spessert, R. and Vollrath L. 1991 Acetylcholine and muscurinic agonists increase synaptic ribbon membranes in the rat pineal. *Neurosci Lett.* **133**, 125–128.
17. Ichimura T. (1992) The ultrastructure of neuronal-pinealocytic interconnections in the monkey pineal. *Microsc. Res. Tech.* **21**, 124–135.
18. Ilnerová H., Backström M., Sääf J., Wetterberg L. and Vangbo B. (1972) Melatonin in the rat pineal gland and serum: rapid parallel decline after light exposure at night *Neurosci. Lett.* **9**, 189–193.
19. Kaku K., Inoue Y., Matsutani A., Okuba M., Hatao K., Kaneko T. and Yanaihara, N. (1983) Receptors for vasoactive intestinal polypeptide on rad dispersed pineal cells. *Biomed. Res.* **4**, 321–328.
20. Kaku K., Tsuchiya M., Tanizawa Y., Okuya Y., Inoue Y., Kaneko T. and Yanaihara N. (1986) Circadian cycles in VIP content and VIP stimulation of cyclic AMP accumulation in the rat pineal gland. *Peptides 7 Suppl.* **1**, 193–195.
21. Kaku K., Tsuchiya M., Matsuda M., Inoue Y., Kaneko T. and Yanaihara N. (1985) Light and agonist alter vasoactive intestinal peptide binding and intracellular accumulation of adenosine 3′,5′-monophosphate in the rat pineal gland. *Endocrinology* **117**, 2371–2375.
22. Kaneko T., Cheng P., Oka H., Oda T., Yanaihara N. and Yanaihara C. (1980) Vasoactive intestinal polypeptide stimulates adenylate cyclase and serotonin N-acetyl transferase activities in rat pineal in vitro. *Biomed. Res.* **1**, 84–87.
23. Kaneko, T. and Kaku, K. (1983) Regulation of pineal function by vasoactive intestinal polypeptide (VIP) and peptides with N-terminal histidine and C-terminal isoleucine (PHI). *Biomed. Res.* (*Suppl*) **4**, 251–254.
24. Kebabian J.W., Zatz M., Romero J.A. and Axelrod J. (1975) Rapid changes in rat pineal B-adrenergic receptor: Alterations in *l*-(^{3}H)alprenolol binding and adenylate cyclase. *Proc. Natl. Acad. Sci. USA* **72**, 3735–3739.
25. Klein D.C. and Berg G.R. (1970) Pineal Gland: stimulation of melatonin production produced by norepinephrine involves cyclic AMP-mediated stimulation of N-acetyltransferase. *Adv. Biochem. Psychopharm.* **3**, 241–263.
26. Klein D.C. and Weller J.L. 1972 Rapid light-induced decrease in pineal serotonin N-Acetyltransferase activity. *Science* **177**, 532–533
27. Klein D.C. and Weller J.L. (1973) Adrenergic-adenosine 3′,5′-monophosphate regulation of serotonin N-Acetyltransferase activity and the temporal relationship of serotonin-N-acetyltransferase activity to synthesis of ^{3}H-N-acetylserotonin and H-melatonin in the cultured rat pineal gland. *J. Pharmac. Exp. Ther.* **186**, 516–527.
28. Klein D.C., Weller J.L. and Moore, R.Y. (1971) Melatonin metabolism: neural regulation of pineal serotonin N-acetyltransferase activity. *Proc. Natl. Acad. Sci. USA* **68**, 2107–3110.
29. Klein D., Yuwiler A., Weller J.L. and Plotkin S. (1973) Postsynaptic adrenergic-cyclic AMP control of the serotonin content of cultured rat pineal glands. *J. Neurochem.* **21**, 1261–1271.
30. Laitinen J.T., Torda, T. and Saavadra J.M. (1989) Cholinergic stimulation of phosphoinositol hydrolysis in the rat pineal gland. *Euro. J. Pharmacol.* **161**, 237–240.
31. Mata M.M., Schrier R.K., Klein D.C., Weller J.L. and Ghiou, G.Y. 1976 On GABA function and physiology in the pineal gland. *Brain Res.* **118**, 383–394.
32. Meunier A.C., Voisin P., Van Camp G., Cenatiempo Y. and Muller J.M. (1991) Molecular characterization and peptide specificity of two vasoactive intestinal peptide (VIP) binding sites in the chicken pineal. *Neuropeptides* **19**, 1–8.
33. Møller M. and Mikkelsen J.D. (1989) Vasoactive intestinal peptide (VIP) and peptide histidine isoleucine (PHI) in the mammalian pineal gland. In *Ad. Pineal Res.* (eds. Reiter, R.J. and Pang, S.F.). Libbey, London 3, 1–10.
34. Møller M., Mikkelsen J.D., Fahrenkrug J. and Korf H.W. (1985) The presence of vasoactive intestinal polypeptide (VIP)-like-immunoreactive nerve fibers and VIP-

receptors in the pineal gland of the mongolian gerbil (*Meriones Unguiculatus*). An immunohistochemical and receptor-autoradiographic study. *Cell Tissue Res.* **241**, 333–340.

35. Møller M., Mikkelsen J.D., Fahrenkrug J. and Korf H.W. (1985) The presence of vasoactive intestinal polypeptide (VIP)-like-immunoreactive nerve fibers and VIP-receptors in the pineal gland of the mongolian gerbil (*Meriones Unguiculatus*). An immunohistochemical and receptor-autoradiographic study. *Cell Tissue Res.* **241**, 333–340.
36. Møller M., Mikkelsen J.D. and Martinet L. (1990) Innervation of the mink pineal gland with neuropeptide Y (NPY)-containing nerve fibers. An experimental immunohistochemical study. *Cell Tissue Res.* **261**, 477–483
37. Moore R.Y. and Sibony P. (1988) Enkephalin-like immunoreactivity in neurons in the human pineal gland. *Brain Res.* **457**, 395–398.
38. Moujir F., Reiter R.J., Rodriguez C. and Yaga, K. (1992) β-Adrenergic and Peptide N-Terminal Histidine and C-Terminal Isoleucine stimulation of N-acetyltransferase activity and melatonin production in the cultured rat pineal gland. *Endocrinology* **130**, 2076–2082.
39. Moujir F., Richardson B.A., Yaga K. and Reiter R.J. (1992) Vasoactive intestinal peptide stimulates N-acetyltransferase and hydroxyindole-O-methyltransferase activities and melatonin production in cultured rat but not in Syrian hamster pineal glands. *J. Pineal Res.* **12**, 35–42.
40. Nonaka K.O., Reiter R.J., Withyachumnarnkul B., Stokkan K.A. and Lerchl A. (1991) Adenosine effects on the rat pineal gland *in vitro*: cyclic adenosine monophosphate levels, N-acetyltransferase, and thyroxine type II 5′-deiodinase activities and melatonin production. *J. Pineal Res.* **11**, 1–6.
41. Nikodijevic O. and Klein D.C. (1989) Adenosine stimulates adenosine 3′,5′-monophosphate and guanosine 3′,5′-monophosphate accumulation in rat pinealocytes: evidence for a role for adenosine in pineal neurotransmission. *Endocrinology* **125**, 2150–2157.
42. Olcese J. (1991) Neuropeptide Y: an endogenous inhibitor of norepinephrine-stimulated melatonin secretion in the rat pineal gland. *J. Neurochem.* **57**, 943–947.
43. O'Dea R.F. and Zatz M. (1976) Catecholamine-stimulated cyclic GMP accumulation in the rat pineal: apparent presynaptic site of action. *Proc. Natl. Acad. Sci. USA* **73**, 3398–3402.
44. Phansuwan-Pujito J.D., Mikkelsen P., Govitrapong P. and Moller M. (1991) Cholinergic innervation of the bovine pineal gland visualized by immunochemical detection of choline acetyltransferase-immunoreactive nerve fibers. *Brain Res.* **545**, 49–58.
45. Pratt B.L. and Takahashi J.S. (1989) Vasoactive intestinal polypeptide and alpha 2-adrenoceptor agonists regulate adenosine 3′,5′-monophosphate accumulation and melatonin release in chick pineal cell cultures. *Endocrinology* **125**, 2375–2384.
46. Reiter R.J., Withyachumnarnkul B., Stokkan K.A. and Lerchl A. (1991) Adenosine effects on the rat pineal gland in vitro: cyclic adenosine monophosphate levels, N-acetyltransferase, and thyroxine type II 5′-deiodinase activities and melatonin production. *J. Pineal Res.* **11**, 1–6.
47. Reuss S. and Schroder H. (1987) Neuropeptide Y effects on pineal melatonin synthesis in the rat. *Neurosci. Lett.* **74**, 158–162.
48. Reuss S., Riemann R. and Vollrath L. (1992) Substance P- and calcitonin gene-related peptide-like immunoreactive neurons in the rat trigeminal ganglion—with special reference to meningeal and pineal innervation. *Acta Histochem.* **92**, 104–109.
49. Romero J.A., Zatz M., Kebabian J.W. and Axelrod, J. (1975) Circadian cycles in binding of ^{3}H Alprenolol to β-adrenergic receptor sites in rat pineal. *Nature* **258**, 435–436.
50. Ronnekleiv O.K., Kelly M.J. and Wuttke W. (1980) Single unit recordings in the rat pineal gland: Evidence for a Habenulo-pineal neural connection. *Exp. Brain. Res.* **39**, 187–192
51. Rosenstein R.E., Chuluyan H.E., Diaz M.C. and Cardinali D.P. (1990a) GABA as a

presumptive paracrine signal in the pineal gland. Evidence on an intrapineal GABAergic system. *Brain Res. Bull.* **25**, 339–344.

52. Rosenstein R.E., Chuluyan H.E. and Cardinali D.P. (1990b) Presynaptic effects of gamma-aminobutyric acid on norepinephrine release and uptake in rat pineal gland. *J. Neural Trans. Gen. Sec.* **82**, 131–140.
53. Rosenstein R.E., Sanjurjo C. and Cardinali D.P. (1989b) Gamma aminobutyric acid uptake, release, and effect on ^{36}Cl—influx in bovine pineal gland. *J. Neural Trans.* **77**, 141–152.
54. Rosenstein R.E., Chuluyan H.E., Pereyra E.N. and Cardinali D.P. (1989a) Release and effect of gamma-aminobutyric acid (GABA) on rat pineal melatonin production *in vitro*. *Cell. Mol. Neurobiol.* **9**, 207–219.
55. Sarda N., Gharib A., Reynaud D., Ou L. and Pacheco H. (1989) Identification of adenosine receptor in rat pineal gland: evidence for A-2 selectivity. *J. Neurochem* **53**, 733–737.
56. Sartin J.L., Bruot B.C. and Orts R.J. (1979) Neurotransmitter regulation of arginine vasotocin release from rat pineal glands in vitro. *Acta Endocrinol.* **91**, 571–576.
57. Semm P., Demaine C. and Vollrath L. (1981) Electrical responses of pineal cells to melatonin and putative transmitters. Evidence for circadian changes in sensitivity. *Exp. Brain Res.* **43**, 361–370.
58. Schon F., Beart P.M., Chapman D. and Kelly J.S. (1975) On GABA metabolism in the gliocyte cells of the rat pineal gland. *Brain Res.* **85**, 479–490.
59. Schrier B.K. and Klein D.C. (1974) Absence of choline acetyltransferase in rat and rabbit pineal gland. *Brain Res.* **79**, 347–351.
60. Shiotani Y., Yamano M., Shiosaka S., Emson P.C., Hillyard C.J., Girgis S. and Macintyre I. (1986) Distribution and origins of substance P (SP)-, calcitonin gene-related peptide (CGRP)-, vasoactive intestinal polypeptide (VIP)- and neuropeptide Y (NPY)- containing nerve fibers in the pineal gland of gerbils. *Neurosci. Lett.* **70**, 187–192.
61. Sugden D. (1990) 5-Hydroxytryptamine amplifies beta-adrenergic stimulation of N-acetyltransferase activity in rat pinealocytes. *J. Neurochem.* **55**, 1655–1658.
62. Tsuchiya M., Kaku K., Matsuda M., Kaneko T. and Yanaihara, N. (1987) Demonstration of receptors specific for N-terminal histidine and C-terminal isoleucine (PHI) using rat PHI and rat dispersed pineal cells. *Biomed. Res.* **8**, 45–51.
63. Uddman R., Alumets J., Hakanson R., Loren I. and Sundler F. (1980) Vasoactive intestinal peptide (VIP) occurs in nerves of the pineal gland. *Experientia* **36**, 1119–1120.
64. Vacas M.I., Sarmiento M.I., Pereyra E.N., Etchegoyen G.S. and Cardinali D.P. (1987) *In vitro* effect of neuropeptide Y on melatonin and norepinephrine release in rat pineal gland. *Cell Mol. Neurobiol.* **7**, 309–315.
65. Vacas M.I., Sarmiento M.I., Pereyra E.N. and Cardinali D.P. (1989) Effect of adenosine on melatonin and norepinephrine release in rat pineal explants. *Acta Physiol. Et Pharmacol. Latinoamer.* **39**, 189–195.
66. Vanecek J., Sugden D., Weller J. and Klein D.C. (1985) Atypical synergistic $\alpha 1$ and β-adrenergic regulation of adenosine 3′,5′-monophosphate in rat pinealocytes. *Endocrinology* **116**, 2167–2173.
67. Weiss B. and Costa E. (1968) Selective stimulation of adenylate cyclase of rat pineal gland by pharmacologically active catecholamines, *J. Pharm. Exp. Therap.* **161**, 310–319.
68. Wheler G.H., Weller J.L. and Klein D.C. (1979) Taurine: stimulation of pineal N-acetyltransferase activity and melatonin production via a beta-adrenergic mechanism. *Brain Res.* **166**, 65–74
69. Wheler G.H. and Klein D.C. (1980) Taurine release from the pineal-gland is stimulated via a beta-adrenergic mechanism. *Brain Res.* **187**, 155–164.
70. Wheler G.H. and Klein D.C. (1979) Cyclic AMP-induced release of [^{14}C]taurine from pinealocytes. *Biochem. Biophys. Res. Commun.* **90**, 22–27.
71. Williams L.M., Morgan P.J., Pelletier G., Riddoch G.I., Lawson W. and Davidson G.R. (1989) Neuropeptide Y (NPY) innervation of the ovine pineal gland. *J Pineal Res.* **7**, 345–353.

72. Yuwiler A. (1983) Vasoactive intestinal peptide stimulation of pineal serotonin-N-acetyltransferase activity: General characteristics. *J. Neurochem.* **41**, 146–153.
73. Yuwiler A. (1983) Light and agonists alter pineal N-acetyltransferase induction by vasoactive intestinal polypeptide. *Science* **230**, 1082–1083.
74. Yuwiler A. (1987) Synergistic action of postsynaptic alpha-adrenergic receptor stimulation on vasoactive intestinal polypeptide—induced increases in pineal N-acetyltransferase activity. *J. Neurochem.* **49,** 806–811.
75. Zhang E.T., Mikkelsen J.D. and Møller M. (1991) Tyrosine hydroxylase- and neuropeptide Y-immunoreactive nerve fibers in the pineal complex of untreated rats and rats following removal of the superior cervical ganglia. *Cell Tissue Res.* **265**, 63–71.

10

The Mammalian Pineal Gland as an End Organ of the Visual System[1]

RUSSEL J. REITER

Department of Cellular and Structural Biology, The University of Texas, Health Science Center at San Antonio, 7703 Floyd Curl Drive, San Antonio, TX 78284–7762, USA

Abstract

Visible light is the major environmental factor controlling the circadian production of melatonin in the mammalian pineal gland. Regardless of the activity pattern a species displays, i.e. diurnal, nocturnal or crepuscular, high melatonin production is always associated with the dark phase of the light:dark cycle. Light has two effects on the circadian melatonin rhythm; it precisely entrains the rhythm to 24 hours and light exposure during the night rapidly depresses high nighttime melatonin levels. The intensity of light required to inhibit nocturnal melatonin production varies widely among species with melatonin synthesis in the pineal gland of nocturnal animals having a much greater sensitivity (lower threshold) to light than that of diurnally active animals. While optic enucleation prevents all actions of light on the 24 hour melatonin cycle, destruction of the majority, if not all, the retinal photoreceptors does not alter the ability of light to synchronize the melatonin rhythm and only slightly increases the threshold of the pineal gland to acute light exposure. Beside luminance or intensity, the specific wavelength of light is important in determining its efficacy as an inhibitor of pineal melatonin synthesis. In general, blue-green wavelengths seem to be most inhibitory to the melatonin forming ability of the pineal. Red light (>600 nm) exposure, however, also suppresses pineal and serum melatonin levels. The ability of red light to achieve this effect is retained in rats with total or near total destruction of the retinal photoreceptors. To date there is no adequate explanation as to the retinal mechanisms involved in mediating the inhibitory effect of light on the pineal gland. It is possible, however, that the classical photoreceptors, i.e. the rods and cones, may not be involved in this process.

Introduction

LIGHT CONSISTS of electromagnetic radiation limited to a defined band of retinally-detected wavelengths referred to as the visible spectrum. Classically, the visible spectrum consists of wavelengths ranging from 400 nm to 760 nm, although some wavelengths shorter than 400 nm in the

[1] Work by the author was supported by grants from the National Science Foundation and from the National Institutes of Health.

ultraviolet (UV) range) and, perhaps under some circumstances and/or in some species, wavelengths longer than 760 nm (in the infrared range) may be detectable by the mammalian retina. This window of visible wavelengths is a very small portion of the total electromagnetic spectrum and the windowing aspect of detection is a feature common to several sensory systems. Besides these limited wavelengths, it has recently been proposed that the eyes may respond to static and extremely low frequency electromagnetic fields[22] which may produce physiological effects similar to those caused by visible light.[28,30]

Since light is a form of energy, it can be quantified in absolute units referred to as ergs. Physiologically, however, what is identified as the "brightness" or "luminosity" of light, which is of major interest to experimentalists working on the functional aspects of the visible spectrum, is not uniquely determined by the energy content of light, but rather it varies according to the specific wavelength. As a result, on a luminosity scale, two independent light sources have an equal intensity when they induce an equal sensation of brightness.

The effects of light on the retina as well as the signal transduction mechanisms in the photoreceptors have been, in part, defined.[12] These mechanisms have been studied primarily in terms of the visual effects of light and less frequently in reference to the functional aspects of light in relationship to circadian and endocrine physiology.[19,38] Although in mammals the detection of light subserving both systems is in the retinas, the neural pathways mediating the physiological consequences are different (Figure 1). Since this report is concerned with the circadian and endocrine (especially pineal melatonin) changes induced by the visible spectrum, the neural pathways connecting the eyes with the suprachiasmatic nuclei (SCN) and eventually with the pineal gland are of special interest.[24] As summarized in Figure 1, the neural pathway from the eyes to the pineal gland is a highly circuitous one; interruption of the connections anywhere between the SCN and the pineal render the gland incapable of producing its hormone melatonin and totally ineffective as an endocrine gland.[23] Besides the innervation of the pineal gland via the peripheral autonomic nervous system, centrally originating fibers also enter the gland through its stalk;[14] however, these connections seem functionally inept in reference to the circadian melatonin rhythm.

Pineal melatonin synthesis and light

What is known of the signal transduction mechanisms related to the production of the key pineal hormone melatonin (*N*-acetyl-5-methoxytryptamine) has recently been reviewed[26] and is summarized in Figure 2. Whereas norepinephrine (NE) is the neurotransmitter primarily responsible for initiating the production of melatonin in the pineal gland, its

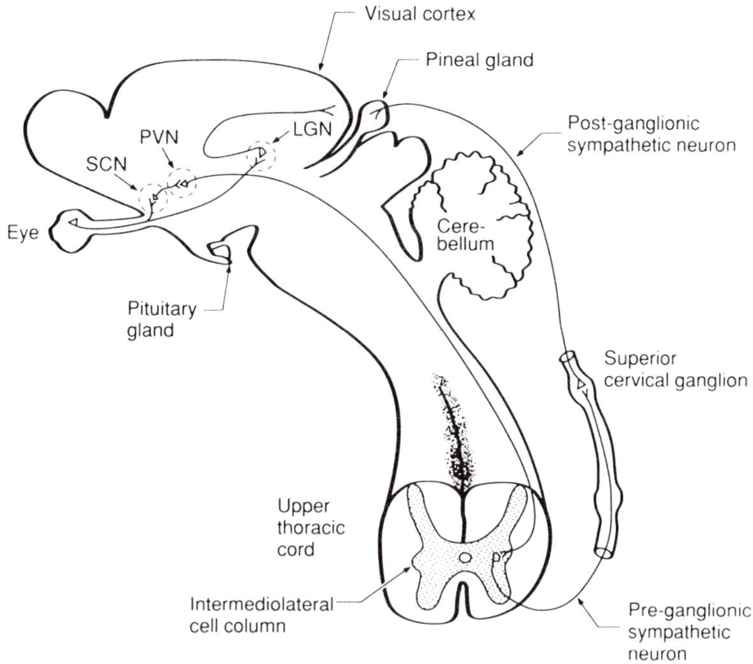

FIG. 1. Schematic representation of the neural connections of the retinas with the visual cortex and with the circadian system via the suprachiasmatic nuclei (SCN) and the pineal gland. The visual cortex receives projections from the eyes by means of ganglion cell axons which synapse in the lateral geniculate nucleus (LGN); neural fibers from the LGN then project to the occipital pole of the cerebral cortex. The eyes are connected to the pineal gland as well with the pathway also involving axons of retinal ganglion cells which synapse in the SCN; the remainder of the pathway includes synapses in the paraventricular nuclei (PVN), the intermediolateral cell column of the upper thoracic cord and the superior cervical ganglia. The postganglionic sympathetic fibers terminate in the pineal gland and end in the proximity of the pinealocytes, the endocrine units of the gland.

interaction with pinealocyte adrenergic receptors is not the only described mechanism for regulating melatonin production. Indeed, pinealocyte membranes contain many other receptors which can modulate the synthesis of melatonin to varying degrees;[10] however, the circadian production and release of melatonin is generally considered to be exclusively related to the peripheral autonomic innervation and its associated catecholaminergic neurotransmitter, NE.

Regardless of the activity pattern a specific animal displays, i.e. diurnal, nocturnal, or crepuscular, peak pineal melatonin production and secretion is always associated with the night. Because of its unique relationship with the dark phase of the light–dark cycle, melatonin has been referred to as the "chemical expression of darkness".[27] In Scandinavia, melatonin could also be thought of as the "troll of the endocrine

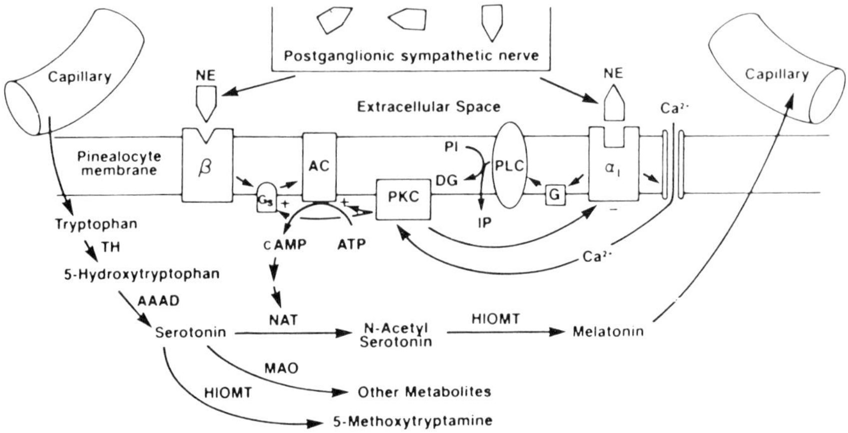

FIG. 2. A simplified schematic representation of the signal transduction mechanisms controlling the production of melatonin in the mammalian pineal gland. The primary neurotransmitter mediating the nighttime rise in melatonin is the catecholamine norepinephrine (NE) which is released from postganglionic sympathetic neurons on command from the suprachiasmatic nuclei in the hypothalamus. The most important interaction of NE is with β-adrenergic receptors in the cell membrane of the pinealocytes; α-adrenergic receptors play a modulatory role in controlling melatonin production. Besides the one visually depicted here, the pinealocytes respond to many other neurotransmitters which act on specific receptors and which either slightly or markedly influence melatonin production. AAAD, l-aromatic amino acid decarboxylase; AC, adenylate cyclase; ATP, adenosine triphosphate; cAMP, cyclic AMP; DG, diacylglycerol; Gs, stimulatory guanine nucleotide binding protein; HIOMT, hydroxyindole-O-methyltransferase; IP, inositol phosphate; MAO, monoamine oxidase; NAT, *N*-acetyltransferase; PI, phosphotidylinositol; PKC, protein kinase C; PLC, phospholipase C; TH, tryptophan hydroxylase; α^1-, α-adrenergic receptor; β,β-adrenergic receptor. From Reiter.

system". For melatonin production, as with the mythical troll, sunlight is synonymous with death.

Melatonin produced in the pineal gland is quickly released into the blood vascular system; the release of melatonin is so rapid that blood levels of the hormone are generally used as an index of the amount being synthesized in the gland at virtually the same time. Remarkably, however, melatonin's release can be further hastened by at least one maneuver, i.e. strenuous exercise. Thus, if rats at night are forced to swim the enzymes controlling melatonin production in the pineal gland remain elevated while the melatonin content of the gland falls to daytime levels. Yet, circulating concentrations of the indole remain high, as during darkness. The best interpretation consistent with these findings is that the newly-synthesized melatonin is very rapidly released (more rapidly than normal), leading to low pineal melatonin levels but high concentrations in the blood.[31]

Once in the blood, melatonin has access to every other fluid in the body and its high lipophilicity, which promotes its ready passage through cell membranes, ensures that it gains admittance to other fluid compartments. Because of this, the circadian melatonin rhythm, typical of the blood, is also present, albeit of somewhat lower amplitude, in other bodily fluids including saliva, ovarian follicular fluid, male seminal fluid, cerebrospinal fluid, amniotic fluid and fluid of the anterior chamber of the eye.[26] Likewise, the chief hepatic metabolite of melatonin, 6-hydroxymelatonin sulfate, exhibits a 24 hour cycle in the urine with high excretion of the compound occurring at night.[6] Thus, all cells with which melatonin comes in contact could use the melatonin message, provided they have the cellular machinery to recognize melatonin changes over time.

Encoded in the melatonin message is both time-of-day (clock) and time-of-year (calendar) information. Thus, daytime levels of melatonin are always low whereas nighttime levels are elevated. Providing the receptor molecules that respond to the message distinguish low (or absence) from high concentrations (or presence) of melatonin, the associated metabolic responses reflect either day or night.

The ratio of light to darkness per 24 hour period varies seasonally and is more exaggerated at higher latitudes (Figure 3). Since the duration of elevated melatonin at night is roughly proportional to the duration of the night, the melatonin signal thereby provides time-of-year-information; this is true both for experimental animals[3,27] as well as for humans.[42,43] In animals, at least, the physiological significance of the annual variation in melatonin secretion is well documented where it is known to control seasonal reproductive capability.[23] Seasonal reproduction in mammals in their natural habitat is an important event since it ensures the young are born at a season most conducive to their survival.

As mentioned above, environmental light is incompatible with high melatonin production. Sunlight as well as artificial light have two obvious effects on the 24 hour melatonin cycle. Thus, light entrains the circadian rhythm of melatonin synthesis and the introduction of light at unusual times, e.g. at night, acutely suppresses melatonin synthesis. The entraining effect of light is readily seen in both experimental animals[1] and humans[43] and is diagrammatically portrayed in Figure 4. The entraining effect of light on the circadian melatonin rhythm is readily apparent in humans living in Antarctica where the 24 hour cycle of melatonin free-runs during the interval when the sun is below the horizon[13] and in blind humans who are incapable of light perception.[16]

Light intensity effects

The intensity of white light required to suppress pineal melatonin production varies greatly among species (Table 1) and may be influenced by the lighting history of the animal.[33] To illustrate the variation among

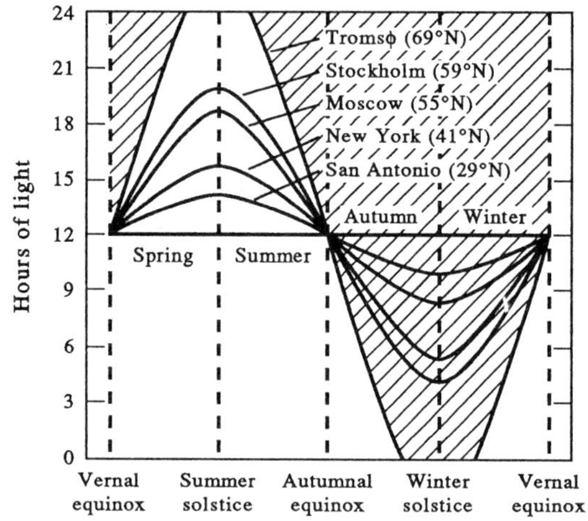

FIG. 3. Diagrammatic representation of seasonal variations in daylength in representative cities at various latitudes. At progressively higher latitudes in both hemispheres, the ratios of daylight hours to nighttime hours become increasingly great. In the most exaggerated case, in the Arctic and Antarctic Circle, long periods of persistent light and darkness are experienced.

the mammals investigated, pineal melatonin production in the albino rat is roughly 3.7×10^6 times more sensitive to inhibition by white light than is the pineal of the Richardson's ground squirrel.[24a] In general, pineal melatonin synthesis in nocturnal animals, who have rod-dominated retinas, exhibits a greater sensitivity to light inhibition when compared with diurnal species who have almost exclusively cone photoreceptors. The one subterranean mammal that has been tested, i.e. the valley pocket gopher (*Thomomys bottae*), has, unexpectedly, a pineal melatonin rhythm that is relatively insensitive to white light. It is estimated that pocket gophers spend 99.9% of their life in underground burrows (with closed openings to the surface) in total darkness. Their activity patterns are seasonally dependent and range from nocturnal, to diurnal, to crepuscular depending on ground temperature conditions near the surface. Despite their almost total lack of exposure to light in their natural habitat, the melatonin rhythm in the pineal gland of wild-captured pocket gophers is easily synchronized by laboratory light:dark cycles (J.H. Sun, R.J. Reiter, K. Yaga and L.C. Manchester, unpublished observations). Information on the photoreceptor composition of the pocket gopher retina is lacking.

It is also apparent that in a real sense the retina-SCN-pineal system can act as a comparator to compare the relative intensity of light to determine what is night and what is day.[18] In the study in question, rats were exposed to a L12:D12 photoperiodic cycle of either darkness alternating with dim light (0.1–0.3 $\mu W/cm^2$) or the dim light alternating with bright light (45–100 $\mu W/cm^2$). In both exposure situations the rhythmic production of

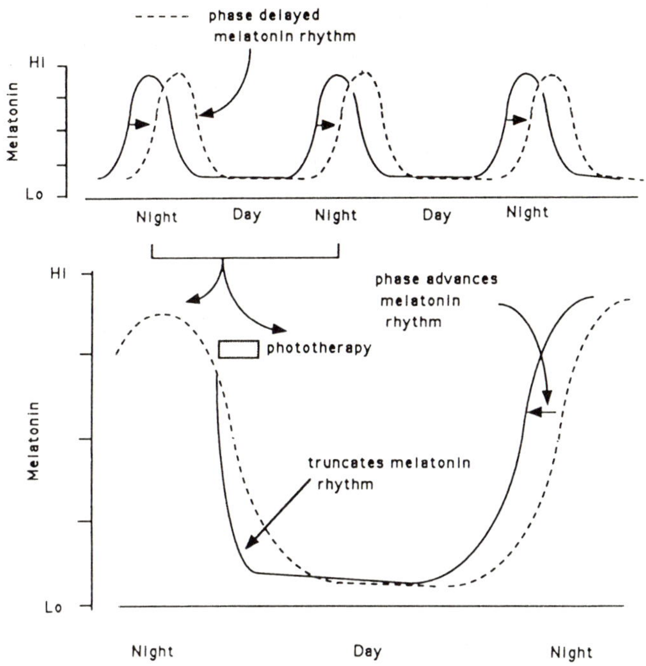

FIG. 4. In the case of an animal or a human where the melatonin rhythm is phase-delayed (upper Figure), it can be phase-advanced (lower Figure) by bright light exposure (in humans, referred to as phototherapy) in either the early morning or possibly throughout the day. Shifting the melatonin rhythm may have psychological and physiological benefits.

melatonin persisted with the maximal levels always being associated with the darkest portion of the photoperiodic cycle. Thus, the dim light was interpreted as either day or night depending on the illumination intensity during the contiguous 12 hour period. The implication is that the system exhibits some plasticity and can adapt to prevailing environmental conditions. This is also supported by the already-mentioned data showing that the lighting history of an animal may influence the sensitivity of the retina-SCN-pineal system to light.[33] It is presumed these adjustments are made at the level of the so-called visual cells in the SCN.[19]

Compared to the sensitivity of the retina-visual cortical system, the retina-SCN-pineal system is relatively insensitive to visible light stimulation. For example, in the Syrian hamster a light irradiance of 0.00001 $\mu W/cm^2$ evokes retinal neuronal activity as indicated by the firing pattern of ganglion cell axons (measured by means of the electroretinogram) in the optic nerve.[34,35] An examination of Table 1 shows that the inhibition of melatonin synthesis in this species has a threshold of approximately 0.1 $\mu W/cm^2$. Therefore, the retina-visual cortical system of the hamster

TABLE 1
Sensitivity of the melatonin rhythm to acute light exposure at night in representative mammals. Listed are the approximate intensities of light required to depress pineal melatonin production; both the irradiance and illumination itensities are given

Species	Irradiance ($\mu W/cm^2$ cool white)	Illumination intensity ftc	lux	Activity pattern
Albino rat (*Rattus norvigecus*)	0.0005	0.0002	0.0017	nocturnal
Cotton rat (*Sigmodon hispidus*)	0.01	0.003	0.034	nocturnal
Syrian hamster (*Mesocricetus auratus*)	0.1	0.031	0.336	nocturnal
Goat (British Saanen strain)	0.4	0.13	1.35	diurnal
Valley pocket gopher (*Thomomys bottae*)	420	131	1,412	diurnal/ nocturnal
13-Lined ground squirrel (*Spermophilus tridecemlineatus*)	925	289	3,109	diurnal
Richardson's ground squirrel (*Spermophilus richardsoni*)	1,850	578	6,220	diurnal

has a sensitivity to light which is 1,000 times greater than that for the retina-SCN-pineal system. Similar relationships hold for the rat.[32] At the other end of the spectrum, it appears that in species such as the Richardson's ground squirrel (Table 1) the retina-visual cortical system may be millions of times more sensitive to light than is melatonin synthesis in the pineal gland.

The suppression of high nocturnal pineal melatonin levels by acute light exposure is very rapid with the melatonin content of the pineal gland falling to 50% of the nighttime levels in less than 10 minutes.[36] Soon after pineal melatonin levels fall, serum levels as well as the concentration of the indole in other bodily fluids are decreased (Figure 5). The duration of light exposure at night required to inhibit pineal melatonin production is very brief. Thus, certainly a 1 sec flash of bright light at night leads to a temporary cessation of pineal melatonin production.[29] The disruption of the circadian melatonin rhythm in mammals by very brief light periods leads to marked physiological effects[9] and probably has significance for fossorial animals in their natural habitat.[25]

Unquestionably, the ability of light to inhibit pineal melatonin production requires the eyes of the animal to be intact. Thus, optically enucleated rats kept under a light:dark cycle no longer exhibit a pineal melatonin rhythm that is either synchronized by or acutely suppressed by light[32] (Figure 6). In these animals the melatonin rhythm free-runs with each animal having a melatonin cycle with a slightly different free-running period. As a result, individual pineals of a group of animals killed at the

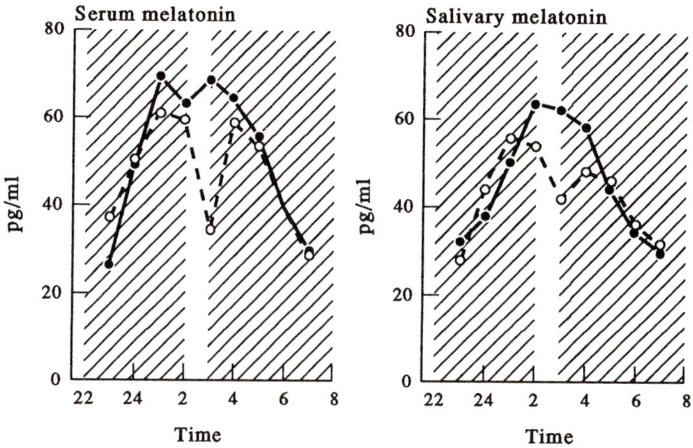

FIG. 5. Serum and salivary melatonin response to a 1 hour exposure to light between 02.00 and 03.00 hours in three adult human males. During the control night (●) melatonin levels were elevated between 02.00 and 03.00 hours; when the night was interrupted by light (○) melatonin levels in both fluids fell. Shaded area represents darkness.

same time during the light:dark cycle may have high, low or intermediate levels of melatonin production with the average melatonin level falling between the expected low daytime and high nighttime values (Figure 6, middle panel). This is quite different from the situation in rats where the photoreceptor cells are lost due to continual light exposure;[15] even in the case of total or near total destruction of the retinal photoreceptors the melatonin rhythm is readily synchronized by the prevailing light:dark environment (Figure 6, bottom panel).

As mentioned above, in optically enucleated rats acute light exposure at night also does not impair the ability of the pineal gland to synthesize its chief hormonal product melatonin[41] (Figure 7, middle panel). Conversely, melatonin synthesis in the pineal gland of rats with light-damaged retinas is still depressed by acute light exposure at night with the threshold of light intensity required to inhibit pineal indoleamine metabolism being only slightly increased over that in rats with normal photoreceptor elements.[41]

Besides lighting history, ambient environmental temperature may be a factor in determining the sensitivity of the melatonin synthesis cycle to light. When Djungarian hamsters (*Phodopus sungorus*) were exposed to reduced temperature during the night the light threshold required to suppress the melatonin forming ability of the pineal gland was increased, i.e. the system was less sensitive to inhibition.[37] It is assumed that the threshold adjustment is made at the level of the hypothalamic SCN since this area of the brain receives sensory projections from the thalamus, which is an important sensory relay center to higher cortical structures.

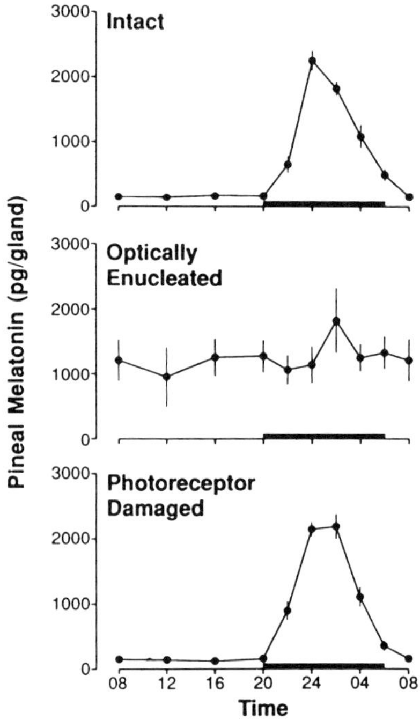

FIG. 6. Pineal melatonin levels over a 24 hour light:dark cycle in rats with an intact retina (top panel), in rats that had been optically enucleated (middle panel) and in rats that had their photoreceptors severely damaged by continual light exposure (bottom panel). Clearly, the pineal melatonin cycle is not synchronized by light in the blinded animals while it is in the rats that have no or few functional photoreceptors remaining in their retinas. Dark bar on the horizontal axis is the daily dark period.

Light wavelength effects

In addition to light intensity, light wavelength is important in determining the suppressive effect of visible electromagnetic radiation on the synthetic activity of the pineal gland. Using the activity of HIOMT, the melatonin-forming enzyme, as an index, Cardinali *et al.*[7] claimed that among the wavelengths tested (red, yellow, blue, green and near ultraviolet (UV-A)), green wavelengths proved most effective in inhibiting the biosynthetic activity of the rat pineal gland. In the Syrian hamster, green wavelengths (half-peak bandwidth = 515–550 nm), but even to a greater extent blue wavelengths (half-peak bandwidth = 435–500 nm), suppress nocturnal pineal melatonin levels.[4] The implication of these findings, as well as those from a similar study in humans where a patternless monochromatic light (λ max 509 nm, with wavelength and irradiance adjustment with glass interference and neutral density filters) was used,[2] is that scotopic mechanisms in the retina are involved in mediating the inhibitory effect of visible light on the pineal gland. Although no definitive claims have been

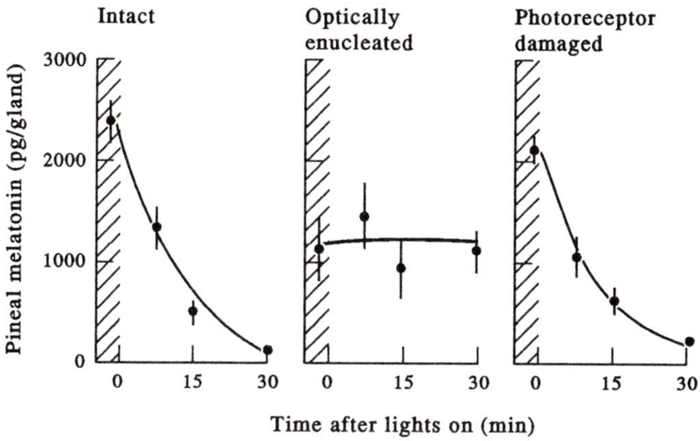

FIG. 7. Responses of the pineal gland to acute light exposure at night in retinal-intact rats (left panel), in optically enucleated rats (middle panel), and in rats that have severely damaged retinas due to prolonged light exposure (right panel). Optically enucleated rats do not respond to light onset with melatonin suppression while those with extensive photoreceptor destruction respond similar to those in rats with intact retinas.

made that the activation of rhodopsin is an important aspect of the light-mediated suppression of melatonin, this is usually tacitly accepted. If true, it would be consistent with observation that very low light intensities readily suppress pineal melatonin in animals with rod-dominated retinas (Table 1).

In a brief report published in 1982, Vanecek and Illnerova[39] claimed that the exposure of albino rats to red light (8 lux, from a 15 W red Tesla photographic bulb) for 30 minutes at night suppressed pineal NAT activity and, by inference, melatonin levels (although they were not measured). With this particular red light source less than 5% of the total emitted light has wavelengths shorter than 600 nm. These findings are of interest because of the widespread use of red background illumination in animal rooms as "safe" lights and because it implies that rat retinal photoreceptors can respond to wavelengths greater than 600 nm; this latter claim could not be definitively made since in fact 5% of the emitted wavelengths were below 600 nm and, in the range of wavelengths under 600 nm, the albino rat pineal gland is very sensitive to light.[41]

To prove whether wavelengths of light greater than 600 nm do suppress the nocturnal levels of not only NAT activity, but also melatonin, rats were exposed to a red light (>600 nm wavelengths only; with the aid of a Kodak 1A Wratten filter) of different intensities. As did Vanecek and Illnerova, we found that at all intensities equal to or greater than 50 $\mu W/cm^2$, not only inhibited pineal NAT, but pineal and serum levels of melatonin also were suppressed (R.J. Reiter, B. Poeggeler, K. Yaga and J.H. Sun, unpublished observations) as a consequence of this exposure. The findings provide unequivocal evidence that nocturnal pineal melatonin production is influenced by wavelengths of 600 nm and above. Furthermore, in the retinas of the same animals we measured the ratio of 11-*cis* to all-*trans*-retinal. Normally, during darkness 85–90% of the retinal aldehyde is in the 11-*cis* form. Whereas white light exposure at night quickly changes this ratio,[12] no alteration in the 11-*cis* to all-*trans* ratio could be measured in the red light exposed rats. Since the isomerization of 11-*cis* retinal is the initial event in the activation of rhodopsin, our findings indicate that the rod photoreceptor is not likely to be involved in the red light-mediated suppression of pineal melatonin production. It is possible that cone cells in the rat retina may respond to red wavelengths although the evidence for this is weak.[8]

We carried these studies one step further in an attempt to determine the role of the rod photoreceptors in detecting red light that was capable of suppressing pineal melatonin production. Rats were placed under constant white light, a procedure known to cause the rapid destruction of the rod photoreceptors.[15] The animals were then returned to cyclic light and during the night they were exposed to >600 nm light (red). Despite the severely damaged retinas with the near total loss of photoreceptors, red light readily inhibited the ability of the pineal gland to reduce melatonin. These findings suggest that rods are not involved in the detection of >600 nm light which acts to inhibit the pineal gland, although it is not incontrovertible evidence since a few cell bodies, if not the outer segments, of the rod photoreceptors may have survived the prolonged continuous light treatment. Also, since the relatively few cone photoreceptors that do exist in the rat retina are less readily destroyed by continual light treatment, it is possible that surviving cones responded to the red wavelengths which depressed the formation of melatonin in the pineal gland.[8] These explanations are not the only tentable ones, however (see Concluding remarks).

Concluding remarks

The discrimination of light intensity or luminance in the retina-SCN-pineal system likely occurs at the level of the SCN.[38] What are referred to as visual cells in the SCN change their firing rate (the majority of the cells increase their discharge rate while the remainder decrease it) in response to sustained illumination.[20] The light-activated visual cells increase their discharge rate as light intensity increases thereby presumably coding for luminance.[21] Interestingly, among the species tested the light intensity required to activate the visual cells varies with the SCN cells of the rat being most light sensitive, followed by those in the Syrian hamster, and finally by those in the ground squirrel (*Spermophilus tridecemlineatus*).[21] The rat and hamster are, of course, nocturnal while the ground squirrel is diurnal. Inspection of Table 1 reveals that the light intensity required to inhibit pineal melatonin production has the same rank order. Although this certainly provides circumstantial evidence for the SCN being the site at which light intensity is discriminated in reference to its ability to inhibit the pineal gland, not all the data support this interpretation. Thus, whereas very brief light pulses (<1 second) inhibit nocturnal melatonin synthesis in the rat pineal, sustained illumination is required to evoke maximal stimulation of SCN visual cell activity.[21]

For the most part, it is usually assumed that pineal melatonin production relies on the ability of retinal photoreceptors to respond to visible electromagnetic illumination. However, there are observations that argue against this. For example, severely damaged retinas with total or near total destruction of the photoreceptors, especially the rods, are still very sensitive to light. Certainly, the possibility exists that very few partially surviving rods could still detect sufficient light to inhibit pineal melatonin production. While this cannot be precluded, a substantial attenuation of light's ability to influence the pineal gland would seemingly be expected. Likewise, it could be argued that rather than rods, surviving cones (which are more likely preserved in eyes with light-damaged retinas[8]), may mediate the inhibitory effects of light on pineal melatonin production. If, in fact, this is a function of cones then it would be expected that pineal melatonin synthesis in diurnally active animals, which have cone-dominated retinas, would be more sensitive to light inhibition than would the pineal metabolic activity of nocturnally active animals which possess rod-dominated retinas. However, just the inverse is the case (Table 1). Also, the three proposed photopic mechanisms in the retina of the rat have peaks of roughly 450 nm, 520 nm and 560 nm.

Whereas there is little doubt that wavelengths of visible light between 460 nm and 580 nm (with peak sensitivity at 495 nm) inhibit pineal NAT activity in the rat and this inhibition may be mediated by a rod photopigment,[5] there is currently no satisfactory explanation for the ability of wavelengths >600 nm (red light) to suppress the melatonin forming ability of the pineal. We have found that the isomerization of 11-*cis* retinal does not occur in the retinas of red light exposed rats even when the light is sufficiently intense to inhibit the ability of the pineal gland to synthesize melatonin (J.H. Sun, R.J. Reiter and K. Yaga, unpublished observations); thus, photoactivation of rhodopsin is clearly not required for red light to act in an inhibitory manner on the pineal gland.

One implication of the information summarized in the previous paragraph is that there may be circumstances under which pineal melatonin synthesis is inhibited by wavelengths between 400 nm and 760 nm where the classical retinal photoreceptors may not be involved. Indeed, the evidence is already quite compelling that light detected by the retinas, but not by rods and cones, may influence the circadian system via the SCN. According to Foster and colleagues,[11] in retinally degenerate mice (C57BL/6J mice homogenous for the *rd* allele: *rd/rd*) light can still entrain their circadian system. These workers advanced the idea that there may exist a yet unidentified group of photoreceptive cells in the mammalian retina that normally function to entrain the circadian system. Considering the possibility of a non-classical photoreceptor in the mammalian eye, it may be noteworthy that the vertebrate retina contains a special class of bipolar cells which has a ciliated dentritic process that has extensive arborizations. These structures, referred to as "Landolt's clubs"[17] have been speculated to be photosensory[40] and they should be given consideration as a photoreceptor which mediates the inhibitory effect of light on pineal melatonin production in mammals.

References

1. Azuma H., Yamatodani A., Yagi A., Nishimura T. and Wada H. (1991) Effect of bright light in the morning on dirurnal variations of pineal indoles in NZBWF1 mice. *Neurosci. Lett.* **131**, 210–212.
2. Brainard G.C., Lewy A.J., Menaker M., Fredrickson R.H., Miller L.S., Weleber R.G., Cassone V. and Hudson D. (1985) Effect of light wavelength on the suppression of nocturnal plasma melatonin in normal volunteers. *Ann. N.Y. Acad. Sci.* **453**, 376–378.
3. Brainard G.C., Petterborg L.J., Richardson B.A. and Reiter R.J. (1982) Pineal melatonin in Syrian hamsters: Circadian and seasonal rhythms in animals maintained under laboratory and natural conditions. *Neuroendocrinology* **35**, 342–348.
4. Brainard G.C., Richardson B.A., King T.S. and Reiter R.J. (1984) The influence of different light spectra on the suppression of pineal melatonin content in the Syrian hamster. *Brain Res.* **294**, 333–339.
5. Bronstein D.M., Jacobs G.H., Gook K.A., Neitz J. and Lytle L.D. (1987) Action spectrum of the retinal mechanism mediating nocturnal light-induced suppression of rat pineal gland N-acetyltransferase. *Brain Res.* **406**, 352–356.
6. Brown G.M., Bar-Or A., Grossi D., Kashur S., Johannson, E. and Yei S.M. Urinary 6-hydroxymelatonin, an index of pineal function in the rat. *J. Pineal Res.* **10**, 141–147.
7. Cardinali D.P., Larin P. and Wurtman R.J. (1972) Control of rat pineal gland by light spectra. *Proc. Natl. Acad. Sci. USA* **69**, 2003–2005.
8. Cicerone, C.M. (1976) Cones survive rods in the light-damaged eye of the albino rat. *Science* **194**, 1183–1185.
9. Earnest D.J. and Turek, F. J. (1983) Effect of one-second light pulses on testicular function and locomotor activity in the golden hamster. *Biol. Reprod.* **28**, 557–563.
10. Ebadi M. and Govitrapong P. (1986) Orphan transmitters and their receptor sites in the pineal gland. In *Pineal Research Reviews* (ed. Reiter, R.J.), Vol. 4, pp. 1–54. Alan R. Liss, New York.
11. Foster R.G., Provencio I., Hudson D., Fiske S., DeGrip W. and Menaker M. (1991) Circadian photoreception in the retinally degenerate mouse (*rd/rd*). *J. Comp. Physiol.* **169**, 39–50.
12. Hargrave P.A. and McDowell J.H. (1992) Rhodopsin and phototransduction: A model system for G protein-linked receptors. *FASEB J.* **6**, 2323–2331.
13. Kennaway D.J. and Van Dorp C.F. (1991) Free-running rhythms of melatonin, cortisol, electrolytes and sleep in humans in Antarctica. *Am. J. Physiol.* **260**, R1137–R1144.
14. Korf H.W. and Møller M. (1984) The innervation of the mammalian pineal gland with special reference to central pinealopetal projections. In *Pineal Research Reviews* (ed. Reiter, R.J.), Vol. 2, pp. 41–86. Alan R. Liss, New York.
15. Lanum J. (1978) The damaging effects of light on the retina. Emperical findings, theoretical and practical implications. *Surv. Ophthal.* **22**, 221–249.

16. Lewy A.J. and Newsome D.A. (1983) Different types of melatonin circadian rhythms in some blind subjects. *J. Clin. Endocrinol. Metab.* **56**, 1103–1107.
17. Locket N.A. (1970) Landolt's club in the retina of the African lungfish, *Protopterus aethiopicus*. *Vision Res.* **10**, 299–306.
18. Lynch H.J., Deng M.H. and Wurtman R.J. (1985) Indirect effects of light: Ecological and ethological considerations. *Ann. N.Y. Acad. Sci.* **453**, 231–241.
19. Meijer, J.H. (1991) Integration of visual information by the suprachiasmatic nuclei. In *Suprachiasmatic Nucleus* (eds. Klein D.C., Moore R.Y. and Reppert S.M.), pp. 107–119. Oxford University Press, Oxford.
20. Meijer J.H., Gross G.A. and Rusak B. (1986) Luminance coding in the circadian pacemaker: The suprachiasmatic nucleus of the rat and hamster. *Brain Res.* **382**, 109–118.
21. Meijer J.H., Rusak B. and Harrington M.E. (1989) Photically responsive neurons in the hypothalamus of the diurnal ground squirrel. *Brain Res.* **501**, 315–323.
22. Olcese J.M. (1990) The neurobiology of magnetic field detection in rodents. *Prog. Neurobiol.* **35**, 325–330.
23. Reiter R.J. (1980) The pineal and its hormones in the control of reproduction in mammals. *Endocr. Rev.* **1**, 109–131.
24. Reiter R.J. (1981) The mammalian pineal gland: Structure and function. Am. J. Anat. 162, 287–313.
24a. Reiter R.J. (1985) Action spectra, dose-response relationships and temporal aspects of light's effect on the pineal gland. *Ann. N.Y. Acad. Sci.* **453**, 215–230.
25. Reiter R.J. (1986) Environmentally determined physiological adjustments made in preparation for hibernation. In *Living in the Cold* (eds. Heller H.C., Musacchia X.J. and Wang L.C.H.), pp. 287–294. Elsevier, New York.
26. Reiter R.J. (1991) Pineal melatonin: Cell biology of its synthesis and of its physiological interactions. *Endocr. Rev.* **12**, 151–180.
27. Reiter R.J. (1991) Melatonin: The chemical expression of darkness. *Mol. Cell. Endocrinol.* **79**, C153–C158.
28. Reiter R.J. (1992) Alterations of the circadian melatonin rhythm by the electromagnetic spectrum: A study in environmental toxicology. *Regul. Toxicol. Pharmacol.* **15**, 226–244.
29. Reiter R.J., Joshi B.N., Heinzeller T. and Nürnberger F. (1986) A single 1- or 5-second light pulse at night inhibits hamster pineal melatonin. *Endocrinology* **118**, 1906–1909.
30. Reiter R.J. and Richardson B.A. (1992) Static magnetic field effects on pineal indoleamine metabolism and possible biological consequences. *FASEB J.* **6**, 2283–2287.
31. Reiter R.J. and Richardson B.A. (1992) Some perturbations that disturb the circadian melatonin rhythm. *Chronobiol. Internat.* **9**, 314–321.
32. Reiter R.J., Sorrentino S. Jr., Ralph C.L., Lynch H.J., Mull D. and Jarrow E. (1971) Some endocrine effects of blinding and anosmia in adult male rats with observations on pineal melatonin. *Endocrinology* **88**, 895–900.
33. Reiter R.J., Steinlechner S., Richardson B.A. and King T.S. (1983) Differential response of pineal melatonin levels to light at night in laboratory-raised and wild-captured 13-lined ground squirrels (*Spermophilus tridecemlineatus*). *Life Sci.* **32**, 2625–2629.
34. Reuter J.H. (1972) A comparison of flash evoked ERG's and ERG's evoked with sinusoidally modulated light stimuli in a number of rodents. *Pflüger's Arch. Gesamte Physiol. Menschen Tiere* **331**, 95–102.
35. Reuter J.H. and Hobbelen J.F. (1978) The effect of exposure to continuous light on the ERG of Syrian hamsters. *Neurosci. Lett.* **7**, 73–77.
36. Rollag M.D., Panke E.S., Trakulrungsi W., Trakulrungsi C. and Reiter R.J. (1980) Quantification of daily melatonin synthesis in the hamster pineal gland. *Endocrinology* **106**, 232–236.
37. Stieglitz A., Steinlechner S., Ruf T. and Heldmaier G. (1991) Cold prevents the light-induced inactivation of pineal N-acetyltransferase in the Djungarian hamster, *Phodopus sungorus*. *J. Comp. Physiol.* **168**, 599–603.
38. Takahashi J.S., DeCoursey P.J., Bauman L. and Menaker M. (1984) Spectral

sensitivity of a novel photoreceptive system mediating entrainment of mammalian circadian rhythms. *Nature* **308**, 186–188.

39. Vanecek J. and Illnerova H. (1982) Night pineal *N*-acetyltransferase activity in rats exposed to white or red light pusles of various intensities and duration. *Experientia* **38**, 1318–1320.
40. Vigh B., Vigh-Teichmann I., Röhlich P. and Oksche A. (1983) Cerebrospinal fluid-contacting neurons, sensory pinealocytes and Landolt's clubs of the retina as revealed by means of an electron microscopic immunoreaction against opsin. *Cell Tissue Res.* **233**, 539–548.
41. Webb S.M., Champney T.H., Lewinski A.K. and Reiter R.J. (1985) Photoreceptor damage and eye pigmentation: Influence on the sensitivity of rat pineal *N*-acetyltransferase activity and melatonin levels to light at night. *Neuroendocrinology* **40**, 205–209.
42. Wehr T.A. (1991) The durations of human melatonin secretion and sleep response to changes in daylength (photoperiod). *J. Clin. Endocrinol. Metab.* **73**, 1276–1280.
43. Wetterberg L., Beck-Friis J. and Kjellman B.F. (1990) Melatonin as a marker for a subgroup of depression in adults. In *Biological Rhythms, Mood Disorders, Light Therapy, and the Pineal Gland* (eds. Schafii M. and Shafii S.L.), pp. 71–95. American Psychiatric Press, Washington.

11

Light Entrainment of Rat and Human Circadian Melatonin Rhythms

HELENA ILLNEROVÁ, LUDMILA SAMKOVÁ and MILENA BURESOVÁ

Institute of Physiology, Czechoslovak Academy of Sciences, 14220 Prague 4, Czechoslovakia

Abstract

The rhythmic production of melatonin in rats is driven by a circadian rhythm of the pineal *N*-acetyltransferase activity. The present study compares light entrainment of the rat pineal *N*-acetyltransferase rhythm with that of the human serum or saliva melatonin rhythm. Exposure to bright light in the early night phase delays the rat NAT rhythm and the human melatonin rhythm as a whole within one day. Exposure to bright light in the late night phase advances the morning rat NAT and human melatonin offset within one day as well; the evening rat NAT onset is, however, not phase shifted at all, whereas the human melatonin onset may or may not be phase advanced. It appears that even in humans the evening melatonin production onset does not necessarily phase shift in parallel with the morning offset. Both the rat NAT rhythm and the human melatonin rhythm are photoperiod dependent; adjustment of the rhythms to a change from long to short days may, however, proceed in a different way in rats than in humans. Altogether, though light entrainment of the circadian rhythm in melatonin production in humans is similar to that in rats, some differences exist as well.

Introduction

THE DAILY rhythm of the pineal hormone melatonin production and secretion in mammals is involved in transduction of the photoperiodic signal and may play a subtle role in the circadian system entrainment.[10,28] In rats, the rhythmic melatonin production is driven by the circadian rhythm of the pineal *N*-acetyltransferase (EC 2.3.1.87) (NAT) activity.[18,23] The NAT rhythm is controlled by a pacemaker located in the suprachiasmatic nuclei of the hypothalamus, similarly as other rhythms are.[22] The NAT rhythm may be reset by exposure of rats to light during their subjective night.[11,12,15] Similarly, chronic and even a single bright light exposure may reset the human melatonin rhythm.[4,5,24,25] The present study compares the light entrainment of the rhythmic melatonin production, i.e. of the NAT rhythm, in the rat, with the entrainment of the human serum or saliva melatonin rhythm in order to show similarities as well as eventual dissimilarities between the rat and the human melatonin rhythm resetting.

NAT activity was determined by a modification of the method of Deguchi and Axelrod, as described elsewhere.[9,13,15] Melatonin was measured by a direct radioimunoassay validated for use in serum and in saliva.[27,31] Unless otherwise specified, rats were maintained under a lighting regime with 12 hours of light and 12 hours of darkness per day (LD 12:12). Resetting of the human circadian pacemaker by artificial light was performed in winter months in order to avoid any interference with bright outdoor light either in the morning or in the evening.

Phase delays

After a 1-minute light pulse in the early night (Figure 1A) or after a delay of the evening darkness onset (Figure 1B), the evening NAT activity onset was phase delayed more than the morning offset during the same night.[16,17] However, within one day, the onset and offset were phase delayed to the same extent, eventually by as much as by 6 hours. Consequently, the whole NAT rhythm was phase shifted within one cycle and its waveform did not change.

After a 3-hour bright light pulse in the early night, the morning saliva melatonin offset was phase delayed by 1–2 hours in four subjects out of four within one day; the evening melatonin onset was phase delayed by 1 hour in three subjects (Figure 2) (Samková and Illnerová, unpublished results). The mechanism of phase delaying of the human melatonin rhythm might thus be similar to the mechanism of phase delaying of the rat NAT rhythm. The magnitude of phase delays of the rat NAT rhythm appeared to be larger than that of the human melatonin rhythm. Chronic bright light treatment of humans in the evening delays the onset of the nighttime melatonin production and the whole rhythm of 6-sulfatoxymelatonin excretion.[20,25]

The melatonin rhythm amplitude increased during the night when the light pulse was applied and the next night in three subjects (Figure 2). Such a rebound effect has been reported for the human serum melatonin rhythm,[1,2] but it has never been observed in the rat NAT rhythm. In the human melatonin rhythm, other factors besides the NAT activity might play a role in the rebound effect, e.g. the pineal hydroxyindole-*O*-methyltransferase activity or concentration of substrates.[11]

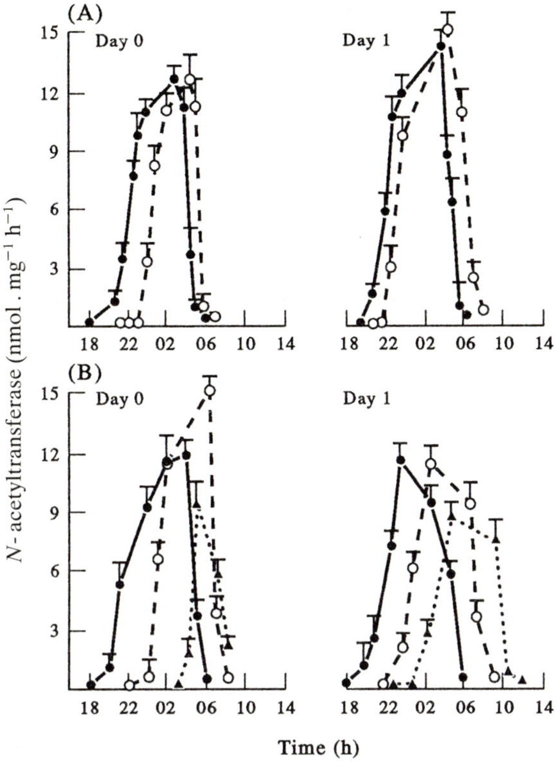

FIG. 1. Delay of the NAT rhythm. Rats maintained in LD 12:12, with lights on from 06.00 to 18.00 hours, were (A) exposed to a 1 minute light pulse at 21.00 hours (dashed line) or (B) subjected to a 4-hour (dashed line) or a 6-hour (dotted line) delay of the evening light offset. Thereafter, the rats were released into darkness and the NAT rhythm was followed during the same night (day 0) and during the next night (day 1). Full lines represent NAT rhythms in control rats undisturbed by light.

Phase advances

After a 1-minute light pulse in the late night (Figure 3A) or after a 3-hour and a 5-hour advance, respectively, of the morning light onset (Figure 3B), the next night only the morning NAT offset was phase advanced whereas the evening NAT onset was not phase shifted at all.[16,17] Consequently, the phase relationship between the onset and the offset became compressed. Apparently, the onset and the offset do not necessarily phase shift in parallel.

After an exposure of human subjects to bright light since 03.00 hours in the morning, the next day the serum melatonin offset was phase advanced by 1–3 hours (Figure 4).[5] In contrary to the rat NAT rhythm, the evening melatonin onset was eventually phase advanced as well though not in all

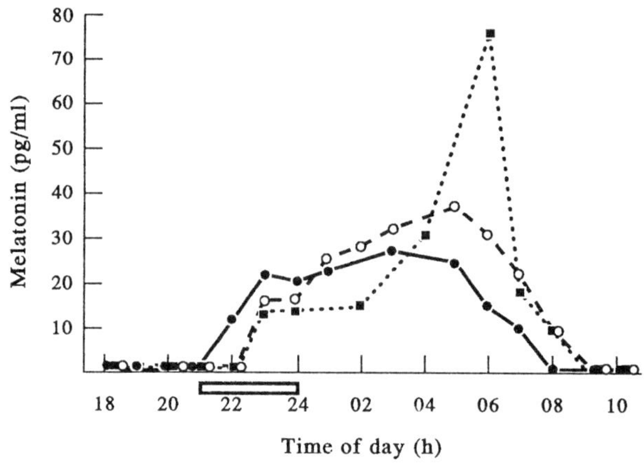

FIG. 2. Effect of a 3-hour bright light pulse in the early night on the saliva melatonin rhythm in a representative subject. Volunteers sampled saliva during the first, control night, when only dim light of an intensity lower than 50 lux was on (full line), during the second night, when they were exposed to bright light of intensity 2,500 lux from 21.00 to 24.00 hours (dotted line) and during the third night, when they again experienced dim light only (dashed line). Open bar under the abcissa denotes the light pulse.

subjects. It appears that even in humans the evening melatonin onset and the morning offset do not necessarily phase shift in parallel. After a 3-hour bright light pulse in the late night, the next day the morning saliva melatonin offset was phase advanced in four subjects out of four by 1–3 hours whereas the evening onset was phase advanced in one subject (Figure 5) (Samková and Illnerová, unpublished results). Chronic morning bright light treatment phase advances the onset of nighttime melatonin production in winter depressive patients and the whole melatonin rhythm pattern during the Antarctic winter.[4,25]

Effect of photoperiod

In the rat, the evening light phase delays the evening NAT onset primarily while the morning light phase advances the morning NAT offset primarily. Consequently, in long summer days the NAT rhythm waveform is compressed due to light intruding into the late evening and early morning hours; in short winter days, the waveform is decompressed.[14] Under all photoperiods equal to 16 hours or shorter, the morning NAT offset occurs about 1 hour before the light onset and is thus locked to the morning light.[11] It appears that the morning light is the entraining agent which synchronizes the rat circadian system with the 24-hour day. The evening light has a strong phase delaying effect only when the morning light is absent. Otherwise, the evening light may serve not as a true entraining agent but rather as a photoperiodic signal. After transition of rats from

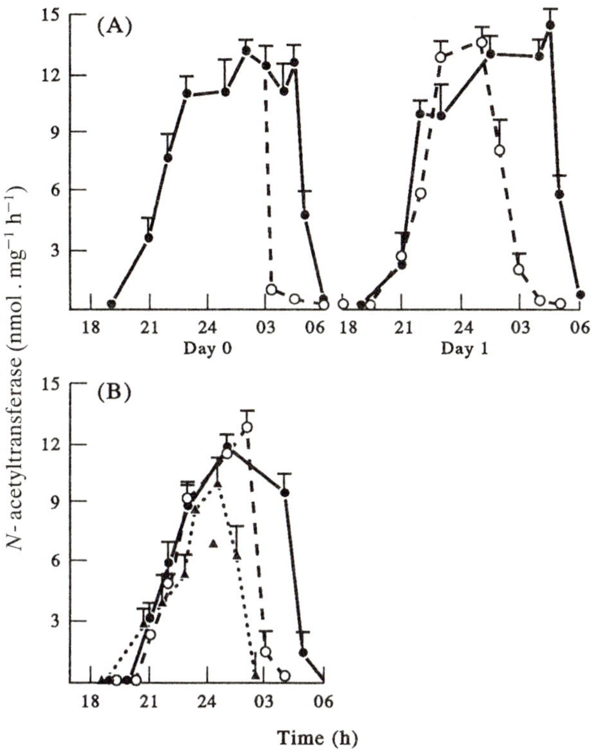

FIG. 3. Effect of a light exposure in the late night on the NAT rhythm. Rats maintained in LD 12:12, with lights on from 06.00 to 18.00 hours were (A) exposed to a 1 minute light pulse at 03.00 hours (dashed line), thereafter released into darkness and the NAT rhythm was followed during the same night (day 0) or the next night (day 1) or (B) subjected to a 3-hour advance (dashed line) and a 5-hour advance (dotted line) respectively, of the morning light onset, and the next day they were released into darkness already at 14.00 hours and the NAT rhythm was followed. Full lines represent NAT rhythms in control rats undisturbed by light.

long to short days, the NAT rhythm waveform extended gradually only (Figure 6).[13] When the dark period was prolonged symmetricaly, the extension proceeded mostly into the morning hours, so that the NAT decline occurred again just 1 hour before the morning light onset (Figure 6A).

In humans, no difference in the plasma melatonin signal duration between summer and winter was observed at 50°N and 35°S latitude, due apparently to shielding of human subjects from bright outdoor light.[19,21,29] However, at 60°N latitude in Sweden and at 68°S latitude in Antarctica, a slight difference between summer and winter was reported.[3,26] During winter months at 50°N, when natural daylight

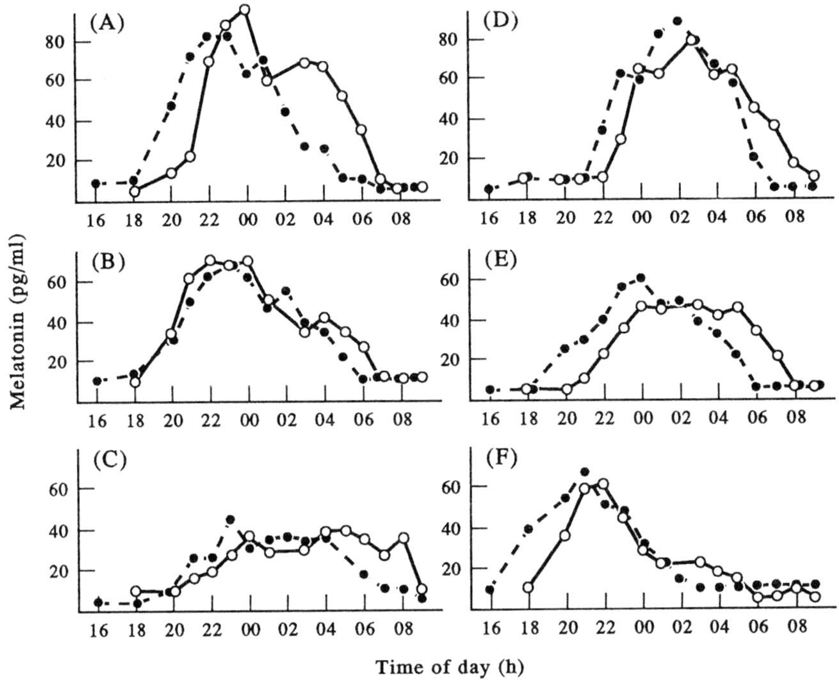

FIG. 4. Effect of a single morning bright light exposure on the human serum melatonin rhythm. Blood was sampled from six volunteers, A–F, during the first control night, when dim light only was on (full lines). During the second night, the subjects were exposed to bright light of intensity 2,000–3,000 lux from 03.00 to 09.00 hours in the morning. During the third night, the subjects again experienced dim light only and blood was sampled (dashed lines). After Ref. 5.

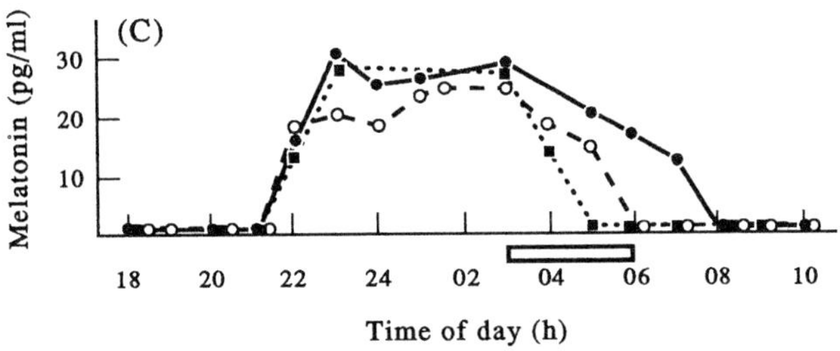

FIG. 5. Effect of a 3-hour bright light pulse in the late night on the saliva melatonin rhythm in a representative subject. Volunteers sampled saliva during the first, control night, when only dim light was on (full line), during the second night, when they were exposed to bright light from 03.00 to 06.00 hours (dotted line) and during the third night when they again experienced dim light only (dashed line). Open bar under the abcissa denotes the light pulse.

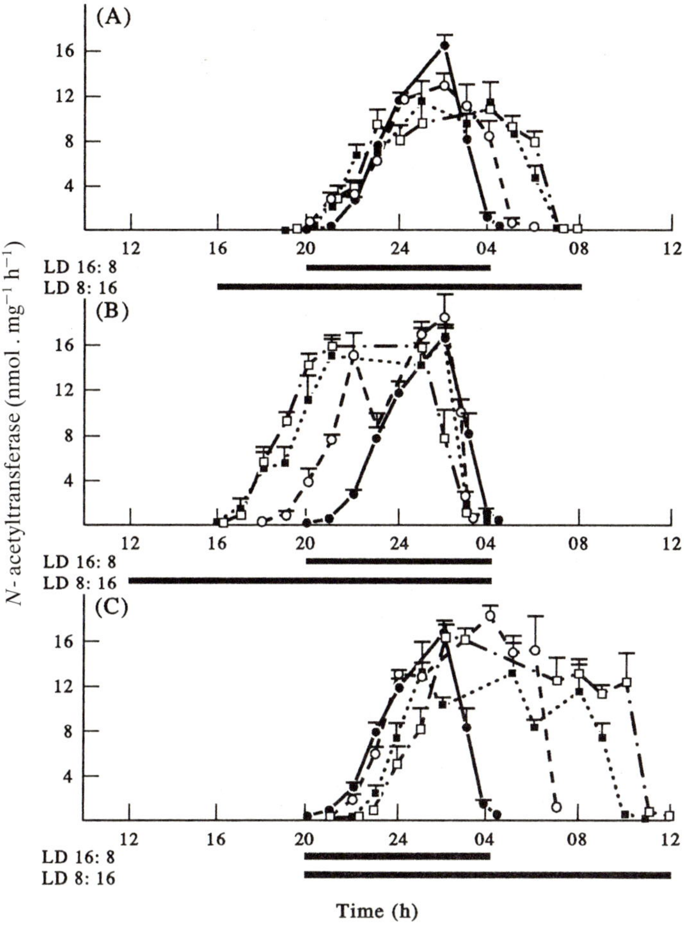

FIG. 6. Extension of the rat NAT rhythm after symmetrical prolongation of the dark period (A), after prolongation of the dark period into the evening hours (B), and after prolongation of the dark period into the morning hours (C). Rats maintained in LD 16:8 were either killed in LD 16:8 (filled circles) or were transferred to LD 8:16 and killed 3 (open circles), 6 (filled squares) and 13 (open squares) days later. Lines below the abcissa denote periods of darkness before (LD 16:8) and after (LD 8:16) transition. After Ref. 13.

resembled LD 8:16, the melatonin signal duration in subjects entrained to artificial long days simulated by the bright light LD 16:8 skeleton photoperiod was shorter by 2.5 hours than the signal in subjects experiencing just short winter days.[6] Hence the nocturnal melatonin signal duration may also provide human beings with the information on day length.

Four-day exposure to the long skeleton photoperiod during winter months, with bright light on from 19.00 to 22.00 hours in the evening and again from 06.00 to 09.00 hours in the morning phase advanced the morning serum melatonin offset in four subjects out of four (Illnerová, Burešová, Dvořáková and Zvolský, unpublished results). Following withdrawal of the bright light treatment, the phase advance of the morning melatonin offset persisted for three subsequent days and eventually the evening melatonin onset became phase advanced as well. In contrary to what happens with the rat NAT rhythm after a change from long to short days,[13] extension of the human melatonin rhythm waveform on short days did not proceed into the morning hours. The subjects were, however, woken each day at 06.00 hours in the morning. The forced sleep-wake schedule might thus maintain the circadian phase advance induced by the long skeleton photoperiod. The prevailing phase advancing effect of the long photoperiod is in agreement with the role the morning light may play in the human circadian system entrainment.[19,24,30] At 50°N and 35°S latitude, the melatonin rhythm in winter is phase shifted towards the morning hours as compared with the summer melatonin pattern, due probably to the fact that in winter humans may experience light of sufficiently strong intensity later in the morning than in summer.[19,21]

Lowering of amplitude

In rats exposed to a 7-hour advance and an 8-hour advance, respectively, of the morning light onset (Figure 7A) or in those exposed to a 4-hour light pulse around the middle of the night (Figure 7B), the next day the morning NAT offset was phase advanced and at the same time the evening onset was phase delayed.[16] The waveform of the NAT rhythm was compressed considerably and the amplitude decreased. Apparently, exposure to light around the middle of the night might have a dual effect on the NAT rhythm, i.e. a phase delaying one on the NAT onset and a phase advancing one on the NAT offset.[12] Eventually, the onset might occur close to or at the time of the offset and the NAT rhythm amplitude diminished.

In humans, a reduction of amplitudes of the plasma cortisol and body temperature rhythms was observed following a repeated exposure of subjects to a long bright light pulse around the time of the temperature minimum.[7,8] Our effort to diminish the melatonin rhythm amplitude in human volunteers by a single 3-hour or even a 7-hour exposure to bright light of intensity 2,500 lux has been unsuccessful till this time, due probably to inappropriate timing, low intensity of light; a single exposure instead of a repeated one or to combination of all the above mentioned reasons.

Conclusions

Following exposure to bright light in the early night, the next day the evening rat NAT and human melatonin onsets are phase delayed and,

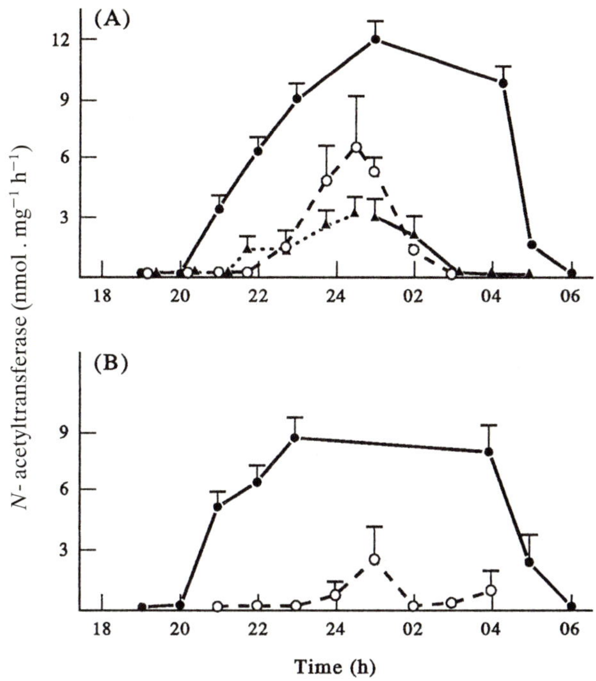

FIG. 7. Lowering of the NAT rhythm amplitude. Rats maintained in LD 12:12, with lights on from 06.00 to 18.00 hours were (A) subjected to a 7-hour advance (dashed line) or to an 8-hour advance (dotted line), respectively, of the morning light onset and the next day they were released into darkness already at 14.00 hours and the NAT rhythm was followed in the subsequent night, or (B) were exposed to a 4-hour light pulse from 23.00–03.00 hours, thereafter released into darkness and the NAT rhythm was followed the next night (dashed line). Full lines represent NAT rhythms of control rats undisturbed by light.

correspondingly, the offsets are phase delayed as well. Following exposure to bright light in the late night, the next day the morning rat NAT offset and the human melatonin offset are phase advanced; the evening rat NAT onset is not phase shifted whereas the human melatonin onset may or may not be phase advanced. Duration of the elevated nocturnal NAT level in rats and duration of the melatonin signal in humans depend on the photoperiod in both species. Following a change from a long to a short photoperiod, the rat NAT rhythm pattern extends mostly into the morning hours. In humans, a long photoperiod causes primarily a phase advance of the morning melatonin offset; such an advance may persist even in short days, due probably to a fixed waking schedule. In humans, similarly as in rats, the evening melatonin production onset does not necessarily phase shift in parallel with the morning offset. In summary, light entrainment of the human melatonin rhythm shows striking similarities with that of the rat pineal NAT rhythm, but some dissimilarities as well.

Acknowledgements

The research was partly supported by the Czechoslovak Academy of Sciences grant number 71118.

References

1. Beck-Friis J., Borg G., Mellgren T., Undén F. and Wetterberg L. (1986) Nocturnal serum melatonin levels following evening bright light exposure. *Clin. Neuropharmacol.* **9**, suppl. 4, 184–186.
2. Beck-Friis J., Borg G. and Wetterberg L. (1986) Rebound increase of nocturnal serum melatonin levels following evening suppression by bright light exposure in healthy men: relation to cortisol levels and morning exposure. *Annals N.Y. Acad. Sci.* **453**, 371–375.
3. Beck-Friis J., Van Rosen D., Kjellman B.F., Ljunggren J.G. and Wetterberg L. (1985) Melatonin in relation to body measures, sex, age, season and the use of drugs in patients with major affective disorders and healthy subjects. *Psychoneuroendocrinology* **10**, 173–183.
4. Broadway J., Arendt J. and Folkard S. (1987) Bright light phase shifts the human melatonin rhythm during the Antarctic winter. *Neurosci. Lett.* **79**, 185–189.
5. Burešová M., Dvořaková M., Zvolský P. and Illnerová H. (1991) Early morning bright light phase advances the human circadian pacemaker within one day. *Neurosci. Lett.* **121**, 47–50.
6. Burešová M., Dvořaková M., Zvolský P. and Illnerová H. (1992) Human circadian rhythm in serum melatonin in short winter days and in simulated artificial long days. *Neurosci. Lett.* **136**, 173–176.
7. Czeisler C.A., Allan J.S., Kronauer R.E., Duffy J.T. (1988) Strong circadian phase resetting in man is affected by bright light suppression of circadian amplitude. *Sleep Res.* **17**, 367.
8. Czeisler C.A., Kronauer R.E., Allan J.S., Duffy J.F., Jewett M.E., Brown E.N. and Ronda J.M. (1989) Bright light induction of strong (Type 0) resetting of the human circadian pacemaker. *Science* **244**, 1328–1333.
9. Deguchi T. and Axelrod J. (1972) Sensitive assay for serotonin N-acetyltransferase activity in the rat pineal. *Anal. Biochem.* **50**, 174–179.
10. Hoffmann K. (1981) The role of the pineal gland in the photoperiodic control of seasonal cycles in hamsters. In *Biological Clocks in Seasonal Reproductive Cycles* (eds. Follet B.K. and Follet D.E.), pp. 237–250. John Wright and Sons, Bristol.
11. Illnerová H. (1988) Entrainment of mammalian circadian rhythms in melatonin production by light. *Pineal Res.* **6**, 173–217.
12. Illnerová H. (1992) Resetting of the mammalian circadian clock through lowering of the amplitude: rat pineal *N*-acetyltransferase rhythm as a model. *Physiol. Res.* **41**, 335–344.
13. Illnerová H., Hoffmann K. and Vaněček J. (1986) Adjustment of the rat pineal *N*-acetyltransferase rhythm to change from long to short photoperiod depends on the direction of extension of the dark period. *Brain Res.* **362**, 403–408.
14. Illnerová H. and Vaněček J. (1980) Pineal rhythm in N-acetyltransferase activity in rats under different artificial photoperiods and in natural daylight in the course of a year. *Neuroendocrinology* **31**, 321–326.
15. Illnerová H., Vaněček J. (1982) Two-oscillator structure of the pacemaker controlling the circadian rhythm of N-acetyltransferase in the rat pineal gland. *J. Comp. Physiol.* **145**, 538–548.
16. Illnerová H. and Vaněček J. (1987) Entrainment of the circadian rhythm in the rat pineal N-acetyltransferase activity by prolonged periods of light. *J. Comp. Physiol. A* **161**, 495–510.
17. Illnerová H. and Vaněček J. (1987) Dynamics of discrete entrainment of the circadian

rhythm in the rat pineal N-acetyltransferase activity during transient cycles. *J. Biol. Rhythms* **2**, 95–108.
18. Illnerová H., Vaněček J. and Hoffmann K. (1983) Regulation of the pineal melatonin concentration in the rat (*Rattus norvegicus*) and in the Djungarian hamster (*Phodopus sungorus*). *Comp. Biochem. Physiol.* **74**, 155–159.
19. Illnerová H., Zvolský P. and Vaněček J. (1985) The circadian rhythm in plasma melatonin concentration of the urbanized man: the effect of summer and winter time. *Brain Res.* **328**, 186–189.
20. Kennaway D.J., Earl C.R., Shaw P.F., Royles P., Carbone F. and Webb H. (1987) Phase delay of the rhythm of 6-sulphatoxymelatonin excretion by artificial light. *J. Pineal Res.* **4**, 315–320.
21. Kennaway D.J. and Royles P. (1986) Circadian rhythms of 6-sulphatoxymelatonin, cortisol and electrolyte excretion at the summer and winter solstices in normal men and women. *Acta Endocrinol.* (*Copenh.*) **113**, 450–456.
22. Klein D.C. and Moore R.J. (1979) Pineal N-acetyltransferase and hydroxyindole-O-methyltransferase: Control by the retinal hypothalamic tract and the suprachiasmatic nucleus. *Brain Res.* **174**, 245–262.
23. Klein D.C. and Weller J.L. (1970) Indole metabolism in the pineal gland. A circadian rhythm in N-acetyltransferase activity. *Science* **169**, 1093–1095.
24. Lewy A.J. (1985) Regulation of melatonin production in humans by bright artificial light: evidence for a clock-gate model and a phase response curve. In *The Pineal Gland. Endocrine Aspects* (eds. Brown G.M. and Wainwright S.D.), Vol. 53, pp. 203–208. Pergamon Press, New York.
25. Lewy A.J., Sack R.L., Miller L.S. and Hoban T.M. (1987) Antidepressant and circadian phase-shifting effects of light. *Science* **325**, 352–354.
26. Makkison I. and Arendt J. (1991) Melatonin secretion in humans on two different antarctic bases (68° and 75°S). *J. Interdisciplinary Cycle Res.* **22**, 149–150.
27. Nowak R., McMillen I.C., Redman J. and Short R.V. (1987) The correlation between serum and salivary melatonin concentrations and urinary 6-hydroxymelatonin sulphate excretion rates: two noninvasive techniques for monitoring human circadian rhythmicity. *Clin. Endocrinol.* **27**, 445–452.
28. Redman J., Armstrong S., Ng K.T. (1983) Free-running activity rhythms in the rat: Entrainment by melatonin. *Science* **219**, 1089–1091.
29. Roenneberg R. and Aschoff J. (1990) Annual rhythm of human reproduction. II. Environmental correlations. *J. Biol. Rhythms* **5**, 217–239.
30. Terman M., Quitkin F.M., Terman J.S., Stewart J.W. and McGrath P.J. (1987) The timing of phototherapy: effects on clinical response and the melatonin cycle. *Psychopharmacol. Bul.* **23**, 354–357.
31. Webley G.E., Mehl H. and Willey K.P. (1985) Validation of a sensitive direct assay for melatonin for investigation of circadian rhythms in different species. *J. Endocrinol.* **106**, 387–394.

12

The Use of Melatonin as a Marker for Circadian Phase and as a Chronobiotic in Blind and Sighted Humans

A. J. LEWY and R. L. SACK

Sleep and Mood Disorders Laboratory, Departments of Psychiatry, Ophthalmology and Pharmacology, Oregon Health Sciences University, Portland, Oregon, 97201, USA

Melatonin assay

MELATONIN IS produced only at night in the dark. Levels of plasma melatonin are quite low during the day (1–5 pg/ml) and increase to about 50–100 pg/ml at night. In order to accurately measure these low levels, we developed the gas chromatographic–negative chemical ionization mass spectrometric (GCMS) assay.[17] The GCMS assay continues to be the "gold standard", because of its sensitivity, specificity and precision. It has helped improve radioimmunoassays (RIAs) over the years, directly through cross-validation and indirectly through the publication of accurate plasma melatonin values. The GCMS assay's future utility lies in a continued role for on-going validation of RIAs established in different laboratories when there is a change in tissue or species, when there is a change in experimental design, or when an unexpected seemingly spurious result needs to be verified. At present, we routinely use this assay for high resolution determination in plasma samples of the endogenous melatonin production circadian rhythm; the GCMS assay also has great potential for the routine determination of the melatonin rhythm in saliva, in which melatonin levels are about one-third those of plasma.[53]

The GCMS assay has been used in the discovery of important findings, such as suppression of human melatonin production by bright light.[30] One implication of this finding was that humans might have circadian and seasonal rhythms that respond to the natural (bright) light–dark cycle and the natural photoperiod, respectively, and that these biological rhythms are relatively unperturbed by the use of ordinary-intensity room light. Prior to this discovery, scientists thought that social cues were the major zeitgeber (time cue) for humans.[54]

Melatonin circadian rhythms in blind people

When we realized that light might be an important zeitgeber for humans, the question naturally arose, what were the consequences for totally blind individuals who had absolutely no light perception and thus could not perceive the light–dark cycle. Initially, we identified 10 individuals who were each studied on one occasion.[18] All of them had melatonin patterns with an approximately 12-hour active phase and 12-hour quiescent phase. While some of these melatonin patterns were in a normal phase relationship with the light–dark cycle, most were markedly phase advanced or phase delayed.

Two of the abnormally phased individuals were studied longitudinally on a weekly basis for 4 consecutive weeks. One of them appeared to be stably entrained, albeit at an abnormal phase. The other individual was shown to be free-running, with an intrinsic period (tau) of about 24.7 hours.

The melatonin onset

It is not necessary to measure the entire nighttime profile when using melatonin to assess circadian phase. Theoretically, any part of the curve can be used to assess phase, as long as the amplitude and the duration (time interval between the onset and the offset) are constant. This appears to be the case in blind people. For most of our studies, the melatonin onset is used because it is the most clearly demarcated part of the melatonin curve and most accurately reflects the phase and period of the melatonin rhythm in blind people.

We prefer the melatonin onset for both theoretical and practical reasons. Theoretically, the melatonin onset is that part of the melatonin curve which is least affected by biochemical factors that could confound the use of melatonin levels as a marker for circadian phase. One such factor is increasing subsensitivity of pineal beta-adrenergic receptors during the night.[41] Furthermore, melatonin levels could also decrease during the night because synthesis precursors become depleted. Consequently, as the night progresses, melatonin levels are marking biochemical events as well as the circadian signal from the endogenous pacemaker (the suprachiasmatic nuclei, or SCN, located in the hypothalamus).

The dim light melatonin onset (DLMO)

In sighted people, the melatonin onset is also the preferred marker for assessment of circadian phase. To avoid suppression of melatonin, it is critical to draw blood under dim light (<30–50 lux), beginning 1 hour before the first draw. We call this measure the dim light melatonin onset, or DLMO. Blood is drawn in the evening as frequently as resolution requires, usually every 30 minutes between 18.00 and 23.00, although sometimes we extend this interval by an hour in either direction.

From our studies of blind people, we know that sleep propensity is phase-locked with the melatonin onset, consistently lagging behind it by 2–3 hours.[34] We also know that when the melatonin onset is within a few hours of sleep onset, sleep tends to be phase-locked with the melatonin onset; that is, they appear to be relatively coordinated.[45] These findings do not necessarily mean that the melatonin rhythm induces sleep, but rather that sleep is strongly influenced by a (free-running) endogenous circadian pacemaker which is marked by the melatonin onset.

Ruling out non-photic effects on the circadian system of humans

In a subsequent study of 11 blind people, 10 had taus greater than 24 hours that ranged between 24.28 and 25.08.[45] One person had a tau less than 24 hours (23.88). This person happened to be the oldest person of this group, which is interesting in that some investigators believe that tau decreases with age.

If tau does change with age, it does so very slowly, because intra-individual variability in tau is small. For example, two free-running blind people studied on a second occasion after an interval of 1 or 2 years had taus unchanged from when they were first studied. An individual who was studied for more than 6 months had a tau of 24.70 with a standard deviation less than the sampling frequency of 1 hour: the regression fit was highly linear ($r = 0.95$, $p < 0.001$). Thus, although inter-individual variability is marked, intra-individual variability is small. Therefore, the study of blind people provides a means for testing non-photic influences on the circadian system. Incidentally, the melatonin onset is phase-locked with the cortisol circadian rhythm and has the same tau.[44]

The marked inter-individual variability in the taus of free-running blind people may explain why some people are "early birds", while others are "night owls". Sighted people are similar, in that DLMO intra-individual variability is small, yet among individuals DLMOs can range between 19.00 and 24.00. Whereas part of this variability is explained by differences in light exposure and sleep times (which can superimpose some structure upon the light-dark cycle), as with blind individuals, some of this variability is intrinsic, in that a person with a greater tau is likely to have a more phase-delayed DLMO when entrained.

Although most totally blind people appear to have free-running circadian rhythms, a few may not be free-running. Some of these people may be entrained to a normal phase position, while others may be entrained to an abnormal phase, and still others may have unstable circadian rhythms. An on-going study in our laboratory is to determine if there actually are blind individuals who appear to be (stably or unstably) entrained. Whereas it is straight-forward if an individual is free-running, stable entrainment is more difficult to demonstrate. One must do studies over several weeks or months to detect a change in circadian phase position if the tau is close to 24 hours. Alternatively, a circadian rhythm with a tau much greater than 24 hours would appear to be entrained every few weeks; these individuals would need to be studied at frequent intervals.

These studies are very important. If all totally blind people are free-running, it is unlikely that activity, food or even non-ocular light perception can affect the circadian system in humans. There is a report of a presumably entrained almost totally blind person; however, this person could suppress melatonin production in response to 6,300 to 8,400 lux light exposure.[32] Also, upon further evaluation of this person's light response, it turned out that some response could be detected on electroretinographic testing. We know of one totally blind person who continues to have a free-running melatonin rhythm, despite daily and vigorous use of a stationary bicycle. We also know of a totally blind shift worker whose circadian rhythms were unaffected after switching from one work shift to another.

Melatonin: from a marker to a chronobiotic

When we started working with melatonin, we regarded it primarily as a biological marker. However, in our efforts to develop ways in which to entrain blind people, we decided to investigate the potential zeitgeber properties of melatonin. We were encouraged in this regard by animal studies.

In animals, the function of the pineal gland and its principle hormone, melatonin, was first associated with an antigonadal effect; thus, melatonin was originally thought to be a reproductive hormone.[40] Then it was discovered that melatonin regulated the seasonal rhythm of estrus as well as other seasonal rhythms. That is, melatonin could be either antigonadal or progonadal, depending on whether the animal was a spring, or autumn breeder, respectively. Consequently, melatonin came to be associated with seasonal rhythms, although there continues to be controversy as to whether it is the duration or the phase of nighttime melatonin secretion that is responsible for communicating the time of the year to animals with seasonal rhythms.[12,51]

A circadian function for the pineal was first shown in birds.[11] In mammals, however, the effects of pinealectomy on circadian rhythms was not robust.[38] Then, in 1983, Redman and Armstrong showed that melatonin could entrain the circadian activity-rest cycle in the rat.[39]

Melatonin and jet lag

Human melatonin-administration studies began in the early 1980s, when melatonin was used by certain peripatetic scientists for self-treatment of jet lag; eventually, a few controlled studies were done.[2,36] However, melatonin's role in helping to synchronize circadian rhythms was not clear.

In these studies, there was some concern that melatonin was helping air travelers primarily through a sedative-hypnotic effect and not through a direct effect on circadian rhythms. Indeed, at the pharmacological doses taken (at least 2 mg po), melatonin can have a soporific effect, and sleep loss is definitely a part of jet lag. Moreover, melatonin can indirectly aid in entraining circadian rhythms, because sleep superimposes dark on the light–dark cycle. Thus, it was thought that melatonin might not have a true zeitgeber effect, at least with respect to most endogenous circadian rhythms. In one further jet lag study,[1] subjects slept better after bedtime melatonin administration. Effects on endogenous melatonin were difficult to interpret, mainly because melatonin data from subjects who did not subjectively improve were eliminated from the analysis.

Phase-advancing effects of melatonin in sighted people

Whether or not melatonin administered in this way was directly resynchronizing circadian rhythms, the use of melatonin in treating jet lag encouraged researchers to test the phase-shifting effects of melatonin. This was first done by Arendt in 1985,[4,56] who gave melatonin at 17.00 hours in order to advance the time when the melatonin signal occurred, yet still be continuous with the endogenous nighttime bout: the results, however, were neither robust nor consistent. Of 12 subjects given melatonin for 1 month in the spring, only two showed phase advances in the endogenous melatonin rhythm. Exogenous melatonin was not administered on the last (sampling) day of the study. This study was repeated in the autumn, in which exogenous melatonin was administered for 3 weeks, including the last (sampling) day. Five out of 11 subjects demonstrated phase advances. Partly because the post-treatment endogenous melatonin levels in most of the remaining subjects could not be discerned from the exogenous levels, it was not clear how many, if any, of the other subjects phase advanced.

The second group to investigate the phase-shifting effects of exogenous melatonin on the endogenous melatonin rhythm was the Claustrat group.[31] There was no effect immediately after administering melatonin (8 mg po) at 22.00, although there was a phase advance 3 days after stopping melatonin treatment. However, the cortisol rhythm did not shift at all.

Finally, Arendt and Wever studied two individuals under free-running conditions and failed to show any effect of melatonin on entrainment.[4,55] In this study, melatonin was given at bedtime during a protocol in which the light–dark cycle was increasingly lengthened. Melatonin had no effect on any of the circadian rhythms measured, except for those of fatigue and alertness. It is no wonder, then, that some of these researchers[4,55] were drawing quite conservative conclusions about the zeitgeber properties of melatonin from their studies. (Despite the negative findings with respect to endogenous physiological rhythms, melatonin did appear to have a consistent effect on sleep.)[4]

Phase-advancing effects of melatonin in blind people

In 1985–1986, we began to study the effects of melatonin (5 mg po) in free-running blind people.[43] We thought that more robust and consistent effects could be demonstrated in individuals who were not influenced by the light–dark cycle. Indeed, we were eventually able to show phase-advancing effects of about 30 minutes per day; both the melatonin onset and the cortisol nadir were phase-shifted to the same extent following exogenous melatonin administration.[46]

Melatonin administration research thus began to focus on blind people. Other groups then began to give melatonin to blind people,[3,35,48] showing at least some effect on the sleep-wake cycle. However, when a physiological marker such as melatonin production was assessed, it was not clear if phase shifts actually occurred.[3,35] This lack of an effect was probably due to the fact that urinary melatonin production was measured, which is not very sensitive to small phase shifts. Consequently, we thought that part of our ability to show robust and consistent phase shifts in blind people was the use of a very sensitive marker for circadian phase, the melatonin onset.

The DLMO and the phase response curve (PRC)

In sighted individuals, the DLMO is also a useful marker for circadian phase. It has been used, for example, to phase type individuals before instituting light therapy.[25] People with phase-advanced DLMOs require a circadian phase delay and therefore are treated with bright light exposure scheduled in the evening. People with phase-delayed DLMOs require a phase advance and are treated with morning bright light. The DLMO not only discriminates between different phase types but also indicates the phase-shifting response to light treatment. Interestingly, the baseline DLMO predicts phase-shifting responses; individuals who are relatively phase delayed at baseline have more of a phase-advance response to morning light and have less of a phase-delay response to evening light; individuals who are phase advanced have more of a phase-delay response to evening light and have less of a phase-advance response to morning light.[19,29] These findings suggest that the DLMO is marking the phase of its phase response curve (PRC) and imply that the DLMO is marking the phase of its underlying circadian pacemaker.

Perhaps a comment would be appropriate here regarding a recent paper published by Shanahan and Czeisler.[50] Although we agree with these investigators that the melatonin circadian rhythm correlates well with the melatonin offset as well as with the melatonin maximum,[49] we do not agree with their statement that the melatonin onset is not as reliable as the fitted melatonin maximum for marking circadian phase shifts, because the high degree of noise in their melatonin RIA appears to have confounded measurement of the low daytime levels needed to determine the melatonin onset. This is not a problem when the melatonin onset is measured by the GCMS assay or by several other RIAs currently in use.

In 1983, we hypothesized a human PRC to bright light,[20] based on the common features of PRCs observed in a variety of animals.[9,37] Essentially, there is a zone of reduced responses during the day. Phase-delay responses occur in the evening and increase in magnitude during the middle of the night; phase advance responses occur in the morning and also increase in magnitude during the middle of the night. We subsequently published data consistent with such a PRC.[25–27] Recently, complete PRCs for light have been described by four groups.[7,13,33,55] One of these PRCs was determined under free-running conditions.[33]

In order not to interfere with sleep, the usual times for scheduling light treatment are in the evening ending about 1 hour before bedtime, and in the morning immediately upon awakening. The phase-shifting responses to bright light have been applied in the treatment of advanced and delayed sleep phase syndrome,[20,28] winter depression (thought to be a phase-delay type of disorder),[25,47] jet lag[8] and maladaptation to shift work.[6,10] Bright light exposure is also being investigated in the treatment of non-seasonal depression[15] and in a few other psychiatric disorders.

The human melatonin PRC

Using the highly resolved DLMO (which is sensitive to small phase shifts), we were able to describe a PRC for melatonin administration in humans.[16,21–24] We administered 0.5 mg of melatonin for 4 days in a placebo-controlled study. For the first week of each trial, subjects received a (cornstarch) placebo. The second week subjects received placebo for 2 days and melatonin for 4 days. There was a DLMO determined at the end of each week.

Ten subjects participated in 30 trials. Subjects were asked to sleep at the same time of week two as that same day of week one. Only one subject failed to keep average sleep time of week two close to (within 1 hour) week one's sleep time. This subject was eliminated from the study.

Most subjects received their medication in two divided doses 1 hour apart. The protocol was modified, however, when capsules were scheduled during sleep for three subjects who received the entire dose in one capsule. Furthermore, placebo administration was eliminated for these trials in order to minimize interference with sleep.

The first DLMO was the baseline DLMO. In addition to serving as a reference point for the post-treatment phase shift, the baseline DLMO was used to demarcate circadian time. For example, if an individual's baseline DLMO was at 21.00 and melatonin was administered at 18.00, then the circadian time of administration was CT 11. This is because, on average, the melatonin onset is usually 14 hours after sleep offset (or "lights on") which in diurnal animals is traditionally designated as CT 0. If another individual's baseline DLMO is at 19.00, then a clock time administration of melatonin at 18.00 converts to a circadian time of CT 13. Between individuals, baseline DLMOs can be as early as 18.00 and as late as midnight. (Within individuals, baseline DLMOs are quite consistent.)

The results indicate a PRC similar in shape to PRCs described for various stimuli in other species. In phase, however, the melatonin PRC is about 12 hours shifted from the PRCs for light. The melatonin PRC has the same phase as a dark-pulse PRC.[5] Indeed, this should not be surprising, since melatonin is produced at night (in the dark) in almost all species. Interestingly, the melatonin PRC more resembles that of the lizard[52] than that of the rat.[39] In the latter species, there is no delay zone and a very limited zone of phase-advance responses (between CT 10 and 12) with no apparent relationship between time of administration and magnitude of phase shift.

The human melatonin PRC appears to have a "dead zone" of reduced responses between CT 14 and CT 20. Between CT 20 and CT 5, there are delay responses but no advance responses. Between CT 5 and CT 14, there are advance responses but no delay responses. Around CT 4–5 there is a cross-over point.

The melatonin PRC explains why previous investigators did not find more consistent or robust results. Melatonin was not administered at the optimal time to induce robust phase advances.[4,31] In the case where phase-delay responses might have been effective, melatonin was not administered at the correct time.[4,55] It is not surprising, then, that some of these investigators were beginning to doubt that melatonin could cause phase shifts in endogenous circadian rhythms.[4,55]

At the very least, our data provide clear evidence for a phase-shifting effect of melatonin. For phase advances, there is a clear linear relationship between the magnitude of the phase shift and the time of administration. There also appears to be a delay zone, although because of the difficulties administering melatonin during sleep, there were fewer trials conducted in this zone. Therefore, we cannot as yet clearly describe the precise shape of the delay zone or where the cross-over point is precisely located.

It is significant that we used a physiological dose of melatonin in this study. The 0.5 mg dose produces levels no greater than 200–300 pg/ml about 1–2 two hours after ingestion. Its half-life is around 40 minutes: exogenous melatonin is removed from the circulation within a few hours. Consequently, exogenous melatonin did not interfere with the next day's DLMO determination. For the modified protocol, the DLMO was determined on the evening of the same day of administration, again without being confounded by exogenous melatonin.

The function of endogenous melatonin production in humans

The use of a physiological dose is also significant because it strengthens the argument that the melatonin PRC suggests a function for endogenous melatonin in humans: melatonin could act to augment entrainment of the endogenous circadian pacemaker by the light–dark cycle. This role of endogenous melatonin is related to the suppressant effect of light. Although it has been argued in some species that the apparent suppression of melatonin by light may in fact be due to instantaneous shifts in either an onset oscillator or an offset oscillator,[14] humans appear to have a separate suppressant effect of light on melatonin production. The suppressant effect of light is instantaneous. On the other hand, shifts of the pacemaker, or at least of its driven rhythms, often require transients; that is, several days may be required before steady-state phase relationships are resumed. (If the pacemaker shifts instantaneously, melatonin probably shifts the driven rhythms that require transients.)

The suppressant effect of light and endogenous melatonin work together in the following way. Suppression of melatonin at the twilight transitions prevents its stimulation of specific portions of the melatonin PRC. For example, while light is stimulating the advance portion of the light PRC as occurs with an earlier dawn, the melatonin offset occurs earlier that day, before the pacemaker that drives the melatonin rhythm has been advanced. Suppression of melatonin in the morning decreases stimulation of the delay portion of the melatonin PRC, thus enhancing advancement of the pacemaker.

After a few days, steady-state entrainment ensues. Thus, augmentation of phase-shifting is probably most important after shifts in the light–dark cycle. Alterations in endogenous melatonin levels may also be more important for changes in the timing of dawn than for dusk, and there may be some differences between advances and delays.

The melatonin PRC is the basis for treating both phase-advance and phase-delay disorders with administration of exogenous melatonin. The melatonin PRC (and the suppressant effect of light) also provide the basis for hypothesizing a function for endogenous melatonin in humans. The melatonin PRC makes clear that time of administration must be taken into account. Whether or not knowledge of a person's baseline DLMO will be important remains to be determined. It is possible that sleep time (most likely sleep offset time) will be a fairly useful marker for circadian time.

Summary

Melatonin has had a very significant role in our understanding of human circadian physiology. The discovery of suppression of melatonin by bright light changed the way in which we think about the regulation of human circadian rhythms. Melatonin may also be important as part of a feed-back loop in this system.

A role for melatonin in humans suggests that bright light treatment of circadian phase disorders might at least partially (particularly initially) shift circadian rhythms through melatonin suppression. For example, it is hoped that the study of beta-blockers in winter depressives[42] will be repeated, this time selectively blocking endogenous melatonin production at a time predicted by the melatonin PRC so as to cause a phase advance.

In conclusion, appropriately timed melatonin administration may be useful in the treatment of circadian phase disorders, alone or in conjunction with appropriately timed bright light exposure. Determining the DLMO may be important in the optimal scheduling of light treatment or melatonin therapy. Melatonin, therefore, continues to be important for assessing as well as treating circadian phase disorders, such as advanced and delayed sleep phase syndrome, winter depression and maladaptation to shift work and transmeridional air travel.

References

1. Arendt J., Aldhous M., English J., Marks V., Arendt J.H., Marks M. and Folkard S. (1987) Some effects of jet-lag and their alleviation by melatonin. *Ergonomics* **30**, 1379–1393.
2. Arendt J., Aldhous M. and Marks V. (1986) Alleviation of jet lag by melatonin: Preliminary results of controlled double blind trial. *Br. Med. J.* **292**, 1170.

3. Arendt J., Aldhous M. and Wright J. (1988) Synchronisation of a disturbed sleep-wake cycle in a blind man by melatonin treatment. *Lancet* **i**, 772–773.
4. Arendt J., Bojkowski C., Folkard S., Franey C., Marks V., Minors D., Waterhouse J., Wever R.A., Wildgruber C. and Wright, J. (1985) Some effects of melatonin and the control of its secretion in humans. In *Photoperiodism, Melatonin and the Pineal* (eds. Evered D. and Clark S.), pp. 266–283. Pitman, London.
5. Boulos Z. and Rusak B. (1982) Circadian phase response curves for dark pulses in the hamster. *J. Comp. Physiol.* **146**, 411–417.
6. Czeisler C.A., Johnson M.P., Duffy J.F., Brown E.N., Ronda J.M. and Kronauer R.E. (1990) Exposure to bright light and darkness to treat physiologic maladaptation to night work. *N. Engl. J. Med.* **322**, 1253–1259.
7. Czeisler C.A., Kronauer R.E., Allan J.S., Duffy J.F., Jewett M.E., Brown E.N. and Ronda J.M. (1989) Bright light induction of strong (Type 0) resetting of the human circadian pacemaker. *Science* **244**, 1328–1332.
8. Daan S. and Lewy A.J. (1984) Scheduled exposure to daylight: A potential strategy to reduce "jet lag" following transmeridian flight. *Psychopharmacol. Bull.* **20**, 566–568.
9. DeCoursey P.J. (1960) Daily light sensitivity rhythm in a rodent. *Science* **131**, 33–35.
10. Eastman C.I. (1986) Bright light in work-sleep schedules for shift workers: Application of circadian rhythm principles. In: *Temporal Disorder in Human Oscillatory Systems* (eds. Rensing U., van der Heiden U. and Mackey M.C.), pp. 176–185. Springer, Heidelberg.
11. Gaston S. and Menaker M. (1968) Pineal Function: The biological clock in the sparrow? *Science* **160**, 1125–1127.
12. Goldman B. and Darrow J. (1983) The pineal gland and mammalian photoperiodism. *Neuroendocrin.* **37**, 386–396.
13. Honma K. and Honma S. (1988) A human phase response curve for bright light pulses. *Jpn. J. Psych. Neurol.* **42**, 167–168.
14. Illnerova H. and Vanecek J. (1982) Two-oscillator structure of the pacemaker controlling the circadian rhythm of *N*-acetyltransferase in the rat pineal gland. *J. Comp Physiol.* **145**, 539–548.
15. Kripke D.F., Mullaney D.J., Savides T.J. and Gillin J.C. (1989) Phototherapy for nonseasonal major depressive disorders. In *Seasonal Affective Disorders and Phototherapy* (eds. Rosenthal N.E. and Blehar M.C.), pp. 167–229. Guildford Press, New York N.Y.
16. Lewy A.J., Ahmed S., Jackson J.M.L. and Sack R.L. (1992) Melatonin shifts human circadian rhythms according to a phase-response curve. *Chronobiol. Int.* **9**, 380–392.
17. Lewy A.J. and Markey S.P. (1978) Analysis of melatonin in human plasma by gas chromatography negative chemical ionization mass spectrometry. *Science* **201**, 741–743.
18. Lewy A.J. and Newsome D.A. (1983) Different types of melatonin circadian secretory rhythms in some blind subjects. *J. Clin. Endocrinol. Metab.* **56**, 1103–1107.
19. Lewy A.J. and Sack R.L. (1989) The dim light melatonin onset (DLMO) as a marker for circadian phase position. *Chronobiol. Int.* **6**, 93–102.
20. Lewy A.J., Sack R.L., Fredrickson R.H., Reaves M., Denney D.D. and Zielske D.R. (1983) The use of bright light in the treatment of chronobiologic sleep and mood disorders: The phase-response curve. *Psychopharmacol. Bull.* **19**, 523–525.
21. Lewy A.J., Sack R.L. and Latham J. (1991) A phase response curve for melatonin administration in humans. *Sleep Res.* **20**, 461.
22. Lewy A.J., Sack R.L. and Latham J.M. (1990) Circadian phase shifting of blind and sighted people with exogenous melatonin administration: Evidence for a phase response curve. *Soc. Light Treatment Biol. Rhythms Abst.* **2**, 22.
23. Lewy A.J., Sack R.L. and Latham J.M. Exogenous melatonin administration shifts circadian rhythms according to a phase response curve. Vth Colloquium of the European Pineal Study Group, Guildford, Surrey, UK, 1990.
24. Lewy A.J., Sack R.L. and Latham J.M. (1991) Melatonin and the acute suppressant effect of light may help regulate circadian rhythms in humans. In *Advances in Pineal Research:* 5 (eds. Arendt J. and Pevet P.), pp. 285–293. John Libbey, London.

25. Lewy A.J., Sack R.L., Miller L.S. and Hoban T.M. (1987) Antidepressant and circadian phase-shifting effects of light. *Science* **235**, 352–354.
26. Lewy A.J., Sack R.L. and Singer C.L. (1984) Assessment and treatment of chronobiologic disorders using plasma melatonin levels and bright light exposure: The clock-gate model and the phase response curve. *Psychopharmacol. Bull.* **20**, 561–565.
27. Lewy A.J., Sack R.L. and Singer C.M. (1985) Immediate and delayed effects of bright light on human melatonin production: Shifting "dawn" and "dusk" shifts the dim light melatonin onset (DLMO). *Ann. NY Acad. Sci.* **453**, 253–259.
28. Lewy A.J., Sack R.L. and Singer C.M. (1985) Melatonin, light and chronobiological disorders. In: *Photoperiodism, Melatonin and the Pineal* (eds. Evered D. and Clark S.), pp. 231–252. Pitman, London.
29. Lewy A.J., Sack R.L., Singer C.M., White D.M. and Hoban T.M. (1988) Winter depression and the phase shift hypothesis for bright light's therapeutic effects: History, theory and experimental evidence. *J. Biol. Rhythms* **3**, 121–134.
30. Lewy A.J., Wehr T.A., Goodwin F.K., Newsome DA and Markey SP. (1980) Light suppresses melatonin secretion in humans. *Science.* **210**, 1267–1269.
31. Mallo C., Zaidan R., Faure A., Brun J., Chazot G. and Claustrat B. (1988) Effects of a four-day melatonin treatment on the 24 h plasma melatonin, cortisol and prolactin profiles in humans. *Acta Endocrinol.* **119**, 474–480.
32. Martens H., Klein T., Rizzo III J.R., Shanahan T.L. and Czeisler C.A. Light-induced melatonin suppression in a blind man. *Soc. Res. Biol. Rhythms*, Amelia Island, Florida, 1992.
33. Minors D.S., Waterhouse J.M. and Wirz-Justice A. (1991) A human phase-response curve to light. *Neurosci. Lett.* **133**, 36–40.
34. Nakagawa H., Sack R.L. and Lewy A.J. (1991) Sleep propensity free-running in parallel with the melatonin rhythm in a totally blind person. *Sleep Res.* **20**, 467.
35. Palm L., Blennow G. and Wetterberg L. (1990) Correction of non-24-hour sleep-wake cycle by melatonin in a blind retarded boy. *J. Neural Transm.* **81**, 17–29.
36. Petrie K., Conaglen J.V., Thompson L. and Chamberlain K. (1989) Effect of melatonin on jet lag after long haul flights. *Br. Med. J.* **298**, 705–707.
37. Pittendrigh C.S. and Daan S. (1976) A functional analysis of circadian pacemakers in noctural rodents. IV. Entrainment: Pacemaker as clock. *J. Comp. Physiol.* **106**, 291–331.
38. Quay W.B. (1968) Individuation and lack of pineal effect in the rat's circadian locomotor rhythm. *Physiol. Behav.* **3**, 109–118.
39. Redman J., Armstrong S. and Ng K.T. (1983) Free-running activity rhythms in the rat: Entrainment by melatonin. *Science* **219**, 1089–1091.
40. Reiter R. (1982) The pineal and its hormones in the control of reproduction. *Endocrine Rev.* **1**, 109–131.
41. Romero J.A., Zatz M., Kebabian J.W. and Axelrod J. (1975) Circadian cycles in binding of ^{3}H-alprenolol to beta-adrenergic receptor sites in rat pineal. *Nature* **258**, 435–436.
42. Rosenthal N.E., Jacobsen F.M., Sack D.A., Arendt J., James S.P., Parry B.L. and Wehr T.A. (1988) Atenolol in seasonal affective disorder: A test of the melatonin hypothesis. *Arch. Gen. Psych.* 145, 52–56.
43. Sack R., Lewy A. and Hoban T. (1987) Free-running melatonin rhythm in blind people: Phase shifts with melatonin and Triazolam administration. In *Temporal Disorder in Human Oscillatory Systems* (eds. Rensing L., an der Heiden U. and Mackey M.), pp. 219–224. Springer, Heidelberg.
44. Sack R.L., Keith L.D. and Lewy A.J. (1989) Cortisol and melatonin rhythms in free-running blind people have identical periods. *Sleep Res.* **18**, 441.
45. Sack R.L., Lewy A.J., Blood M.L., Keith L.D. and Nakagawa H. (1992) Circadian rhythm abnormalities in totally blind people: incidence and clinical significance. *J. Clin. Endocrinol. Metab.* **75**, 127–134.
46. Sack R.L., Lewy A.J., Blood M.L., Stevenson J. and Keith L.D. (1991) Melatonin administration to blind people; phase advances and entrainment. *J. Biol. Rhythms.* **6**, 249–261.
47. Sack R.L., Lewy A.J., White D.A., Singer C.M., Fireman M.J. and Vandiver R. (1990) Morning versus evening light treatment for winter depression. Evidence that the

therapeutic effects of light are mediated by circadian phase shifts. *Arch. Gen. Psychiat.* **47**, 343–351.

48. Sarrafzadeh A., Wirz-Justice A., Arendt J. and English J. (1990) Melatonin stablises sleep onset in a blind man. In *Sleep '90* (ed. Horne J.A.), pp. 51–54. Pontenagel Press, Bocam, Germany.
49. Shanahan T.L. Evaluation of the twenty-four hour plasma melatonin pattern as a marker of the human circadian system. (Bachelor of Science Scholar of the College Thesis). Boston College, 1990.
50. Shanahan T.L. and Czeisler C.A. (1991) Light exposure induces equivalent phase shifts of the endogenous circadian rhythms of circulating plasma melatonin and core body temperature in men. *J. Clin. Endocrinol. Metab.* **73**, 227–235.
51. Stetson M.H., Sarafidis E. and Rollag M.D. (1986) Sensitivity of adult male Djungarian hamsters (*Phodopus sungorus sungorus*) to melatonin injections throughout the day: Effects on the reproductive system and the pineal. *Biol. Reprod.* **35**, 618–623.
52. Underwood H. (1986) Circadian rhythms in lizards: Phase response curve for melatonin. *J. Pineal Res.* **3**, 187–196.
53. Vakkuri O., Leppaluoto J. and Kauppila A. (1985) Oral administration and distribution of melatonin in human serum, saliva and urine. *Life Sci.* **37**, 489–95.
54. Wever R.A. (1979) *The Circadian System of Man. Results of Experiments under Temporal Isolation*. Springer, New York.
55. Wever R.A. (1989) Light effects on human circadian rhythms. A review of recent Andechs experiments. *J. Biol. Rhythms* **4**, 161–184.
56. Wright J., Aldhous M., Franey C., English J. and Arendt J. (1986) The effects of exogenous melatonin on endocrine function in man. *Clin. Endocrinol. (Oxf.)* **24**, 375–382.

SECTION III

Light Influence on Circadian Rhythms

13

Overview: Unmasking Temperature Data Obtained in Field Conditions

J. WATERHOUSE,[1] G. COSTA,[2] M. HARMA,[3] P KNAUTH,[4] D. MINORS,[1]
U. SITKA[5] and D. WEINERT[5]

[1]*Department of Physiological Sciences, University of Manchester, Manchester M13 9PT, UK*
[2]*Istituto di Medicina del Lavoro, Universita degli Studi di Verona, 37134 Verona, Italy*
[3]*Institute of Occupational Health, Laajaniityntie 1, SF-01620 Vantaa, Finland*
[4]*IIP, Universitat Karlsruhe, Hertzstrasse 16, D-7500 Karlsruhe 21, Germany*
[5]*Institute for Zoologie, Universitat Halle, Domplatz 4, 0-4020 Halle, Germany*

Abstract

Under normal circumstances, masking effects due to an individual's lifestyle and environment give misleading information when the phase of the endogenous oscillator(s) is inferred from overt circadian rhythms. Methods to overcome this problem are often not feasible in field studies. However, we have developed a series of methods that can be used in such conditions and which give a better estimate of the endogenous oscillator than using raw, masked data. These methods have been assessed for validity in laboratory-based studies and also used in several studies in the field where the amount and quality of data collected can be less.

Results confirm the view that the use of masked data leads to an overestimate of the rate of adjustment of the endogenous oscillator to time-zone transitions and shiftwork. They also indicate that masking factors might contribute to the phase stability that normally is measured in nychthemeral circumstances.

The methods require further testing but open up the possibility of assessing more accurately than before the process of adjustment of the body clock to changes in the field.

Masking and some ways of dealing with it

THE OVERT circadian rhythms measured in most variables are a mixture of endogenous and exogenous components; the endogenous component is due to the internal oscillator(s)—the body clock—and the exogenous component is caused by aspects of our environment and lifestyle—and also referred to as "masking". As a result, measured circadian rhythms may be only poor reflections of the internal oscillators, particularly with regard to the phase and amplitude of their output. Accurate measurements of the internal oscillators—for example, of their response to a change in sleep/activity schedule, of the day-by-day variation in their timing in nychthemeral circumstances, or of their development in the neonate—will all need recordings to be made in the absence of these masking influences, or at least for the record to be corrected (unmasked or purified) in some way.

Several successful approaches have been used, amongst which are:

A. The rhythm can be measured under circumstances in which the masking influences are decreased sufficiently to become unimportant. For plasma melatonin concentration, rectal temperature and some other variables, this has been achieved. With melatonin, the dim light melatonin onset method of Lewy has been successful,[7] as has a version of the constant routine method for rectal temperature, as refined by Czeisler's group.[1] For other variables, however, including those of mental performance, of hormones influenced by plasma glucose,[14] and of cardiovascular variables such as heart rate and blood pressure, it is not certain that the exogenous influences have been removed fully or that the effects of sleep loss have not begun to exert another disturbing effect. Clearly the constant routine method is unsuitable both for the study of several successive days and in most field conditions.

B. The rhythm has been measured under circumstances in which masking influences are present but attempts have been made to separate the endogenous and exogenous effects mathematically. Such methods have proved to be fruitful in studies involving time-zone transitions and irregular sleep/activity schedules.[5,6,12] These methods have the disadvantages, however, that they can require large amounts of data and the biological interpretation of the mathematical description of the data (particularly the exogenous component) can be complex.

C. We have developed a set of unmasking methods[10] that, at least for rectal temperature data, enable an estimate to be made of the phase of the endogenous component and yet overcome some of the difficulties discussed above. Thus:

(a) the rhythm can be measured in nychthemeral circumstances;
(b) there is good evidence that the exogenous component is directly related to the sleep/activity cycle;
(c) successive days can be studied;
(d) the method can be used with missing data.

Unmasking methods: the basic concept

The basic concept is that a measured rhythm consists of exogenous and endogenous components that act additively.[2,3,10] The endogenous component, whose phase is sought, is assumed to be described by "normative endogenous data". (These data are the mean rectal temperature rhythm from many subjects studied in the supine position and corrected for any effects of sleep, when masking effects will be minimal.) The shape and timing of the exogenous component depend upon several factors, including the individual's sleep/activity cycle which can be estimated in more than one way, see below. The estimated rhythm will be the sum of these components and the method consists of changing the phase of the endogenous component until the sum of it and the exogenous component approximates most closely to the rhythm that has been measured on the day in question. This approximation is assessed by finding the phase lag of the endogenous component which maximizes the cross-correlation coefficient between the estimated and measured circadian rhythms. Alternatively, the phase lag of the endogenous component which minimizes the summed squared deviations between the estimated and measured rhythms can be used; this method is required if the size of the masking effect is being assessed, see later.

In the following overview, some of the uses to which the basic concept and its variants have been put will be outlined.

Testing the basic method in the laboratory environment

We first tested the method with a simulated time-zone transition.[4] The laboratory enables constant routines to be carried out also, though only on selected rather than consecutive days. Nevertheless, this will enable a comparison to be made between shifts measured by the new and the "gold-standard" methods.

The method is applicable because the sleep/activity cycle is fairly reproducible from day to day in comparison with the expected shift of the endogenous component.

The stages of the analysis would be as follows:

(a) The control days are pooled to give an average nychthemeral day.
(b) The normative endogenous curve is subtracted to give the exogenous profile for that subject.
(c) This profile is then adjusted fully to the new local time after the time-zone transition.
(d) Sets of test data are produced by combining (c) with the normative endogenous data shifted by different amounts. In our studies we used hourly data points, so the shifts were in hourly increments from 0 to 23.
(e) The set of test data which correlates most highly with the measured rhythm on a particular day is found; the shift of the normative endogenous data which contribute to this set of test data is taken as a measure of the shift of the endogenous component on that day.

Such a method was used for 8 successive days after a simulated time-zone transition in seven subjects. The results indicated a progressive adjustment to the new time zone, but one that was significantly less than that measured conventionally (by cosinor analysis). Moreover, the shifts were not significantly different from those that were measured in independent experiments in which the shift of the endogenous component was estimated by constant routines. That is, the method appears to produce an estimate of the phase of the endogenous component even though nychthemeral data are being used.

However, this method makes an important assumption, namely it assumes that the profile of the exogenous component is, overall, constant from day to day. In many cases, particularly in the field, this will not be so, and a more versatile method is required.

The purification method: dealing with variable patterns of sleep and activity

This, based upon the ideas of Wever,[17] we have called the "purification" method. Here the temperature record is corrected or "purified" in accord with categories which represent the major type of activity during the previous hour. In one of our studies,[11] the categories used were: "sleeping"; "lying down but awake"; "sitting"; "active". The method then consisted of the following stages:

(a) Categorize each temperature on the basis of the activity in the previous hour (see above).
(b) Raise all temperatures in the "sleeping" category by 0–1.0°C in 0.1°C steps.
(c) Leave temperatures in the "lying down but awake" category unchanged (there is assumed to be no masking in this posture).
(d) Lower all temperatures in the "sitting" category by 0–1.0°C in 0.1°C steps.
(e) Repeat stage (d) for each of the other categories.
(f) Consider all the purified data sets produced by a combination of the above and compare each with normative endogenous data shifted in hourly increments from 0–23. This enables the "best" combination of purification factors and shift of normative endogenous data to be found—and hence the shift of the endogenous component on that day. "Best" can be assessed by finding the highest cross-correlation coefficient as before, but finding the minimum summed squared deviations between normative and purified data sets is an alternative that we normally use.

We have tested the purification method upon a group of 10 nurses.[11]

Each nurse was studied for at least 2 rest days (when she lived conventional hours) before working up to 6 consecutive night shifts. Rectal temperature was measured throughout and hourly readings were used in the analysis. The nurses also kept a diary of their sleep and different categories of activity, as described above. It was established that adjustment of the endogenous component was progressive but incomplete (as compared with the delay in midsleep), and that it was slower than that assessed by cosinor analysis of raw (masked) data.

That is, such a purification method indicates that there is an overestimate of phase adjustment when unpurified data are used. Whether the phase adjustment calculated by the purification method is "correct" is not known since it was not feasible for the nurses to undergo constant routines. However, in an earlier laboratory-based simulation of nightwork,[9] rectal temperature data were analyzed by other methods that also attempted to correct for masking effects. Not only did these results show that shifts using corrected data were less than those using uncorrected (or masked) data, but also these smaller shifts were close to those estimated by constant routines. That is, the evidence points to the results being a better assessment of the phase of the endogenous component when purified rather than raw data are used.

Assessing nychthemeral rhythms using both methods

The aim of developing the new methods was to be able to use them in field conditions. Since neither of the methods just described involves sleep loss, either can be used with data collected on an extended number of consecutive days to assess how reproducibly the circadian rhythm is phased from day to day.

With subjects living in an isolation chamber, in which activity is restricted by lack of space to a mainly sedentary existence, the activity profile is very similar each day and therefore the first method (see the section on testing the basic method) can be used. We have done this with a group of 11 subjects.[16] The analysis consisted of the following stages.

(a) All data were pooled to give the mean nychthemeral rhythm.
(b) The normative endogenous data were subtracted to give the mean exogenous profile.
(c) This profile was taken from each day's data in turn, to estimate the endogenous component on that day.
(d) The phase of this component was compared with normative endogenous data.

As a control, the phase of the masked data was assessed each day by carrying out a comparison with normative data in the same way but without having first removed the exogenous component. Results indicated that there was entrainment to 24 hours, but that there was a greater phase variation between days when the endogenous component was assessed (that is, the exogenous component had been removed) than when the unchanged raw data were assessed.

In a related experiment,[16] two subjects kept detailed activity logs for a period of 23 consecutive days whilst they lived with their families. The purification method (see the section on the purification method) was more appropriate with these data, the normal five categories of activity being entered into the logs. Again results showed that there was entrainment to the 24-hour day and that the day-by-day phase variation was greater when purified rather than raw data were used. (Incidentally, if this purification method was used with the 11 subjects in the isolation unit then, again, the purified data showed more daily phase variation than did the raw data.)

The full implications of these results remain to be determined but they indicate that masking effects during nychthemeral circumstances have the effect of stabilizing the phase of the measured circadian rhythm in comparison with the phase of the endogenous component. The results might also enable some estimate to be made of the phase variability of the entrained endogenous component. This seems an important factor when the process of adjustment and the relationship described by the phase response curve are considered.

Using the models in less ideal field conditions

The rectal temperature data described so far have been collected from volunteers in the laboratory or, in the cases of the nurses and two subjects just described, from individuals who were able and willing to record rectal temperature and details of their lifestyle for extended periods of time. In most studies performed in the field, one or more factors prevent such "ideal" collection of data. Some of these will now be considered.

Changing the number of categories of activity

The nychthemeral study upon two subjects living with their families made use of logs in which five types of activity (sleeping, lying but awake, sitting, standing, active) were used; the study using volunteers in the isolation unit used only two categories (sleeping and sitting); and the study using nurses used four categories (sleeping, lying but awake, sitting or relaxed, active or working). The choice of categories reflects in part what is convenient to the subjects and appropriate for their lifestyles, but it raises the question whether more categories, which means more purification, also means that the shift estimates are significantly improved. Closely related to this is the question of whether an activity log assessed by an objective means, say a wrist accelerometer, is equally suitable.

Our studies with the nurses doing shiftwork (above) indicated that the greater the number of categories, the better was the purification achieved. ("Better" in this context has two meanings: first it means a smaller value for the minimum summed squared deviations; second, it means a shift of the endogenous component that is less than that of raw data by a larger amount). However, the use of as few as two categories (sleeping and waking) produced some improvement over the use of masked data. (In this last case we are very close to the original process of purification as used by Wever.[17])

The record from a wrist accelerometer could also be used, the hourly counts being used as a measure of activity. Purification using categories based on these counts produced results that were indistinguishable from those using an activity log and the same number of categories.

Recently we have analyzed data from 19 nurses who wore rectal probes and whose activity record only distinguished between "sleep"; "leisure" and "work". Results, Figure 1, compare the phase change between the control day (normal routine) and the 24 hours from the end of the first night shift to the end of the second night shift, using raw and purified data. The overestimate of the amount of adjustment when using raw data is clearly evident, as is the fact that adjustment of the endogenous component was rather inconsistent and, on average, very small. (Indeed it was only marginally significantly different from zero. Thus: for raw data, the delay was 6.47 ± 0.63 hour (mean $\pm$ SE); for purified data, the delay was 2.11 ± 1.10 hour, $p < 0.10$.)

Estimation of rhythm changes with missing data

Assessments of circadian rhythms in the field not only have to deal with problems caused by masking but also with those due to missing data. There is a general resistance to the use of a rectal probe (though we must reiterate our thanks to the nurses who were honourable exceptions and whose data we have summarized above). In practice, therefore, recordings during sleep will be missing and even those taken during the subjects' waking phase cannot be taken frequently.

A fairly typical outcome would be that the protocol would consist of measurements of *oral* temperature made every 2 or 3 hours when awake (during leisure and work time) and single readings taken on retiring and immediately after waking but before rising. We have recently investigated the adjustment of a group of 13 nurses during two successive night shifts using the protocol just described. The process of purification needed some minor changes due to differences in the measurements. These were:

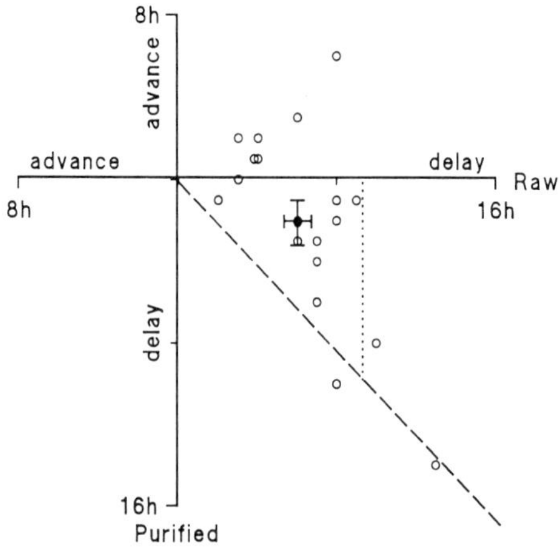

FIG. 1. Rectal temperature. Comparison of shifts of circadian rhythm between control day and night duty using raw and purified data. ○, individual nurses; ●, mean value ± SE for group; – – – –, line of equal change; dotted line, mean delay of midsleep

(a) Each temperature was assigned to the nearest clock hour. (This is required as the normative data are hourly). When the reading was taken on the half hour, then it was assigned to both adjacent hours.
(b) Each temperature was assigned a category (in this case "sleeping"; "lying but awake"; "sitting"; "standing"; "working"). "Sleeping" was the category assigned to the single reading obtained immediately on waking.
(c) The purification process was unchanged except that: (i) the summed squared deviation was calculated only for those hours when a temperature reading was available; (ii) the normative endogenous data set were decreased in value slightly in order to enable better alignment with oral temperatures which are slightly lower than those obtained from the rectum.

As before, the average delay of the rhythm assessed from raw data (4.15 ± 1.01 hour) was greater than that of the endogenous component assessed after purification (1.23 ± 0.96 hour). An example of raw and purified temperature rhythms from one subject is shown in Figure 2.

Whilst the results confirm previous findings, and therefore indicate that the present methods have at least some value in this type of field study, there is a caveat. As with any analysis of a circadian rhythm, as the numbers of data points per cycle falls, so the estimate of phase will be influenced more by any extreme result. In the present case, the single value designated "sleeping" necessarily assumes considerable importance when the processes of purification and phase assessment are carried out. One outcome is that there is rather more inter-individual variation in phase

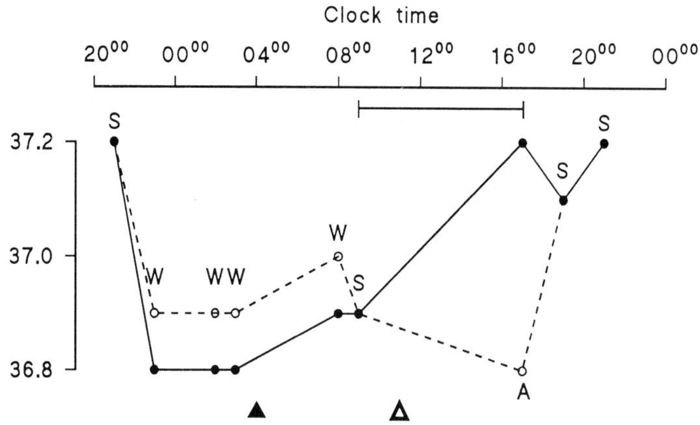

FIG. 2. Raw, ○ – – ○, and purified, ●—●, oral temperature data for single subject. |——|, time of sleep; S, sitting; W, awake; A, asleep. Estimated times of minimum shown for, △, raw data and, ▲, purified data.

estimates than when 24 data points per cycle are available. Even so, it is worth pointing out that the mean values for the purification factors (sleeping, sitting etc) were not different from those found when complete data sets were used in other studies.

A further test of the purification method—assessing amplitude

In the studies so far we have always assumed that the amplitude of the endogenous component of an individual's rhythm was approximated by that of normative endogenous data. The test of this assumption—to compare normative endogenous data with those obtained during a constant routine—cannot be carried out in the field conditions we are investigating, but it must be remembered that the normative data were obtained with protocols that approximated to such conditions in healthy subjects. With aged subjects[15] and, particularly, infants,[8] however, there is evidence to suggest that the amplitude of the endogenous component is decreased; our method would need modification if it were to be applied to such subjects, therefore.

We have performed a series of measurements upon healthy, full-term babies during the first week of extra-uterine life (day 2) and 4 weeks later. Skin and rectal temperature were measured hourly for 24 hours and a record of "activity" was kept, this being divided into: "deeply asleep"; "lightly asleep"; "awake, but lying passively"; "awake and moving arms and legs"; "awake and crying". The category "awake, but lying passively" was assumed to be unaffected by masking, and so was not purified. The analysis proceeded as before except for two changes.

A. The amplitude of the normative endogenous data could be changed. This is achieved by altering K in the equation:

$$Y_t = K(E_t - M_E) + M_E,$$

where

Y_t is the amplitude-corrected value for normative endogenous data at time t;

E_t is the uncorrected value for normative endogenous data at this time—this is the value used in the other methods;

M_E is the mean value for normative endogenous data.

With this transformation, a value of K equal to zero implies a normative endogenous component with no amplitude and equal to the mean value, a value of unity gives values equal to those normally used in our other studies; other values of K can be tried systematically during the purification process.

B. The mean value for the raw data was set equal to M_E and all individual values were adjusted appropriately. This is necessary to superimpose the mean values for normative and raw data, particularly in the present case as the temperature of an infant might differ from that of an adult. Misalignment of normative and raw data might lead to incorrect assessments of the size of masking effects in any study—but it does not alter the estimate of phase of the endogenous component—but this could be particularly misleading in the present case, as masking effects would be predicted to be small on account of activity being so limited in an neonate. For this reason also, the size of the masking effects in all categories was estimated to the nearest 0.05°C rather than 0.1°C.

The results from the 11 babies showed some differences in amplitude and the size of masking effects between day 2 and week 4. Masking effects due to activity were generally small at both ages, but those due to sleep increased between the first and fourth week, Figure 3. The former result no doubt reflects the small amount of physical activity involved and "crying" as a category was too infrequent for a quantitative assessment of the masking effects of this to be made. The increased masking effect of sleep might indicate some consequence of the process of establishing sleep infrastructure that is occurring at this age (see reference 8, for example). The amplitude of the endogenous component of rectal temperature did not change significantly between day 2 and week 4, all amplitudes on both occasions being less than values predicted for an adult, Figure 4. Such a result would corroborate several findings on the process of development of circadian rhythmicity during the weeks immediately after birth,[8,13] though these other studies have been upon masked data. By contrast the amplitude of the endogenous component of skin temperature (which correlated highly with rectal temperature in most babies) showed a significant increase in amplitude between day 2 and week 4. Taken

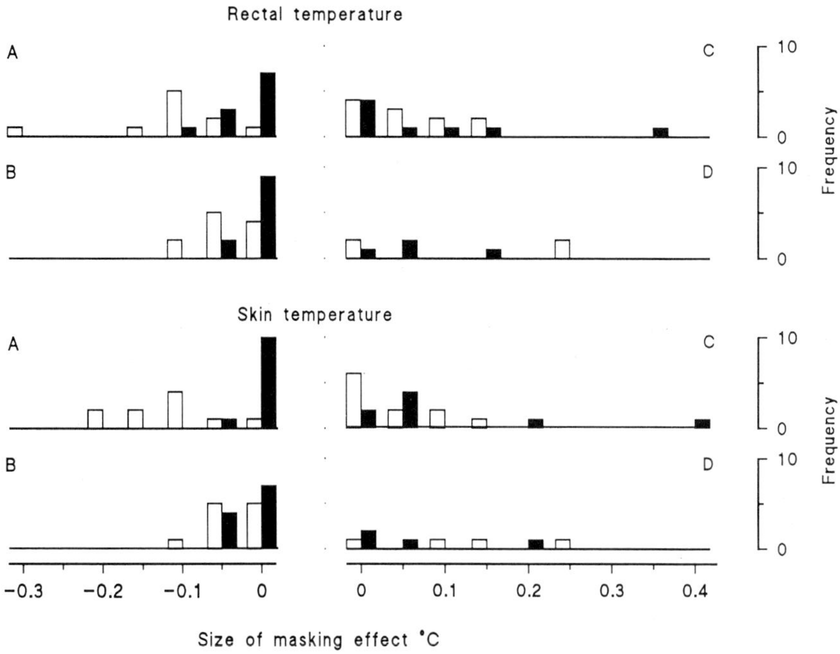

FIG. 3. Frequency distributions of masking effects for rectal and skin temperature in 11 babies. Closed bars, at 2 days; open bars, at 4 weeks. A, deeply asleep; B, lightly asleep; C, awake and moving arms and legs; D, awake and crying.

together, these results suggest that the development of the circadian rhythm of temperature regulation is taking place by the end of the first month after birth.

Concluding comments

We do not claim that the present methods give estimates of the phase and amplitude of the endogenous component of the circadian rhythm of body temperature that are as accurate as those given by constant routines or by sophisticated mathematical analyses of data obtained under nychthemeral conditions.

We accept that the present methods are based upon assumptions about the constancy of shape of normative endogenous data, the additive nature of endogenous and exogenous components, and the ability to estimate the exogenous component from activity records. We also accept that the methods have not been justified in all the experimental circumstances in which they have been applied. There are also the problems that validating a set of assumptions in one circumstance does not validate them in another, and that further difficulties might arise when combining assumptions. We have outlined the evidence, both here and elsewhere,[10]

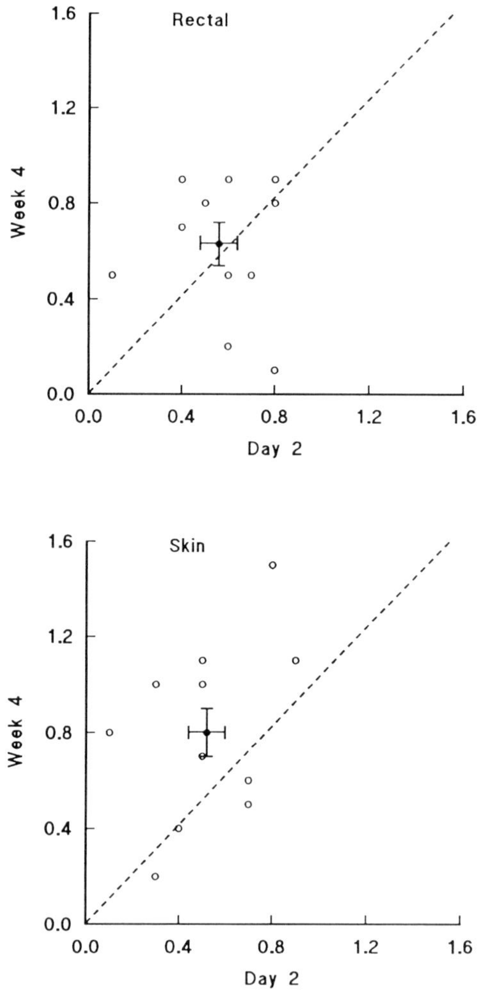

FIG. 4. Comparison of estimated amplitude of endogenous component of temperature rhythm at 2 days and 4 weeks of age in 11 babies. Results expressed as fraction of amplitude of normative data in adults. ○, individual babies; ●, mean value ± SE for group. – – – –, line of equal change.

which supports these assumptions and the results which have been obtained by the different methods.

Clearly, more validation is required, but we suggest that the methods offer the prospect of gaining some insight into the endogenous component of the circadian rhythm of body temperature in field circumstances, that is, in circumstances where the experimental demands of the laboratory cannot be met for ethical or practical reasons. The results and preliminary analyses reported and overviewed here confirm that results obtained with masked data can be misleading and that theoretical concepts developed in the laboratory can be applied to the field. It is hoped that this cross-fertilization between theory and practice, both in experimentation and analysis, will continue to provide evidence relevant to an understanding of circadian rhythms mechanisms.

References

1. Czeisler C.A., Kronauer R.F., Allan J.S., Duff J.F., Jewett M.E., Brown E.N. and Ronda J.M. (1989) Bright light induction of strong (type 0) resetting of the human pacemaker. *Science* **244**, 1328–1333.
2. Folkard S. (1988) Circadian rhythms and shiftwork: Adjustment or masking? In *Trends in Chronobiology* (eds. Hekkens W.Th., Kerkhof G.A. and Rietveld W.J.), pp. 173–182. Pergamon Press, Oxford.
3. Folkard S. (1989) The pragmatic approach to masking. *Chronobiol. Int.* **6**, 55–64.
4. Folkard S., Minors D.S. and Waterhouse J.M. (1991) "De-masking" the temperature rhythm after simulated time-zone transitions. *J. Biol. Rhythms* **6**, 81–91.
5. Gander P.H., Kronauer R.E. and Graeber R.C. (1985) Phase shifting two coupled circadian pacemakers: implications for jet lag. *Am. J. Physiol.* **249**, R704–R719.
6. Gundel A. and Spencer M.B. (1992) A mathematical model of the human circadian system and its application to jet lag. *Chronobiol. Int.* 9, 148–159.
7. Lewy A.J. and Sack R.L. (1989) The dim light melatonin onset as a marker for circadian phase position. *Chronobiol. Int.* **6**, 93–102.
8. Mills J.N. (1975) Development of circadian rhythms in infancy. *Chronobiologia* **2**, 363–371.
9. Minors D.S. and Waterhouse J.M. (1968) Effects upon circadian rhythmicity of an alteration to the sleep-wake cycle: problems of assessment resulting from measurement in the presence of sleep and analysis in terms of a single shifted component. *J. Biol. Rhythms* **3**, 23–40.
10. Minors D.S. and Waterhouse J.M. (1992) Investigating the endogenous component of human circadian rhythms: a review of some simple alternatives to constant routines. *Chronobiol. Int.* **9**, 55–78.
11. Minors D.S. and Waterhouse J.M. (1993) Separating the endogenous and exogenous components of the circadian rhythm of body temperature during night work using some "purification" models. *Ergonomics* **36**, 497–507.
12. Minors D.S., Nicholson A.N., Spencer M.B., Stone B.M. and Waterhouse J.M. (1986) Irregularity of rest and activity: studies on circadian rhythmicity in man. *J. Physiol.* **381**, 279–296.
13. Sitka U., Nagel F., Rumler W. and Weinert D. (1984) Entwicklung der Zirkadian periodik der Korpertemperatur bei Neugeborenen. *Dt. Gesundh. Wesen.* **39**, 1334–1339.
14. Van Cauter E., Blackman J.D., Roland D., Spire J-P., Refetoff S. and Polansky K.S. (1991) Modulation of glucose regulation and insulin secretion by circadian rhythmicity and sleep. *J. Clin. Invest.* **88**, 934–942.
15. Vitiello M.V., Smallwood R.G., Avery D.H., Pascauly R.A., Martin D.C. and Prinz P.N. (1986) Circadian temperature rhythms in young adult and aged men. *Neurobiol. Aging* **7**, 97–100.
16. Waterhouse J.M., Minors D.S. and Folkard S. (1993) Estimating the endogenous component of the circadian rhythm of rectal temperature in humans undergoing normal sleep/activity schedules. *J. Interdiscipl. Cycle Res.* **24** (in press).
17. Wever R. (1985) Internal interactions within the human circadian system: the masking effect. *Experientia* **41**, 332–342.

14

Some Effects of Light and Melatonin on Human Rhythms

JOSEPHINE ARENDT

School of Biological Sciences, University of Surrey, Guildford, Surrey GU2 5XH, UK

Abstract

In a natural environment, body rhythms are adapted to the regular alternation of night and day, and the slower seasonal cycles. Current social organization (inter alia) leads to specific mismatches between our internal physiology and our environment. Shift-work and rapid travel across several time-zones leads to forced desynchronization of internal rhythms from the external environment and from each other, with consequent problems of behavior (e.g. sleep), physiology (e.g. gut function) and performance (e.g. accident rate). Similar disorganization of daily rhythms is seen in the aged, some blind subjects and in certain pathological situations, such as delayed sleep phase insomnia, some psychiatric disorders, possibly some cancers and other pathology.

The pineal hormone melatonin conveys information concerning light-dark cycles to body physiology in animals and may play a similar role in humans. Its rhythmic secretion provides an excellent circadian marker rhythm and has been extensively used for the investigation of the control of circadian rhythms in humans, in particular with regard to the effects of light. It has become clear that the human clock can be reset by suitable application of bright light.

A substantial number of recent investigations suggest that rhythm disturbance can be alleviated or minimized by pharmacological and non-pharmacological methods. Light provides a non-pharmacological approach to the alleviation of rhythm disturbance. Similarly, there is good evidence that the pineal hormone melatonin can reset the hypothalamic clock and that in humans it has resynchronizing effects with regard to a number of circadian rhythms including the sleep-wake cycle. Alleviation of rhythm disturbance in jet-lag, delayed sleep phase insomnia and selected blind subjects underlines the potential wide therapeutic possibilities for this natural hormone.

Introduction

MOST IF not all circadian rhythms in mammals are generated in the suprachiasmatic nucleus (SCN) and entrained (synchronized) to the 24 hour day primarily by light-dark cycles.[1,2] Factors (zeitgebers) other than light–dark cycles which are involved in the entrainment of circadian rhythms in animals include behavioral imposition, for example forced activity and rest, social and nutritional (rhythmic feeding) cues, temperature variations, the pineal hormone melatonin and in humans knowledge of clock time. In a normal environment these factors operate in concert. The extent to which any one of these can act as a principal zeitgeber in the absence of the others is species dependent and in general not well defined. Sufficient intensity and duration of white light is, however, sufficient for entrainment to 24 hours in most species.

During the last decade the ability of light of suitable duration and intensity to manipulate human circadian rhythms has become evident. Melatonin provides an endogenous circadian marker rhythm generated in the SCN and relatively easily assessed in human plasma, urine and saliva.[3] It has been extensively used in these investigations. It is normally secreted during the dark phase of the day and is suppressed by sufficient intensity of light at night.[3,13] In addition to its usefulness as an indicator of circadian status, melatonin may well play an important role in the organization of human circadian rhythms. There is now good evidence that exogenous melatonin can manipulate some aspects of human circadian function[3] and that *in vitro* it is able to modify directly the phase of the (rodent) SCN.[4] This review aims to summarize some recent work.

Effects of light

Melatonin secretion in relation to daylength

In most species melatonin secretion is related to the length of the night: the longer the night the longer the duration of secretion. This has been particularly well demonstrated in sheep where melatonin rises within a few minutes of lights off and in photoperiods of more than around 14 hours of light does not decline until lights on.[5,6] The duration of melatonin secretion is a critical feature of its physiological role in photoperiodic time measurement.[5,6]

For many years it has proved difficult to demonstrate a change in duration of melatonin secretion with daylength in humans even in polar regions, although very small changes have been reported. The most consistent observation is that human melatonin profiles show a phase change from winter to summer, with earlier secretion in summer compared to winter.[7,8] If, however, humans are kept strictly in darkness for 14 hours per day for a period of 2 months, the melatonin secretion pattern expands to cover almost the entire dark period and concomitantly, in extended periods of 14 hours of light, the rhythm contracts to less than 10 hours.[9] Thus in appropriate experimental conditions humans show a duration change like animals.

Our behavior, particularly that of urban populations, is such that in "normal" circumstances it is likely that we percieve only very small changes in daylength during the year. For example in the UK it is fairly typical to rise around 07.00 hours (well after dawn in the summer) and to be inside a dwelling from around 19.00 hours (well before dusk in the summer) and thus not exposed to natural sunlight for the entire photoperiod in summer.

Light suppression of melatonin secretion

Even a brief exposure to white or full spectrum light of sufficient intensity at night will rapidly suppress melatonin production.[10] The amount of light required to suppress melatonin secretion either during the night or at the beginning and the end of the night varies considerably from species to species. A most important factor involved in this suppression is the previous exposure to light, with evidence from laboratory animals that prolonged maintenance in light of the intensity of natural light leads to decreased sensitivity.[11] It is likely that light sensitivity varies acutely, increasing in the course of the night, probably through an increase in retinal sensitivity.

Humans require substantial amounts of light to suppress melatonin secretion at night. Early experiments suggested that domestic intensity light (100–500 lux) did not suppress human melatonin, and the inference was made that either we are resistant to such intensities or that human melatonin could not be suppressed by light.[12] In a classic experiment, however, Lewy[13] showed conclusively that if sufficient intensity of white light (2,500 lux for 2 hours during the night between 02.00 and 04.00 hours) was used human melatonin could be suppressed to basal (daytime) levels. This observation has been of very considerable importance, not only for an understanding of the control of human melatonin secretion but also for a general appreciation of the role of light in human physiology, in particular its importance in the control of human rhythms.

In fact partial suppression of melatonin secretion in humans can be affected by 2–300 lux applied for 30 minutes.[14] Using this protocol a time of day and time of year variation in sensitivity is suggested by recent work in Antarctica—a region where long term exposure to domestic intensity light is possible during the winter in the absence, for prolonged periods, of natural sunlight. As might be expected sensitivity increases in the course of the night and was greater in winter than in summer.[15] In healthy subjects there are very substantial individual variations in sensitivity which may be both genetically and environmentally determined.

Entrainment of the melatonin rhythm with light

Clear evidence of the importance of light–dark for the entrainment of melatonin was demonstrated in humans[16] where, following an inversion of the light–dark cycle, the urinary melatonin rhythm adapts to the new photoperiod over a period of several days, finally assuming the same phase relationship with the new light dark cycle as was present originally with the old light dark cycle. If humans are subjected to a forced phase shift of 24 hour time cues, such as in rapid flight across several time zones, readjustment of the phase of melatonin secretion takes on average approximately 1 day per hour of shift but with substantial individual differences.[17]

Light pulses are conventionally used to investigate the control of circadian rhythms according to a protocol known as the phase response curve (PRC) developed from observations on activity rest cycles.[18] This approach has been used extensively in the rat by Illnerova[19] to investigate the circadian control of melatonin production with evidence for separate control of evening onset and morning offset of melatonin by two distinct oscillators—E and M.

A single short daily pulse of light has not to the authors knowledge been investigated in free-running humans with regard to melatonin. A week of daily exposure to bright light in the morning advances the onset of the evening melatonin rise (measured in dim light to prevent suppression) by up to 2.5 hours and a week of exposure to the same amount of light in the evening delays the onset by 1 hour even when the sleep wake cycle is kept constant.[20]

Recent work designed to clarify phase shifting mechanisms in general in humans has used constant routine conditions to evaluate phase and amplitude before and after the use of 2 to 3 daily (5–6 hours) applications of bright (approximately 8,500 lux) light to induce phase shifts.[21] Here subjects are kept semi-recumbent in constant dim light and awake for 36 hours or more with identical hourly snacks. Such procedures abolish the "masking" effects of light dark alterations, sleep, activity, social contacts and intermittent meals and allow a pure assessment of circadian phase and amplitude. Bright light-induced phase shifts in the core temperature rhythm were accompanied by precisely equivalent shifts of the plasma melatonin rhythm.[21] Melatonin is only "masked" to a major extent by ambient light and provides probably the purest indicator of central biological clock status provided that light exposure is controlled.

Skeleton photoperiods

"Skeleton" light pulses at dawn and dusk, with otherwise continuous darkness, are read as a full photoperiod by animals and have been used to investigate control of human melatonin rhythms. In Antarctica, specifically on bases with an imposed structure of daily activity, human melatonin rhythms show a more marked tendency to delay in winter compared to summer, than in temperate zones, presumably due to looser coupling of the circadian system to 24 hours in the dim domestic intensity light.[8] One hour of light (2,500 lux) in the morning and in the evening for 6 weeks is sufficient to re-establish an earlier phase position comparable to summer.[8] Pulses of 500 lux were insufficient, hence we may assume that in these particular conditions of dim light conditioning, the intensity of light experienced in the morning and the evening required to entrain human rhythms lies between these limits.

There is no reason to suppose that the intensity of light required for melatonin suppression and the intensity of light required for entrainment are the same in a particular individual, but they may be related.

Circadian rhythms in the blind

An unknown proportion of blind people have abnormal circadian rhythms.[22] Reports of free-running sleep-wake, cortisol and melatonin rhythms exist, however, to what extent these are representative of the total blind population is not known. In a small number of registered blind people in the UK reporting sleep disturbance and willing to provide body fluids for the assessment of hormones, the majority showed free-running melatonin with tau varying from 23.8 to 25 hours.[23] Others had a synchronized rhythm with a delayed phase relationship to the 24 hour day. In other studies normal, free-running and phase delayed rhythms have been reported.[22]

Thus it would appear that for some individuals time cues other than light will maintain synchronization, albeit with the weak coupling evident from a delayed phase, whereas for others the light–dark cycle is essential. Whether these differences are genetic, environmental, related to the specific pathology of the blindness or a combination of factors remains an open question. Moreover it is possible that some visually blind people still have hypothalamic light perception via the retino-hypothalamic projection, and could use this information for the entrainment of circadian rhythms.

Social zeitgebers in dim light

Further evidence for the relative importance of different zeitgebers for the entrainment of melatonin again comes from studies in Antarctica. On the British base of Halley (75°S) the sun does not rise for 3 months in the winter and does not set for 3 months in the summer. The maximum light intensity during the winter is around 500 lux. Sleep logs and melatonin assessment throughout the year show that the vast majority of personnel on the base of Halley remain synchronized to 24 hours at all times.[8,24] In view of the low intensity light in winter the prediction was that rhythms would desynchronize from the 24 hour day and "free-run" at their own endogenous period. In fact it is rare to observe such a phenomenon on British bases. Rhythms generally remain entrained at all seasons with, however, a clear delay in the timing of sleep at weekends especially in the winter, and a delay in the timing, and exceptionally a small increase in the duration, of melatonin in winter compared to summer, similar to changes found in temperate latitudes.[8,24]

In contrast, Kennaway and colleagues[25] have recently reported a study of four Greenpeace volunteers overwintering in Antarctica, all of whom showed free-running sleep-wake and melatonin rhythms throughout the period of sundown. This group were aware of clock time, maintained daily radio contact (a social zeitgeber) and even intermittent personal contact with other bases, however their daily routine was not otherwise structured. They were free to sleep and eat as they wished, in contrast to British bases where the daily routine is structured.

Some interesting conclusions can be drawn comparing these studies. Evidently in the absence of a strong light–dark cycle behavioral impositions can serve to maintain the integrity of the circadian system. Without this stricture, daily time cues and knowledge of clock time do not suffice. The delayed, but entrained, rhythms in winter reflect the weak nature of the time cues. In such circumstances coupling between the pacemaker (rhythm generating system) and the zeitgeber is loosened.

For most purposes the imposed social routine enforces synchronization in combination with the dim light–dark zeitgeber and undoubtedly contributes to well-being and performance. The neural and/or biochemical mechanisms whereby imposed behavior influence the circadian system remain to be determined.

Non-24 hour light–dark cycles

The limits of entrainment of melatonin by light–dark have been explored in humans kept in environmental isolation. Imposition of a light–dark cycle (domestic intensity light less than 1,000 lux) steadily increasing in length from 24 to 29 hours by 10 minutes a day leads to a phenomenon known as fractional desynchronization. The sleep wake cycle remains entrained up to 29 hours but more strongly endogenous variables such as temperature, cortisol and melatonin uncouple from the zeitgeber at different imposed daylengths. Melatonin breaks away at approximately 27 hours[26] and free-runs with a free-running period identical to that of temperature. If, however, bright light (3–4,000 lux) is used, all variables remain coupled up to at least 29 hours and the upper limit of entrainment has yet to be fully evaluated with this intensity of light.[27]

Role of the pineal in the control of mammalian circadian rhythms

Until quite recently, received opinion was that the pineal did not have a role in the mammalian circadian system. This conclusion was essentially based on one approach to function: the lack of effect of pinealectomy on free-running rest activity cycles. However, as long ago as 1970 Quay[28] reported that when rats are subjected to forced phase shifts of the light–dark cycle, pinealectomy increased the rate of reentrainment. He even suggested that rate of adaptation to phase shift could be a diagnostic approach to the detection of destructive lesions of the pineal in humans. Redman *et al.*[29] have confirmed this work and very recently Cassone and coworkers (1992)[30] have shown that pinealectomy of hamsters in constant light leads to a major disruption of the circadian system.

Effects of melatonin on circadian rhythms

Early work

In the late 1970s accumulated experience suggested that administration of melatonin to mammals might provide more rewarding information concerning its role in circadian rhythms than pinealectomy. Only when the importance of timing in the physiology of melatonin was appreciated, rather than a pharmacological emphasis on dose, did the design of experiments produce coherent and interpretable results.

Two groups working simultaneously and independently with rats and humans, and observing the circadian effects of timed daily melatonin administration, showed for the first time that melatonin had entraining and phase shifting effects respectively in mammals. Redman and coworkers[31] injected melatonin at 24 hour intervals into free-running rats and obtained entrainment of the rest activity cycle when saline control injections were usually ineffective. Arendt and coworkers[26,32,33] gave daily melatonin orally in the late afternoon to human volunteers and obtained a phase advance of the endogenous melatonin rhythm, of evening fatigue and of prolactin morning offset.

Timed administration of melatonin in humans

Daily feeding of low dose (2 mg) melatonin in the late afternoon advanced the timing of evening self rated fatigue, the endogenous melatonin rhythm and the morning decline of prolactin compared to placebo. The duration and quality of sleep were not significantly altered but there was a non-significant advance of morning wake up time. The evening rise of melatonin was advanced more than the morning decline.

There were no significant effects on self rated mood, or on LH, FSH, testosterone, cortisol, growth hormone or T4. The effect on fatigue was not immediately apparent but became significant in the group as a whole after 4 days. No deleterious effects were reported by the subjects. This study, conducted twice over 4 weeks on two different occasions in spring and autumn showed that in low doses melatonin has chronobiotic effects in humans at least with respect to sleep, its own rhythm and prolactin.[26,32,33] It indicated that such a daily dose is well tolerated over 4 weeks. Importantly it showed no significant reproductive effects and indeed the dose is well below that needed to influence human reproduction as a combined contraceptive. It clearly suggested that melatonin would be useful for phase shifting and synchronizing strategies at least for phase advances. Since this initial study a substantial volume of work has served to confirm this prediction.

Amongst the conditions likely to benefit from the ability to manipulate human circadian rhythms are jet-lag, shift work, insomnia with delayed or advanced phase, blindness, old age and possibly some psychiatric conditions.

Jet lag and shift work

Melatonin has been used in the treatment of jet lag in both real life and simulation conditions.[17,34–37] Field studies show that self rated jet lag can be reduced on average by 50% with appropriately timed treatment both westwards and eastwards. The improvement is greater with larger numbers of time zones. Using advance shifts over eight time zones with pre-shift administration of melatonin in the late subjective day and post shift administration at the desired bedtime, the subjective impressions are reinforced by improved latency and quality of sleep, greater daytime alertness and more rapid resynchronization of melatonin and cortisol rhythms. In simulation studies using advanced phase shift and a similar administration protocol, effects on sleep are not significant but there is a marked increase in the rate of re-entrainment of both hormonal and electrolyte rhythms and an immediate effect of lowering body temperature which persists as more rapid re-entrainment.

Neither the dose nor the timing of melatonin administration has been optimized for different time zone shifts although a phase response curve for melatonin would serve to indicate preferred treatment times. To date a minority (less than 10%) of subjects feel significantly worse after melatonin.[35] Individual circadian status is probably a major determinant of response in that undesirable changes in direction of entrainment may occur with inappropriate timing of treatment. Moreover, unpredictable exposure to bright light can theoretically act in opposition to the desired result. Bright light exposure needs to be titrated during treatment for jet lag. An ambulatory technique for instantaneous assessment of individual circadian phase position would be of great interest.

There is very little published work on the use of melatonin in shift work although exposure to bright light during the night is clearly beneficial to night shift workers.[38] One study has found improved sleep and increased daytime alertness in night shift workers receiving melatonin at the desired bedtime during a night shift week compared to placebo and baseline conditions.[39] Here there was a suggestion that one performance task (visual search speed) was adversely affected by melatonin. This is a very important consideration if melatonin is to be used on members of the workforce. There are no reported deleterious effects on performance in jet lag studies but acute studies using much higher doses (240 mg) describe a brief decline in visual reaction time, accompanied by an increase in fatigue ratings but with no effect on memory.

Recently, moderately bright light treatment during the night has been used to force phase delays in human volunteers kept otherwise in a normal environment. Melatonin was used in the post phase shift adaptation to induce phase advances. In the course of this study no deleterious effects on performance were seen.[40] It will be essential to assess thoroughly performance effects of melatonin in the work place and on specific tasks before its use can be recommended.

Sleep disturbance in the blind

Free-running blind subjects often find this condition difficult to tolerate, with intermittent problems particularly of sleepiness and poor performance during the day and poor night time sleep, when their pacemaker is out of phase with the environment.

Only a small number of studies have been reported, however it is clear that some blind subjects reporting sleep problems derive benefit from melatonin ingestion at desired bedtime.[41,23] One effect is a stabilization of sleep onset, sometimes with significant improvement of some other sleep parameters such as quality and duration. In some sufficiently long term experiments it is clear that a synchronization of the sleep wake cycle has occurred under melatonin. The successful treatments have included both free-running subjects with a period greater than 24 hours and others with abnormal phase delay. Insufficient data are available to determine in which blind people melatonin is likely to be useful as yet. As melatonin may have sites of action within the visual system the integrity of such targets may be compromised in some blind people.

Whether synchronization of sleep wake is accompanied by synchronization of other circadian variables in these circumstances is a question of much interest. One subject with a previously free-running sleep wake cycle and whose sleep has been stabilized for a number of years showed persistent free-run in melatonin, cortisol and temperature rhythms.[42] In spite of this his well being was restored with improvement of mood and daytime alertness and from this one case it appears that if the sleep problem is allayed, the behavior of other major circadian variables may be irrelevant. It also suggests that behavioral changes do not necessarily entrain all of the pacemaker in the absence of any light–dark cycle. There may be enormous individual differences however. In the work of Lewy and colleagues[43] and this volume, other blind subjects show clear phase shifts and occasionally entrainment of the endogenous melatonin rhythm during melatonin treatment.

Delayed sleep phase insomnia

Another situation in which phase delay forms an integral part of a clinical problem is delayed sleep phase insomnia (DSPI). Patients cannot sleep at the socially acceptable time of night and delay sleep onset until the early hours of the morning, sleeping through much of the day. This condition is endemic in teenagers and in the student population where it is usually self limiting. The obligation to earn a living provides sufficient motivation to overcome the problem. In true DSPI subjects the aberrant phase position cannot be overcome with consequent inability to work and socialize at usual times. This condition has been successfully treated with bright light in the early morning to induce phase advances of the clock. In others evening melatonin (5 mg) will also advance sleep time significantly.[44]

A PRC for melatonin

Lewy and coworkers (45 and this volume) have described a PRC for melatonin in people. They report phase advances in the late subjective day and small phase delays during the early morning. The data are approximately 12 hours out of phase with the PRC to light and are consistent with the long held view that melatonin can act like darkness both in circadian and seasonal rhythms.

Melatonin as an endogenous time cue

The physiological role of melatonin in circadian organization in humans can be considered as follows. With a free-running period (tau) greater than 24 hours the phase advancing effects of morning light and the phase advancing effects of evening melatonin may combine with other zeitgebers as yet incompletely defined to maintain 24 hour synchrony. The reverse situation can evidently be postulated for the rare individual with tau less than 24 hours. With this in mind it would obviously be advantageous to combine controlled bright light (artificial or natural) and melatonin treatments in the management of rhythm disorder. Preliminary work indicates that this approach will yield a rich harvest.

If the maintained night time levels of melatonin are important for sleep maintenance then a slow release preparation given in the evening should prove of greater benefit to sleep duration than a rapidly metabolized dose. Good slow release preparations of melatonin for human use have only been developed recently and there is a lack of information in this area.

The interest in melatonin as an endogenous thermoregulatory factor has been reinforced by studies designed to uncover the acute effects of melatonin and light in humans. There is an inverse relationship between acute suppression of melatonin and acute elevation of body temperature by bright light.[46] These acute and opposite effects of bright light and melatonin may represent the primary manifestation of a phase shift in the making, in terms of brain metabolic events.

References

1. Rusak B. and Zucker I. (1979) Neural regulation of circadian rhythms. *Physiol. Rev.* **59**, 449–526.
2. Moore-Ede M.C., Czeisler C.A. and Richardson G.S. (1983a) Circadian timekeeping in health and disease. Part 1. Basic properties of circadian pacemakers. *N. Eng. J. Med.* **309**, 469–476.
3. Arendt J. (1988) Melatonin. *Clin. Endocrinol.* **29**, 205–229.
4. McArthur A.J., Gillette M.U. and Prosser R.A. (1991) Melatonin directly resets the rat suprachiasmatic circadian clock in vitro. *Brain Res.* **565**, 158–161.
5. Tamarkin K., Baird C.J. and Almeida O.F.X. (1985) Melatonin: a coordinating signal for mammalian reproduction. *Science* **227**, 714–720.
6. Arendt J. (1986) Role of the pineal gland and melatonin in seasonal reproductive function in mammals. *Oxf. Rev. Reprod. Biol.* **8**, 266–320.
7. Illnerova H., Zvolsky P. and Vanecek J. (1985) The circadian rhythm in plasma melatonin concentration of the urbanised man, the effect of summer and winter time. *Brain Res.* **328**, 186–189.
8. Broadway J., Arendt J. and Folkard S. (1987) Bright light phase shifts the human melatonin rhythm during the Antarctic winter. *Neurosci. Letts.* **79**, 185–189.
9. Wehr T.A. (1991) The durations of human melatonin secretion and sleep respond to changes in daylength (photoperiod). *J. Clin. Endocrinol. Metab.* **73**, 1276–1280.
10. Illnerova H. (1988) Entrainment of mammalian circadian rhythms in melatonin production by light. *Pineal Res. Revs.* **6**, 173–217.
11. Reiter R.J. (1985) Action spectra, dose-response relationships and temporal aspects of light's effects on the pineal gland. In *The Medical and Biological Effects of Light* (eds. Wurtman R.J., Baum M.J. and Potts J.R. Jr.), Vol. 453, pp. 243–268. Annuals of the New York Academy of Sciences.
12. Arendt A. (1978) Melatonin assays in body fluids. *J. Neural Trans. Suppl.* 13, 265–278.
13. Lewy A.J., Wehr T.A., Goodwin F.K. *et al.* (1980) Light suppresses melatonin secretion in humans. *Science* **210**, 1267–1269.
14. Bojkowski C., Aldhous M., English J., Franey C., Poulton A.L., Skene D.J. and Arendt J. (1987) Suppression of nocturnal plasma melatonin and 6-sulphatoxymelatonin by bright and dim light in man. *Horm. Metab. Res.* **19**, 437–440.
15. Owen J. and Arendt J. (1992) Melatonin suppression in human subjects by bright and dim light in Antarctica : time and season dependent effects. *Neurosci. Letts.* In press.
16. Lynch H.J., Jimerson D.C., Ozaki Y., Post R.M., Bunney Jr. W. E. and Wurtman R.J. (1978) Entrainment of rhythmic melatonin secretion in man to a 12-hour phase shift in the light-dark cycle. *Life Sci.* **23**, 1557–1564.

17. Arendt J., Aldhous M., English J. *et al.* (1987) Some effects of jet lag and their alleviation by melatonin. *Ergonomics* **30**, 1379–1393.
18. Elliot J.A. (1981) Circadian rhythms, entrainment and photoperiodism in the Syrian hamster. In *Biological Clocks in Seasonal Reproductive Cycles* (eds. Follett B.K. and Follett D.E.), pp. 203–218. Wright, Bristol.
19. Illnerova H. (1991) The Minds Clock. In *Suprachiasmatic Nucleus. The Minds Clock* (eds. Klein D.C., Moore R.Y. and Reppert S.M.), pp. 197–219. Oxford University Press, New York, Oxford.
20. Lewy A.J., Sack R.L. and Singer C.M. (1985) Treating phase typed chronobiologic sleep and mood disorders using appropriately timed bright artificial light. *Psychopharmacol. Bull.* **21**, 368–372.
21. Shanahan T.L. and Czeisler C.A. (1991) Light exposure induces equivalent phase shifts of the endogenous circadian rhythms of circulating plasma melatonin and core body temperature in man. *J. Clin. Endocrinol. Metab.* **73**, 227–235.
22. Lewy A.J. and Newsome D.A. (1983) Different types of melatonin circadian secretory rhythms in some blind subjects. *J. Clin. Endocrinol. Metab.* **56**, 1103–1107.
23. Aldhous M.E. and Arendt J. (1991) Assessment of melatonin rhythms and the sleep wake cycle in blind subjects. In *Proceedings of the European Pineal Society, Guildford, 1990. Advances in Pineal Research* (eds. Arendt J. and Pevet P.), Vol. 5, pp. 307–311.
24. O'Conor R.A.A. and Arendt J. (1992) A year in Antarctica: sleep and mood studies on Halley Base, 75oS. In *Proceedings of the European Society for Chronobiology*. J. Interdisciplinary Cycle Research, Leiden, Holland, 1992. In press.
25. Kennaway D.J. and Van Dorp C.F. (1991) Free-running rhythmns of melatonin, cortisol, electrolytes and sleep in humans in Antarctica. *Am. J. Physiol.* **260**, R1137–R1144.
26. Arendt J., Bojkowski C., Folkard S., Franey C., Minors D.S., Waterhouse J.M., Wever R.A., Wildgruber C. and Wright J. (1985) Some effects of melatonin and the control of its secretion in man. In *Ciba Foundation Symposium 117. Photoperiodism, Melatonin and the Pineal* (eds. Evered D. and Clark S.), pp. 266–283. London, Pitman.
27. Wever R.A. (1989) Light effects on human circadian rhythms: a review of recent Andechs experiments. *J. Biol. Rhythms* **4**, 161.
28. Quay W.B. (1970) Precocious entrainment and associated characteristics of activity patterns following pinealectomy and reversal of photoperiod. *Physiol. Behav.* **5**, 1281–1290.
29. Redman J.R., Armstrong S. and Ng K.T. (1983) Pinealectomy: increased sensory input or reduced homeostatic mechanism? *Proc. Int. Union Physiol. Sci.* **15**, 41.
30. Cassone V. (1992) The pineal gland influences rat circadian activity rhythms in constant light. *J. Biol. Rhythms* **7**, 27–40.
31. Redman J., Armstrong S. and Ng K.T. (1983) Free-running activity rhythms in the rat: entrainment by melatonin. *Science* **219**, 1089–1091.
32. Arendt J., Borbely A.A., Franey C. and Wright J. (1984) The effect of chronic, small doses of melatonin given in the late afternoon on fatigue in man: a preliminary study. *Neurosci. Lett.* **45**, 317–321.
33. Wright J., Aldhous M., Franey C., English J. and Arendt J. (1986) The effects of exogenous melatonin on endocrine function in man. *Clin. Endocrinol.* **24**, 375–382.
34. Arendt J., Aldhous M. and Marks V. (1986) Alleviation of jet-lag by melatonin: preliminary results of controlled double-blind trial. *Br. Med. J.* **292**, 1170.
35. Skene D.J., Aldhous M. and Arendt J. (1989) Melatonin, jet-lag and the sleep-wake cycle. In *Proceedings of the European Sleep Congress* (eds. Horne J. and Karger S.), Jerusalem.
36. Samel A., Wegman H.M., Vejvoda M. and Maas H. (1991). Influence of melatonin treatment on human circadian rhythmicity before and after a simulated 9 hour time shift. *J. Biol. Rhythms* **6**, 235–248.
37. Nickelsen T., Lang A. and Bergau L. (1991) The effect of 6-, 9- and 11-hour time shifts on circadian rhythms: adaptation of sleep parameters and hormonal patterns following the intake of melatonin or placebo. In *Advance Pineal Research* (eds. Arendt J. and Pevet P.), Vol. 5, pp. 303–306.
38. Czeisler C.A., Johnson P.J., Duffy J.F., Brown E.N., Ronda J.M. and Kronauer R.E.

(1990) Exposure to bright light and darkness to treat physiologic maladaptation to night-work. *N. Engl. J. Med.* **322**, 1253–1259.

39. Folkard S., Arendt J. and Clark M. (1993) Can melatonin improve shiftworkers' tolerance of the night shift? Some preliminary findings. *Chronobiol. Int.* (in press).
40. Deveson S., Deacon S., Arendt J., Totterdell P. and English J. (1992) Simulation of circadian rhythm disturbances associated with jet lag and shift work using moderately bright light. In *Proceedings of the Pineal Satellite Symposium.* International Congress of Endocrinology. Elsevier, Paris, France. In press.
41. Arendt J., Aldhous M. and Wright J. (1988) Synchronisation of a disturbed sleep-wake cycle in a blind man by melatonin treatment. *Lancet* **1**, 772–773.
42. Folkard S., Arendt J., Aldhous M. and Kennett H. (1990) Melatonin stabilises sleep onset time in a blind man without entrainment of cortisol or temperature rhythms. *Neurosci. Lett.* **113**, 193–198.
43. Sack R.L., Lewy A.J. and Hoban T.M. (1987) Free-running melatonin rhythms in blind people: phase shifts with melatonin and Triazolam administration. In *Temporal Disorder in Human Oscillatory Systems* (eds. Rensing L.V. an der Heiden V. and Mackey M.C.), pp. 219–224. Springer, Heidelberg.
44. Dahlitz M., Alvarez B., Vignau J., English J., Arendt J. and Parkes J.D. (1991) Delayed sleep phase syndrome response to melatonin. *Lancet* **i, 337**, 1121–1124.
45. Lewy A.J., Saeeduddin A., Latham Jackson J. and Sack R.L. (1992) Melatonin shifts human circadian rhythms according to a phase-response curve. *Chronobiol. Int.* **9**, 380–392.
46. Badia P., Myers B. and Murphy P. (1992) Melatonin and thermoregulation. In *Melatonin: Biosynthesis, Physiological Effects, and Clinical Applications* (eds. Reiter R.J. and Yu H.S.). CRC Press, Boca Raton, FL. In press.

15

Understanding the Use of Light to Control the Circadian Pacemaker in Humans

RICHARD E. KRONAUER and CHARLES A. CZEISLER

Harvard University, Division of Applied Sciences, Cambridge, MA 02138 and Laboratory for Circadian and Sleep Disorders Medicine, Harvard Medical School, Brigham and Women's Hospital, Boston, MA 02115 USA

Abstract

Bright light can make large changes in the amplitude of circadian rhythms as well as shift their phase. When amplitude is much reduced, the phase shifting ability of light is greatly enhanced, in a manner which is most readily comprehended by means of vector diagrams. To obtain large phase shifts, light is used most effectively when the stimulus reduces amplitude at the beginning of the stimulus episode. Stimulus strategies which embody strong (Type 0) resetting can be very powerful, but to work they require careful adherence to a regimen of extended light exposures. A complete inversion of rhythm phase (a 12 hour phase shift) is possible with 10,000 lux episodes of 8 hours duration or of 6 hours duration on two successive days. Briefer exposures producing a succession of small shifts (cumulative type 1) can achieve the same result but may require several more days.

Quantifying pacemaker response to light

ONE OF the most revealing demonstrations of the action of light on the suprachiasmatic nuclei (SCN pacemaker) was given by Inouye and Kawamura.[8] Electrodes measured the circadian rhythm of mass neuronal discharge before and after 3 hours of light application to a rat's retinae (Figure 1). The circadian pattern of discharge is nearly sinusoidal, maximal by day and minimal by night (corresponding to metabolic activity for all mammalian species, whether the animal is nocturnal or diurnal[20,21]). For reasons which we will develop, it is most useful to use the nadir of neuronal activity as a timing reference for stimuli. Although light was applied for 3 hours it is clear that the effect on the rhythm was essentially complete in the first few minutes. This interpretation agrees with the data of Nelson and Takahashi[17] who showed that extending light exposure to hamsters beyond about 5 minutes produced very little increase

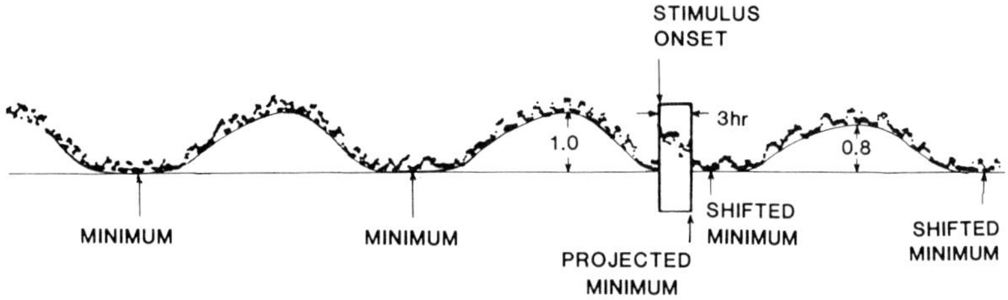

FIG. 1. Mass neuronal discharge in the suprachiasmatic nucleus of a rat. A 3 hour light stimulus was applied to the retina beginning 3 hours before the projected minimum neuronal activity. In response to the stimulus the minimum was delay-shifted by 2 hours and the rhythm amplitude was reduced by 20%.

in phase shifting. Such time-saturation of circadian responsiveness may be a protective mechanism for systems with high sensitivity (for example, 5 minutes of 250 lux light can shift the hamster rhythm by 2 hours). Other less sensitive mammals such as the Polynesian rat[6] and humans appear to respond in a continuous, integrating way to time-extended stimuli. As Figure 1 shows, the effective stimulus occurred about 3 hours before the nadir of neuronal activity. The light produced two changes of significance: The rhythm was phase-delayed by about 2 hours (30 phase degrees or 0.5 radian) and reduced in amplitude by about 20%.

It is tempting to think that an assay of SCN electrical activity reveals directly the circadian pacemaker function. However, since the blocking of neuronal sodium channels with tetrodatoxin silences the electrical activity without eliminating the rhythm itself,[22] the neuronal firing must be regarded as "hands" of the pacemaker clock, indicative of underlying chemical rhythms in a manner yet to be quantified. It is reasonable to anticipate a simple relationship between these "hands" and the chemical cycle itself. Other commonly used human markers of pacemaker function, such as the endogenous components of core body temperature, plasma cortisol and plasma melatonin may have more complex relationships to the pacemaker.

In appraising SCN electrical activity we have repeatedly referred to rhythm amplitude in order to emphasize an important feature of circadian function which is too often ignored. The direct physiological significance of amplitude has yet to be worked out. Some writers have linked reduced amplitude to human depression[5,24] or to a reduced circadian modulation of alertness.[9] In insects there seems to be a connection between amplitude and geographic latitude.[19] Regardless of how important, physiologically, the amplitude of pacemaker rhythm is found to be, it indisputably plays a crucial role in our ability to manipulate effectively the phase of that rhythm, as we shall elaborate below. In particular, strong phase resetting of the type which Winfree[28] called "Type 0", requires a stimulus capable of reducing amplitude to zero!

Rhythm assessment

The example of Figure 1 presents a protocol paradigm for all experiments which seek to quantify the effect of light on the pacemaker. The structure of that experiment entails first an assessment of the state of the pacemaker (its rhythmic amplitude and phase), next a quantified stimulus intervention (specified strength and duration of light) at a chosen phase of the endogenous rhythm, and finally an assessment of the resulting state of the pacemaker. A full rhythm cycle is needed to appraise each state.

In human experiments, assessment of initial and final rhythm states is effected through subsidiary variables (e.g. temperature, cortisol, melatonin). Temperature has been most widely adopted due to the ease with which it can be measured. However, the endogenous circadian temperature pattern is ordinarily "masked" by comparably large temperature swings evoked by posture change, activity, sleep, showering, etc. The endogenous rhythm is best "unmasked" by the Constant Routine (CR) procedure[2] which requires monitoring and control of the subject with a strictness usually achieved only in the laboratory. Repeated CR performed on control subjects maintained under low light levels (less than 20 lux) show amplitude variations in endogenous temperature of about 27% (SD). These can be decomposed into intersubject (intrinsic) variability of about 20% and intrasubject variability (i.e. "noise" in the CR assessment) of about 18%. The "noise" produces errors in circadian phase assessment of about ± 0.9 hour (SD). Comparison between subjects is improved when amplitudes are normalized relative to each subject's intrinsic value.

Any physiological variable which can be sampled frequently (e.g. hourly) over an extended time has the potential to give data on both amplitude and phase of a circadian rhythm. Blood plasma cortisol has been used to corroborate pacemaker phase shifts inferred from temperature data[4] and as an alternative measure of amplitude in experiments designed to reduce greatly the pacemaker amplitude.[10] Lewy and co-workers have used plasma melatonin as a phase marker via the Dim Light Melatonin Onset (DLMO) procedure.[15] Since Shanahan and Czeisler[23] have shown that the timing of the peak of nocturnal melatonin in any subject is closely related to both onset and offset timing, it is possible that melatonin can also be used as a marker of rhythm amplitude. Hopefully, salivary or urinary measures of cortisol and melatonin may be able to replace invasive blood samples, which would greatly facilitate the use of these hormones as circadian markers.

Masking effects on hormonal rhythms have been studied very little, beyond a very strong, acute effect of light on melatonin production.[14,16] Of course, blood sampling under CR conditions automatically controls for possible masking effects due to those variables which are held constant during the CR.

Reporting resetting results

Rodent studies have been greatly simplified by the existence of a sharply defined activity onset (wheel running or drinking) which provides an easily monitored marker of circadian phase. Consequently, the more difficult task of assessing rhythm amplitude in animals has largely been ignored. Unfortunately, the partial information offered by phase alone has significantly narrowed the perspective of much research on animal rhythms.

When one ignores the amplitude of the pacemaker rhythm, only the timing of the cycle remains to be considered. The effect of a brief light stimulus can be described by the magnitude and direction of the phase shift which it produces, typically plotted as a function of the phase of the cycle at which it is applied. This is the Phase Resetting Curve (PRC) introduced by Hastings and Sweeney.[7] The experiment shown in Figure 1 generates one point of a PRC. Taking the minimum of neural activity as the phase reference (phase = 0), the effective stimulus in Figure 1 was applied 21 hours after the minimum. We denote this as the initial phase, ϕ_{init}, at which the stimulus was applied, with respect to the unshifted rhythm ($\phi_{\text{init}} = 21$). The phase shift produced was a 2 hour delay (or -2 hours according to convention). "Final phase", ϕ_{final}, is defined as the phase at which the stimulus occurs relative to the *shifted* rhythm. In this experiment, $\phi_{\text{final}} = 19$ hours. Phase shift is typically designated as $\Delta\phi$; clearly $\Delta\phi = \phi_{\text{final}} - \phi_{\text{init}}$.

When we consider rhythm amplitude (designated $\mathbf{A}$), both the experiments and their reporting become far more complex. We must select initial amplitude, $\mathbf{A}_{\text{init}}$, as well as the stimulus timing, ϕ_{init}, and we must measure $\mathbf{A}_{\text{final}}$ as well as ϕ_{final}. Two functions:

$$\mathbf{A}_{\text{final}} = f_1(\mathbf{A}_{\text{init}}, \phi_{\text{init}})$$

$$\phi_{\text{final}} = f_2(\mathbf{A}_{\text{init}}, \phi_{\text{init}})$$

are needed to report the results. We have found that a graphical presentation helps greatly in organizing and interpreting them.

Consider the polar diagram of Figure 2. The endogenous rhythm amplitude* is represented by the radius from the center (designated $\oplus$), while the circadian phase (at which a stimulus is applied) is represented by the angle measured clockwise in hours from the indicated zero phase. Once around the circle is 24 hours (i.e. 1 hour $= 15°$). (We are ignoring here the approximately 1–3% difference between intrinsic rhythm period and 24 hours.) As suggested earlier, we will normalize rhythm *amplitude* for each animal or human subject to the value seen under reference conditions. There is a long tradition in rodent research to consider the free-run

*The endogenous rhythm exists within the SCN neurons, presumably in the form of intracellular chemical oscillations.

condition in darkness or very dim light as the reference state. The rat reported in Figure 1 was in the reference state before the light stimulus. Consequently, $\mathbf{A}_{\text{init}} = 1$.

The experiment of Figure 1 is represented by the initial and final vectors indicated in Figure 2. The initial phase is 21 hours (or −3 hours), describing the time in the cycle at which the stimulus is applied. The final state vector shows the reduced amplitude of 0.8 and a phase of 19 hours, which describes the time in the newly-altered rhythm at which the stimulus was applied. The vector drawn from the initial state to the final state describes the change produced by the stimulus (a delay shift, −2 hours).

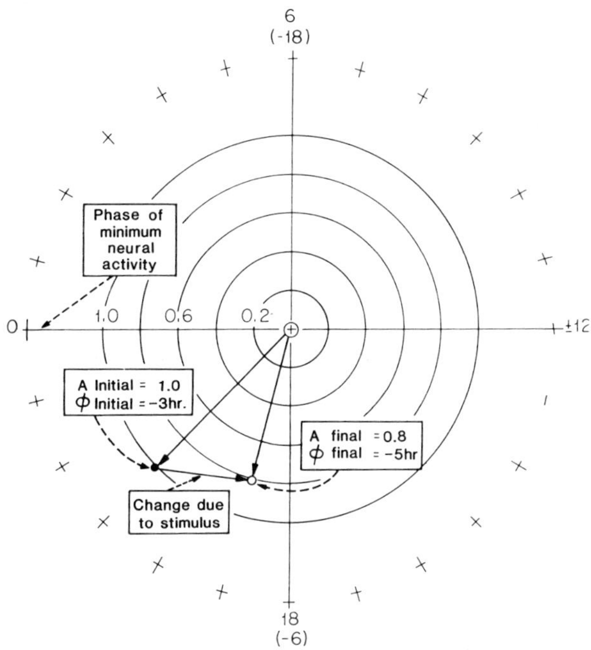

FIG. 2. Schematic representation of the data of Figure 1. The filled circle represents the initial circadian amplitude and initial circadian phase at which the stimulus occurs. The open circle represents the final (reduced) circadian amplitude and the final (earlier) circadian phase at which the stimulus occurred. Since the endogenous rhythm was *delay-shifted* 2 hours by the stimulus, the timing of the stimulus was *advanced* 2 hours in relative circadian phase.

Human resetting results

In what follows we shall describe an idealized or schematic picture of the effect of bright light (i.e. 10,000 lux) on the human pacemaker. The representation is intended to be realistic and corresponds quantifiably to the results of several key experiments, which will be cited where relevant.

At the same time the schematic picture contains many details which must be extrapolated from the limited available data and which will require many more experiments either to be corroborated or modified. To begin, we must observe that, for humans, bright light durations of several hours are needed to produce measurable phase and amplitude changes. To compare stimuli of different durations, experiments have shown it useful to use the *midpoint* of the light episode for the phase reference, as did Gander and Lewis.[6] By so doing, critical points of the response (i.e. stimulus phases at which the direction of the phase shift, or the sign of $\Delta\phi$, changes) are very nearly independent of stimulus duration.

Since human subjects do not tolerate protracted darkness, the reference condition must be modified from the conventional one for noctural rodents. We have therefore lowered ambient light exposure to the range of 10–20 lux to reduce its effect on the human pacemaker. For experimental convenience we find it useful to impose a 24 hour sleep/wake pattern. This is close to free-run periods and insures both regular sleep duration and internal synchrony. Together these constitute our reference condition.

Figure 3 shows the estimated response of a human subject to a single 5 hour episode of 10,000 lux. We assume that the subject is a typical young male placed in the reference condition by being entrained to a standard laboratory protocol in which sleep is scheduled from 24.00 to 08.00 and waking hours are spent in a room of about 20 lux. Given these conditions, the minimum of the endogenous temperature rhythm would be at about 05.30 and $\mathbf{A}_{init}=1$. For the stimulus depicted in Figure 3 the subject is awake during at least a part of his normal sleep time, and bright light (10,000 lux) is viewed from 04.00 to 09.00 (centered at 06.30). Thus, $\phi_{init}=1.0$ hours. The final state is shown as $\mathbf{A}_{final}=0.56$ and $\phi_{final}=2.4$ hours.

The 5 hours of bright light has *advanced* the pacemaker rhythm by 1.4 hours, since the timing of the center of the light episode (06.30) now occurs 2.4 hours after temperature minimum (i.e. the next minimum will now be found at 04.06, which is 1.4 hours earlier than 05.30, where it would have occurred without the light stimulus). The pacemaker responds to the stimulus progressively during the 5 hours of light application (from 04.00 to 09.00). The stimulus interval actually encompasses the timing of the minimum in both the “initial state” (05.30) and “final state” (04.06). Consequently, an experimental appraisal of the change would require comparison of the “initial” minimum (located approximately 24 hours before the stimulus) and the “final” minimum (located approximately 24 hours after the stimulus). The vector describing the change from the initial to the final state in Figure 3 shows that the reduction of amplitude (to about half its reference value) is a stronger effect of light than is the phase change.

Figure 4 shows schematically the effect of 5 hours of 10,000 lux when

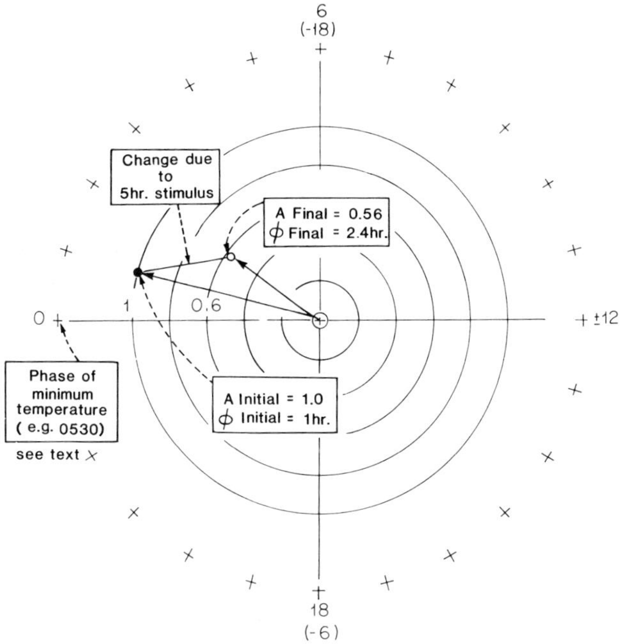

FIG. 3. Schematic representation of the effect of a single 5 hour episode of bright light on the human circadian pacemaker. The filled circle represents the initial circadian amplitude and initial circadian phase (referenced to the prestimulus rhythm) at which the mid-point of the stimulus occurs. If the reference circadian phase (endogenous temperature minimum) were at 05.30, as chosen, application of bright light from 04.00 to 09.00 (centered at 06.30) corresponds to the chosen $\phi_{initial} = 1$ hour. A 5 hour episode timed in this way is presumed to produce a 1.4 hour phase advance of the endogenous rhythm and a reduction of amplitude to 0.56. Relative to this shifted rhythm the stimulus is now centered 2.4 hours after the (new) cycle reference phase. Consequently, ϕ_{final} is 2.4 and $\mathbf{A}_{final}$ is 0.56. These are represented by the open circle.

applied at all phases of the pacemaker cycle. The filled circles represent the initial states. They are all on the unit circle, indicating that the subject was in the reference condition for each hypothetical experiment, and they are located at the integer phase points. The corresponding *final* states are indicated by open circles and identified by the value of the corresponding *initial* phase. Most of the vector changes have been drawn in. The dashed line connecting the final states is the *resetting contour*, showing all possible final states which can be reached from the reference condition ($\mathbf{A}_{init} = 1$) by using the 5 hour stimulus. The resetting contour shown is symmetrical about the horizontal axis, indicating that advancing phase shifts and delaying shifts are equally attainable. When the 5 hour bright episode is centered over the temperature minimum ($\phi_{init} = 0$) the phase shift is zero so $\phi_{final} = 0$. However, the amplitude is reduced to 0.4, which is consistent with related human data as we shall describe below. There is an important implication for animal experiments in which only phase shifting is measured: zero phase shift may very well correspond to strong amplitude reduction.

Figure 4 shows amplitude to be somewhat enhanced for ϕ_{init} between 7 and −7, typically about 20%. Whether light does enhance amplitude during these hours (between about 12.00 and 22.00 for a typical subject entrained to the normal daily routine) must be regarded as conjectural at this time; the requisite experiments have yet to be performed. In fact, identifying changes in amplitude of this size is very difficult, since the standard deviation in the measured amplitude of the endogenous temperature rhythm (in the CR) is almost as large as the hypothesized change.

An interesting feature of Figure 4 is that the principal effect of stimuli timed close to the reference phase (O) is to generate vector changes directed rightward (towards ±12). When timed exactly at reference phase, no change of phase is produced. Rightward vector changes correspond to delay shifts when they occur before reference phase and advance shifts when they occur after. Takahashi[26] has shown that the expression of FOS and JUNB in the SCN of hamsters is induced by light during the animal's subjective night at very much the same levels whether the light occurs in the delay portion of the PRC or the advance portion. If we draw an analogy between the rightward displacement produced by light in Figure 4 and the expression of FOS or JUNB we find a possible explanation for how a single chemical drive would be interpreted as either a delay shift or an advance shift depending on other chemical processes in the oscillatory system.

Because of the cumulative effects of light, we can approximate the resetting produced by bright episodes shorter than 5 hours simply by scaling down the vector change proportionally to stimulus duration. For example, a 2 hour episode will produce about 40% as much change, so the vector length would be reduced by about 60%. However, the same principle cannot be applied to episodes much longer than 5 hours. The reason can be simply deduced from Figure 4. If we proceed around the circadian cycle, we find that for about half of the cycle the rhythm amplitude is reduced by light while for the other half-cycle it is increased. Similarly, phase is advanced for half of the cycle while for the opposite half it is delayed. Any desired effect (e.g. phase advance) is limited to a "window of opportunity" and furthermore, near the edges of that window the effect is small. Thus, 12 hours of stimulus can, at best, produce little more effect than 9 hours, while 9 hours may produce only about 40% more than 5 hours. A stimulus which lasts more than 12 hours can only produce *less* effect than the best 12 hours episode. The essential fact is that there are inescapable limits to the effects which 10,000 lux can produce within any circadian cycle. To produce stronger changes it is necessary to wait until

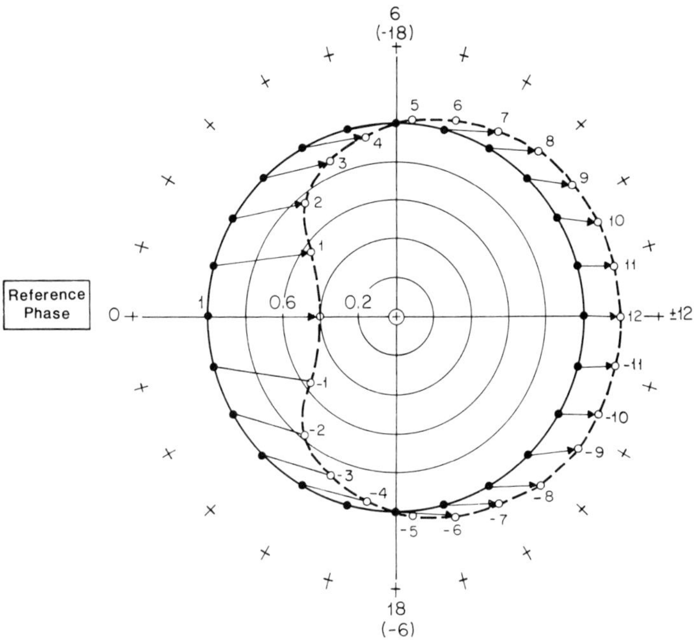

FIG. 4. The schematic representation of the response of the human circadian system to a single 5 hour episode of bright light centered at all different phases of the endogenous rhythm. Conventions are the same as in Figure 3. Filled circles represent the initial circadian amplitude (always the reference value of 1) and initial circadian phase at which the mid-point of the stimulus occurs. Open circles represent the final circadian amplitude and phase at which the stimulus occurs. The numbers by the open circles identify the corresponding *initial* phase. The closed-circle and open-circle pair identified as "1" show the effect examined in detail in Figure 3. The dashed line drawn through the open circles is the resetting contour which summarizes the effects of the 5 hour bright light stimulus for any phase of application. A *clockwise* displacement from a filled circle to the corresponding open circle indicates that the stimulus has *advanced* the endogenous rhythm, while *counterclockwise* displacement indicates *delay*. Advance and delay shifts are symmetrical in this construction.

the desired "window of opportunity" comes back again on the next cycle and applies further stimulus there.

Repetitive stimuli

Once we admit the option of applying light on successive circadian cycles, the possible combinations are innumerable. From the standpoint of experimental protocol it is clear that unless we are prepared to re-assay the system after the first light episode, the second episode can only be timed relative to the pacemaker assay which preceded the first episode. In such a case, by far the simplest protocol is to time the second, third and subsequent episodes with the same relation to the original assay as the first episode, i.e. all episodes at the same clock hour. With the same objective of simplifying the protocol, all episodes should embody the same strength of light and have the same duration. Five hours is a convenient duration since it is readily tolerated by subjects and avoids the "diminishing returns" of much longer episodes. These then constitute the protocol standards which we will use as the basis for our further schematic representations.

In the 24 hours between the centers of the stimulus episodes, the pacemaker is presumed to "run free" and express its intrinsic drives. The most significant drive is that to return rhythm amplitude to its reference value, $\mathbf{A}=1$. The simplest mathematical representation of a self-sustaining rhythm generator is the van der Pol oscillator, which we have used to represent the pacemaker on various occasions.[12,13] Only a single non-dimensional parameter, the "stiffness" of the oscillator, is needed to characterize the amplitude recovery process. We have estimated this stiffness by comparing the duration of a single 10,000 lux episode which will essentially reduce **A** to zero (an 8 hour episode) with the duration of two equal episodes required to produce the same reduction of **A** (two 5.5 hour episodes).[10] When two episodes are used (spaced 24 hours apart) there is one full cycle of amplitude recovery intervening. The resulting estimate for stiffness is 0.13.[12] Using this value of stiffness, it is a straightforward matter to calculate for the van der Pol oscillator the spontaneous change of amplitude in one cycle, $\Delta\mathbf{A}$, as a function of the amplitude at the start of the cycle, $\mathbf{A}_{\mathrm{init}}$. These are shown in Figure 5. When $\mathbf{A}_{\mathrm{init}}$ is less than 1 the amplitude increases ($\Delta\mathbf{A}$ is positive). Amplitude decreases when $\mathbf{A}_{\mathrm{init}}$ is greater than 1. Spontaneous increases as large as 0.18 can occur in one cycle.

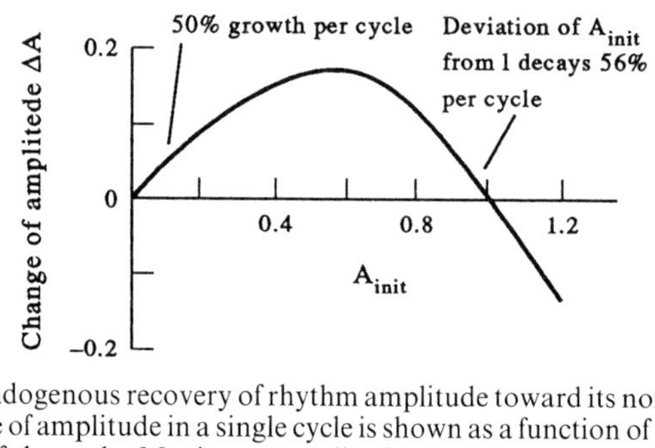

FIG. 5. The endogenous recovery of rhythm amplitude toward its nominal value of 1. The change of amplitude in a single cycle is shown as a function of the amplitude at the start of the cycle. Maximum amplitude growth is 18% of nominal for an initial amplitude of about 0.6. For initial amplitudes larger than 1, amplitude decreases in each cycle.

The second intrinsic drive of the pacemaker is embodied in its intrinsic period, τ, which is typically longer than 24 hours. In our report of the phase resetting potential of bright light,[4] the data suggest that the pacemaker period is longer than 24 hours by $\sim$0.2 hour (which is considerably closer to 24 hours than previous estimates[27]). Consequently, in the 24 hour interval between stimuli the pacemaker will drift in a delaying sense by 0.2 hours typically. Figure 6 shows schematically the combination of effects to be expected in a 2-cycle protocol where stimulus episodes are centered 24 hours apart. Between the "final state" for the first episode (open circle) and the "initial state" for the second episode (filled square) there is an amplitude recovery and a delay phase shift of the rhythm. Then the second episode is applied, inducing the second light effect.

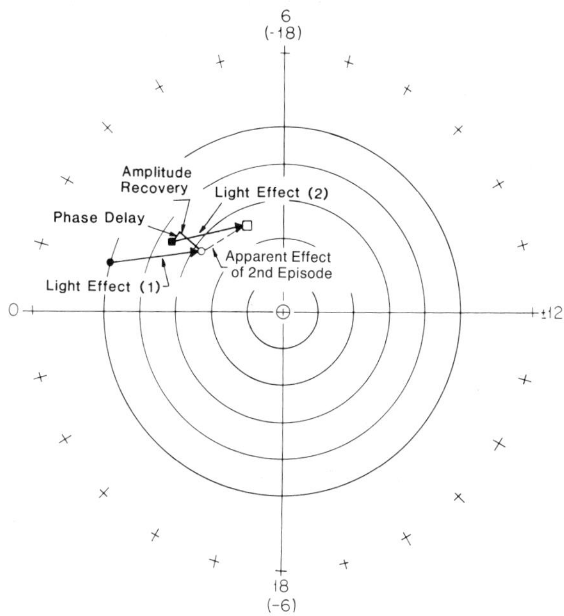

FIG. 6. A schematic representation of the endogenous circadian effects operating between successive bright episodes. (Conventions are the same as in Figure 3.) These include amplitude recovery and a phase delay shift due to an endogenous period, τ, which is larger than 24 hours. The "apparent" effect of a second episode is altered by these processes.

Figure 7 shows the schematic resetting contour for single 5 hour episodes of 10,000 lux (open circles, taken from Figure 4) along with the schematic resetting contour for two repeated episodes (open squares). The most significant feature is the extremely low amplitude produced when these episodes are timed to occur precisely over the temperature minimum. Related to this amplitude reduction is an extreme sensitivity of phase shift to the timing of the stimuli. If the two episodes are centered only 30 minutes away from the temperature minimum the phase shift will be more than 3 hours of advance or delay, depending on the sense in which the episodes are offset. The delay shifts due to τ are quite small by comparison, but they do make the resetting contour slightly asymmetric.

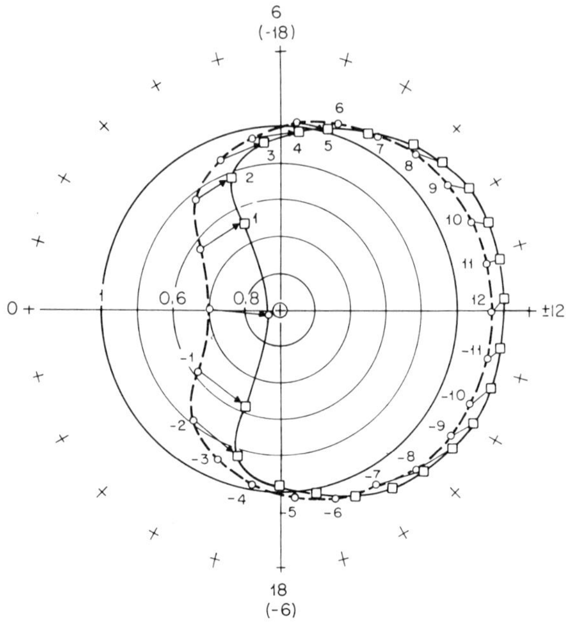

FIG. 7. A schematic representation of the "apparent" amplitude and phase changes produced by a second 5 hour episode of bright light, applied at the same clock time as the first episode but 24 hours later. As in Figures 3 and 4, the open circles represent the endogenous amplitude and the circadian phase of the midpoint of the bright light after the first episode. The open squares represent the endogenous amplitude and circadian phase of the midpoint after the second episode. The numbers indicate the circadian phase of the midpoint relative to the initial (prestimulus) rhythm shown in Figure 4 as filled circles. The open-circle and open-square sequence labeled by the number "1" corresponds to the combination of effects diagrammed in Figure 6. All "apparent" changes represented here combine amplitude recovery and phase delay (due to endogenous τ) along with the direct effect of the second bright light episode.

By waiting for the "window of opportunity" to reappear again, a third episode can be applied. During the intervening cycle another recovery of amplitude will occur and another small phase shift due to the intrinsic period, τ. Figure 8 shows the effective transition for the resetting contour from its configuration for two 5 hour episodes found in Figure 7 (open squares) to the configuration for three successive 5 hour episodes (open triangles). This transition is very special. Whereas the resetting contours for one and two episodes each encircled the "origin" of the polar diagram (the "phaseless" point, $\mathbf{A}=0$), for three episodes the origin lies *outside* the contour. The significance of this distinction is best understood by comparing just the phase aspect of the resetting process as shown in the phase resetting curves of Figure 9.

The abscissa in Figure 9 represents the 24 hour range of initial phases (at

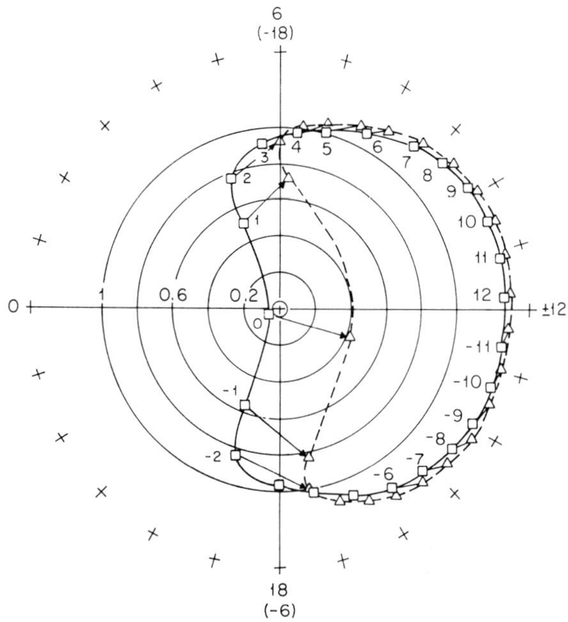

FIG. 8. A schematic representation of the "apparent" amplitude and phase changes produced by a third 5 hour episode of bright light applied at the same clock time as the first and second episodes but on the third circadian cycle. The open squares are positioned as in Figure 7. The open triangles represent the endogenous amplitude and circadian phase of the episode midpoint after the third episode. As in Figures 4 and 7 the numbers beside open-square and open-triangle sequences indicate the circadian phase of the episode midpoint relative to the initial (prestimulus) rhythm. (To avoid clutter many numbers and change vectors have been omitted.) For number 11 we observe that the "apparent" change has no net phase component. This is because the phase advance produced by the bright light exactly matches the phase delay due to endogenous τ. We also observe that the dashed line joining the open triangles (the resetting contour for the entire 3-episode stimulus) *does not* encircle the origin (which represents zero-amplitude of the endogenous rhythm). The solid line joining the open squares (the resetting contour for a stimulus consisting of only two bright light episodes) does enclose the origin. This transition thus represents Type 0 resetting.

which the first stimulus episode is applied) while the ordinate represents the final phase, after one stimulus episode, or two episodes or three episodes as the case may be. For either one episode or two episodes, for which the resetting contours in Figure 6 or Figure 7 enclose the origin, the final phase advances exactly one full cycle when the initial phase advances one cycle. That is, the *average* slope of phase resetting is unity. It is for this reason that Winfree[28] called the process "Type 1 resetting". In contrast, for three episodes, where the resetting contour in Figure 8 does not enclose the origin, all of the ϕ_{final} lie in a band of about 12 hours width and there is *no* overall advance of final phase: the average slope of phase resetting is zero.

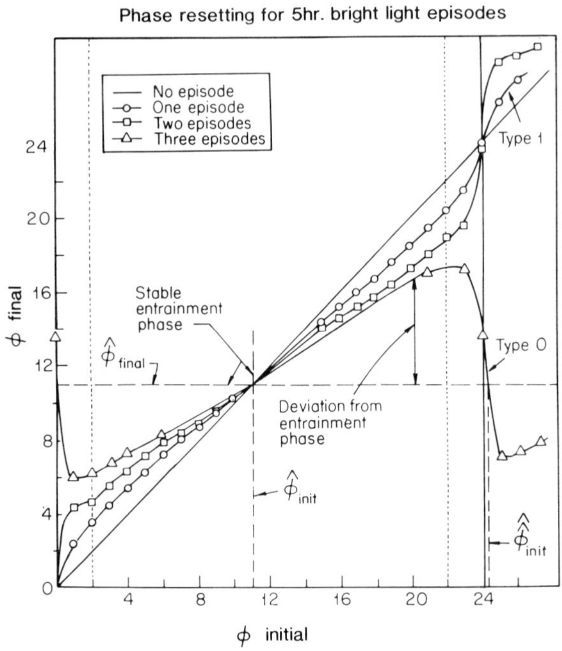

FIG. 9. A summary of the phase relationships produced by stimuli consisting of one episode of bright light, two episodes or three episodes, taken from Figures 4, 7 and 8. The abscissae correspond to the circadian phase of the midpoint of the 5 hour episodes relative to the prestimulus circadian rhythm. Second and third episodes are given at the same clock time as first episodes. The ordinates correspond to the circadian phase of these episodes after the designated number of episodes have been given. The response curves corresponding to one and two episodes are "Type 1" since ϕ_{final} increases by one cycle for an increase in ϕ_{init} of one cycle. Three episodes produce "Type 0" response since there is zero overall increase in ϕ_{final} for an increase in ϕ_{init} of one cycle. All response curves intersect at $\phi_{init} = 11$. This special condition, designated $\hat{\phi}_{init}$, is where $\phi_{final} = \phi_{init} + \delta$ (where δ is just the advance required to offset the delay produced by endogenous τ in each cycle). This is the stable entrainment condition for which $\hat{\phi}_{final} = \hat{\phi}^{initial}$. The other special condition designated $\hat{\hat{\phi}}_{init}$ corresponds to a timing of the 3-episode stimulus which will bring the circadian rhythm to stable entrainment in precisely those three episodes (by inducing a phase shift of about 11 hours).

Winfree called this "Type 0 resetting". In Figure 9, if we look at the range $2 < \phi_{init} < 22$ we find the resetting curves for one, two and three episodes progress in an orderly sequence of stronger phase shifts, although three episodes do not quite produce three times the shift of one episode. However, if we look at ϕ close to zero (or close to 24, which is the same thing) we see that by progressing from two to three episodes the slope of the resetting curve changes from steeply positive (about $+10$) to steeply negative (about -10). It is for these ϕ_{init} that Type 0 resetting is essentially different from Type 1 resetting. If we consider the stimulus to be centered very soon after the reference phase so that ϕ_{init} is, say, 0.25 hour we find that after the first light episode the phase is advanced about 1 hour, after the second episode (at the same clock time as the first episode) the phase is advanced (overall) about 3 hours, but after the third episode it is advanced (overall) almost 9 hours. The extraordinarily strong phase-shifting effect of the third episode is due to a great reduction of amplitude produced by the first two episodes.

Type 0 resetting can occur *only* if the cumulative effect of individual light episodes is able to reduce the amplitude to zero. Figure 5 shows that the maximum growth of amplitude (for van der Pol "stiffness", $\mu=0.13$) in a single cycle is approximately 0.18. Thus, any light stimulus which can reduce amplitude by more than 0.18 in a single episode is capable of producing Type 0 resetting (although the number of repetitions required may be very large). Weaker stimuli can only give Type 1 resetting. An important corollary to these observations is that if the single episode has sufficient strength to reduce amplitude to zero by N repetitions, then the resetting will certainly be Type 0 for all protocols in which the stimulus is repeated more than N times, and will certainly be Type 1 when the repetitions are fewer than N.

The distinction is shown qualitatively in Figure 10. For simplicity, endogenous τ is assumed to be exactly 24 hours, and also only half of the phase resetting curve is shown. Figure 10a portrays the weak stimulus situation for which the endogenous recovery of amplitude is strong enough to prevent the stimulus from driving amplitude to zero. No matter how many times the stimulus episode is repeated there will always be a band of ϕ_{init} close to zero for which phase shifts are small. Figure 10b shows the situation in which N repetitions can produce zero amplitude. The resetting pattern bifurcates at N repetitions. When the repetitions, n, are fewer than N, the resetting curves all pass through the point $\phi_{\text{init}}=0$, $\phi_{\text{final}}=0$ (much as they do for Figure 10a) but when $n>N$ the resetting curves all pass through the point $\phi_{\text{init}}=0$, $\phi_{\text{final}}=12$.

When undertaking to classify experimental resetting data as Type 1 or Type 0 the protocol must be designed to provide maximum discriminability in the critical region near $\phi_{\text{init}}=0$. The data points must be dense enough so that the sense in which the phase resetting curve is cyclically closed is unambiguous (despite "noise" in the assessment of each point). This means also that the total stimulus strength must be weak enough so that the band near $\phi_{\text{init}}=0$, for which phase shifts would be small if the resetting were Type 1, is clearly defined. For the human resetting we have reported,[4] the data points were principally clustered in the critical region while the choice of a three-episode stimulus was the fewest repeats of 5 hour episodes which could produce Type 0.

The transition between Type 1 and Type 0 resetting is traditionally

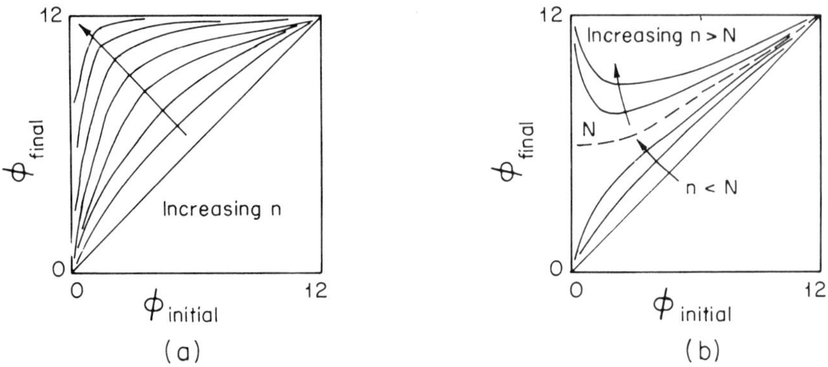

FIG. 10. Schematic representation of phase resetting curves for increasing number of repeated stimulus episodes. (a) The individual stimulus episode is weaker than the maximum endogenous recovery of amplitude between episodes. The resetting is Type 1 however many the repetitions. (b) The individual stimulus episode is strong enough to reduce the amplitude to zero in N repetitions. The resetting is Type 1 for $n<N$, bifurcates when $n=N$ and is Type 0 for $n>N$.

discussed in terms of a single (brief) stimulus of increasing strength in which the strength increase may come about through brighter light or a longer episode.[18,28] Here our stimulus consists of a patterned repetition of separated episodes; increasing overall stimulus strength is obtained by increasing the number of episodes. Without amplitude recovery, Type 0 resetting will ultimately be found for any repeated stimulus episode which reduces amplitude, however weak that reduction may be for each repetition.

Strogatz[25] reviewed the mathematical arguments which show how repeated applications of Type 1 resetting can never produce Type 0 resetting—a result which may superficially seem to contradict the presentation here. However, Strogatz also observed that if a sequence of resettings gives Type 0 as the final result then one member of the sequence must be Type 0. In Figures 4 and 7 the resettings are each Type 1 since both the "initial" and "final" contours in each figure encircle the "phaseless point" (the origin). It is in Figure 8 that the resetting (due to the third bright light episode) is Type 0, and this establishes Type 0 for the overall sequence of three episodes. Strogatz discussed amplitude reduction and subsequent endogenous recovery in terms of two limiting cases: rapid recovery and slow recovery. The human response to 5 hour bright light epsiodes which we have depicted is intermediate between these. Significant amplitude recovery does occur between the first and second episodes, but the total reduction achieved by these episodes is strong enough to win the competition. Once the amplitude is close to zero, the recovery per cycle is presumed to be small (Figure 5).

Entrainment and re-entrainment

All of the resetting curves of Figure 9 intersect at $\hat{\phi}_{\text{init}} \cong 11$ hours implying that neither the second nor third stimulus produces any net phase shift. In fact, they each produce a phase advance of 0.2 hour which just offsets the daily phase delay drift due to intrinsic $\tau = 24.2$ hours. The pacemaker is seen to be stably entrained by daily 5 hours bright episodes when they are centered 11 hours after temperature minimum. However, a person who has $\tau = 24.4$ hours requires a daily phase advance of 0.4 hour for entrainment to an imposed 24 hour rhythm. A daily 5 hour bright light episode will provide this if centered at 10 hours after the circadian reference phase. In daily life the patterning of the light stimulus is dictated primarily by the social constraints of employment and recreation which are tied to solar time. Thus stimulus timing is primarily independent of a person's individual τ. In that case, persons with longer τ will have their reference circadian phase at a later clock hour. If the total daily stimulus approximately matches the strength of the 5 hours of bright light discussed here, the clock time of the reference phase would be $\frac{1}{2}$ hour later for each 0.1 hour increase of τ. For weaker stimuli the effect of τ on entrainment phase would be even greater.

Entrainment is the condition for which $\phi_{\text{init}} = \phi_{\text{final}}$, after the drift due to intrinsic τ has been taken into account. The horizontal line in Figure 9 at $\hat{\phi}_{\text{final}} = 11$ hours is a representation of the entrained condition equivalent to $\hat{\phi}_{\text{init}} = 11$ hours. This horizontal line has a second intersection with the three-episode resetting curve at $\hat{\hat{\phi}}_{\text{init}} \cong 24.2$ hours (or equivalently $\hat{\hat{\phi}}_{\text{init}} = 0.2$ hours). The significance of this second intersection is that three episodes critically timed in this precise way, relative to the initial state of the pacemaker, will theoretically advance the pacemaker by 10.8 hours exactly so that no further phase adjustment is required for stable entrainment.

A basic approach to the re-entrainment needed to cope with transmeridional jet travel is to separate the phase shifting process from that of entrainment. One or more bright light episodes are timed to develop the required phase change rapidly and conveniently. Once the new phase has been approximately achieved, the natural stimuli in the new environment are relied upon to complete and maintain entrainment. The bright light episodes chosen for this scenario are often much different from either $\hat{\phi}_{\text{init}}$ or $\hat{\hat{\phi}}_{\text{init}}$.

In the preindustrial ages artificial light was a scarce commodity and so the entrainment of the circadian pacemaker was necessarily governed by natural light which is centered about noon. This would have placed the pacemaker reference phase shortly after midnight, say 01.30. We have found that present-day young adult men (18–31 years) have their reference phase close to 06.45 ± 1.3 hours SD while persons of age 60–85 show reference phase of 04.55 ± 2.0 hour SD.[3] These present-day entrainment phases appear to have two principal causes. The first is a reduced light exposure during daytime hours, compared to that in an agrarian setting. Kripke and coworkers have found[11] surprisingly low daily exposures, even in the favorable climate of San Diego, California. The second is the extension of illuminated hours well beyond sundown. Indoor light, though more than 50 times lower in physical intensity than typical outdoor light appears to have a significant cumulative effect.

There appears to be a strongly "compressive" nonlinear relationship between the physical measure of light (e.g. lux) and the effect it exerts on the pacemaker. We have proposed that the effect is proportional to the cube-root of the lux measure.[12] In laboratory studies where human subjects are allowed to select their own sleep/wake schedule (human "free-run" protocol) they traditionally have been allowed to turn on room lights during waking hours (150–200 lux). The free-run period could often be about 25.5 hours with the pacemaker remaining entrained to the sleep/wake rhythm. That is, the pacemaker could be drawn-off from its natural period by more than an hour.[13] Our current data indicate that entrainment of the pacemaker to such a different period was effected solely by the self-selected dark/light cycle which was coincident with the sleep/wake cycle. Thus it appears that about 17 hours of 100–200 lux can shift the pacemaker phase by as much as 1 hour per day.

Conclusions

When multiple episodes of light are used on successive days, analysis of experimental data is made more difficult by the endogenous rhythmic processes which intervene between episodes. Before spontaneous internal desynchrony was reported by Aschoff in 1965[1] it was thought that the free-run period which characterized internally synchronized records (approximately 25.5 hours) represented the intrinsic pacemaker period. From a study of free-run records in which internal desynchrony was seen, it was recognized that with synchrony the true pacemaker period was being compromised by the endogenous sleeep/wake rhythm. Wever[27] ascribed to the pacemaker the period 24.85 ± 0.3 (SD) hours. Kronauer *et al.*[13] pointed out that pacemaker periods observed even in desynchrony could be somewhat high due to a residual effect of the sleep/wake cycle. The phase shifting results of Czeisler *et al.*[4] on which many features of the present discussion are based, indicated that the daily phase delay attributable to pacemaker period were in fact considerably smaller than Wever's estimate, being only about 0.2 hours per cycle rather than 0.85 hours.

In the vector presentation used in Figure 6, the 0.2 hours delay drift is seen to be small compared to the recovery of amplitude which occurs in a single cycle. Since Type 0 resetting is a powerful method of achieving rapid phase change, and since Type 0 resetting requires large amplitude reduction, endogenous amplitude recovery has emerged as having as strong an impact on phase resetting as phase drift due to endogenous τ. We conclude that to utilize the Type 0 strategy the effect of *individual* bright light episodes must be strong: amplitude recovery makes Type 0 inaccessible if light exposure is casual or weak.

Finally, we wish to emphasize that the resetting contours of Figures 4, 7 and 8 are idealizations based upon limited data. They are the result of a simplified model[12] with parameters adjusted to match the data reasonably well. As experiments are refined and repeated the details of the resetting contours may change but we expect the essential features to remain.

Acknowledgments

We thank Megan E. Jewett for her thoughtful criticism of this manuscript. This research analyzed herein was supported in part by NIA 1-R01-AG06072, NIA 1-P01-AG-09975, NIMH 5-R01-MH45130, NASA NAG 9-524 and NIH DRR GCRC5-M01-RR00888. Richard E. Kronauer is a Gordon McKay Profesor of Mechanical Engineering at Harvard University. Charles A. Czeisler and Richard E. Kronauer serve as consultants to Light Sciences, Inc., Brookline, Massachusetts.

References

1. Aschoff J. (1965). Circadian rhythms in man. *Science* **148**, 1427–1432.
2. Czeisler C.A., Brown E.N., Ronda J.M., Kronauer R.E., Richardson G.S. and Freitag W.O. (1985). A clinical method to assess the endogenous circadian phase (ECP) of the deep circadian oscillator in man. *Sleep Res.* **14**, 295.
3. Czeisler C.A., Dumont M., Duffy J.F., Steinberg J.D., Richardson G.S, Brown E.N., Sánchez R., Ríos C.D. and Ronda J.M. (1992). Association of sleep-wake habits in older people with changes in output of circadian pacemaker, *The Lancet*, **340**, 933–936.
4. Czeisler C.A., Kronauer R.E., Allan J.S., Duffy J.F., Jewett M.E., Brown E.N. and Ronda J.M. (1989). Bright light induction of strong (type 0) resetting of the human circadian pacemaker. *Science* **244**, 1328–1333.
5. Czeisler C.A., Kronauer R.E., Mooney J.J., Anderson J.L. and Allan J.S. "Biologic rhythm disorders, depression, and phototherapy. A new hypothesis", *Psychiatric Clinics of North America*, Vol. 10, No. 4, "Sleep Disorders", pp. 687–709. W.B. Saunders Co, Philadelphia, Dec. 1987.
6. Gander P.H. and Lewis R.D. (1983). Phase resetting action of light on the circadian rhythm of *Rattus Exulans. Am. J. Physiol.* **245**, R10–R17.
7. Hastings J.M. and Sweeney B.W. (1958). *Biol. Bull.* **115**, 440–458.
8. Inouye S.T. and Kawamura H. (1982). Characteristics of a circadian pacemaker in suprachiasmatic nucleus. *J. Comp. Physiol.* **146**, 153–160.
9. Jewett M.E. and Czeisler C.A. (1992). The role of endogenous circadian amplitude in the modulation of alertness and performance: An investigation of the singular region. *Sleep Res.* **21**, 378.
10. Jewett M.E., Kronauer R.E. and Czeisler C.A. (1991). A light-induced suppression of endogenous circadian amplitude in humans. *Nature* **350**, 59–62.
11. Kripke D.F. and Gregg L.W. (1990). Circadian effects of varying environmental light.

In *Medical Monitoring in the Home and Work Environment* (eds. Miles L.E. and Broughton R.J.), pp. 187–195. Raven Press, New York.
12. Kronauer R.E. (1990). A quantitative model for the effects of light on the amplitude and phase of the deep circadian pacemaker, based on human data. *Sleep '90* (*Proceedings of the Tenth European Congress on Sleep Research, Strasbourg*). (ed. Horn J.A.), Bochum: Pontenagel Press, 306–309.
13. Kronauer R.E., Czeisler C.A., Pilato S., Moore-Ede M.C. and Weitzman, E.D. (1982). Mathematical model of the human circadian system with two interacting oscillators. *Am. J. Physiol.* **242**, R3–R17.
14. Lewy A.J., Wehr T.A., Goodwin F.K., Newsome D.A. and Markey S.P. (1980). Light suppresses melatonin secretion in humans. *Science* **210**, 1267–1269.
15. Lewy A.J., Sack R.L. and Singer C.M. (1985). Immediate and delayed effects of bright light on human melatonin production: Shifting "dawn" and "dusk" shifts the dim light melatonin onset (DLMO). *Ann. NY Acad. Sci.* **453**, 253–259.
16. McIntyre I.M., Norman T.R., Burrows G.D. and Armstrong S.M. (1989). Human melatonin suppression by light is intensity dependent. *J. Pineal Res.* **6**, 149–156.
17. Nelson D.E. and Takahashi J.S. (1991). Sensitivity and integration in a visual pathway for circadian entrainment in the hamster (*Mesocricetus Auratus*). *J. Physiol.* **439**, 115–145.
18. Peterson E.L. (1980). Phase-resetting a mosquito circadian oscillator. *J. Comp. Phyisol.* **138**, 201–211.
19. Pittendrigh C.S., Kyner W.T. and Takamura T. (1991). The amplitude of circadian oscillations: Temperature dependence, latitudinal clines, and the photoperiodic time measurement. *J. Biol. Rhythms* **6**(4), 299–313.
20. Reppert S. and Schwartz W. (1983). Maternal coordination of the fetal biological clock in utero. *Science* **220**, 969–971.
21. Schwartz W.J., Davidsen L.C. and Smith C.B. (1980) In vivo metabolic activity of a putative circadian oscillator, the rat suprachiasmatic nucleus. *J. Comp. Neurol.* **189**, 157–167.
22. Schwartz W.J., Gross R.H. and Morton, M.T. (1989) The suprachiasmatic nuclei contain a tetrodotoxin resistant circadian pacemaker. *Proc. Natl. Acad. Sci.* **84**, 1694–1698.
23. Shanahan T.L. and Czeisler C.A. (1991). Light exposure induces equivalent phase shifts of the endogenous circadian rhythms in circulating plasma melatonin and core body temperature in men. *J. Clin. Endocrinol. Metab.* **73**(2), 227–235.
24. Souetre E., Salvati E., Belugou J.-L., Pringuey D. *et al.* (1989) Circadian rhythms in depression and recovery: Evidence for blunted amplitude as the main chronobiological abnormality. *Psychiatry Res.* **28**(3), 263–278.
25. Strogatz S.H. (1990). Commentary: Interpreting the human phase response curve to multiple bright-light exposures. *J. Biol. Rhythms* **5**(2), 169–174.
26. Takahashi J.S. Biological Rhythms: From Gene Expression to Behavior. This volume, pp. 3–20.
27. Wever R.A. (1979). *The Circadian System of Man.* Springer, New York.
28. Winfree A.T. (1979). *The Geometry of Biological Time.* Springer, New York.

16

Sleep/Wake Regulation

TORBJÖRN ÅKERSTEDT[1] and SIMON FOLKARD[2]

[1]*IPM & Stress Research, Karolinska Institute, Stockholm, Sweden*
[2]*MRC Social and Applied Psychology Unit, Sheffield, UK*

A MAJOR reason for having the present symposium is that misalignment frequently occurs between behavior and the circadian system and that this seriously affects the individual. In particular, one observes disturbed alertness patterns that lead to excessive sleepiness, disturbed sleep, and a loss of capacity to function properly. This is seen in clinical populations as demonstrated in other contributions to the present meeting. It also occurs, however, in healthy populations, and then as perfectly "normal", non-pathological phenomena. In particular, we can observe the effects in relation to shift work and transmeridian jet travel. This presentation will give an overview of such effects and their causes, and will propose a practical mathematical model for the prediction of sleepiness/alertness.

Shift work

A number of studies have reported that shift workers feel **sleepy** during particularly the night shift.[41] Recently, this has been verified with ambulatory EEG methods. Figure 1 shows the hypnogram of a control room worker falling asleep three times during a night shift.[31] This type of behavior occurred in 25% of the subjects and was not condoned by management or unions. In fact, it was a reason for dismissal. Interestingly, the sleep incidents were in most cases not recognized by the subjects themselves—a circumstance that constitutes an additional problem. Similar patterns of sleep intrusions into the waking EEG have been demonstrated for train drivers[29] as well as for truck drivers.[19] In both cases spectral power density in the alpha (8–12 Hz) and theta (4–7 Hz) bands reached high levels towards the early morning and were closely correlated with self-rated sleepiness. High levels of EEG alpha and theta

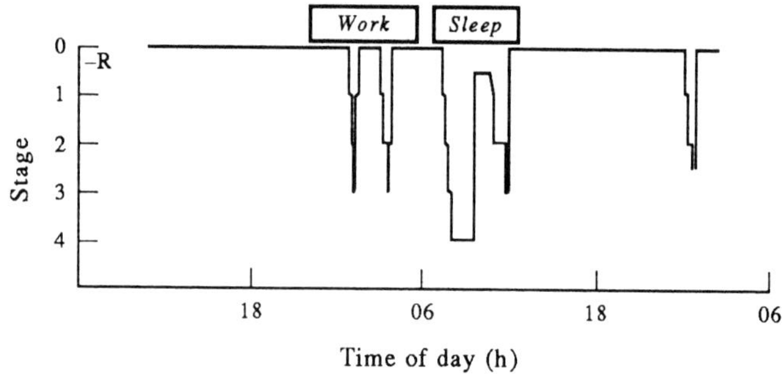

FIG. 1. 24 hour hypnogram involving a night shift for a process operator. Sleep stages 0–4 and REM are indicated.

activity are essentially incompatible with purposeful interaction with the external world.[25,30,33]

If sleepiness is excessive during work one would expect an accompanying performance degradation and an increased accident risk at work. This appears to be the case, indeed. Thus, gauge reading errors of process operators[2] and response latency of switchboard operators increase during the night shift.[5] More importantly, the risk of single vehicle road accidents reaches a pronounced peak in the morning hours[17] and night shift sleepiness has been identified as a major contributor to many of the recent nuclear reactor accidents and incidents.[22] Much research remains to be done in this area, however.

The effects of shift work on **sleep** have been studied in a number of EEG studies. Essentially, the results show that sleep is shortened by 2–4 hours in connection with night shift and morning shift work.[11,28,31,32] Mostly the sleep loss affects the total amounts of stage 2 and REM sleep. Slow wave sleep (SWS = stages 3 and 4) does not seem to be much affected. Interestingly, sleep after the night shift is often spontaneously terminated, leaving a certain feeling of recuperation.[45] The sleep before a morning shift, however, is terminated by the alarm clock (or similar arrangements) under considerable resistance. The latter leaves the sleeper with a feeling not being refreshed by sleep. The afternoon shift, in contrast, is usually perceived as very comfortable with respect to sleep or wakefulness.

Jet lag

Jet lag is a problem similar to that of shift work. It usually involves out-of-phase feelings of excessive sleepiness and sleep disturbances after transmeridian flights across three or more time zones.[36]

After a westward flight across, for example, eight time zones, alertness and performance will be reduced in the afternoon and evening of the new time zone.[6,23,25,34,37] After an eastward flight alertness and performance will be reduced from the early morning to the beginning of the night. Interestingly, the performance effects correspond to those seen in connection with blood alcohol levels of 0.05%.[1]

With respect to sleep, a westward flight involves a rapid sleep onset but an early (local time) awakening. Both progressively approach normality across a week.[8,24,26,35] Eastward flights mainly involve problems of initiating sleep and of difficulties awakening.[8,26] Frequently, the first sleep after arrival is quite normal, whereas the second sleep is disturbed, followed by subsequent, gradual recovery.

The mechanism of disturbed sleep

The reason for the sleep disturbances in shift work and jet travel seem rather straightforward. Controlled laboratory studies (under optimal sleep conditions) show that sleep that is displaced towards later times of night gradually decreases in length with increasing displacement (see Figure 2).[42] But only up to a point—from noon, when sleep length is between 4 and 5 hours, a lengthening occurs, reaching a maximum of 11 hours after a bedtime around 19.00 hours. This then, clearly shows that the short post-night shift sleep of the shift worker, or the post-flight sleep of the transmeridian jet traveller, is a part of a general pattern of greatly reduced daytime sleep propensity, i.e. a **time of day** pattern. External influences, such as noise, need not be invoked to explain shortened day time sleep, although such influences may exacerbate the problems. Interestingly, also REM sleep shows a morning peak in duration.[7,18,38,42]

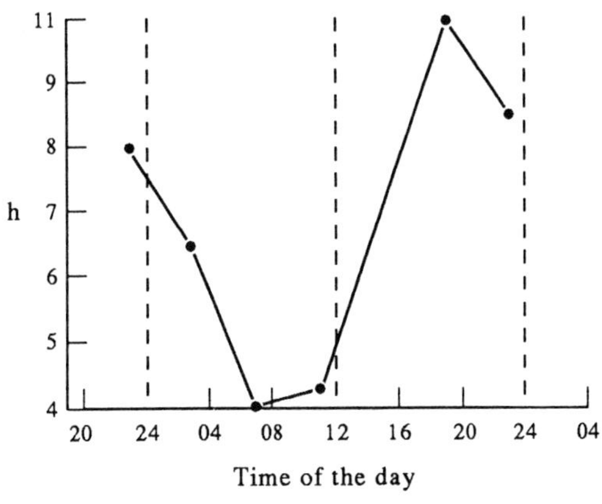

FIG. 2. Mean ± SE of TST following different bedtimes. $N = 6$.

Isolation studies have demonstrated that the time of day pattern is highly correlated with the phase of the rectal temperature rhythm, such that high or rising portions are associated with a low sleep propensity as well as with a low REM propensity.[6,39] Thus, the phase of rising body temperature is associated with strong sleep offset tendencies (not necessarily causal). Also sleep latency is influenced by circadian factors such that it reaches a minimum in the early morning.[20,23,27] Even a very gradual phase advance of bedtime will increase sleep latency.[44]

It should be emphasized that the circadian influence on sleep and wakefulness is dependent on the rate of adjustment of circadian rhythmicity to phase shifts of the sleep/wake pattern. With respect to shift work the adjustment is only marginal.[40] Even in permanent night workers the circadian phase may only be delayed by 1 or 2 hours. The adjustment to time zone shifts is approximately 1 hour per day after eastward flights and slightly more after westward flights.[37] Note, however, that masking effects may cause an impression of faster adjustment.[21] As is discussed in several other contributions to the present symposium, light plays a major role in the adjustment process.

The circadian influence on sleep may be sizeable, but it should be strongly emphasized that sleep is equally strongly regulated by **homeostatic** influences. Thus, for example, a daytime sleep started at 11.00 hours will be differentially long depending on the amount of sleep during the prior night.[43] The resulting TST may vary from 4.5 hours, after no prior night sleep, via 3.5 hours after 2 hours of night sleep and 2.8 hours after 4 hours of night sleep to 2 hours after a full night's sleep.

The homeostatic influence is also clearly evident in the amount of deep sleep—stage 3 and 4 sleep (SWS). Both the sleep stage measure (SWS) and the computer derived measure (slow wave activity—SWA) are almost entirely dependent on the amount of prior waking[4,18,34] or the amount of prior sleep.[43] If SWS is suppressed (by e.g. noise), sleep is spontaneously extended until the SWA needed has been recovered.[15]

Borbély and others have combined the two main regulatory influences into a "two-process model of sleep regulation".[3] Together (approximately 50% each), homeostatic and cirdadian regulation account for most of the effects of both shift work and transmeridian travel on sleep.

Mechanism of excessive sleepiness

Inspired by the sleep model we have constructed a practical model for sleepiness regulation. Using published data from experiments with life on a 22.0 hour day,[9] sleep deprivation[10–14] and other studies, we have found that subjective sleepiness (or alertness) is predictable using three processes: S, C and W. Space does not permit here a discussion of the derivation of the model but the reader is referred to Folkard and Åkerstedt.[10] Here will be described the main traits of the model.

Process C

This represents sleepiness due to circadian influences and has the general sinusoidal form indicated below (Figure 3 and Table 1), where M = amplitude (in arbitrary units on a scale from 1 to 21), p = acrophase (in decimal hours), and t = time of day (in decimal hours). To the right, model default parameters have been inserted. Our attempts to model empirical data suggest that the phase estimate may have to be delayed to somewhat later hours in students, shift workers or other groups in which evening type preferences dominate.

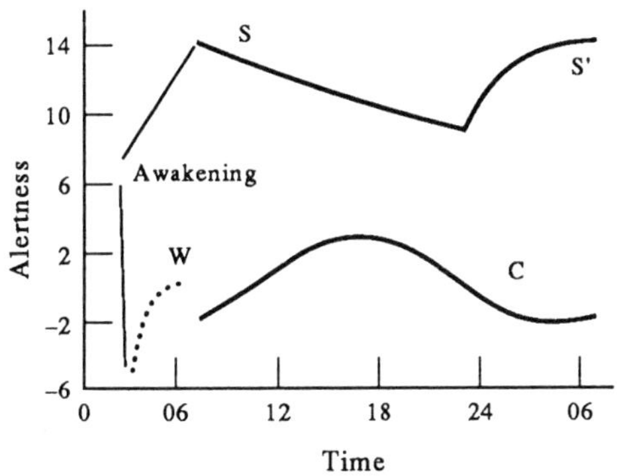

FIG. 3. The three components S, C and W of the alertness regulation model (plotted in hourly intervals).

TABLE 1
Mathematical functions of the model of alertness regulation

$$C = M\cos(t-p)\pi/12 \quad \text{or} \quad C = 2.5\cos(t-16.8)\pi/12$$
$$S = (S_a - L)e^{-0.0353t} + L \quad \text{or} \quad S = (14-2.4)e^{-0.0353t} + 2.4$$
$$S' = U - (U - S_r)e^{-0.381t} \quad \text{or} \quad S' = 14.3 - (14.3 - 7.96)e^{-0.381t}$$
$$W = 5.72e^{-1.51t}$$
$$A = C + S \text{ (or } S') - W$$
$$S' + C' = U$$

Process S

This is an exponential function of the time since awakening. Maximum alertness is reached upon awakening and alertness initially falls rapidly but levels out and gradually approaches an asymptote.The rate of decrease of the exponential function corresponds to 3.5% of the previous value per hour. S_a = value of S at awakening, L = lower asymptote, t = time since awakening. At sleep onset process S is reversed and called S' and recovery occurs as an exponential function that initially increases at a very rapid rate but subsequently levels off towards an upper asymptote. $S_r = S$ value at retiring; U = upper asymptote. The rate of increase of the exponential function is 28.7% per hour, i.e. much steeper than the decay during the day.

The final component is the wakeup **process W**, or sleep inertia, after *forced* awakenings. This function is also exponential but with an even steeper initial decrease—the rate of decrease is 62.7% per hour, i.e. already after the first hour most of the inertia has dissipated. W is subtracted (in effect) from the $S+C$ level. Spontaneous awakenings constitute a special problem since the exact time of awakening often is not availabe, and since the time between the first formal awakening and the time of finally rising might be considerable. Tentatively, we assume that such a process might take approximately 1 hour.

The estimated alertness (or sleepiness) is then expressed as the arithmetic sum of the three functions above. Using the model we have found it possible to predict with accurracy the variation of subjective sleepiness under conditions of sleep deprivation and various experimental or naturally occurring alternations of sleep/wake behavior.[10] Figure 4 shows an example of prediction of latency to stage 1 sleep during a 30 minute vigilance test performed every 2 hours in a study of 64 hours of continuous activity. The study involved 12 subjects under conditions of isolation from external synchronizers.[16] The circadian acrophase was set at 20.48 hours. The covariation between the curves is very close, reaching an r^2 of 0.70. Note the line at 7 units, denoting the "critical level". This level was determined after laboratory experiments and represents the level at which EEG-signs of sleep will start appearing within 5 minutes in a waking subject with open eyes focusing on a spot on the wall.

Figure 4 also shows the prediction of (mean) sleepiness in 20 workers in connection with night shift work.[31] Again the variance accounted for is very high. EEG-defined sleep incidents start to occur around and below level 7.

Sleep and the alertness model

In order to be comprehensive the sleepiness model also needs to be able to predict sleep. In a previous study we have demonstrated that *SWA* during sleep is predicted by component S',[41] that is SWA follows the same pattern as does S'.

We have also recently started to apply the model to **sleep termination** (unpublished). This application is based on the assumption that sleep termination occurs when $S+C$ reach a certain level of alertness (see

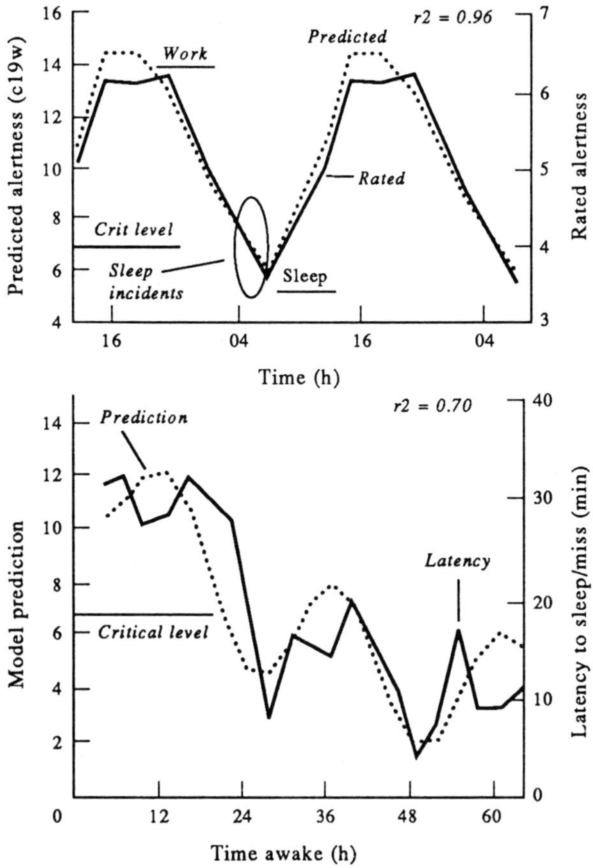

FIG. 4. Comparison of model prediction and empirical data. Above: rated alertness and sleep incidents in night shift workers. Below: sleep latency data (group mean) during a 64 hour sleep deprivation experiment.

Figure 5). Furthermore, we assume that the value in question is the upper asymptote, since it may be argued that S' would reach close to its asymptote if circadian factors would not interfere. Hence, we propose that sleep termination occurs when $S'+C=U$ (the latter is the upper asymptote of S').

Figure 5 illustrates the application of the model. We assume 16 hours of waking preceeding each of the bedtimes indicated. Alertness is regained during sleep according to process S'—up to the point where circadian influences stop the process. In the afternoon this interference will occur very early and in the early morning it will occur much later. We also assume that the influence (amplitude) of C is less during sleep. We therefore sought the optimum amplitude of C by modelling empirical sleep length data from two earlier studies of displaced sleep.[42,43] This was done

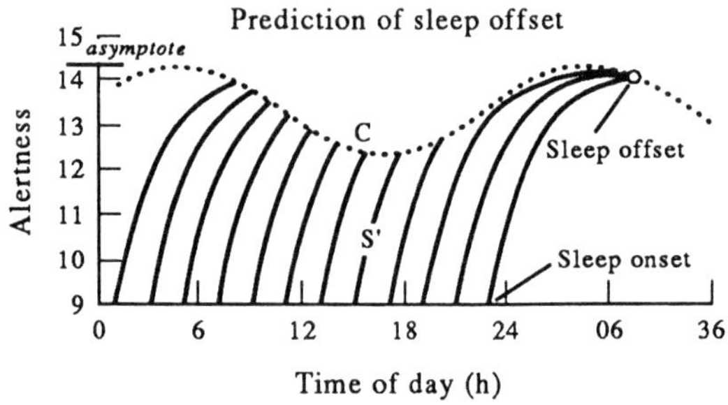

FIG. 5. Use of the alertness regulation model for prediction of sleep length. See text for further information.

by varying the amplitude of C in the function $S'+C=U$. The results showed that optimum (96%) prediction could be obtained by using an amplitude of 1. This yielded a regression coefficient of 1.01 (through the origin) for the prediction of sleep length by the model. The r^2 reached 0.98 for group mean sleep lengths. This new, sleep terminating, version of C has been named C'.

Final comments

This presentation has wanted to emphasize that alertness and sleep are severely affected by phase shifts of sleep wake behavior but that appreciation of the mechanism and provisions for countermeasures make it necessary to consider not only circadian rhythmicity, but also homeostasis. The latter has a tendency to be left out in connection with work on rhythms and light.

References

1. Klein K.E., Wegmann H.M. and Hunt B.I. (1972) Desynchronization of body temperature and performance circadian rhythm as a result of outgoing and homegoing transmeridian flights. *Aerospace Med.* **43**, 119–132.
2. Bjerner B., Holm Å. and Swensson, Å. (1955) Diurnal variation of mental perfomance. A study of three-shift workers. *Br. J. Ind. Med.* **12**, 103–110.
3. Borbély A. (1982) Sleep regulation: Circadian rhythm and homeostasis. In *Sleep. Clinical and Experimental aspects* (eds. Ganten D. and Pfaff D.), pp. 83–104. Springer, Berlin.
4. Borbély A.A., Baumann F., Brandeis D., Strauch I. and Lehmann D. (1981) Sleep deprivation: effects on sleep stages and EEG power density in man. *Electroenceph. Clin. Neurophysiol.* **51**, 483–493.
5. Brown R.C. (1949) The day and night performance of teleprinter switchboard operators. *Occup. Psychol.* **23**, 121–126.
6. Czeisler C.A., Weitzman E.D., Moore-Ede M.C., Zimmerman J.C. and Knauer R.S. (1980) Human sleep: its duration and organization depend on its circadian phase. *Science* **210**, 1264–1267.
7. Czeisler C.A., Zimmerman J.C., Ronda J.M., Moore-Ede M.C. and Weitzman E.D. (1980) Timing of REM sleep is coupled to the circadian rhythm of body temperature in man. *Sleep* **2**, 329–346.

8. Dement W.C., Seidel W.F., Cohen S.A., Bliwise N.G. and Carskadon M.A. (1986). Sleep and wakefulness in aircrew before and after transoceanic flights. *Aviat. Space Environ. Med.* **57**, B14–28.
9. Folkard S., Hume K.I., Minors D.S., Waterhouse J.M. and Watson, F.L. (1985) Independence of the circadian rhythm in alertness from the sleep/wake cycle. *Nature* **313**, 678–679.
10. Folkard S. and Åkerstedt T. (1991) A three process model of the regulation of alertness and sleepiness. In *Sleep, Arousal and Performance: Problems and Promises* (eds. Ogilvie R. and Broughton R.), pp. 11–26. Birkhäuser, Boston.
11. Foret J. and Lantin G. (1972) The sleep of train drivers: An example of the effects of irregular work schdules on sleep. In *Aspects of Human Efficiency. Diurnal Rhythm and Loss of Sleep* (ed. Colquhoun W.P.), pp. 273–281. The English Universities Press, London.
12. Fröberg J., Karlsson C.G., Levi L. and Lidberg L. (1975) Circadian variations of catecholamine excretion, shooting range performance and self-ratings of fatigue during sleep deprivation. *Biol. Psychol.* **2**, 175–188.
13. Fröberg J.E. (1977) Twenty-four hour patterns in human performance, subjective and physiological variables and differences between morning and evening active subjects. *Biol. Psychol.* **5**, 119–134.
14. Fröberg J.E., Karlsson C.G., Levi L. and Lidberg L. (1975) Psychological circadian rhythms during a 72-hour vigil. *Försvarsmedicin.* **II**, 192–201.
15. Gillberg M., Anderzén I. and Åkerstedt T. (1991). Recovery within daytime sleep after slow wave sleep suppression. *Electroencephal. Clin. Neurophysiol.* **78**, 267–273.
16. Gillberg M. and Åkerstedt T. (1981) Possible measures of "sleepiness" for the evaluation of disturbed and displaced sleep. In *Night and Shift Work: Biological and Social Aspects* (eds. Reinberg A., Vieux N. and Andlauer P.), pp. 155–160. Pergamon Press, Oxford.
17. Hamelin P. (1987) Lorry driver's time habits in work and their involvement in traffic accidents. *Ergonomics* **30**, 1323–1333.
18. Hume K.I. and Mills J.N. (1977) Rhythms of REM and slow-wave sleep in subjects living on abnormal time schedules. *Waking and Sleeping* **1**, 291–296.
19. Kecklund G. and Åkerstedt T. (in press). Sleepiness in long distance truck driving: an ambulatory EEG study of night driving. *Ergonomics.*
20. Lavie P. (1986) Ultrashort sleep-waking schedule. III. "Gates" and "forbidden zones" for sleep. *Electroenceph. Clin. Neurophysiol.* **63**, 414–425.
21. Minors D.S. and Waterhouse J.M. (1989) Masking in humans: The problems and some attempts to solve it. *Chronobiol. Intern.* **6**, 29–53.
22. Mitler M.M., Carskadon M.A., Czeisler C.A., Dement W.C., Dinges D.F. and Graeber R.C. (1988) Catastrophes, sleep and public policy. Concensus Report. *Sleep* **11**, 100–109.
23. Monk T.H. and Moline M.L. (1989) The timing of bedtime and waketime decisions in free-running subjects. *Psychophysiology* **26**, 304–310.
24. Nicholson A.N., Pascoe P.A., Spencer M.B., Stone B.M. and Green R.L. (1986). Nocturnal sleep and daytime alertness of aircrew after transmeridian flights. *Aviat. Space Environ. Med.* **57**, B42–52.
25. O'Hanlon J.F. and Beatty J. (1977) Concurrence of electroencephalographic and performance changes during a simulated radar watch and some implications for the arousal theory of vigilance. In *Vigilance* (ed. Mackie R.R.), pp. 189–202. Plenum Press, New York.
26. Sasaki M., Kurosaki Y., Mori A. and Endo S. (1986) Patterns of sleep-wakefulness before and after transmeridian flight in commercial airline pilots. *Aviat. Space Environ. Med.* **57**, B29–42.
27. Strogatz S.H., Kronauer R.E. and Czeisler C.A. (1987) Circadian pacemaker interferes with sleep onset at specific times each day: role in insomnia. *Am. J. Physiol.* **253**, 172–178.
28. Tilley A.J., Wilkinson R.T. and Drud M. (1981) Night and day shifts compared in terms of the quality and quantity of sleep recorded in the home and performance measured at work: a pilot study. In *Night and Shift Work. Biological and Social Aspects* (eds. Reinberg A., Vieux N. and Andlauer P.), pp. 187–196. Pergamon Press, Oxford.

29. Torsvall L. and Åkerstedt T. (1987) Sleepiness on the job: continuously measured EEG changes in train drivers. *Electroenceph. Clin. Neurophysiol.* **66**, 502–511.
30. Torsvall L. and Åkerstedt T. (1988) Extreme sleepiness: quantification of EOG and spectral EEG parameters. *Int. J. Neurosci.* **38**, 435–441.
31. Torsvall L., Åkerstedt T., Gillander K. and Knutsson A. (1989) Sleep on the night shift: 24-hour EEG monitoring of spontaneous sleep/wake behavior. *Psychophysiology* **26**, 352–358.
32. Torsvall L., Åkerstedt T. and Gillberg M. (1981) Age, sleep and irregular work hours: a field study with EEG recording, catecholamine excretion, and self-ratings. *Scand. J. Work Environ. Health* **7**, 196–203.
33. Valley V. and Broughton R. (1983) The physiological (EEG) nature of drowsiness and its relation to performance deficits in narcoleptics. *Electroenceph. Clin. Neurophysiol.* **55**, 243–251.
34. Webb W.B. and Agnew J.H.W. (1971) Stage 4 sleep: influence of time course variables. *Science* **174**, 1354–1356.
35. Wegmann H.M., Gundel A., Naumann M., Samel A., Schwartz E. and Vejvoda M. (1986) Sleep, sleepiness and circadian rhythmicity in aircrews operating on transatlantic routes. *Aviat. Space Environ. Med.* **57**, 53–64.
36. Wegmann H.M. and Klein K.E. (1985) Jet-lag and aircrew scheduling. In *Hours of Work* (eds. Folkard S. and Monk T.H.), pp. 539–552. John Wiley, Chichester.
37. Wegmann H.M., Klein K.E., Conrad B. and Esser P. (1983) A model for prediction of resynchronization after time-zone flights. *Aviat. Space Environ. Med.* **54**, 524–527.
38. Zulley J. (1980) Distribution of REM sleep in entrained 24 hour and free running sleep—wake cycles. *Sleep* **2**, 377–389.
39. Zulley J., Wever R. and Aschoff J. (1981) The dependence of onset and duration of sleep on the circadian rhythm of rectal temperature. *Pflügers Arch.* **391**, 314–318.
40. Åkerstedt T. (1985) Adjustment of physiological circadian rhythms and the sleep-wake cycle to shift work. In *Hours of Work* (eds. Monk T.H. and Folkard S.), pp. 185–198. John Wiley, Chichester.
41. Åkerstedt T. (1988) Sleepiness as a consequence of shift work. *Sleep* **11**, 17–34.
42. Åkerstedt T. and Gillberg, M. (1981). The circadian variation of experimentally displaced sleep. *Sleep* **4**, 159–169.
43. Åkerstedt T. and Gillberg, M. (1986) A dose-response study of sleep loss and spontaneous sleep termination. *Psychophysiology* **23**, 293–297.
44. Åkerstedt T., Hume K.I., Minors D.S., Waterhouse J.M. and Folkard S. (1992) Sleep on a shortening day/night schedule. *Electroencephalogr. Clin. Neurophysiol.* **82**, 102–111.
45. Åkerstedt T., Kecklund G. and Knutsson A. (1991) Spectral analysis of sleep electroencephalography in rotating three-shift work. *Scand. Work Environ. Health* **17**, 330–336.

17

Hormone Rhythms and Sleep

J. SAINI, G. BRANDENBERGER, C. SIMON and M. FOLLENIUS

Laboratoire de Physiologie et de Psychologie Environnementales, INRS/CNRS, 21, rue Becquerel, 67087 Strasbourg Cedex, France

Abstract

Recent studies suggest that the temporal relationships between hormone rhythms and sleep are more complex than previously believed. To clarify the situation, an acute shift in the sleep period was used to dissociate sleep from circadian influences.

The results demonstrated that hormones from a wide range of endocrine systems, namely the anterior pituitary, hydromineral regulating and pancreatic systems all show temporal associations with sleep. For certain hormones the occurrence of sleep was the major factor determining the 24-hour profile, either being related to sleep as a whole or to the internal sleep structure. Prolactin was influenced by the presence of sleep rather than sleep quality, with an immediate shift in the nocturnal rise as soon as sleep was displaced. A significant pulse was observed at the time of habitual sleep suggesting a possible weak circadian component to the profile which is dominated by the sleep-wake cycle. The profiles of GH and PRA were also determined by sleep, but more precisely by the internal sleep structure. GH showed some association with SWS, particularly with the first SWS episode of both night and day sleep. As for prolactin an "anamnestic" pulse was observed during the night of sleep deprivation, possibly reflecting intrinsic rhythmicity. In contrast to a specific sleep stage, PRA oscillations were associated with sleep stage alternance. The acute shift in sleep confirmed that the increase in nocturnal PRA is sleep related and not circadian in nature. Glucose and insulin oscillations were also amplified during sleep but were not related to the sleep cycles. For other hormones, such as TSH and cortisol, temporal associations with sleep stages were superimposed on a strong circadian rhythm. Sleep onset had an inhibiting effect on TSH and declining levels of TSH and cortisol pulses coincided with deep sleep.

These temporal associations probably reflect the close proximity of the central controlling mechanisms involved in sleep and hormone secretion. In opposition ANP proved to be independant of sleep and circadian rhythmicity confirming peripheral mechanical regulation.

In summary, these results highlight the complexity of hormone and sleep interactions in that no two hormones relate to sleep in exactly the same manner and provide a basis for further research including manipulations of the light–dark cycle.

Introduction

Over the last 20 years a large body of experimental evidence has accumulated concerning the description of temporal associations between hormones and sleep in man. These relationships involve endocrine systems with very different functions, including hydromineral regulation, glucose metabolism, lactation and growth. Mullen in 1983,[21] classified hormones into three categories; (1) those dependent on sleep; (2) those not influenced by sleep; (3) those related to a specific sleep stage. The current use of precise and sensitive radioimmunoassays, frequent blood sampling methods and deconvolution models to determine hormone secretion rates from plasma concentrations is providing evidence that these relationships are more complex than originally believed.

Many hormones show a 24 hour rhythm, which may be driven by an internal clock and termed "circadian" or synchronized to the sleep-wake cycle or derived from a combination of the two. Superimposed on the 24 hour profile, ultradian oscillations may be associated with the internal sleep structure so that any given hormone can exhibit more than one relationship with sleep and recent research suggests that this is frequently the case rather than the exception. Hormones such as adrenocorticotropic hormone (ACTH) and cortisol considered for a long time to be circadian in nature may have a sleep related component.[1,16] Others, for example growth hormone (GH) and prolactin thought to be solely sleep dependent appear to possess an inherent rhythmic component.[34] For renin, a key enzyme in the angiotensin-aldosterone hydromineral regulatory system, a close association exists between the ultradian oscillations of plasma renin activity (PRA) and the rapid-eye movement (REM)—nonREM sleep cycles. PRA levels are in the increasing phase during NREM sleep and decrease as sleep becomes lighter(3). These investigations have also established the consequential relationship between aldosterone[20] and sleep, but as yet the absence of a circadian rhythm in PRA has not been confirmed.[5] Controversy persists as to the presence or absence of a circadian rhythm for the recently identified hormone, atrial natriuretic peptide (ANP), but recent data suggests that ANP shows no associations with internal sleep structure.[15]

A variety of methods have been used to investigate sleep–hormone interactions, including: complete or partial sleep deprivation; acute or chronic shifts in sleep time or complete sleep-wake reversal as occurs with transmeridian travel or shift work; administration of pharmacological agents which perturb either sleep or hormone secretion and finally the study of specific pathologies which manifest either endocrine or sleep disturbances. This paper presents current knowledge of the relationships between hormones of different endocrine systems, namely the anterior pituitary, hydromineral and pancreatic systems, and presents findings using one approach, an acute shift in sleep time, to dissociate circadian and sleep related influences.

Experimental procedures

Six healthy male subjects with regular sleep wake patterns underwent two 24-hour studies (23.00–23.00 hours), following a night of habituation. One study was considered as a reference period in which the subjects slept at night (23.00–07.00 hours) and the other an experimental session in which the sleep period was shifted by 8 hours so the subjects slept during the day (07.00–15.00 hours).

From 8 hours before and throughout the study the subjects received continuous enteral nutrition, to avoid the acute effects of meals on hormone profiles. Blood sampling began at 23.00 hours and was pumped continuously from an indwelling intravenous catheter until the end of the experiment. Samples were collected at 10 minute intervals for subsequent plasma hormone analysis. Sleep was recorded during sleep times and examined in parallel with the hormone profiles.

Analysis of hormone profiles

Plasma hormone concentrations were measured using radioimmunoassay methods as previously detailed.[3,15,16,26] To evaluate hormone pulsatility an objective pulse detection program (ULTRA) was employed. The general principle of the ULTRA algorithm is to eliminate all pulses of plasma concentration for which either the increment or the decrement does not exceed a pre-defined threshold, based on measurement error.[29] As plasma hormone concentrations reflect not only secretion rates but also rates of clearance and redistribution, a mathematical deconvolution model, described by Eaton *et al.*[12] was used to estimate hormone secretory rates from plasma concentrations.

Results

Pituitary hormones

Prolactin

The nocturnal prolactin increase was displaced by the acute shift in sleep time (Figure 1). Neither the sleep induced rise nor the levels observed during the waking periods differed significantly in the two conditions, suggesting that the prolactin rhythm is solely determined by the sleep-wake pattern. In addition no differences were found in the number or amplitude of the sleep associated oscillations. However, a significant prolactin pulse was identified in all subjects, during the night in which sleep was deprived, at the time that habitual sleep occurred (Figure 1). The mean maximal value of this peak was at 3.15 hours ± 19 minutes and is perhaps indicative of some innate circadian rhythmicity.

Analysis of ascending and descending phases of prolactin plasma and secretory pulses in relation to the different sleep stages, confirmed that neither slow-wave sleep (SWS) nor nocturnal awakenings were associated with any trend in prolactin. For REM sleep, a temporal association was

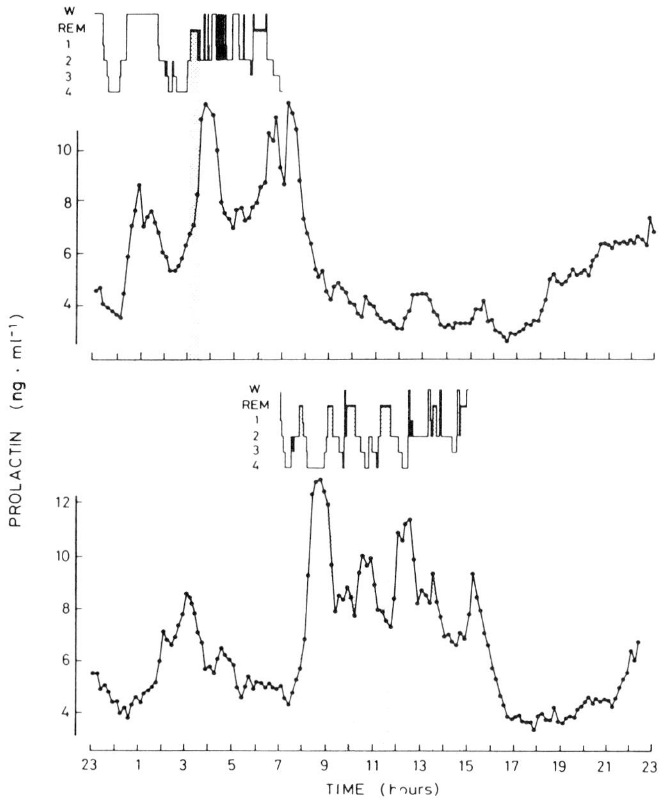

FIG. 1. Individual 24-hour plasma prolactin profiles with nighttime sleep (top) and daytime sleep (bottom). Shaded areas represent REM sleep.

apparent in that 91% of plasma and 77% of secretory pulses were in a descending phase at the onset of REM sleep ($p<0.001$, $p<0.02$ respectively). Prolactin secretion was then seen to increase following 20 ($p<0.05$) and 30 ($p<0.02$) minutes of REM sleep.

Growth Hormone

Plasma and secretory GH profiles consisted of low basal levels interrupted by episodic pulses. During nighttime and daytime sleep a large pulse occurred soon after sleep onset. The amplitude of this pulse varied widely between individuals but was not altered by the time of sleep, giving a mean maximum concentration of 20.04 ± 4.76 $ng \cdot ml^{-1}$ at 01.10 hours $\pm$ 16 minutes for night sleep and 19.52 ± 5.59 $ng \cdot ml^{-1}$ at 09.00 hours $\pm$ 15 minutes for day sleep. As for prolactin a significant pulse was detected in four out of the six subjects during the night of sleep deprivation at the habitual time of the nocturnal pulse (Figure 2). The number of plasma pulses

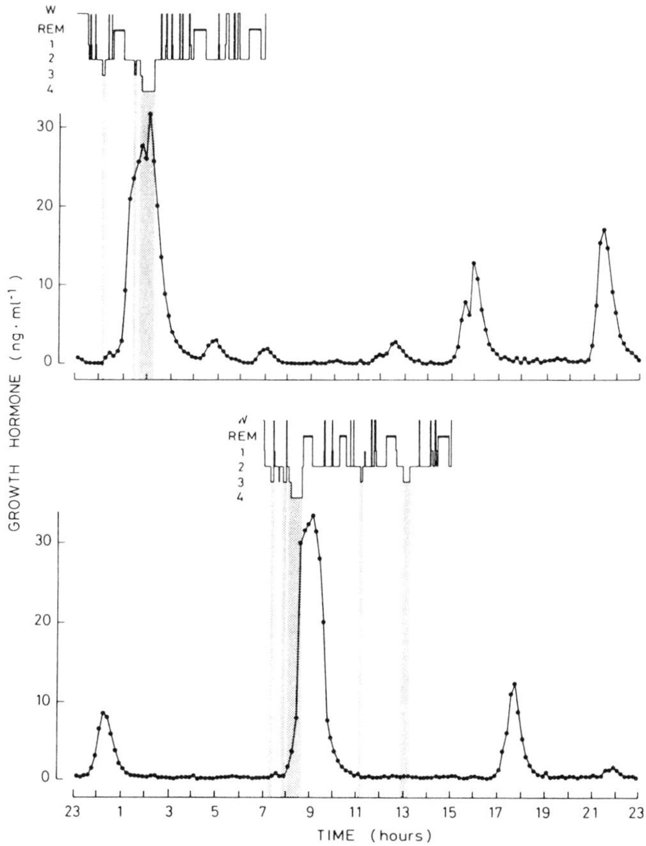

FIG. 2. Individual 24-hour plasma GH profiles with nighttime sleep (top) and daytime sleep (bottom). Shaded areas represent SWS.

for the six subjects during night and day sleep was 21 and 25 pulses respectively corresponding to 35 and 41 secretory pulses. In relation to sleep stages, 59% and 68% of plasma and 46% and 63% of secretory pulses of the night and day time sleep periods were associated with SWS episodes and for SWS, 74% of night and 92% of day time episodes were associated with a significant GH secretory pulse.

Thyrotropin (TSH)

Under basal conditions, plasma TSH concentrations showed a circadian rhythm with a slow rise during the evening, a peak around the time of sleep onset and a subsequent decrease during the night hours and low values during the day (Figure 3). Sleep deprivation during the night preceding day sleep was associated with significantly higher levels, and a decline as soon as sleep onset occurred. In addition to a sleep onset influence on the TSH

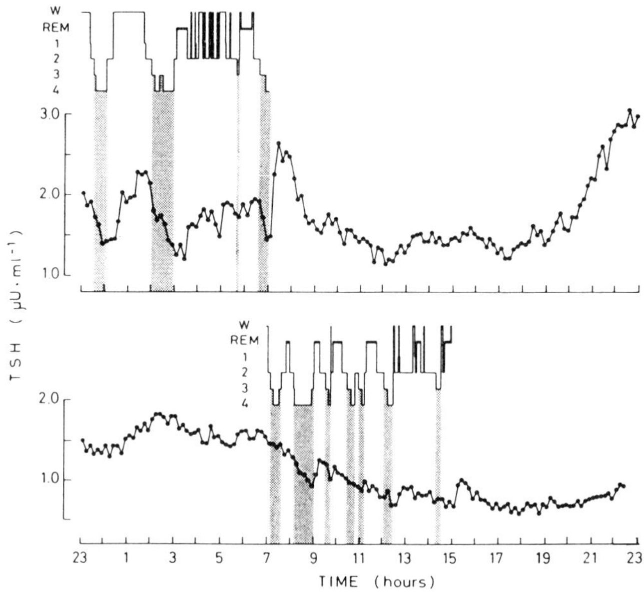

FIG. 3. Individual 24-hour plasma TSH profiles with nighttime sleep (top) and daytime sleep (bottom). Shaded areas represent SWS.

rhythm the temporal association which has recently been described between SWS episodes and declining plasma TSH levels was confirmed.[18]

ACTH and Cortisol

The shift in sleep time had very little influence on the cortisol rhythm (Figure 4), reflecting the dominant circadian rhythmicity of the controlling pituitary hormone ACTH. ACTH is rarely in the ascending phase at the time of REM sleep onset[16] and as cortisol pulses lag behind those of ACTH the number of pulses in the descending phase at REM sleep onset is reduced. During day sleep the major SWS episodes coincided with a period of enhanced cortisol secretion, allowing the analysis of cortisol secretory episodes concomitantly with a greater number of SWS episodes than for night sleep. In confirmation of a prior experiment,[13] using partial sleep deprivation, SWS episodes tended to occur when adrenal cortical activity was diminishing (Figure 4).

Hydromineral hormones

Renin-angiotensin-aldosterone system

The profiles of PRA, a key enzyme in the angiotensin-aldosterone system, in a subject who had regular REM-NREM sleep cycles during night and daytime sleep are presented in Figure 5. Pulse analysis revealed no significant difference in the number, duration or amplitude of the oscillations under the

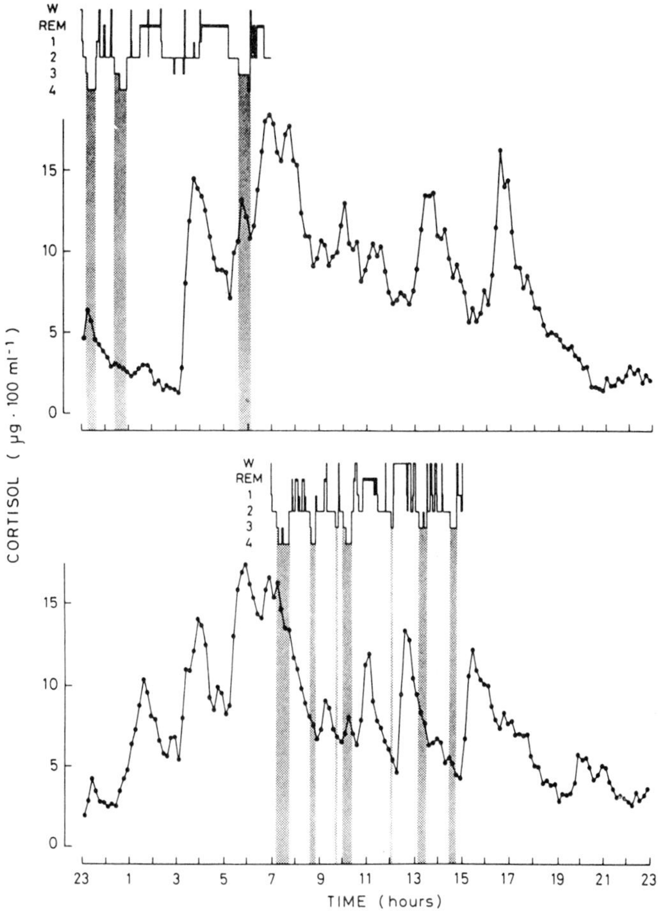

FIG. 4. Individual 24-hour plasma cortisol profiles with nighttime sleep (top) and daytime sleep (bottom). Shaded areas represent SWS.

two conditions, but in each case their absolute amplitude was 1.5 times greater during sleep compared to waking periods. The previously described relationship of increasing PRA levels with the occurrence of NREM sleep and decreasing levels coinciding with REM sleep persisted during daytime sleep. Throughout the waking periods, PRA oscillations existed but were damped and irregular. The PRA rhythm appeared to be determined by sleep and independent of intrinsic circadian rhythmicity.

ANP

ANP fluctuated throughout the 24 hours showing between two and five transient increases, which was not altered by a shift in sleep time (Figure 6).

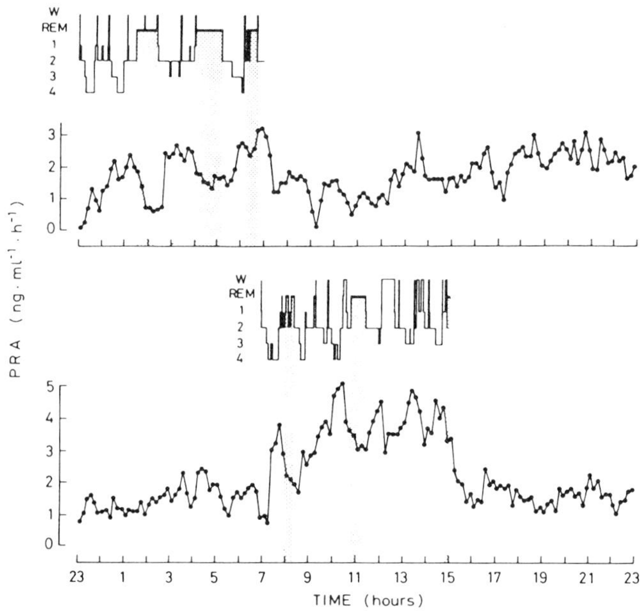

FIG. 5. Individual 24-hour PRA profiles with nighttime sleep (top) and daytime sleep (bottom). Shaded areas represent REM sleep.

No differences were found between either the mean level, 38.0 ± 3.7 pg · ml^{-1}, for the normal sleep wake cycle and the shifted sleep cycle, 39.9 ± 3.2 pg · ml^{-1} or the concentrations during the sleep or wake periods. Under standardized conditions, the profile of ANP was unaffected by sleep and seemed to possess no circadian rhythmicity.[14]

Pancreatic hormones

Glucose and insulin oscillate regularly throughout the 24 hours with a periodicity of about 80 minutes with a greater amplitude following meals.[26] Under the steady state condition obtained by administration of continous enteral nutrition, plasma insulin and glucose demonstrated larger oscillations during both night and day time sleep. The mean amplitude of plasma glucose was 0.26 ± 0.03 and 0.28 ± 0.02 g · l^{-1} for the respective 8 hour sleep periods. These values were significantly higher ($p < 0.01$) than those obtained for the corresponding waking periods (mean values: 0.14 ± 0.02 and 0.19 ± 0.02 g · l^{-1}). For plasma insulin the pulse amplitude was 16.6 ± 1.7 uU · ml^{-1} and 22.4 ± 2.7 uU · ml^{-1} for night and day sleep also differing significantly ($p < 0.01$) from corresponding periods during which the subjects were awake (11.5 ± 1.2 uU · ml^{-1} and 12.6 ± 1.2 uU · ml^{-1} respectively). It appeared that sleep itself amplified the pulses, although

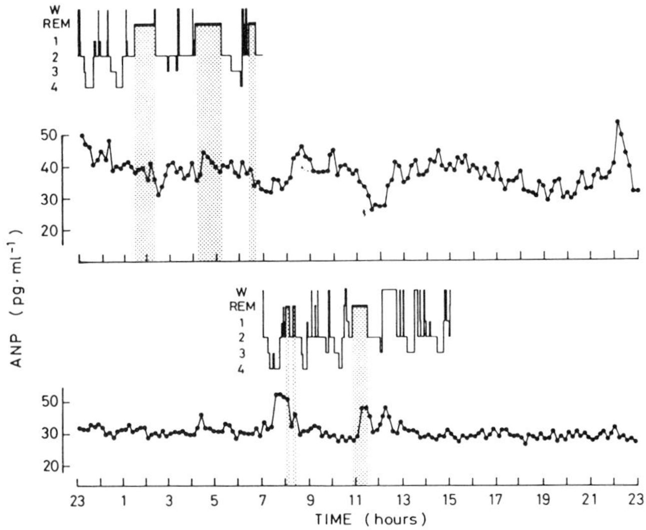

FIG. 6. Individual 24-hour plasma ANP profiles with nighttime sleep (top) and daytime sleep (bottom). Shaded areas represent REM sleep.

despite their similar periodicity, could not be related to REM-NREM sleep cycles (Figure 7).

Discussion

An acute shift in sleep time is one of a variety of methods that can be used to investigate relationships between hormone rhythms and sleep. The findings from this study along side current data highlight the complexity of these interrelationships, with hormones from different endocrine systems, with different controlling mechanisms exhibiting temporal relationships with sleep.

Temporal associations were present between basal rhythms of all the hypothalamo-pituitary hormones studied and sleep, which is a reflection of their central control, although each hormone related to sleep in a unique manner. Hormones such as prolactin and GH, which have a certain amount of structural homology, are secreted by adjacent pituitary acidophil cells and show increased plasma concentrations during exercise and other conditions of stress,[9] have very different relationships with sleep.

Prolactin is classically known to be influenced by sleep as a whole with sleep deprivation abolishing the nocturnal prolactin rise and a shift in sleep immediately displacing the nocturnal secretory episode.[24] Findings of Desir *et al.*[10] have challenged the concept that the prolactin rise is solely sleep dependent, firstly because the sleep associated acrophase was advanced after a westward and delayed after an eastward flight and secondly, as we observed, following an acute shift in sleep, an "anamnestic" pulse was present in 9 out of 10 profiles at the time of pre-travel sleep.

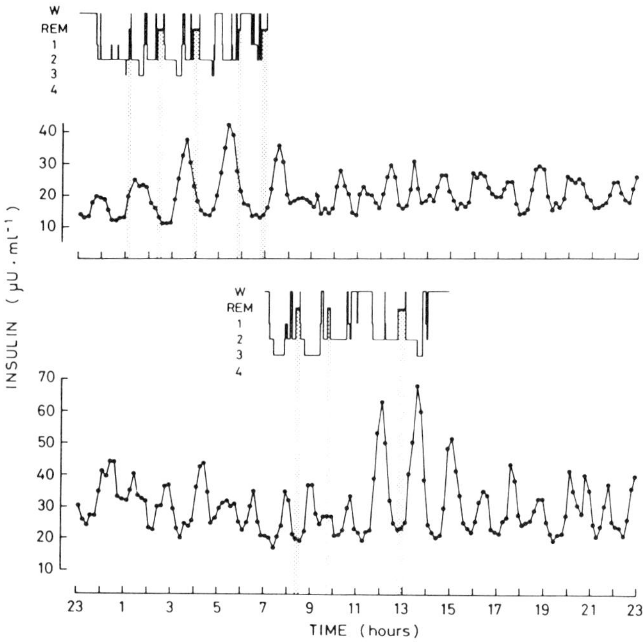

FIG. 7. Individual 24-hour plasma insulin profiles with nighttime sleep (top) and daytime sleep (bottom). Shaded areas represent REM sleep.

Reports have been contradictory concerning the relationship between nocturnal prolactin oscillations and sleep structure.[23,32] Analysis of the prolactin pulses, as previously described, were rarely in the ascending phase at REM sleep onset,[16] but also demonstrated that pituitary prolactin secretion is reduced at this time and then subsequently increases particularly if the REM sleep episode is of long duration.

In contrast to prolactin, where sleep as a whole is the major determinant of the 24-hour profile, many studies have described a temporal relationship between GH and SWS, especially between the major secretory pulse occurring soon after sleep onset and the first SWS episode. Discordance in this temporal relationship has, however, been reported.[2] Plasma GH levels may begin to increase even before sleep onset or be displaced after the SWS episode.[28] Deconvolution methods which permit an accurate definition of the onset and offset of GH secretion reveal an increased number of secretory pulses, with a single plasma pulse often reflecting more than one secretory episode. The number of GH secretory pulses observed to coincide with a SWS episode was lower than that reported by Van Cauter *et al.*[33] but when the percentage of SWS episodes was considered the degree of concordance was similar. This was true for both night and day sleep and in both cases GH secretion increased in all subjects during the first SWS episode. Recent findings suggest that SWS facilitates GH secretion but SWS is not obligatory for its occurrence.[31] As for prolactin, a significant GH secretory pulse was observed in most subjects during the night of sleep deprivation at the time of the habitual sleep associated pulse, again possibly representing an innate circadian component of the hormone rhythm.[34]

Other anterior pituitary hormones, such as ACTH and TSH, show strong circadian rhythmicity. For TSH an inhibitory effect of sleep onset is superimposed on the intrinsic rhythm as when sleep is deprived elevated TSH levels are observed.[22] Decreasing levels of TSH and cortisol pulses were associated with SWS,[13,18] suggesting that some mechanisms regulating SWS may modulate their secretion or conversely increased hormone concentrations prevent SWS occurrence. The decline in plasma cortisol appears to slightly precede SWS onset indicating that it may be enhanced cortisol activity that prevents sleep deepening rather than an effect of sleep on cortisol secretion.[13]

The temporal associations between hormones under hypothalamo-pituitary regulation and sleep are probably a reflection of the intimacy of sleep generating processes and the central control of endocrine secretion. A number of neuroendocrine peptides and transmitters, including noradrenaline, dopamine and 5-hydroxytryptamine have all been shown to affect sleep.[11]

PRA, reflecting angiotensin-aldsoterone activity demonstrates a clear relationship with the internal sleep structure. The acute shift in sleep confirmed that increased nocturnal PRA levels is a sleep related phenomenon and not circadian in nature. The amplitude as well as the frequency of the oscillations was dependent on the regularity of the REM-NREM sleep cycles, with increasing PRA levels coinciding with NREM and decreasing levels with REM sleep. Modulating renin release, either using a β-blocker to suppress or a low sodium diet to stimulate renin release, only alters the magnitude of the pulses and does not disturb the relationship with the sleep cycles.[4] The importance of the internal sleep structure in maintaining rhythmicity of PRA is supported by the nocturnal profiles of patients with various sleep disturbances such as obstructive sleep apnea and narcolepsy. In these cases, PRA profiles reflect all the irregularities of the sleep pattern, but as soon as sleep quality is restored by treatment and regular REM-NREM sleep patterns are re-established, PRA rhythmicity returns and the amplitude of the oscillations increases.[17] A "renin-releasing factor" located in the hypothalamus has been identified which under serotoninergic innervation possibly acts as a central control of renin release.[35] At present, it is not known whether this partially characterized peptide is a candidate for the common factor involved in PRA and sleep stage alternance.

Aldosterone, the active hormone of this hydromineral regulating system, is also under the influence of ACTH and, as for ACTH, levels are low early on in the night and increase towards the morning. Stimulating and suppressing PRA reveal the relative influence of the two hormone systems on aldosterone secretion. When PRA is stimulated aldosterone oscillations reflect those of PRA with a delay of 20 minutes and when PRA is depressed the oscillations follow those of ACTH by 10 minutes. Thus the relationship of the aldosterone rhythm to sleep is dependent on that of PRA and ACTH.[20]

The diurnal variations of ANP have often been compared to the renin-angiotensin-aldosterone system activity, both being sodium and volume dependent. The acute shift in sleep established that, contrarily to PRA, sleep regulatory processes seem not to influence its release,[15] neither did ANP profile show any circadian rhythmicity.[14] This data substantiates the argument that, unlike PRA in which central mechanisms are clearly involved, ANP is released in response to peripheral mechanical factors.

A number of studies indicate that the time of day influences the regulation of blood glucose levels, with higher levels obtained in the afternoon or evening compared with the morning, following either oral or intravenous glucose administration, as well as in response to meals.[19] In normal subjects glucose tolerance reaches a minimum around mid-sleep, suggesting circadian rhythmicity.[25] A recent study, monitoring glucose and insulin levels during a constant glucose infusion suggest that both circadian and sleep modulate the regulation of glucose.[30]

Both glucose and insulin oscillate regularly throughout the 24 hours and in normal subjects show a concomitancy of about 90%, suggesting interdependence. However, results from noninsulin dependent diabetic patients question the hypothesis of an intrinsic insulin-glucose regulatory loop as compared to control subjects these patients have regular glucose oscillations, but less regular, less frequent and smaller amplitude insulin pulses showing a poor association with those of glucose.[27] Oscillations of glucose and insulin secretion were amplified in this study, both during nighttime and daytime sleep, reflecting either central mechanisms or changes in glucose metabolism during sleep.

In conclusion, these results demonstrate the diverse relationships between hormones and sleep. Sleep may be the major factor determining the basal 24-hour rhythm, either by being temporally associated with sleep as a whole, for example prolactin, or linked to the internal sleep structure, such as PRA which reflects sleep stage alternance, and GH which is specifically associated with SWS. For other hormones, the relationship with sleep is superimposed on a strong circadian rhythm, as for ACTH, cortisol and TSH. Melatonin, the pineal hormone is also dominated by a circadian rhythm but recent reports do not favor the presence of any direct relationship between melatonin secretion and sleep.[7] Both melatonin and cortisol as well as the circadian rhythm of body temperature can be entrained by light exposure,[6,8] however, the influence of the light–dark cycle on other hormone rhythms has been little studied. The present data provides a basis for further investigations into this area.

References

1. Born J., Kern W., Bieber K., Fehm-Wolfsdorf G., Schiebe M. and Fehm H.L. (1986) Night time plasma cortisol secretion is associated with specific sleep stages. *Biol. Psychiatry* **21**, 1415–1424.
2. Born J., Muth S. and Fehm H.L. (1988) The significance of sleep onset and slow wave sleep for the nocturnal release of growth hormone and cortisol. *J. Endocrinol. Invest.* **8**, 381–391.
3. Brandenberger G., Follenius M., Simon C., Ehrhart J. and Libert J.P. (1988) Nocturnal oscillations in plasma renin activity and REM-NREM sleep cycles in humans: a common regulatory mechanism? *Sleep* **11**, 242–250.
4. Brandenberger G., Krauth M.O., Ehrhart J., Libert J.P., Simon C. and Follenius M. (1990) Modulation of episodic renin release during sleep in humans. *Hypertension* **15**, 370–375.
5. Brandenberger G., Simon C. and Follenius M. (1987) Night day differences in the ultradian rhythmicity of plasma renin activity. *Life Science* **40**, 2325–2330.
6. Broadway J., Arendt J. and Folkard S. (1987) Bright light phase shifts the human melatonin rhythm during Antarctic Winter. *Neuroscience Lett.* **79**, 185–189.
7. Claustrat B., Brun J., Garry P., Roussel B. and Sassolas G. (1986) A once repeated study of nocturnal plasma melatonin patterns and sleep recordings in six normal young men. *J. Pineal Res.* **3**, 301–310.
8. Czeizler C., Kronauer R., Allan J., Duffy J., Jewett M., Brown E. and Ronda J. (1989) Bright light induction of strong (type 0) resetting of the human pacemaker. *Science* **244**, 1328–1333.
9. Delitala G., Tomasi P. and Virdis R. (1987) Prolactin, growth hormone and thyrotropin-thyroid hormone secretion during stress states in man. *Ballieres Clin. Endocrinol. Metab.* **1**, 391–414.
10. Desir D., Van Cauter E., L'Hermite M., Refetoff S., Jadot C., Caufriez A., Copinschi G. and Robyn C. (1982) Effects of "Jet Lag" on hormonal patterns. III. Demonstration of an intrinsic circadian rhythmicity in plasma prolactin. *J. Clin. Endocrinol. Metab.* **55**, 849–857.
11. Drucker-Colin R. and Valverde C. (1982) Endocrine and peptide functions in the sleep-waking cycle. In *Sleep Clinical and Experimental Aspects* (eds. Ganten D. and Pfaff D.) Current topics in Neuroendocrinology, pp. 37–82. Springer, Berlin, Heidelberg, New York.
12. Eaton R.P., Allen R.C., Schade D.S., Erickson K.M. and Standefer J. (1980) Prehepatic insulin production in man; Kinetic analysis using peripheral connecting peptide behavior. *J. Clin. Endocrinol. Metab.* **51**, 520–528.
13. Follenius M., Brandenberger G., Bandesapt J., Libert J.P. and Ehrhart J. (1992) Nocturnal cortisol release in relation to sleep structure. *Sleep* **15**, 21–27.
14. Follenius M., Brandenberger G. and Saini J. (1992) Lack of diurnal rhythm in plasma atrial natriuretic peptide. *Life Sciences* **51**, 143–149.
15. Follenius M., Brandenberger G. and Simon C. (1988) Nocturnal fluctuations of plasma atrial natriuretic peptide in normotensive and hypertensive men. *Life Sciences* **42**, 1635–1639.
16. Follenius M., Brandenberger G., Simon C. and Schlienger J.L. (1988) REM sleep in humans begins during decreased secretory activity of the anterior pituitary. *Sleep* **11**, 546–555.
17. Follenius M., Krieger J., Krauth M.O., Sforza F. and Brandenberger G. (1991) Obstructive sleep apnea treatment: peripheral and central effects on plasma renin activity and aldosterone. *Sleep* **14**, 211–217.
18. Goichot B., Brandenberger G., Saini J., Wittersheim G. and Follenius M. (1992) Nocturnal plasma thyrotropin variations are related to slow wave sleep. *J. Sleep Research* **1**, 186–190.
19. Jarrett R. (1979) Rhythms in insulin and glucose. In *Endocrine Rhythms* (ed. Krieger D.), pp. 247–258. Raven Press, New York
20. Krauth M.O., Saini J., Follenius M. and Brandenberger G. (1990) Nocturnal

oscillations of plasma aldosterone in relation to sleep stages. *J. Endocrinol. Invest.* **13**, 727–735.

21. Mullen P.E. (1983) Sleep and its interaction with endocrine rhythms. *Br. J. Psychiat.* **142**, 215–220.
22. Parker D.C., Pekary A.E. and Hershman J.M. (1987) Effect of 64 hour sleep deprivation on the circadian waveform of thyrotropin (TSH): further evidence of sleep-related inhibition of TSH release. *J. Clin. Endocrinol. Metab.* **64**, 157–161.
23. Parker D.C., Rossman L.G. and Vanderlaan E.F. (1974) Relation of sleep-entrained human prolactin release to REM-nonREM cycles. *J. Clin. Endocrinol. Metab.* **38**, 646–651.
24. Sassin J.F., Frantz A.G., Kapen S. and Weitzman E.D. (1973) The nocturnal rise of human prolactin is dependent on sleep. *J. Clin. Endocrinol. Metab.* **37**, 436–440.
25. Shapiro E.T., Tillil H., Polonsky K.S., Fang V.S., Rubenstein A.H. and Van Cauter E. (1988) Oscillations in insulin secretion during constant glucose infusion in normal man: Relationship to changes in plasma glucose. *J. Clin. Endocrinol. Metab.* **67**, 307–314.
26. Simon C., Brandenberger G. and Follenius M. (1987) Postprandial oscillations of plasma glucose, insulin and C-peptide in man. *Diabetologia* **30**, 769–773.
27. Simon C., Brandenberger G., Follenius M. and Schlienger J.L. (1991) Alteration in the temporal organisation of insulin secretion in Type 2 (non-insulin-dependent) diabetic patients under continuous enteral nutrition. *Diabetologia* **34**, 435–440.
28. Steiger A., Herth J. and Holsboer F. (1987) Sleep-electroencephalography and the secretion of cortisol and growth hormone in normal controls. *Acta Endocrinol.* **116**, 36–42.
29. Van Cauter (1981) Quantitative methods for the analysis of circadian and episodic hormone fluctuations. In *Human Pituitary Hormones: Circadian and Episodic Variations* (eds. Van Cauter E. and Copinschi G.), pp. 1–25. Martinus Nijhoff, The Hague.
30. Van Cauter E., Blackman J., Roland D., Spire J-P., Refetoff S. and Polonsky K. (1991) Modulation of glucose regulation and insulin secretion by circadian rhythmicity and sleep. *J. Clin. Invest.* **88**, 934–942.
31. Van Cauter E., Caufriez A., Kerkhofs M., Van Onderbergen A., Thorner M.O. and Copinschi G. (1992) Sleep awakenings and insulin-like growth factor-1 modulate the growth hormone (GH) secretory response to GH-releasing hormone. *J. Clin. Endocrinol. Metab.* **74**, 1451–1459.
32. Van Cauter E., Desir D., Refetoff S., Spire J.P., Noel P., L'Hermite M., Robyn C. and Copinschi G. (1982) The relationship between episodic variations of plasma prolactin and REM-NonREM cyclicity is an artefact. *J. Clin. Endocrinol. Metab.* **54**, 70–75.
33. Van Cauter E., Kerkhofs M., Caufriez A., Van Onderbergen A., Thorner M. and Copinschi G. (1992) A quantitative estimation of growth hormone secretion in normal man: reproducibility and relation to sleep and time of day. *J. Clin. Endocrinol. Metab.* **74**, 1441–1450.
34. Van Cauter E. and Refetoff S. (1985) Multifactorial control of the 24-hour secretory profiles of pituitary hormones. *J. Endocrinol. Invest.* **8**, 381–391.
35. Van de Kar L.D., Urban J.H., Brownfield M.S. and Simons W.H. (1987) Partial characterization of a renin-releasing factor from plasma hypothalamus. *Hypertension* **9**, 598–606.

18

Biological Rhythms and Sleep Disorders in Man: The Delayed Sleep Phase Syndrome

J. VIGNAU, M. DAHLITZ, J. ARENDT, J. ENGLISH and J.D. PARKES

University Department of Neurology, King's College Hospital and Institute of Psychiatry, London SE5 9RS, UK
School of Biological Sciences, University of Surrey, Guildford, UK
University of Lille, France

Abstract

The delayed sleep phase syndrome is a form of insomnia with inability to fall asleep at conventional times but with normal sleep at a delayed clock time in respect of light–dark cycles, social, economic and family demands. The syndrome is the most common intrinsic disorder of human sleep-wake rhythmicity with a reported population prevalence rate of between 0.1 and 0.4%. In some clinics it accounts for 2–5% of all subjects with a primary complaint of evening insomnia. The syndrome is familiar in about one quarter of all subjects, more common in males than females and usually present around puberty. It needs to be distinguished from motivated sleep-phase delay associated with psychiatric problems, psychological factors or school avoidance behavior in adolescence. The delayed sleep phase syndrome is life-long and may involve several biological rhythms coupled to the sleep-wake cycle including core temperature, alertness and melatonin rhythms.

The plasma melatonin pulse duration and amplitude is normal, with a slight delay in low-light melatonin onset as compared with subjects with a normal sleep-wake schedule in a similar environment. Light suppression (moderate intensity: 600 lux at 00.01–00.30) of melatonin secretion is of normal amplitude. Melatonin 5 mg po will advance sleep-phase but the response is not maintained on stopping treatment and depends on the exact timing of melatonin administration in relation to the endogenous melatonin cycle. The phase-response curves of sleep-wake cycles, body temperature and salivary melatonin profile after oral melatonin are described.

Treatment strategies include the use of chronotherapy with a round-the-clock successive 2 hour delay in scheduled sleep onset time, phototherapy (10,000 lux for 30 minutes at 08.00 for 14 days), and melatonin 5 mg po at 22.00. Also, vitamin B12 treatment has been reported effective in some but not all subjects.

The cause of the delayed sleep phase syndrome is uncertain, but both genetic and environmental factors are involved. There is no evidence of intra-uterine or perinatal involvement and the reported incidence of breast feeding is normal. Childhood development is also normal. Both paternal and maternal inheritance have been reported. One mechanism, as shown by the occasional progressive delay in sleep phase may involve the near 25-hour period of the endogenous pacemaker. Also, there may be a failure of normal entrainment mechanisms in the advance phase of the sleep-wake cycle.

Introduction

IN THE London *Spectator* of 26 April 1711, there is a report of a group of philosophers at Cambridge and Oxford Universities called "the Loungers".[1] Also, W.B. Yeats in "The Trembling of the Veil" refers to the long literary tradition of a delayed sleep phase syndrome[1]

> "When I had first gone to see him in 1889 at the Charlotte Street house I had called at about 5 in the afternoon but the manservant told me he was not yet up, adding with effusion: 'He is always up for dinner at 7' This habit of breakfasting when others dined had been started by insomnia, but he came to defend it for its own sake."[1]

Clinical features

The delayed sleep phase syndrome was first described by Weitzman *et al.* in 1981 in a group of insomniac patients who complained of inability to fall asleep at a socially acceptable time but were able to sleep normally at a clock-time independent of social, economic and family demands.[2] The subjects comprised about 7% of the total population presenting at a Sleep Disorders Clinic with insomnia. Delayed sleep phase often began in childhood and on the owl-lark, "morning-evening" scale derived by Horne and Östberg, all the subjects were extreme evening types.[3] Thirteen of 30 subjects had a psychiatric disorder but the nature and distribution of this was no different from other insomniacs. Almost every subject had taken many different sedatives without success.

In the last 10 years about 75 subjects with the delayed sleep phase syndrome have been reported.[4–15] The age of onset varies from infancy to the 6th decade and is most often at adolescence.

The sex incidence was equal in the 30 subjects originally reported by Weitzman *et al.*[2] but most reports have stressed male predominance. The syndrome is similar in children and adults, characterized by habitual sleep onset and wake times several hours later than desired. Children and adolescents have a poor school record and often truancy whilst adults often have many short periods of employment, marital and financial problems and increasing social isolation.[10]

A recent Norwegian epidemiological study suggests a fairly high population prevalence rate for the delayed sleep phase syndrome of between 0.1 and 0.4%.[16] Possible latitude variation was not recorded in this study.

The delayed sleep phase syndrome has been distinguished from a *motivated* sleep phase syndrome in adolescents, where school avoidance, psychological and psychiatric factors are usually very prominent with more normal sleep patterns during school holidays.[5]

Details of subjects with the delayed sleep phase syndrome seen in a 5-year period at the King's College Hospital/Maudsley Hospital Sleep Disorders Clinic are shown in Table 1.

Diagnostic criteria

We have adopted the following diagnostic criteria for the delayed sleep phase syndrome from those of Weitzman *et al.*, Thorpy and the American Sleep Disorders Association Classification.[2,10]

(1) Complaint of inability to fall asleep and wake spontaneously at the desired clock-time.
(2) A phase delay of the major sleep episode in relation to the desired time for sleep.
(3) Symptoms present for at least 12 months.
(4) No evidence of any medical, psychological, environmental or psychiatric factor sufficient to explain the above symptoms.
(5) Absence of any anatomical lesion on head CT scan; and normal polysomnographic findings apart from an atypical time of sleep onset and waking.

Technological problems in the study of the delayed sleep phase syndrome

In our investigation of subjects with the delayed sleep phase syndrome we have met a number of problems with technology. These have included:

(1) Difficulties with accurate core temperature continuous measurement. Available systems need to be refined. Few patients can tolerate an indwelling rectal probe for long periods and these are often displaced during sleep periods. This technology is not very suitable for use in the home environment. There have been frequent problems in our experience with the CorTemp gut capsule broadcasting system—in particular with the signal reception via flimsy antenna devices.
(2) Continuous melatonin sampling from plasma or saliva raises formidable problems in the home or ward environment over the long term. Salivary samples interrupt sleep; attempts to use a dental sucker without arousing the subject have not been successful in our experience. Intravenous cannula require constant monitoring, the use of heparin to keep patent and may need re-inserting into different veins. In the long term, urinary melatonin metabolite sampling may be the only practical technique but will not allow a detailed analysis of melatonin rhythms.

TABLE 1
Clinical features of 14 cases of delayed sleep phase syndrome

							HLA Type		Sleep–wake Parameters Decimal time			IQ	
Case	Age	Sex	Age at puberty	Family history		Developmental/ environmental factors	DR	DQ	Sleep onset mean	Wake mean	Alertness acrophase	Performance	Verbal
1	14	M	8	14	a	Prolonged labour	1,x	1,x	01.48	11.90	19.22	80	75
2	39	M	33	14		EBV[1]	2,6	1,x	07.95	11.81	03.43	125	103
3	18	M	16	13	b	Premature birth EBV[1]	3,x	*	08.10	17.15	20.19	103	74
4	61	M	45	13			4,13	1,3	04.05	15.87	04.60	126	92
5	22	M	12	15		Birth hypoxia	1,2	1,3	01.17	08.12	01.93	119	76
6	20	M	8	15			2,11	1,3	02.68	11.00	15.75	101	70
7	42	M	27	16	c		2,x	1,x	03.25	10.80	20.72	108	93
8	58	M	53	14		Influenza[2]	1,7	1,2	05.30	17.30	23.50	126	112
9	22	M	16	12			1/10,4	1,3	00.22	12.00	10.18	NT	NT
10	14	M	10	11	c		8,4	3,x	03.00	12.28	20.11	NT	NT
11	17	M	25	15		Prolonged labour	4,6	1,3	01.48	12.04	13.59	111	99
12	39	M	8	16	d		2,3	1,2	01.55	08.01	04.46	126	86
13	71	M	18	12	d		2,6	1,x	02.00	09.00	23.05	NT	NT
14	28	F	18	13	adopted	Tonic-clonic seizures, age 12	*	*	05.00	16.10	NT	NT	86

a Family history of daytime sleepiness.
b Family history of narcolepsy and cataplexy. c Father and son. d Father and son.
[1] Onset of delayed sleep phase syndrome following Epstein–Barr viral infection.
[2] Onset after influenza-like illness.
* Failed HLA typing.
NT Not tested.
Sleep-wake parameters given in decimal time, i.e. 01.30 = 01.50. Sleep–wake data derived from mean 28-day sleep logs. Acrophase: time of subjective highest alertness self rating (computing data by cosinor analysis: from mean 28-day self recording).

(3) Motor activity recording with actimetry is a simple technique but will not separate for example sleep from quiet immobility.
(4) There are considerable difficulties in determining temperature phase shifts over a short time span if these are minor. For example, it is very difficult to evaluate the possible effect of a single small oral dose of melatonin on the temperature rhythm over a single 24-hour cycle.
(5) Practical patient considerations. Few subjects will tolerate detailed prolonged circadian studies in their normal environment and the use of a chronobiology laboratory, with, e.g. a 40-hour protocol, is almost certainly necessary for analytical studies.

Delayed, advanced and non-24 hour sleep-wake phase syndromes

As well as the delayed sleep phase syndrome, a number of less common rhythmic and arrythmic sleep-wake patterns have been described including the rare advanced sleep phase syndrome. Non-24-hour sleep-wake patterns are sometimes found in blind or mentally retarded individuals.[17,18] In these conditions, the exact sleep-wake pattern may be determined largely by life-style, social and other time cues and different patterns of sleep phase may occur in the same individual. In the delayed sleep phase syndrome prevalence study in Norway referred to above, no subject with advanced sleep phase was identified.

Pathophysiology of the delayed sleep phase syndrome

The clinical presentation of the delayed sleep phase syndrome, with normal sleep architecture although at an unusual time as well as the normal quality of wakefulness suggest a chronobiological disorder with preserved mean levels and amplitudes of sleep-wake and sleep-wake coupled biological rhythms including those of the temperature and melatonin cycle: in contrast to an expected delay in temperature and melatonin cycles accompanying the delay in sleep onset and wake times, alertness and motor activity rhythms. It should be noted that in some blind subjects, temperature and melatonin rhythms are not always closely coupled to the sleep-wake cycle.

It has been shown that oral melatonin causes a minor sleep phase advance in the delayed sleep phase syndrome although the effect of melatonin on sleep-coupled rhythms has not been determined. The actions of melatonin would be expected to depend on the exact timing of melatonin administration in relation to the endogenous melatonin cycle and to show a definite phase-response curve. Present data suggest that the response to melatonin is greatest in the 6 hour period prior to the endogenous melatonin peak.[19] However, in the delayed sleep phase syndrome it has been suggested there may be a diminshed phase advance capability.[7]

The delayed sleep phase syndrome and melatonin

(a) Endogenous melatonin rhythmicity

Detailed information on plasma melatonin and urinary excretion of the melatonin metabolite aMT6s as clock markers in the delayed sleep phase syndrome in the free-running condition is not available. In a fixed 24 hour light:dark wake:bed schedule, Dahlitz *et al.* showed that plasma melatonin urinary 6-sulphatoxy melatonin profiles, pulse duration and peak concentrations were within the normal range for men aged 20–60 living under normal light:dark environmental conditions. However there was a minor delay in plasma melatonin acrophase as compared with normal subjects (mean 04.43 ± 0.67 v 03.15 ± 1.00: decimal times ± 1 SD: groups not identical in number and age).[15]

(b) Effect of exogenous melatonin

Exogenous melatonin 5 mg at 22.00 given 5 hours prior to the estimated mean time of sleep onset in the delayed sleep phase syndrome results in a mean advance in sleep onset time of a little over 60 minutes, and also results in an advance in wake time without prolongation of the total sleep time. It has not been established if this results from a specific phase-advancing rather than minor hypnotic effect of melatonin on sleep—or indeed whether these two actions are separate. However, in most normal subjects, melatonin 5 mg does not have a reliable hypnotic effect. In the delayed sleep phase syndrome, the advance is not maintained when melatonin is discontinued.[15] The magnitude of the phase shift in sleep onset time in the delayed sleep phase syndrome is closely comparable to that produced by the same dose of melatonin (5 mg) in normal subjects after a simulated time shift. (A single individual has been described with a delay, not advance, in sleep phase with melatonin.[20]) A sleep phase shift of similar magnitude occurs in blind subjects.[19]

The magnitude of phase shift in endogenous melatonin rhythmicity induced by exogenous melatonin 5 mg may be of slightly greater magnitude than the advance in sleep phase produced by the same dosage of melatonin at the same phase of the endogenous melatonin cycle. Arendt *et al.* reported a 1–3 hour phase advance in the rhythm of endogenous melatonin in 5 of 11 normal subjects given melatonin 2 mg.[21]

(c) Light suppression of melatonin

Although human light sensitivity is less than that reported in many other animal species, a short low intensity light exposure will result in partial suppression of human melatonin production or release.[22] The degree of melatonin suppression by light (600 lux—this may be above the threshold to detect individual variation in sensitivity) found in the delayed sleep phase syndrome falls within the range reported in normal subjects at 00.00–00.30, and there is no evidence from this data of anatomical involvement of retinal-pineal-melatonin mechanisms in the delayed sleep phase syndrome although the phase-response curve to light has not been determined.[23] Bojkowski *et al.* (1987) reported that 300 lux induced differential suppression (from 0–100%) in normal subjects[24] (see Table 2).

Phase-response curve to melatonin in delayed sleep phase syndrome

Subjective response

Melatonin 5 mg was given single blind as to drug time in four subjects (subjects no 2, 8, 14, Table 1 and one additional female subject) with the delayed sleep phase syndrome at 06.00, 12.00, 18.00 and 00.00 under a fixed light-dark wake-bed schedule repeated at 1 week intervals on four separate occasions. Two subjects, both female, described a subjective 1–2 hour advance in sleep phase following melatonin administration, one at 06.00, one at 18.00 but not at other times. The two male subjects gave no preference for time of melatonin administration. In a study of blind and sighted subjects, Lewy *et al.* found a mean sleep phase advance of 38 minutes in the sighted group following melatonin 5 mg, the maximum advance occurring when melatonin administration preceded endogenous melatonin onset by about 6 hours, less phase advance when the interval was greater than 6 hours.[19]

Temperature rhythm in delayed sleep phase syndrome

The original description of the delayed sleep phase syndrome made no mention of the characteristics of other rhythms although Wagner *et al.* subsequently described a late phase of temperature nadir in a single subject with the delayed sleep phase syndrome.[7]

In a study in temporal isolation, six subjects with the delayed sleep phase syndrome had the nadir of temperature in the last 2 hours of sleep compared with one of eight normal subjects.

The circadian rhythm of body temperature is coupled to the timing of REM sleep as well as melatonin excretion in man, the body temperature

TABLE 2

Light suppression of melatonin in the delayed sleep phase syndrome

Time	20.00	21.00	22.00	23.00	00.00	00.30	01.00	01.30	02.00	03.00
Constant dark 23.00–08.00	5±5	11±6	16±7	18±8	28±10	29±10	29±12	34±10	37±13	39±13
600 lux 00.00–00.30	2±2	9±3	10±4	14±5	18±7	25±10	16±6	22±9	31±12	38±15

LIGHT

Mean Plasma melatonin (±1 SEM:pg/ml) $n=9$

Delayed sleep phase syndrome

Conditions of constant dark, 23.00–08.00; compared with 600 lux light administration at 0.00–00.30.

low corresponding to the melatonin acrophase.[25–27] A slight delay in the phase of core body temperature from that of normal subjects in the natural environment would be expected to accompany delayed sleep phase periods in the delayed sleep phase syndrome. This may be the case despite problems of masking as well as of possible differences in correlations between spontaneous, and planned, sleep times.

Treatment strategies in the delayed sleep phase syndrome

For successful treatment, the delayed sleep phase syndrome should be distinguished from the motivated sleep delay syndrome. However, despite the diagnostic criteria outlined, it can be difficult to separate the two conditions.

Whatever approach to management is used, good subject compliance is essential and an exact treatment protocol is necessary. Without both factors, success is uncommon. Treatment is rarely practical in the home environment; in particular, phase advance schedules are not practical at home in young children. See Table 3 for various treatment strategies.

TABLE 3
Treatment strategies

Strategy	Overall results in our experience
Chronotherapy Around the clock phase delay with successive 2 hour delay in bedtime until the new desired sleep phase is reached	In adolescents, success rate about 50%. New sleep phase is retained for 3–8 months.
Phototherapy 10,000 lux (Wolff Lighting Systems) for 30 minutes at 08.00 for 14 days. This requires enforced arousal at 08.00.	Variable response in adolescents; ? better results in adults
*Vitamin B*12 100—1000 micrograms daily: oral or sc: for 4 weeks at 08.00 (requires enforced arousal at 08.00)	No effect
Melatonin 5 mg po for 2–40 weeks at 22.00	Partial advance in sleep phase maintained during treatment period
Morning CNS stimulants (amphetamines) with *Evening hypnotics* (benzodiazepines)	Successful in 1 adult subject; unsuccessful in others

Conclusion

The delayed sleep phase syndrome has a fairly characteristic phenotype. It has been recognized with increasing frequency in the last decade. In our experience it is sometimes difficult to disentangle the role of social, environmental, psychological, psychiatric factors. There is increasing evidence for phase delay in a number of biological rhythms coupled to the sleep–wake cycle including melatonin and temperature although the phase of alertness and motor activity rhythms during wakefulness may depend largely on environmental factors.

Although the syndrome is likely to be multi-factorial, about one quarter of cases have an affected first degree relative and genetic factors may be involved. The anatomical integrity of the retinal-pineal pathways appears to be intact as indicated by normal light suppression of the melatonin rhythm although the physiological function of this pathway has not yet been established. However, there may be a low response in the advance portion of the phase-response curve or subjects may have an unusually long endogenous cycle. No treatment is universally successful. Chronotherapy with high motivation and exact schedule may be the best initial option, followed by a trial of light or melatonin if the first approach is unsuccessful. Perhaps the combination of light with melatonin may prove to be the most effective remedy.

Acknowledgments

J.V. was supported by a grant from the British Council, M.D. by the King's College Hospital Joint Research Committee. We gratefully acknowledge the helpful review of D. Minors and J. Waterhouse and the assistance of J. Grimshaw.

References

1. Letter to *The Spectator*, April 26, 1711 (1987) "Stenua nos exercet inertia" cited in: Wirz-Justice A. and Pringle C. The non-entrained life of a young gentleman at Oxford. *Sleep* **10**, 57–61.
2. Weitzman E.D., Czeisler C.A., Coleman R.M., Spielman A.J., Zimmerman J.C., Dement W., Richardson G. and Pollak C.P. (1981) Delayed sleep phase syndrome. A chronobiological disorder with sleep-onset insomnia. *Arch. Gen. Psychiatry* **38**(7), 737–46.
3. Horne J.A. and Östberg O. (1976) A self-assessment questionnaire to determine morningness-eveningness in human circadian rhythms. *Int. J. Chronobiol.* **4**, 97–110.
4. Czeisler C.A., Richardson G.S., Coleman R.M. *et al.* (1981) Chronotherapy: resetting the circadian clock of patients with delayed sleep phase syndrome. *Sleep* **4**, 1–21.
5. Ferber R. and Boyle M.P. (1983) Delayed sleep phase syndrome versus motivated sleep phase delay in adolescents. *Sleep Res.* **21**, 239.
6. Ferber R. and Boyle M.P. (1983) Phase shift dyssomnia in early childhood. *Sleep Res.* **12**, 242 (abstract).
7. Wagner D.R., Motira M.C., Pollak C.P. and Czeisler C.A. (1986) Entrained sleep and temperature rhythms in delayed sleep phase syndrome. *Sleep Res.* **15**, 179 (abstract).

8. Pelayo R.P., Thorpy M.J. and Glovinsky P. (1988) Prevalence of delayed sleep phase syndrome among adolescents. *Sleep Res.* **17**, 392.
9. Kamgar-Parsi B., Wehr T.A. and Gillin J.C. (1983) Successful treatment of human non-24-hour sleep-wake syndrome. *Sleep* **6**(3), 257–264.
10. Thorpy M.J., Korman E., Speilman A.J. and Glovinsky P.B. (1988) Delayed sleep phase syndrome in adolescents. *J. Adolesc. Health Care* **9**, 22–27.
11. Ozaki N., Iwata T., Itoh A., Ohta T., Okada T. and Kasahara Y. (1989) A treatment trial of delayed sleep phase syndrome with triazolam. *J. Neurol. Neurosurg. Psychiat.* **43**, 51–55.
12. Ozaki N., Iwata T., Itoh A., Kogawa S., Ohta T., Okada T. and Kasahara Y. (1988) Body temperature monitoring in subjects with delayed sleep phase syndrome. *Neuropsychobiology* **20**, 174–177.
13. Okawa M., Mishima K., Nanami T., Shimizu T., Iijima S., Hishikawa Y. and Takahashi K. (1990) Vitamin B12 treatment for sleep-wake rhythm disorders. *Sleep* **13**, 15–23.
14. deBeck T.W. (1990) Delayed sleep phase syndrome—criminal offence in the military? *Milit. Med.* **155**, 14–15.
15. Dahlitz M.J., Alvarez B., Vignau J., English J., Arendt J. and Parkes J.D. (1991) Delayed sleep phase syndrome response to melatonin. *Lancet* **337**, 1121–1124.
16. Schrader H., Bovin G. and Sand T. (1992) Prevalence of delayed and advanced sleep sleep phase syndrome. *J. Sleep Res.* **1** Suppl. 1 415, 208 (abstract).
17. Kamei R., Hughes L., Miles L. and Dement W. (1979) Advanced sleep-wake syndrome studied in a time-isolation facility. *Chronobiology* **6**, 115.
18. Weber A.L., Cary M.S., Connor N. *et al.* (1980) Human non-24-hour sleep-wake cycles in an everyday environment. *Sleep* **2**, 347–354.
19. Lewy A.J., Sack R.L. and Latham K.L. (1990) Exogenous melatonin phase shifts circadian rhythms according to a phase response curve. Proceedings of European Pineal Study Group University of Guildford, 2–7 September, Abstract 021.
20. Samel A., Wegmann H.M. and Vejvoda M. (1992) Circadian adaptation of the simulated time shifting effect of melatonin. *J. Sleep Res.* **1** Suppl. 1 205 (abstract 410).
21. Arendt J., Bojkowski C., Folkard S. *et al.* (1985) Some effects of melatonin and the control of its secretions in humans. In *Photoperiodism, Melatonin and the Pineal Ciba Foundation Symposium 117*, pp. 266–283. Pitman, London.
22. McIntyre I.M., Norman T.R., Burrows E.G.D. and Armstrong S.M. (1989) Quantal melatonin suppression by exposure to low intensity light in man. *Life Sci.* **45**, 327–332.
23. Vignau J., Dahlitz M., Parkes J.D., Alvarez B., Arendt J. and English J. (1992) The delayed sleep phase syndrome: light suppression of melatonin. *J. Sleep Res.* **1** Suppl. 1 243 (abstract 485).
24. Bojkowski C., Aldhous M., English L., Franey C. *et al.* (1987) Suppression of nocturnal plasma melatonin and 6-sulphatory melatonin by bright and dim light in man. *Hormone Metabol. Res.* **18**, 437–440.
25. Czeisler C.A., Zimmerman J.C., Ronda J.M. and Moore-Ede M.C. (1980) Timing of REM sleep is coupled to the circadian rhythm of body temperature in man. *Sleep* **2**, 329–346.
26. Lewy A.J. and Sack R.L. (1989) The dim light melatonin onset is a marker for circadian phase position. *Chronobiol. Int.* **6**, 93–102.
27. Wever R.A. (1986) Characteristics of circadian rhythms in human functions. *J. Neural. Transmis.* **21**, 323–374.

SECTION IV

Circadian, Menstrual and Seasonal Light Regulated Rhythms

19

The Influence of Age, Sex, Height, Weight, Urine Volume and Latitude on Melatonin Concentrations in Urine from Normal Subjects: A Multinational Study

LENNART WETTERBERG,[1] GÖRAN EBERHARD,[2] LARS VON KNORRING,[3] M. A. KOHAN,[4] EVA YLIPÄÄ,[5] WOLFGANG RUTZ,[6] TROND BRATLID,[7] VERENA LACOSTE,[8] CHRISTOPHER THOMPSON,[9] HIROSHI YONEDA,[10] NORMAN E. ROSENTHAL,[11] MICHAEL McGUIRE,[12] ALESSANDRO POLLERI,[13] MARGA FREEDMAN,[14] D. J. MORTON,[15] JENNY REDMAN,[16] JOANNE D. BERIGANNAKI,[17] COLIN SHAPRIO,[18] HELEN DRIVER[18] and ARTHUR YUWILER[19]

[1]Karolinska Institute, Department of Psychiatry, St Göran's Hospital, Stockholm, S-11281 Sweden; [2]University of Lund, Department of Psychiatry, Malmö Hospital, Sweden; [3]University of Uppsala, Department of Psychiatry, Uppsala; [4]Mare del Plata, Republica Argentina; [5]Department of Psychiatry, St. Göran's Hospital, Stockholm, Sweden; [6]Psykiatriska kliniken, Visby lasarett, Visby, Sweden; [7]Åsgård sykehus, N-9010 Åsgård, Norway; [8]Psychiatrische Universitäts Klinik Basel, 4025 Basel, Wilhelm Klein-Strasse 27, Switzerland; [9]Department of Psychiatry, 1q Royal South Hants., Hospital, Graham Road, Southampton, Hants S09 4EE, England; [10]Department of Neuropsychiatry, Osaka Medical College, Daigakumachi 2–7, Takatsuki City, Osaka 569, Japan; [11]Outpatient Services, Clinical Psychobiology Branch, NIMH, Room 4S-239 Building 10, 9000 Rockville Pike, Bethesda, MD 20892, USA; [12]NPI, UCLA, 760 Westwood Plaza, Los Angeles, CA 90024, USA; [13]University of Genoa, Department of Endocrinology and Metabolic Sciences, Chair of Internal Medicine, Viale Benedetto XV, 6, 16132 Genova, Italy; [14]Departemento. de Fisiologia, Facultad de Medicina, Universidad de Buenos Aires, Paraguay 2155 7o Piso, C.P. 1121 Buenos Aires, Argentina; [15]Harare University, Faculty of Medicine, P O Box MP 167, Mt Pleasant, Harare, Zimbabwe; [16]Department of Psychology, Monash University, Victoria, Australia; [17]Athens University Medical School, Department of Psychiatry, Eginition Hospital, Athens, Greece; [18]Department of Psychiatry, Toronto Western Hospital, 399 Bathurst Street, Toronto, M5T Canada and Department of Physiology, University of Witwatersrand, Johannesburg, South Africa; [19]Department of Psychiatry and Biobehavioral Sciences, University of California at Los Angeles, Los Angeles, CA 90024, USA and VA Medical Center, West Los Angeles (Brentwood Division) Los Angeles, CA 90073, USA

Introduction

THE DAILY rhythmic rise and fall of melatonin formation is driven by the nightly release of norepinephrine from sympathetic terminals innervating the pineal gland.[20] This rhythm reflects the free-running activity of a central pacemaker in the suprachiasmatic nucleus which is synchronized to the daily 24 hour cycle by light information passed to it by the retina.[24]

Because melatonin formation involves the stimulated metabolism of one monoamine by the transmitter actions of another in accordance with the coupling between an endogenous, free-running pacemaker and light, it has been studied in a number of clinical conditions as a possible marker for endogenous rhythmicity and for noradrenergic activity. Syndromes associated with possible metabolic dysynchrony such as depression,[5,9,10,18,21,22,23,25,34,36,40,42] seasonal affective disorder,[11,19,29] premenstrual syndrome,[16,37,38,42] and degenerative diseases of the elderly,[31,32,33] have been given particular attention. Age and sex distributions in these populations often differ from that in the general population and defining the influences of sex and age on melatonin metabolism and distribution is of obvious importance to study design and data interpretation.

The association between melatonin and age has been studied in both animals and man with the general consensus that melatonin production shows a post-pubertal decline with age[7,14,17,26,27,28,30,32,35] although this relationship has not been seen in all studies.[2,5,12,43] There is less information on gender differences in melatonin production in man although Touitou *et al.*[35] reported slightly higher plasma melatonin in women and Wetterberg *et al.*[41] found a similar relationship in one population of control but not another. These differences may be due to the statistical vagaries associated with small populations or to environmental or seasonal influences although seasonal effects in man may be modest.[15,30]

A recent worldwide multinational study on seasonal and geographic influences on human melatonin production provided multiple urine samples from a large population of geographically dispersed normal subjects. In that study, monthly measures of nighttime urinary melatonin concentration were obtained throughout the year on 370 normal subjects at 19 centers in 14 countries distributed on 5 continents at latitudes from 31°01′ South to 69°40′ North. Complete demographic and related data was also available on 321 of the 370 subjects. A value for "yearly" average nighttime urinary melatonin concentration could therefore be calculated for each individual which would be more representative than one obtained on a random sample at some unspecified time of the year. Further, because sampled population consisted of healthy subjects of varying ethnicity living in widely different climatic, dietary, social and geographic circumstances, the group data should be widely generalizable.

The present paper reports on the relationship between yearly nighttime melatonin concentration and sex, age, height, weight and urine volume on the 321 subjects for whom data was complete on these measures. Finally, the relationship between geographic location and melatonin was examined.

Methods

Subjects

Subjects were 321 healthy volunteer university students and faculty (165 men and 156 women), aged 18–62 at medical centers throughout the world for whom information on sex, age, height, weight, urine volume and urinary melatonin concentration was available. Written informed consent was obtained from all subjects. Volunteers were medically and psychiatrically screened to exclude subjects with somatic or mental disease and subjects taking medications. Use of oral contraceptives was permitted provided they were employed consistently throughout the study period.

Sample collection

Urine samples were collected on the first Wednesday of each month (+ one day) for a period of 12–16 months. Subjects were given graduated plastic beakers to measure urine volume and plastic vials for storage. They were instructed to empty the bladder at bedtime (usually 10–11 p.m.) on the night of the collection period, discard the urine, and record the specific time of voiding. Thereafter the total urine produced during the night, including the first morning urination (at approximately 7 a.m.), was collected into the graduated plastic beaker, the total volume and time of any nocturnal voiding recorded, and a portion of the total urine poured into the plastic bottle. The bottle was then taken in the morning to a collection point where bottles were frozen and stored at $-20°C$. Samples were not taken from subjects traveling more than 300 miles (500 km) north or south of their residence during the month preceding sampling. At the end of the study, all urine samples were transported to Stockholm, Sweden on dry ice, where they arrived in frozen condition, and were retransferred to $-20°C$ freezers until assayed. Previous studies have shown that urinary melatonin is stable under these conditions of storage and transfer.[37] Previous studies have also shown that urinary melatonin concentration in samples collected in this way correlates highly with the 2 a.m. peak value of serum melatonin ($N = 64$; $r = 0.8$; $p < 0.001$).[1]

Analytical procedure

Urine was analyzed for melatonin using a specific radioimmunoassay developed for use with urine and blood samples (Wetterberg *et al.*, 1978).

The assay had a sensitivity of 0.01 nmoles/l. Interassay variability was 4.8% for melatonin levels above 0.15 nmol/l ($N=60$). Melatonin values were expressed as concentration (nM).

Statistics

Correlations, scatter plots and estimates of statistical significance of correlations were computed using the BMDP programs 6D, 1R and 2R, while 2V was used for the analysis of gender difference covaried for age. For each subject the data used was the mean of all available monthly values for that individual. The log of urinary melatonin concentration was used for computations of statistical significance because the log, but not the concentration itself, was normally distributed. $p<0.05$ was considered significant.

Results

Demographic data on the subjects at the 19 centers are presented in Table 1. Mean ages ranged from 30 (Genoa) to 44 (Umea). Subjects at the most southerly site, Mar De Plata Argentina, consisting of one male and five females, were the shortest (mean height = 160 cm) and lightest (61 kg) while those at Visby Sweden, consisting of four males and six females, were the tallest (177 cm) and heaviest (77 kg). The correlations between age and melatonin at each center is shown in Table 2 and in Figure 1. Overall, the correlation coefficient was 0.19 for males, females, and both groups combined ($p<0.001$ in each case). Correlation coefficients between age and melatonin at individual sites were negative in 15 and positive in 4 of the cases. The positive correlations were generally small (0.02, 0.04, 0.08 and 0.23) while negative correlations were generally more robust. Four of the negative correlations were statistically significant. There did not appear to be a relationship between latitude and the magnitude or sign of the correlation coefficient. However, as seen in Table 3, there was a statistically significant relationship between latitude and melatonin for the whole sample ($R=0.202$, $p<0.001$), for subjects in the northern hemisphere alone ($N=241$, $R=0.157$, $p=0.015$) and there was a similar, but nonsignificant, correlation for subjects in the Southern hemisphere ($N=80$, $R=0.197$, $p=0.81$). There was no significant relationship between melatonin and longitude ($R=-0.07$; NS).

Table 3 also presents correlations between melatonin and height, weight and urine volume. Height was not significantly correlated with melatonin ($r=-0.004$, 0.018 respectively) for males or females, but was significantly correlated for the whole group (-0.114; $p<0.04$). Male body weight ($R=-0.17$; $p=0.03$), but not female body weight (-0.14; NS), correlated

TABLE 1

Demographic data for the 19 centers participating in the study

CNTR	Location	Latitude °/min	*N*	Age years	Height cm	Weight kg	U vol ml	Melatonin pm/l
24	Svalbard, Norway	77 00N	19	38±07	175±10	72±14	328±087	328±087
09	Tromso, Norway	69 40N	19	42±10	174±08	72±12	459±153	273±140
07	Pajala, Sweden	67 11N	20	39±10	168±09	69±13	413±128	387±095
01	Umea, Sweden	63 50N	20	44±08	170±08	65±11	379±149	282±108
05	Stockholm, Sweden	59 20N	25	36±08	173±08	71±13	410±141	337±214
08	Visby, Sweden	57 05N	10	33±07	177±08	77±17	429±115	271±109
03	Malmo, Sweden	55 36N	20	34±09	171±08	66±13	387±129	272±126
12	London, England	51 30N	13	38±11	170±13	69±14	455±144	165±098
11	Basel, Switzerland	47 38N	20	31±07	172±10	64±09	398±144	333±111
20	Genoa, Italy	44 25N	15	30±05	176±06	70±13	379±116	218±196
18	Bethesda, USA	38 54N	09	35±10	174±06	69±10	405±218	201±128
26	Athens, Greece	37 58N	22	34±05	170±12	68±12	308±102	271±124
17	Osaka, Japan	34 40N	10	32±03	172±08	65±11	380±161	274±088
19	Los Angeles, USA	34 04N	19	39±10	174±12	72±13	388±135	261±143
23	Harare, Zimbabwe	17 15S	17	42±10	175±09	74±12	521±123	144±063
22	Johannesburg, S. Africa	25 10S	17	28±06	170±09	66±12	338±087	338±123
21	Buenos Aires, Argentina	34 37S	21	34+09	169±10	69±12	380±160	263±097
25	Melbourne, Australia	37 45S	19	32±10	173±08	68±12	454±169	257±115
06	Mar de Plata, Argentina	38 01S	6	37±08	160±07	61±15	399±138	150±073
All			321	36+09	172±09	69±12	398±142	274±120

Values are means+SE Abbreviations are: CNTR=Center, °/min=degrees and minutes of latitude, *N*=number of subjects and Uvol=urine volume.

TABLE 2
Correlation between age and melatonin for the 19 centers participating in the study

CNTR	City	Country	Latitude	Male	Female	N	R	$B \times 100$	M
24	Svalbard	Norway	77 00N	12	7	19	−0.22	49	−4.8
09	Tromso	Norway	69 40N	11	8	19	−0.32	45	−4.3
07	Pajala	Sweden	67 11N	8	12	20	+0.23	30	+2.1
01	Umea	Sweden	63 50N	5	15	20	−0.14	36	−1.8
05	Stockholm	Sweden	59 20N	14	11	25	−0.40*	75	−11.2
08	Visby	Sweden	57 05N	4	6	10	−0.03	28	−0.4
03	Malmo	Sweden	55 36N	7	13	20	+0.08	23	+1.1
12	London	England	51 30N	7	6	13	−0.60*	37	−5.3
11	Basel	Switzerland	47 38N	11	9	20	−0.16	41	−2.6
20	Genoa	Italy	44 25N	12	3	15	−0.51*	79	−19.0
18	Bethesda	USA	38 54N	3	6	9	+0.04	18	−0.0
26	Athens	Greece	37 58N	12	10	22	−0.11	36	−2.5
17	Osaka	Japan	34 40N	9	1	10	−0.44	73	−14.1
19	Los Angeles	USA	34 04N	11	8	19	−0.40	48	−5.6
23	Harare	Zimbabwe	17 15S	9	8	17	−0.56*	29	−3.3
22	Johannesburg	S. Africa	25 10S	8	9	17	−0.24	46	−5.0
21	Buenos Aires	Argentina	34 37S	13	8	21	−0.03	27	−0.3
25	Melbourne	Australia	37 45S	8	11	19	+0.02	25	+0.0
06	Mar de Plata	Argentina	38 01S	1	5	6	−0.21	22	−2.1
Men						165	−0.19**	34	−2.6
Women						156	−0.19**	42	−3.1
Both						321	−0.19**	38	−2.8

Abbreviations are: CNTR = center number; N = number of subjects, R = Pearson correlation coefficient; $B \times 100$ = intercept in pM times 100; M is the slope in nM/year; * = $p < 0.05$; ** = $p < 0.001$.

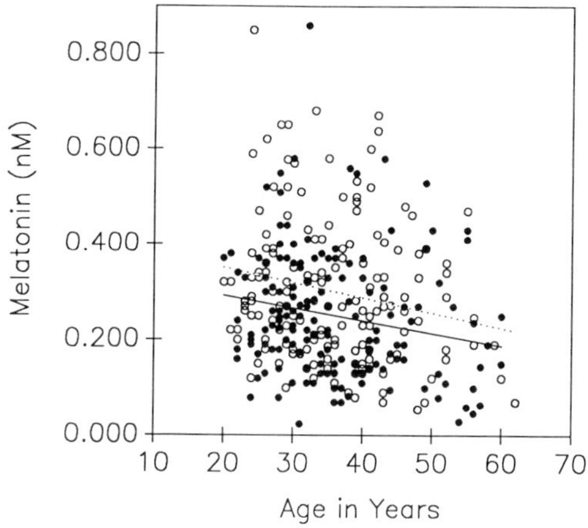

FIG. 1. Scatter diagram of urinary melatonin concentration for males (closed circles) and females (open circles).

TABLE 3
Correlations between melatonin concentration and height, weight, urine volume, latitude and longitude

	Male		Female		Total	
Melatonin vs:	*R*	(*p*)	*R*	(*p*)	*R*	(*p*)
Height	−0.004	(NS)	−0.018	(NS)	−0.114	(0.04)
Weight	−0.166	(0.03)	−0.143	(NS)	−0.230	(<0.001)
Urine V	−0.234	(0.002)	−0.239	(0.003)	−0.246	(<0.001)
Latitude	—		—		0.202	(<0.001)
Longitude	—		—		−0.068	(NS)

R=correlation coefficient, *p*=probability, NS=not significant.

with melatonin while weight and melatonin were correlated for the entire group ($r=-0.23$; $p<0.001$).

Correlations between age and the other variables are shown in Table 4. The correlation between log melatonin and age was even stronger than that between melatonin and age. Normalizing melatonin to body weight did not materially alter the relationship between melatonin concentration and age or between log melatonin and age. Melatonin content in the urine of males correlated with age but not in the urine of females or the melatonin content for both groups combined. As expected, urine volume was strongly negatively correlated with melatonin concentration for males ($R=-0.23$, $p=0.002$), females ($R=-0.24$, $p<0.003$) and both groups together ($R=-0.246$, $p<0.001$) (Table 3).

Gender differences are shown in Table 5. As might be expected, the sexes differ in height and weight. They also differ in melatonin concentration. For this reason an analysis of variance with sex as covariate was carried

TABLE 4
Correlations between Age and melatonin concentration, height, weight and urine volume

	Male		Female		Total	
Age vs:	*R*	(*p*)	*R*	(*p*)	*R*	(*p*)
Melatonin (nM)	−0.193	(0.01)	−0.190	(0.02)	−0.188	(<0.001)
Log melatonin (nM)	−0.245	(<0.001)	−0.237	(0.001)	−0.245	(<0.001)
Melatonin (nM)/kg wt	−0.240	(0.002)	−0.212	(0.008)	−0.204	(<0.001)
Log melatonin (nM)/kg wt.	−0.177	(0.023)	−0.186	(0.02)	−0.178	(<0.001)
Melatonin (pmoles)	−0.149	(0.06)	+0.040	(NS)	−0.55	(NS)
Height	−0.075	(NS)	−0.097	(NS)	−0.061	(NS)
Weight	−0.243	(<0.01)	−0.147	(NS)	−0.147	(0.01)
Urine V	−0.145	(NS)	−0.335	(<0.001)	−0.237	(<0.001)

R = Correlation coefficient, *p* = probability, NS = not statistically significant.

TABLE 5
Gender differences in melatonin, body height and weight, and in urine volume

	Male		Female				
	Mean	SD	Mean	SE	*F*	*Df*	*p*
Height	178	7	165	7	298	1,319	<0.0001
Weight	77	10	60	9	257	1,319	<0.0001
Melatonin	0.250	0.129	0.300	0.151	10	1,319	0.0016
U volume	409	139	386	145	2	1,319	0.2 NS
log Melatonin	−0.665	0.250	−0.580	0.018	10	1,319	0.0017

out confirming the effect of age on melatonin ($Df = 1, 1, 318$, $F = 12$, $p = 0.0006$) with age adjusted mean melatonin concentrations of 0.250 ± 0.129 for males and 0.300 ± 0.151 for females.

Discussion

Our finding of a negative correlation between age and nighttime urinary melatonin concentration is consistent with other reports in the literature[5,14,17,26,27,28,30,32,35] and seems generalizable because of the diverse population sampled. The magnitude of the correlation in this study, however, is considerably smaller than reported elsewhere (e.g. 0.47 by Sack *et al.*[30] measuring urinary 6-hydroxymelatonin and 0.38 by Sharma *et al.*[32] measuring the acrophase of the blood melatonin rhythm), perhaps due to environmental influences on the measures or to the differences in methodology. The unexpected correlation between latitude and melatonin could be interpreted as supporting an environmental influence on melatonin although a genetic influence on melatonin from geographically separated genetically different populations could also explain the results.

As seen in Table 1 the magnitude of the age-melatonin correlation varied considerably between centers and correlations between age and melatonin concentration in these small populations only reached statistical significance in 4 of the 19 participating centers (Stockholm, London, Genoa and Harare). It approached significance in two others (Los Angeles and Osaka).

The negative correlation between body height and urinary melatonin found in this study corresponds to the correlation found by Beck-Friis *et al.*[5] between height and plasma melatonin and by Sack *et al.*[30] between height and urinary 6-hydroxymelatonin. However, it was much weaker and was only seen for the whole population but not for males or females alone.

The inverse relationship between melatonin concentration and overnight urine volume was expected as a natural consequence of the free passage of melatonin from blood into the renal tubule. Since blood melatonin peaks around 2 a.m., urine passing into the bladder at this time is also high and concentration is diluted by urine collecting before and after that period in accordance with the lowered blood levels at those times.

In agreement with Ferrier *et al.*[12] (who did not, however, find an age-melatonin correlation) and Arendt,[3] we found a significant correlation between melatonin and weight. However, normalizing melatonin to body weight did not materially affect the correlation of melatonin with age indicating that age correlate is not a simple reflection of age-related weight changes.

Our finding of a gender difference in urinary melatonin production which persisted after age correction, supports and extends the observation by Touitou *et al.*[35] of higher blood melatonin in aged women than in aged men. This does not appear to be accounted for by gender differences in body mass.

The mechanism accounting for the relationship between age and melatonin is not clear. Age-linked losses of noradrenergic neurones,[8] decreases in catecholamine metabolism[13] or a fall in the pineal enzyme, hydroxy-O-methyltransferase, which converts *N*-acetylserotonin to melatonin[4] could each contribute to the relationship. A more practical issue is that of evaluating the contributions of sex, age, height and weight to the results of studies on the clinical significance of melatonin production. While we find a consistent and statistically significant relationship between melatonin and age, the magnitude of the relationship is small accounting for less than 5% of the variance. This may account for the the lack of agreement between studies on the linkage between age and melatonin. Therefore, these results with normals suggest that age would have a significant but small effect in studies comparing populations provided only that the quantitative relationship between age and melatonin is not altered by clinical state. Whether this proviso is true, however, needs to be established.

Similar considerations apply to assessing the influence of gender on studies of melatonin. Females do have significantly higher urinary melatonin concentrations than males after age correction but again the differences are relatively small so that incomplete sex matching of experimental and control populations is unlikely to result in significant population differences when *N*'s are small but may lead to statistically significant, but physiologically trivial differences when *N*'s are large. Obviously good experimental design requires close matching for both age and sex.

References

1. Almay B., von Knorring L. and Wetterberg L. (1987) Melatonin in serum and urine in patients with idopathic palin syndrome. *Psychiatry Res.* **22**, 179–191.
2. Arendt J., Wirz-Justice A. and Bradtke J. (1977) Annual rhythm of serum melatonin in man. *Neurosci Lett.* **7**, 327–330.
3. Arendt J., Hampton S., English J., Kwasowski P. and Marks V. (1982) 24-hour profiles of melatonin, cortisol, insulin, C-peptide and GIP following a meal and subsequent fasting. *Clin. Endocrinol.* **16**, 89–95.
4. Axelrod J. and Weissbach H. (1960) Enzymatic O-methylation of N-acetylserotonin to melatonin. *Science* **131**, 1312.
5. Beck-Friis J., von Rosen D., Kjellman B.F., Ljunggren J.G. and Wetterberg, L. (1984) Melatonin in relation to body measures, sex, age, season and the use of drugs in patients with major affective disorders and healthy subjects. *Psychoneuroendocrinol.* **9**, 261–277.
6. Birau N. (1981) Melatonin in human serum: Progress in screening investigation and clinic. In *Melatonin: Current Status and Perspectives* (eds. Birau N. and Schloot W.), pp. 297–326. Pergamon Press, Oxford.
7. Bojkowski G.T and Arendt J. (1990) Factors influencing urinary 6-sulfatoxymelatonin, a major melatonin metabolite, in normal human subjects. *Clin. Endocrinol.* **33**, 435–444.
8. Bondareff W., Mountjoy C.Q. and Roth M. (1981) Selective loss of neurones of origin of adrenergic projection to cerebral cortex (nucleus locus coeruleus) in senile dementia. *Lancet* **1**, 783–784.
9. Brown R., Kocsis J.H., Caroff S., Amsterdam J., Winokur A., Stokes P.E. and Frazer, A. (1985) Differences in nocturnal melatonin secretions between melancholic depressed patients and control subjects. *Am. J. Psychiat.* **142**, 811–816.
10. Claustrat B., Chazot G., Brun J., Jordan D. and Sassolas G. (1984) A chronobiological study of melatonin and cortisol secretion in depressed subjects: Plasma Melatonin, A biochemical marker in major depression. *Biol. Psychiat.* **19**, 1215–1228.
11. Esterbrook M.R. and Stiasny S. (1978) Psychiatric disorder, hospital admission and season. *Arc. Gen Psychiat.* **35**, 769–771.
12. Ferrier I.N., Arendt J., Jonstone E.C. and Crow T.J. (1982) Reduced nocturnal melatonin secretion in chronic schizophrenia: Relationship to body weight. *Clin. Endocrinol.* **17**, 181–187.
13. Finch C.E. (1973) Catecholamine metabolism in the brains of ageing male mice. *Brain Res.* **52**, 261–276.
14. Grinevich Y.A. and Lebunetz I.F. (1986) Melatonin, thymic serum factor and cortisol levels in healthy subjects of different age and patients with skin melanoma. *J. Pineal Res.* **3**, 263–275.
15. Griffiths P.A., Folkard S., Bojkowski C., English J. and Arendt J. (1986) Persistent 24-hr variations of urinary 6-hydroxymelatonin sulfate and cortisol in Antarctica. *Experentia* **15**, 430–432.

16. Hariharasubramanian N., Nair N.P.V. and Pilapil C. (1984) Circadian rhythm of plasma melatonin and cortisol during the menstrual cycle. In *The Pineal Gland: Endocrine Aspects* (eds. Brown G.M. and Wainwright, S.D.), pp. 31–36. Pergamon Press, Oxford.
17. Iguchi H., Kato K.J. and Ibayashi, H. (1982) Age-dependent reduction serum melatonin concentration in healthy human subjects. *J. Clin. Endocrinol. Metab.* **55**, 27–29.
18. Jimerson D.C., Lynch H.J., Post R.M., Wurtman R.J. and Bunney W.E. (1977) Urinary melatonin rhythms during sleep deprivation in depressed patients and normals. *Life Sci.* **20**, 1501–1508.
19. Kevan S. (1980) Perspectives on season of suicide: A review. *Soc. Sci. Med.* **14**, 369–378.
20. Klein D.C., Weller J.L. and Moore R.Y. (1971) Melatonin metabolism: neural regulation of pineal serotonin N-acetyltransferase activity. *Proc. Natl. Acad. Sci. U.S.A.* **68**, 2107–3110.
21. McIntyre I., Judd F., Norman T. and Burrows G. (1986) Plasma melatonin concentrations in depression. *Aust. NZ Psychiat.* **20**, 381–383.
22. Mendlewics J., Branchey L., Weinberg U., Branchey M., Linkowski P. and Weitzmann E.D. (1980). The 24 hour pattern of plasma melatonin in depressed patients before and after treatment. *Commun. Psychopharmacol.* **4**, 49–55
23. Miles A. and Philbrick D.R.S. (1988) Melatonin and Psychiatry. *Biol. Psychiat.* **23**, 405–425.
24. Moore R.Y., Heller A., Wurtman R.J. and Axelrod J. (1967) Visual pathways mediating pineal response to environmental light. *Science* **155**, 220–223
25. Nair N.P.V., Hariharasubramanian N. and Pilapil C. (1984) Circadian Rhythm of plasma melatonin in endogenous depression. *Prog. Neuropsychopharmacol. Biol. Psychiat.* **8**, 715–718.
26. Nair N.P.V., Hariharasubramanian N., Pilapil C., Isaac I. and Thivundayil J.X. (1986) Plasma melatonin, an index of brain aging in humans? *Biol. Psychiat.* **21**, 141–150.
27. Pang S.F., Tsang C.W., Hong G.X., Yip P.C., Tang P.L. and Brown G.M. (1990) Fluctuation of blood melatonin concentrations with age: result of changes in pineal melatonin secretion, body growth, and aging. *J. Pineal Res.* **8**, 179–92.
28. Reiter R.J., Craft C.M., Johnson L.Y., King T.S., Richardson B.A., Vaughan G.M. and Vaughan M.K. (1981) Age -associated reduction in nocturnal pineal melatonin levels in female rats. *Endocrinol.* **109**, 1295–1297.
29. Rosenthal N.E., Sack D.A. and Jacobsen F.M. (1986) Melatonin in seasonal affective disorder and phototherapy. *J. Neural Trans.* (Suppl.) **21**, 257–267.
30. Sack R.L., Lewy A.L., Herb D.L., Vollmer W.M. and Singer C.M. (1986) Human melatonin production decreases with age. *J. Pineal Res.* **3**, 379–388.
31. Sandyk R., Anninos P.A. and Tsagas N. (1991) Age related disruption of circadian rhythms: possible relationship to memory impairment and implication for therapy with magnetic fields. *Int. J. Neurosci.* **59**, 259–62.
32. Sharma M., Palacios-Bois J., Schwartz G., Iskandar H., Thakur M., Quirion R. and Nair N.P. (1989) Circadian rhythms of melatonin and cortisol in aging. *Biol. Psychiatry.* **25**, 305–19.
33. Skene D.J., Vivien-Roels B., Sparks D.L., Hunsaker J.C., Pevet P., Ravid D. and Swaab D.F. (1991) Daily variation in the concentration of melatonin and 5-methoxytryptophol in the human pineal gland: Effect of age and Alzheimer's disease. *Brain Res.* **528**, 170–174.
34. Stewart J.H. and Halbreich U. (1989) Plasma melatonin levels in depressed patients before and after antidepressant medication. *Biol. Psychiat.* **25**, 33–38.
35. Touitou Y., Fevre-Montange M., Proust J., Klinger E. and Nakache T.P. (1985) Age- and sex- associated modification of plasma melatonin concentrations in man. Relationship to pathology, malignant or not, and autopsy findings. *Acta Endocrinol.* **108**, 135–144.
36. Thompson C., Franey C., Arendt J. and Checkley S.A. (1988) A comparison of melatonin secretion in depressed patients and normal subjects. *Br. J. Psychiat.* **152**, 260–265
37. Webley G.E. and Leidenberger, F. (1986) The circadian pattern of melatonin and its

positive relationship with progesterone in women. *J. Clin. Endocrinol. Metab.* **63**, 323–328.

38. Wetterberg L., Arendt J., Paunier L., Sizonenko P., van Donselaar W. and Heyden T. (1976) Human serum melatonin changes during the menstrual cycle. *J. Clin. Endocrinol. Metab.* **42**, 185–199.
39. Wetterberg L., Eriksson O., Friberg Y. and Vangbo B. (1978) A simplified radioimmunoassay for melatonin and its application to biological fluids. Preliminary observations on the half-life of plasma melatonin in man. *Clin. Chim. Acta.* **86**, 169–177.
40. Wetterberg L., Beck-Friis J., Aperia B. and Petterson U. (1979) Melatonin/cortisol ratio in depression. *Lancet* **2**, 1361
41. Wetterberg L., Aperia B., Gorelick D.A., Gwirtzman H.E., McGuire M.T., Serafetinides E.A. and Yuwiler A. (1992) Age, alcoholism, and depression are associated with low levels of urinary melatonin. *J. of Psychiat. Neurosci.* **17**, 215–224.
42. Wirz-Justice A. and Arendt J (1979) Diurnal, menstrual cycle and seasonal indole rhythms in man and their modification. In *Biological Psychiatry Today* (eds. Obiols J., Ballus C., Gonzales-Monclus E. and Pujol E.), pp. 294–302. Elsevier/North Holland, Amsterdam.
43. Wirz-Justice A. and Richter R. (1979) Seasonality in biochemical determinations: A source of variance and a clue to the temporal incidence of affective illness. *Psychiat. Res.* **1**, 53–60.

20

Individual Variations of Rhythms in Morning and Evening Types with Special Emphasis on Seasonal Differences

VERENA LACOSTE[1] and LENNART WETTERBERG[2]

[1]*Psychiatric University Clinic, Wilhelm Klein-Str. 27, CH-4025 Basel, Switzerland*

[2]*Department of Psychiatry, St. Göran's Hospital, S-11281 Stockholm, Sweden*

THE PRESENT communication focuses on human rhythm patterns and their modification by the individual "diurnal type" (evening-/morning-), and thus relates to a respectable Scandinavian tradition: Swedish scientists such as Öquist, Östberg, Fröberg, Patkai, Torsvall and Akerstedt among others, have essentially contributed to the progress in various areas of diurnal typology including the development of self-assessment instruments[11,32,34,41] and shift-work concerns.[7,33,34,41]

Since recognition of the existence of so called "morning" and "evening" active persons by Kleitman 1939,[18] various synonymous terms have appeared in chronobiological literature, ranging from the more abstract "subjective circadian phase position" to the poetical terms like larks and owls (Table 1). Commonly morning (M-) and evening (E-) types are regarded as two extremes of a continuum on which the intermediate (I-) ones represent the largest category.[17] But, for example according to Wendt,[45] morningness-eveningness self-reports are more complex and would be more appropriately described within a multidimensional model.

In search for factors which might influence the time structure of a human individual, measurable in biological rhythm characteristics such as phase position, amplitude and waveform, up to now individual, social and environmental sources have been proposed (Table 2). Based on circadian rhythm studies,[17] Kerkhof concluded that the morningness-eveningness dimension is the most important factor associated with reliable interindividual differences.

TABLE 1
Terms used in literature to describe the morningness-eveningness dimension

1. Diurnal type
2. Subjective circadian phase position
3. Time-of-day preference
4. Circadian activity type
5. Circadian chronotype
6. Morning/evening type
7. Early/late sleeper
8. Lark/owl

TABLE 2
Factors which contribute to the variability of rhythm patterns

Individual
- Age and gender
- Height and weight
- Ethnicity
- Personality traits
- Diurnal type (morningness-eveningness)
- Activity load and sleep
- Illness related features

Social
- Work conditions and vacation
- Altered sleep/wake schedules (e.g. shift work, transmeridian flights)
- Dietary constituants and meal timing
- Psychosocial "stressors"

Environmental
- Climate
- Weather conditions
- Light and temperature
- Geographical latitude
- Altitude

Here we will briefly summarize literature dealing with morningness-eveningness as related to the circadian system and circadian rhythms, and then in a second part present new data that indicate an effect of the morningness-eveningness factor on seasonal rhythms.

Morningness-eveningness and the human circadian system

Diurnal typology is mainly based on the individual circadian phase position of spontaneous sleep-wake rhythms and subjective alertness[12,17,34,42] and, as an external validation criterion, body temperature.[12] M-types generally retire and arise earlier and show less variable sleep length and awakening time as compared to E-types and oral temperature peaks approximately 2 hours earlier.

It was suggested that interindividual differences in subjective circadian phase position could be related to the autonomous intrinsic frequency of the circadian rhythm of an individual, which generally deviates from the 24-hour-rhythm of external zeitgebers.[1,2,12,17] Thus, M- and E-types could mirror differences whith regard to the timing of their biological oscillators, whereby according to Foret and Benoit,[6] the morningness-eveningness index seems to reflect the sleep/wake oscillator rather than the temperature oscillator of an individual. In favor of such a view is the observation that circadian sleep/wake cycles in human temporal isolation paradigms exhibit individual differences with a range of 16- to 56-hour-days.[43] Furthermore, in animal studies the dependency of phase position upon the free-running circadian activity period length has clearly been demonstrated.[1,2,10] Alternatively differences in circadian phase positions between M- and E-types could be a consequence of their habits (e.g. sleep times), a view supported by results from Fröberg.[7] Whether morningness-eveningness is of endogenous or exogenous origin still remains to be clarified. Although the two explanations not mutually exclude one another and interactive effects of both are most plausible.

Morningness-eveningness and circadian rhythms

Even in the 19th century interindividual differences in sleep patterns, namely depth of sleep, have been attributed to individual diurnal type related features (see the waking experiments of Michelson 1891,[29] Figure 1). But morningness-eveningness related literature expanded only in the seventies of the actual century, in parallel to the increasing interest in circadian rhythms. Differences between M- and E-types with regard to circadian parameters have been observed in various behavioral, psychophysiological and biological measures and are reviewed in detail by Kerkhof.[17] The majority of studies and the most replicable M-E differences concern the phase or the timing of the rhythms. All studies agree that the maximum value for the M-types precedes that for E-types. Regarding the magnitude of the difference, a metaanalysis showed that for body temperature the average M-E difference is 121 minutes (11 studies), for the alertness rhythm 171 minutes (7 studies) and for sleep-time 80 minutes (4 studies). The scarce results on amplitude and shape lack consistency.

Morningness-eveningness has also been connected with personality and coping styles,[27a,28,31] adaptability to night- and shift-work[4,7,9,33,41] and jet-lag.[46] There is general agreement that E-types show greater tolerance

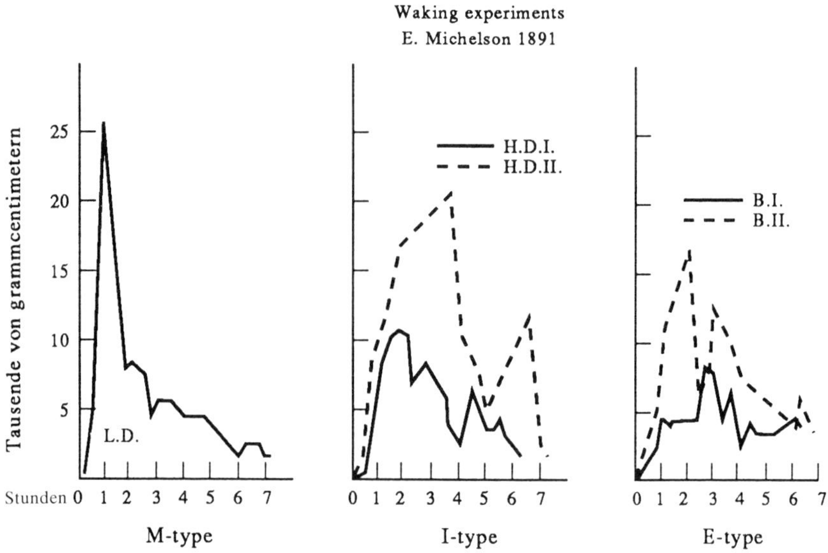

FIG. 1. Individual sleep patterns ("depth" of sleep) evaluated by arousal stimuli presented randomly across the night, as attributed to the morning, intermediate and evening types, respectively. Ordinate: intensity of acoustic arousal stimuli Abscissa: hours after sleep onset. Dashed line: modification of the individual pattern by external events. From Michelson.[29]

to shift- and night-work as compared to M-types. M-types appear to have more difficulty to adjust to westward, E-types to eastward transmeridian flights.

A further condition associated with changing external time cue schedules is season, raising the question about a possible link between morningness-eveningness and seasonal adaptation, an issue however not considered so far.

Morningness-eveningness as related to seasonal rhythms and subjective seasonality

Participation in the "Season Around the World" multinational project (Wetterberg *et al.*, 1993, Chapter 19, this volume) rendered it possible to address the question whether seasonal rhythm patterns and/or "seasonality" are modified by the individual diurnal type. In this purpose the local Basel study group was investigated in more detail, applying in addition to the standard protocol a Morningness-Eveningness-Questionnaire across the year.

Study

Subjects and methods

According to a German version of the Horne-Östberg Morningness-Eveningness-Questionnaire[11] administered to 20 (11m, 9f) healthy volunteers aged between 23 and 52 years (mean: 32.0 ± 7.4) and living in Basel (47.3°N; 7.4°E) under normal daily routine lighting conditions, M-, E- and I-type groups of individuals were formed and compared with regard to the following measures assessed over 14 months:

(1) Nighttime melatonin and cortisol/urine (RIA method) and overnight urine volume.
(2) Self-ratings (100 mm Visual Analogue Scales, VAS) of mood and sleep including: sadness, elation, fatiguability, broken and disturbed sleep, sleep need and difficulties falling asleep.
(3) The degree of diurnality (morningness-eveningness) using the Horne-Östberg index, corresponding to the summary score of the 19 item Horne-Östberg Questionnaire.[11]
(4) In addition, at the end of the investigation period the Seasonal Pattern Assessment Questionnaire (SPAQ),[36] designed to assess the extent to which each subject had previously been affected by seasonal changes, was administered. For analysis the seasonality change score was used.

As to the statistical analysis, considering the small N's, particularly in the extreme groups (M- and E-) priority was given to an explorative approach and a descriptive presentation. Complementary statistical analysis comprises: 1-way ANOVA for repeated measures, 1- and 2-way ANOVA for group comparison and group × time interaction, as well as Pearson product-moment correlations.

Results

Test-retest-reliability of the Horne-Östberg index (H-O index) ranged depending upon the test-retest interval (1–13 months) between 0.67 and 0.96, mean: 0.84 ± 0.08) and thus can be considered as adequate. However, since it has been shown that the H-O Scale undergoes seasonal variation[20] (Figure 2), we considered the average value of the 14 investigations obtained from each individual for typology. Accordingly, the diurnal types were normally distributed, 12 subjects falling in the I-, 4 in the M- and 4 in the E-categories, respectively.

As shown in Table 3, the subgroups are comparable with regard to their mean age, height, weight and body surface area but differ in their sex distribution and as a trend in the seasonality score. A subsequent separate analysis for male and female subjects revealed no further sex difference,

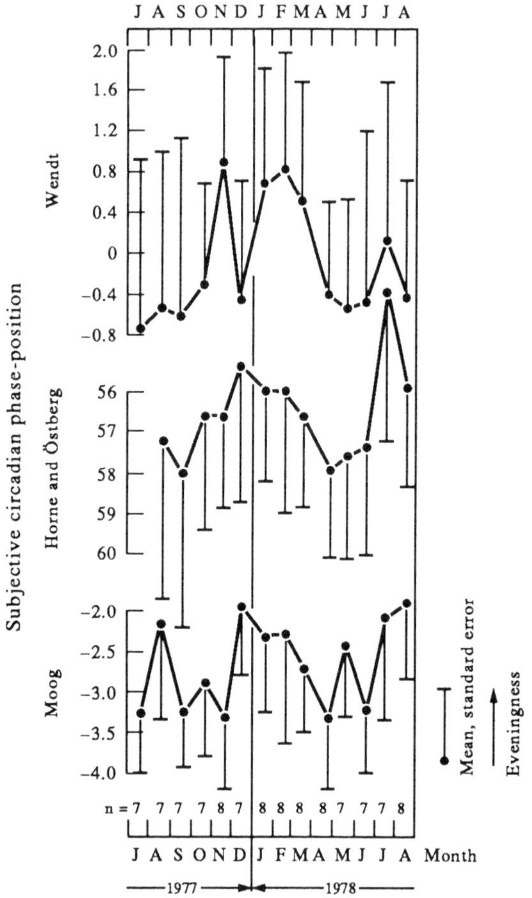

FIG. 2. Mean monthly changes in the degree of morningness-eveningness in eight healthy men and women (Marburg/Germany, 50°N) from three different questionnaires. Comparison showed that only the Horne-Östberg-Scale and that of Wendt shifted significantly across the year. From Klöppel.[19]

apart from the H-O index ($p < 0.01$) and with exception of urine volume being higher in males ($p = 0.04$).

Yearly means of psychobiological measures in M-, I- and E-types

Significant group differences concern mainly fatiguability and sleep items, but also sadness and cortisol, as can be seen in Table 4. Sleep measures as well as fatiguability and sadness generally scored higher in the extreme groups (M- and E-) particularly in the E-types. Cortisol was highest in the M-group, melatonin in the E-group, the latter however without reaching

TABLE 3
Subject characteristics by diurnal type subgroups

	M-type $N=4$	E-type $N=4$	I-type $N=12$	p-level
Sex ratio (m/f)	3/1	1/3	7/5	
Age (years)	35.0 (12.0)	32.5 (7.1)	30.8 (4.6)	0.58
Heigth (cm)	175.3 (10.2)	165.0 (6.0)	172.8 (10.3)	0.29
Weight (kg)	65.3 (11.1)	60.5 (10.5)	66.0 (10.3)	0.66
Body surface area (m^2)	1.79 (0.21)	1.64 (0.15)	1.78 (0.19)	0.39
H-O index (pts)[a]	62.7 (2.7)	36.2 (4.9)	51.9 (3.2)	0.00
Seasonality—score (pts)[b]	5.50 (3.7) mean SD	8.75 (3.3)	5.92 (1.1)	0.09

[a] Summary score from the Horne-Östberg 19-item Morningness-Eveningness Questionnaire (MEQ),[11] reflecting the subjective circadian phase position and serving as the basis for diurnal typology.

[b] As derived from the Seasonal Pattern Assessment Questionnaire (SPAQ) from Rosenthal *et al.*,[36] representing the sum of six items (amount of sleep per day, social activity, mood, weight, appetite and energy level) for which the individuals rated the degree they considered themselves to be affected throughout the year.

significance, provided winter values are considered (December: $p<0.01$). That group differences were greater in winter as compared to summer applies to all measures investigated, except elation.

Seasonal rhythm patterns of psychobiological measures in M-, I- and E-types

Figure 3 demonstrates that the subjective circadian phase position (H-O index) varies throughout the year in the E-group and to a lesser extent in the M-group, while it remains stable in the I-group.

With regard to mood and sleep impairment, in addition to the higher levels (yearly means), M- and E-types also exhibit larger seasonal variations than the I-types, as illustrated in Figure 4. The items difficulties falling asleep and broken and disturbed sleep show statistical trends ($p<0.10$) for a group × time of year interaction.

Regarding melatonin (Figure 5), it is the only variable studied with a statistically significant seasonal rhythm ($p<0.0001$). The seasonal pattern is similar in all diurnal type subgroups, with low values during the summer half and high levels during the winter half of the year, although the E-type shows a more pronounced winter increase resulting in an almost significant group × time of year interaction (July/December: $p=0.05$).

With regard to cortisol (Figure 6), it is notable that peak values are reached in March (M-type) and October (E-type), respectively.

A differential seasonal pattern was also observed in urine volume: while

TABLE 4

Yearly means and standard deviations of self-rated mood and sleep (100 *mm Visual Analogue Scale*) *and physiological measures by diurnal type subgroups*

Variable	M-type $N=4$	E-type $N=4$	I-type $N=12$	p-level
Mood				
Sadness (mm)	33.0 (10.8)	51.5 (4.5)	25.5 (16.9)	0.049
Elation (mm)	47.9 (5.4)	36.2 (1.6)	50.7 (16.5)	0.701
Fatiguability (mm)	43.3 (20.4)	66.1 (11.4)	31.8 (17.5)	0.009
Sleep				
Sleep need (mm)	45.6 (18.3)	56.3 (7.1)	28.7 (17.3)	0.019
Broken and disturbed sleep (mm)	33.4 (10.8)	58.8 (4.5)	20.3 (16.9)	0.002
Difficulties falling asleep (mm)	27.3 (8.9)	55.9 (32.8)	14.5 (3.9)	0.006
Physiological				
Melatonin (nmol/l)	0.33 (0.15)	0.41 (0.11)	0.32 (0.10)	0.360
Cortisol (nmol/l)	201.0 (200)	138.6 (44)	126.2 (60.1)	0.034
Urine volume (8 hours overnight)	336.0 (95.3)	236.4 (101.9)	463.9 (186.8)	0.107

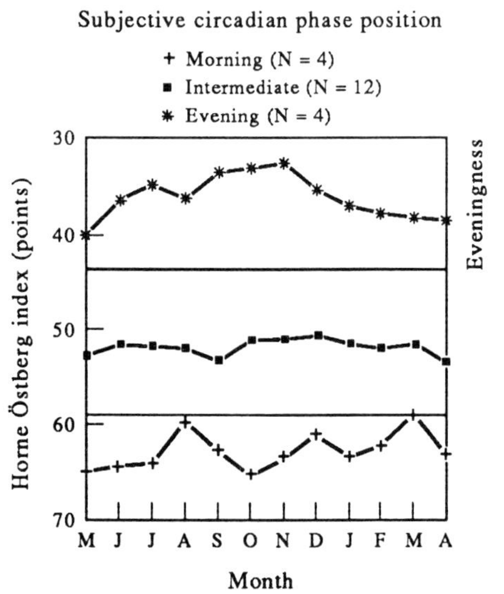

FIG. 3. Seasonal pattern (monthly means) of the Horne-Östberg index separately for the diurnal type subgroups. The horizontal lines represent the cut-off points on which classification is based. Lower values indicate increasing eveningness.

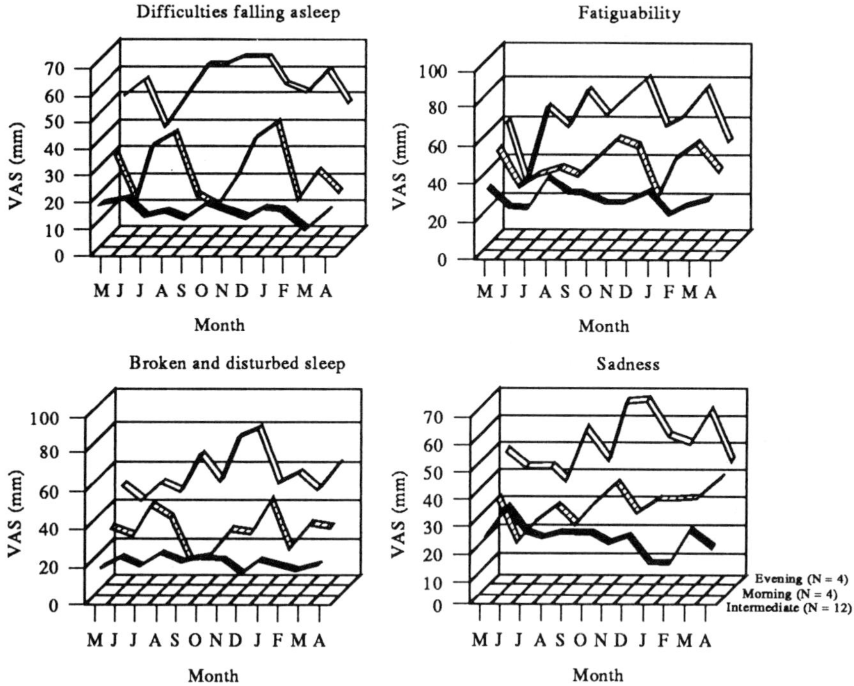

FIG. 4. Seasonal patterns (monthly means) of self-rated mood and sleep (100 mm Visual Analogue Scale) by diurnal type subgroups. Upper curve: E-type, bottom: I-type, in between: M-type.

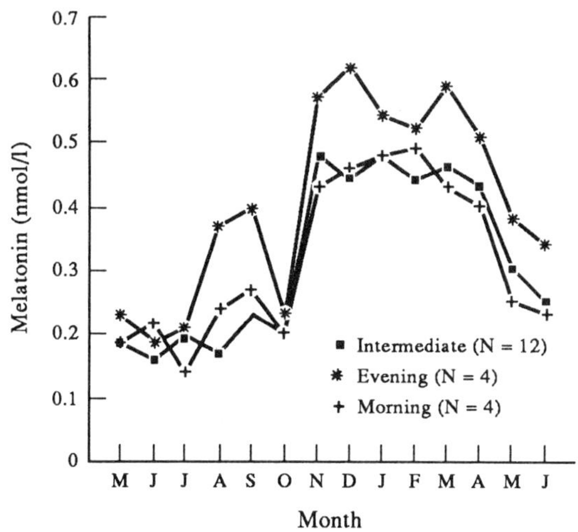

FIG. 5. Seasonal pattern (monthly means) of nighttime urinary melatonin by diurnal type subgroups.

I-types showed high levels in winter, M- and E-types reached minimal values at this time of the year, where group differences become significant on the 1% level (data not shown).

Correlations

As indicated by the coefficients of correlations listed in Table 5, the H-O index correlated negatively with fatiguability and sleep complaints. A low negative correlation was obtained between the H-O index and the seasonality score (Figure 7a), becoming significant when winter values were considered (December: $r = -0.58$, $p < 0.01$).

The seasonality score itself showed more substantial relations, being highly positively correlated with fatiguability, sleep complaints, particularly difficulties falling asleep (Figure 7b) but also with sadness and wintertime melatonin (December: $r = 0.64$, $p < 0.01$). Furthermore, the degree of seasonality appears to increase with age.

Group differences were found in the temporal relationship between melatonin and cortisol, the correlations being positive in the M- ($r = 0.71$, $p < 0.01$), negative ($r = -0.41$, $p = 0.10$) in the E-type, as described elsewhere.[23]

Discussion

According to our results morningness-eveningness appears to be a relatively stable individual characteristic that nevertheless can vary in degree across the seasons. This is in agreement with reports from literature demonstrating a high reliability independent of the population studied and the questionnaire used.[14,16,40,41]

In a longitudinal study applying concomitantly three different morningness-eveningness questionnaires, two out of them—that of Horne and Östberg and that of Wendt—have shown to undergo seasonal variations with shifts towards an eveningness phase position in winter[19] (Figure 2). In the present study a similar seasonal pattern has been observed in the E-type group (Figure 3). Parallel to the within subject studies, in a previous transverse population study a likewise varying distribution of self-estimated M- and E-types in function of season has been noticed.[21]

A winter phase delay in circadian rhythms of various vegetative functions,[19] but also of hormones such as melatonin[5,13] has been described. Thus, seasonal variation in subjective circadian phase position may have a physiological basis in the annually changing phase position of circadian rhythms.

As concerns age, we did not find an association between morningness and increasing age, as has systematically been reported.[28,32,41] But in agreement with epidemiological findings issued from a survey in the general

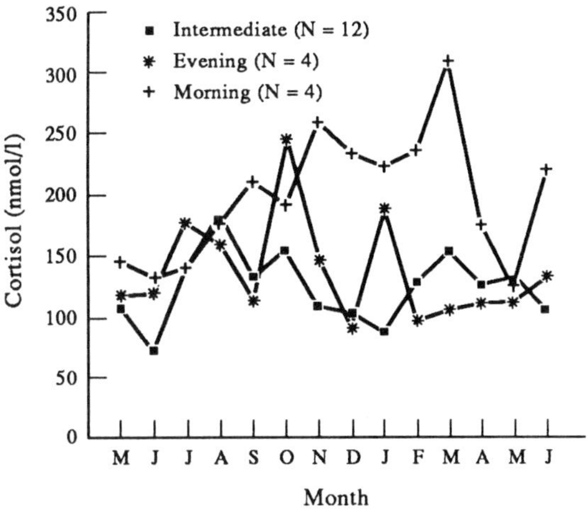

FIG. 6. Seasonal pattern (monthly means) of nighttime urinary cortisol by diurnal type subgroups.

TABLE 5
Pearson's product-moment correlations of diurnality and seasonality with psychobiological measures ($N=20$)

	H-O index[a]	Seasonality score
	r	*r*
Age (years)	0.19	−0.45
Height (cm)	0.20	−0.12
H-O index[a]		−0.37(*)
Seasonality score (pts)	−0.37(*)	
Sadness (mm)	−0.36	0.53*
Elation (mm)	0.27	0.06
Fatiguability (mm)	−0.43*	0.61**
Sleep need (mm)	−0.21	0.54**
Broken and disturbed sleep (mm)	−0.46*	0.54**
Difficulty falling asleep	−0.43*	0.71***
Melatonin (nmol/l)	−0.37(*)	0.39(*)
Cortisol (nmol/l)	0.34	0.13
Urine volume (8 hours, overnight)	0.20	−0.17

[a] A high H–O index is associated with morningness.
(*) $p<0.10$.
* $p<0.05$.
** $p<0.01$.
*** $p<0.001$.

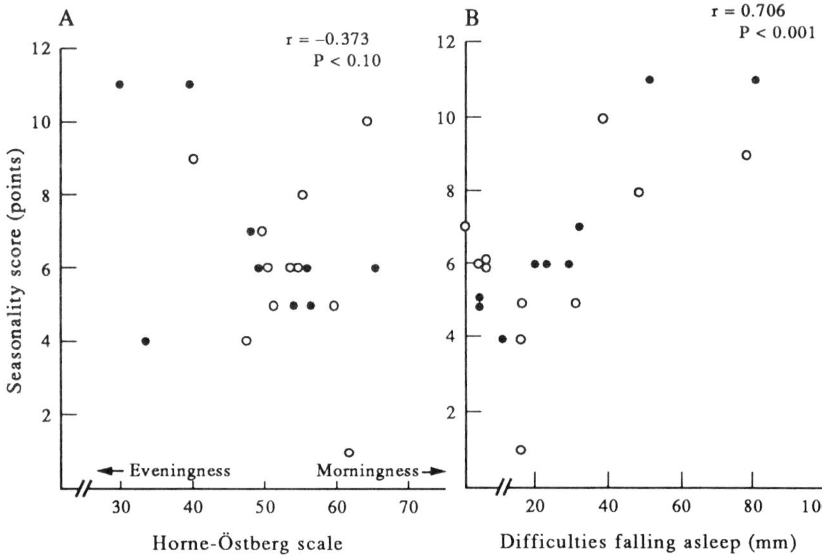

FIG. 7. Morningness-eveningness (Horne-Östberg index) and "difficulties falling asleep" plotted in function of the degree of subjective seasonality as derived from the SPAQ in 20 healthy subjects. Pearson correlation coefficient = *r*. Open circles: men, closed circles: women

population in Montgomery Country/Maryland,[15] we could establish a negative correlation between subjective seasonality and age.

Our results, although preliminary considering the small sample studied, are in support of a view assuming that morningness-eveningness contributes to seasonal variability. M-, I- and E-types have shown to differ in objective (seasonal rhythms) and subjective (seasonality score) measures and indirectly, in seasonal sensitivity and/or seasonal adaptation. That there is a link between diurnality and various aspects of seasonality is not surprising, since the circadian system is closely related to the seasonal system by the interdiurnal changes in daylight over the year. Also the lacking or reduced rhythmicity observed in the I-type corresponds somewhat to expectation. The comparably large seasonal variations seen in the E-type are in agreement with one of the rare studies available, reporting a greater number of significant seasonal rhythms in E-types as compared to M-types[19] (Figure 8). In the same investigation it was shown that, when only weekends were considered, the number of significant rhythms substantially increased in both M- and E-types, suggesting that social constraints may partially be responsible for "a-seasonality" in man.

In the present study M- and E-types were rather close together deviating both in the same direction from the I-types, at least as far as psychometric measures (levels) are concerned. In most work dealing with diurnal

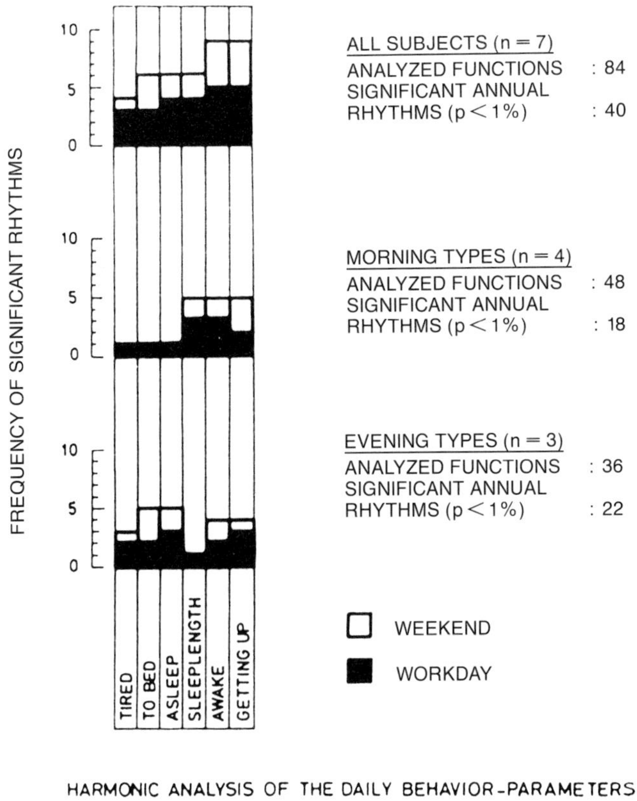

FIG. 8. Frequency of significant (1% level) annual rhythms in daily behavior-parameters separately for M- and E-types. From Klöppel.[19]

typology however, M- and E-types were opposed one to another. Given the period length of a free-running intrinsic rhythm is shorter with regard to the 24-hour environment in M-types, (tending to phase advance) and longer in E-types (tending to phase delay) it is conceivable that M- and E-types have more difficulties to remain synchronized with the geophysical environment and become easier disturbed in their equilibrium as reflected by the higher levels of sadness, fatiguability and sleep complaints. Thus, including the I-type in group comparison appears to be essential and may provide complementary information.

Another important finding is that diurnal type related differences in self-rated mood and sleep as well as in physiological measures were greatest in winter. Likewise, the individual values within the whole group scattered wider in winter as compared to summer. A similar distribution across the year has previously been described for total metanephrines[47] and can tentatively be interpreted as a consequence of the weak or lacking zeitgeber effect of winter light conditions in temperate zones.

Among the three diurnal type subgroups, the E-type appears to be particularly susceptible to seasonal changes in the exogenous light/dark cycle, with prevailing difficulty to adapt to winter conditions, as suggested by findings summarized in Table 6. It is noteworthy that the E-type not only scored highest in subjective seasonality and exhibited largest seasonal amplitudes in self-rated mood and sleep, but also showed highest levels of nighttime melatonin in comparison to M- and I-types. This is in accordance with clinical findings from a depressed population,[3] relating low nocturnal melatonin to diminished rhythmicity in depressive symptomatology. While there is some support from literature for an association between eveningness and day-time fatigue,[7,38] an association between eveningness and increased seasonality however, has not yet been described.

TABLE 6
Synopsis of characteristics associated with an eveningness-typical subjective circadian phase position

Mood	• Sadness↑ • Fatiguability↑	this study
Sleep	• Sleep complaints↑ (especially difficulties falling asleep)	this study
Physiology	• Melatonin/urine (overnight)↑	this study
Seasonality	• Subjective Seasonality score (SPAQ) • Objective Amplitude (annual) Pattern: unimodal with winter peaks	this study
Gender distribution	• Predominantly women	
Age	• High proportion of young subjects	41
Psychopathology	• Neuroticism	31, 41
	• Anxious and neurotic personality	29
	• Cyclothymic personality Catell's 16 PF	32
	• High proportion of E-types in bipolar depression (free interval)	30

Clinical issues

Two chronobiological disorders affected by season are of particular interest with regard to eveningness: the so-called midwinter insomnia (MI) occurring frequently in the far north (e.g. in 23.7% of the general adult population in Tromsö[26]) and winter depression, a form of seasonal affective disorder (SAD) described by Rosenthal and coworkers,[37] with a varying prevalence rate in function of geographical latitude.[35] MI as well as SAD manifest increased day-time drowsiness and sleepiness during the dark period in winter, and MI is characterized by a predominantly initial type of insomnia, ranging in degree from moderate difficulty in getting asleep, to an almost total inability to sleep during the whole night.[27] In addition, both are illness conditions, where a tendency to circadian phase delay due to the lack of regulatory influence of normal day light is presumed, and both respond favorably to therapy with artificial bright light.[25,27] As summarized in Table 6, the E-type appears to share some common characteristics with MI and SAD, such as the high degree of subjective seasonality, the unimodal seasonal pattern with regard to sadness, fatiguability and sleep complaints (initial insomnia)—all peaking around winter—as well as the predominance of female subjects. That in the E-type the latest subjective circadian phase position (maximum eveningness) occurred in late autumn/winter and coincided with the maxima of sadness, fatiguability and difficulties falling asleep add indirect support to the hypothesis that SAD and MI may reflect underlying circadian phase delay of some or all circadian rhythms. In accordance with our findings, in a recent review about seasonal rhythms in healthy subjects, it was shown that changes in metabolic processes, increased sleep and circadian phase delay occurs somewhat physiologically in winter, and that winter impairment of mood, sleep/-wake- and vegetative functions is not restricted to MI and SAD, but occurs in an attenuated form in normals as well.[22]

In opposite to the phase delay hypothesis of SAD, a phase advance hypothesis, based a.o. on earlier peaking of the 24-hour body temperature and early morning awakening, has previously been formulated for major affective disorder,[8] being of interest with regard to a morningness phase position, where an analogous phase advance in circadian rhythms is presumed.[17] According to our results, morningness appears to be related to high nocturnal cortisol, corresponding to elevated cortisol levels found in major affective disorder,[24] but not in SAD.[39] This fact, together with the observation that extreme M-types are strikingly often seen in melancholic depressed patients during their symptom free interval,[30] brings the M-type near to this clinical category of affective disorder.

Whether an extreme phase position, particularly eveningness, would predispose to seasonal variations and sleep and mood pathology must be corroborated by further studies. Such future research might include the evaluation of distribution patterns of M-, I- and E-types with regard to MI and SAD as well as to the Delayed Sleep Phase Syndrome (DSPS)—another chronobiological sleep disorder but with a chronical course[44]—and also with regard to male and female subjects and in function of geographical latitude.

In conclusion, the "diurnal type" appears to modify aspects of seasonality in healthy volunteers. Whether this is due to differences in the structure of the circadian system of M-, I- and E-types, or to a differential response to exogenous zeitgebers through an association between personality traits and cognitive and social factors affecting entrainment of circadian rhythms to the sleep/wake cycle, remains to be clarified. Nevertheless, our findings suggest, that including diurnal typology in chronobiological studies, especially those focusing on seasonal concerns, is a promising area of research, with important implications for sleep and mood disorders and providing new insights into mechanisms relating the circadian and the seasonal system in man.

References

1. Aschoff J. (1991) *Why should we rise with the lark? In Pathways in Applied Immunology* (eds. Messmer M. and Stein M.), pp. 127–137. Springer, Berlin-Heidelberg.
2. Aschoff J. (1965) The phase-angle difference in circadian periodicity. In *Circadian Clock* (ed. Aschoff J.), pp. 262–276. North-Holland, Amsterdam.
3. Beck-Friis J., Kjellman B.F., Aperia B., Unden F., Ljungren J.G. and Wetterberg L. (1985) Serum melatonin in relation to clinical variables in patiens with major depressive disorder and a hypothesis of a low melatonin syndrome. *Acta Psychiat. Scand.* **71**, 319–330.
4. Breithaupt H., Hildebrandt G.H., Döhre S., Josch R., Sieber U. and Werner M. (1978) Tolerance to shift of sleep, as related to the individuals circadian phase position. *Ergonomics* **21**, 767–774.
5. Broadway J.W., Arendt J. and Folkland S. (1987) Bright light phase-shifts the human melatonin rhythm during the antarctic winter. *Neurosci. Lett.* **79**, 185–189.
6. Foret J. and Benoit O. (1981) Individual factors, sleep characteristics and circadian evolution of body temperature. In *Sleep 1980*. 5th Eur. Congr. Sleep Res. Amsterdam, pp. 195–197. Karger, Basel.
7. Fröberg J.E. (1977) Twenty-four hour patterns in human performance, subjective and physiological variables and differences between morning and evening active subjects. *Biol. Psychol.* **5**, 119–134.
8. Goodwin F.K., Wirz-Justice A. and Wehr T. (1982) Evidence that the pathophysiology of depression and the mechanisms of action of antidepressant drugs involve alterations in circadian rhythms. *Adv. Biochem. Psychopharmacol.* **31**, 1–11.
9. Hildebrandt G.H., Breithaupt H., Döhre S., Stratmann I. and Werner M. (1977) Arbeitsphysiologische Bedeutung und Bestimmung der circadianen Phasentypen. *Z. Arb. Wiss.* **31**, 98–102.
10. Hoffmann K. (1963) Zur Beziehung zwischen Phasenlage und Spontanfrequenz bei der endogenen Tagesperiodik. *Z. Naturforsch.* **188**, 154–157.
11. Horne J.A. and Östberg O. (1976) A self-assessment questionnaire to determine morningness-eveningness in human circadian rhythms. *Int. J. Chronobiol.* **4**, 97–110.
12. Horne J.A. and Östberg O. (1977) Individual differences in human circadian rhythms. *Biol. Psychol.* **5**, 179–190.
13. Illnerova H., Zvolsky P. and Vanecek J. (1985) The circadian rhythm in plasma melatonin concentration of the urbanized man: the effect of summer and winter time. *Brain Res.* **328**, 186–189.
14. Ishihara K., Miyashita A. and Inugami M. (1986) Morningness-eveningness. The results of investigations by the Japanese version of the morningness-eveningness questionnaire. (engl. abstract). *The Japanese Journal of Psychology* **57**, 87–91.
15. Kasper S., Wehr T.A., Bartko J.J., Gaist P.A. and Rosenthal N.E. (1989) Epidemiological findings of seasonal changes in mood and behaviour. *Arch. Gen. Psychiatry* **46**, 823–833.
16. Kerkhof G.A. (1984) A Dutch questionnaire for selection of morning- and evening-type

persons (engl. abstract). *Nederlands-Tijdschrift-voor-de Psychologie- en haar Grensgebieden* **39**, 281–294.

17. Kerkhof G.A. (1985) Interindividual differences in the human circadian system: A review. *Biol. Psychol.* **20**, 83–112.
18. Kleitman N. (1939) *Sleep and Wakefulness*. University of Chicago Press, Chicago (revised and enlarged edition 1963).
19. Klöppel H.-B. (1980) Circannuale Änderungen der circadianen Phasenlage des Menschen. PhD dissertation. University of Marburg/Lahn.
20. Klöppel H.-B. (1982) Circannual changes of the circadian phase position in man. In *Biological Adaptation* (eds. Hildebrand G.H. and Hankel H.), pp. 137–145. Georg Thieme, Stuttgart/New York.
21. Lacoste V. and Wirz-Justice A. (1987) Seasonality in personality dimensions. *Psychiat. Res.* **21**, 181–183.
22. Lacoste V. and Wirz-Justice A. (1989) Seasonal variation in normal subjects: an update of variables current in depression research. In *Seasonal Affective Disorder and Phototherapy* (eds. Rosenthal N.E. and Blehar M.), pp. 167–229. Guilford Press, New York.
23. Lacoste V. and Wetterberg L. (1992) Is there a seasonal relationship between urinary melatonin and cortisol? *Experientia* **48**, A92.
24. Larson A.J., Newman D.C., Lin S.C. and Maruta T. (1988) Cortisol in major depression. Biol. *Psychiatry* **24**, 957–958.
25. Lewy A.J., Sack R.L., Fredrickson R.H., Reaves M., Denney D. and Zielske D.R. (1983) The use of bright light in the treatment of chronobiological sleep and mood disorders: the phase response curve. *Psychopharmacol. Bull.* **19**, 523–525.
26. Lingjaerde O., Bratlid T. and Hansen T. (1985) Insomnia during the "dark period" in northern Norway. *Acta Psychiat. Scand.* **71**, 506–512.
27. Lingjaerde O., Hansen T. and Gotestam K.G. (1986) Seasonal affective disorders and midwinter insomnia in the far north: studies on two related chronobiological disorders in Norway. *Clin. Neuropharmacol.* **9** (Suppl.) 4, 187–189.

27a. Matthew G. (1988) Morningness-eveningness as a dimension of personality: trait, state and psychophysiological correlates. *Europ. J. of Personality* **2**, 277–293.

28. Maccaci L., Zani A., Rocchetti G. and Lucioli R. (1986) The relationship between morningness-eveningness, aging and personality. *Person. indiv. Diff.* **7**, 911–913.
29. Michelson E. (1899) Untersuchungen über die Tiefe des Schlafs. *Psychol. Arbeiten* **2**, 84–117.
30. Middelhoff H.D. (1967) Tagesrhytmische Schwankungen bei endogenen Depressionen im symptomfreien Intervall und während der Phase. *Arch. Psychiatrie u. Z. ges. Neurol.* **209**, 315–339
31. Mura LE.L. and Levi A. (1986) Relationship between neuroticism and circadian rhythms. *Psychol. Reports* **58**, 298.
32. Öquist O. (1970) Kartläggning av individuella dygnsrytmer. Thesis at the Dept. of Psychology. University of Göteborg.
33. Östberg O. (1973) Interindividual differences in circadian fatigue patterns of shift workers. *British Journal of Industrial Medicine* **30**, 341–351
34. Patkai S.G. (1971) The diurnal rhythm of adrenaline secretion in subjects with different working habits. *Acta Physiol. Scand.* **81**, 30–34.
35. Potkin S.G., Zetkin M. and Stamenkovic V. (1986) Seasonal affective disorder: prevalence varies with latitude and climate. *Clin. Neuropharmacol.* **9** (Suppl.) 4, 181–183.
36. Rosenthal N.E., Bradt G.H. and Wehr T.A. (1984) *Seasonal Pattern Assessment Questionnaire*. National institute of Mental Health, Bethesda ML.
37. Rosenthal N.E., Sack D.A., Gillin C., Lewy A.J., Goodwin K., Müller D.A., Newsome D.A. and Wehr T. (1984) Seasonal affective disorder. A description of the syndrome and preliminary findings. *Arch. Gen. Psychiatry* **41**, 72–80.
38. Sano K. (1987) Circadian rhythm of CFF in morningness and eveningness children. *Neurosciences* **13**, 298–300.
39. Skwever R.G., Jacobson F.M., Duncan C.C., Kelly K.A., Sack D.A. and Tamarkin L. (1989) Neurobiology of seasonal affective disorder and phototherapy. In *Seasonal*

Affective Disorders and Phototherapy (eds. Rosenthal N.E. and Blehar M.), pp. 311–332. Guilford Press, London/New York.

40. Sverko B. and Fabulic L. (1985) Stability of morningness- eveningness: retest changes after seven years (engl. abstract). *Revij za psihologiju* **15**, 71–78.
41. Torsvall L. and Akerstedt (1980) A diurnal type scale: construction, consistency and validation. *Scand. Environ. Health* **6**, 283–290.
42. Webb W.B. and Bonnet M.H. (1978) The sleep of the "morning" and "evening" types. *Biol. Psychol.* **7**, 29–35.
43. Wehr T.A. and Goodwin F.K. (1981) Biological rhythms and psychiatry. In *American Handbook of Psychiatry*. Vol. 7 *Advances and Directions* (eds. Arieti S. and Brodio M.K.), pp. 46–74. Basic Books Inc. Publishers, New York.
44. Weitzman E.D., Czeisler C.A., Coleman R.M., Spielman A.J. and Dement W. (1981) Delayed sleep phase syndrome. A chronobiological disorder with sleep-onset insomnia. *Arch. Gen. Psychiatry* **38**, 737–746.
45. Wendt H.W. (1977) Population, sex and constitution in typologies based on individual circadian rhythms. *J. interdiscipl. Cycle Res.* **8**, 286–290.
46. Winget C.M., DeRoshia C.W. and Holley D.C. (1985) Circadian rhythms and athletic performance. *Med. Sci. Sports Exerc.* **17**, 498–516.
47. Yamamoto T., Takeuchi Y., Baba M. and Tanaka M. (1976) Seasonal variation of urinary excretion of total metanephrines. *Clinica Chemica Acta* **68**, 241–244.

21

Light Regulation of the Menstrual Cycle

DANIEL F. KRIPKE*

Department of Psychiatry, UCSD and the Sam and Rose Stein Institute for Research on Aging, La Jolla, CA 92093–0667, USA

*Supported by MH00117, by HL40930, by the Department of Veterans Affairs, and by a gift from Apollo Light Systems, Inc.

Abstract

The 1967 report of Edmund Dewan that a bedside 100 watt bulb could stabilize irregular menstrual cycles was initially greeted with skepticism. Recent studies of light treatments of humans have indicated that both bright and dim light have real biological effects. To examine the validity of Dewan's report, our group has treated two samples of women with irregular menstruation with randomly-assigned 100 watt bedside night lights or dim red light placebos. In both groups, the 100 watt treatments significantly decreased the duration of menstrual cycles and reduced their variability. Effective were treatments both from days 10–14 and from days 13–17 of the menstrual cycle (counting the first bleeding as day 1). The night light effect on long cycles in oligomenorrheic women seems to be a genuine and rather powerful biological effect and not a result of suggestion.

Use of light to regulate menstrual cyclicity may have clinical value both for rhythm contraceptive methods and for treatments of infertility. Further study of night light effects on the menstrual cycle may expand our understanding of reproductive endocrinology. Extensive clinical testing is needed to determine the optimal intensities, timing, and durations of night light treatments for producing endocrine benefits.

Light regulation of the menstrual cycle

IN 1967, Edmund Dewan reported that by simulating moonlight with nocturnal light exposures, the menstrual cycles of women could be brought nearer to the lunar cycle of 29.5 days.[9] Dewan argued that such dim light might be used to augment natural contraceptive methods. Dewan published several follow-on reports,[10,11] and a feminist popularization called "Lunaception" appeared,[16] but for 20 years, Dewan's work was virtually ignored by the scientific community. Perhaps the idea that mere moonlight had biological effects on humans strained the credulity of the time.

In the intervening years, more evidence has accumulated that circadian systems of plants, nocturnal rodents,[22] and other organisms are indeed sensitive to light exposures in the range of moonlight. Furthermore, through effects on the circadian timing system, in nocturnal rodents, dim light exposures are able to trigger a variety of reproductive endocrine responses including seasonal reproduction.

Whereas it once seemed accepted that human beings were immune to the photic effects which regulate lower organisms, pioneering work of Wetterberg, Lewy, and colleagues demonstrated that very bright light does indeed influence human circadian and melatonin systems.[19,26] In the 1980s, we have witnessed an expanding science of phototherapy, inspired by the concept that illumination of at least 1,500–2,500 lux is required for melatonin suppression and related effects in humans. Meanwhile, the field has tended toward experiments with very bright illumination levels. Terman *et al.*[25] found that phototherapeutic effects of 10,000 lux for 30 minutes exceeded effects of 3,000 lux for 30 minutes and approximated effects of 2,500 lux for 2 hours. Meanwhile, Wever reported that strong circadian synchronizing effects required exposures up to 4,000 lux for as much as 8 hours.[27] Czeisler and colleagues[8,13] obtained surprisingly strong Type-0 circadian phase-shifting effects from 7–12,000 lux exposures for 5 hours on 3 consecutive days. Such data have encouraged some investigators to work with 5,000–10,000 lux illuminations, even though there is some data in monkeys suggesting a possibility of eye damage with prolonged exposures to 10,000 lux.[24]

In retrospect, the notion of a threshold of suppression of 1,500–2,500 lux was based on a small study of only a few cases.[19] Subsequent studies showed that melatonin suppression in humans is a graded effect produced by graded intensities and durations of illumination.[6,21] Indeed partial human melatonin suppression can be produced by 500 lux,[18] 300 lux,[4] or even 17 lux[5] in particular circumstances. Nevertheless, little study has been done of effects on human melatonin of relatively dim illumination.

The intensity/response relationship for human circadian phase shifting has likewise been examined only vaguely. Without strong empirical evidence, Kronauer[15] suggested that circadian phase shifting might be related to the cube root of illumination intensity. If a cube root rule is correct, 8,000 lux should be only twice as powerful as 1,000 lux. Further, 8,000 lux might be only five times as powerful as 64 lux and only 20 times as powerful as 1 lux. A logarithmic rule might produce relationships of similar magnitude, e.g. 10,000 lux would be four times as powerful as 10 lux. Whether longer durations compensate for dimmer illuminations by cube, logarithmic or more linear rules is essentially unknown. From such considerations, the possibility that Dewan's report could be correct became more plausible.

First San Diego replication: experimental procedures

At UCSD, Lin *et al.*[20] decided to replicate Dewan's reports with placebo-control similar to designs we had used for studies of phototherapy of depression. Women were recruited who reported long (or irregular) menstrual cycles averaging 46–54 days. To focus on women who were likely to have oligomenorrhea of functional hypothalamic origin, we excluded women with grossly abnormal body weights, those with any appearance of androgenizing syndromes, and those with any history of major psychiatric or endocrine disorders. Each woman was followed prospectively for one untreated menstrual cycle. In the following cycle, randomized women were treated with a 100 watt light bulb placed by the side of the bed, approximately 1 m from the head, on the nights following the 13th–17th days after the start of menstruation. To assure that the women had some exposure to this treatment, which measured 235 lux, we asked these women to read for 0.5 hour in bed. Randomized controls used a dim red photographic safe light which produced only 1.7 lux by photopic measurement. A red light placebo was chosen, because in previous phototherapy studies, we had found that subjects commonly believed that dim red light was the active treatment, thus balancing expectations.

First San Diego replication: results

In this study, we were able to study seven subjects who received the 100 watt light and nine who received the dim red placebo. All of the women exposed to the 100 watt bulb reduced their menstrual cycle durations: from an average of 45.7 days at baseline to 33.1 days during the treated cycle. Placebo had no significant effect, thus, differences between the 100 watt treatment and placebo were robustly significant. One problem with this study was that eight subjects assigned to the 100 watt bulb had dropped out, apparently either because the light disturbed their sleep or because of negative expectations. Another problem was failure to formally balance pre-sleep behaviors and expectations. Thus, further replication was needed.

Second San Diego replication: experimental procedures

In our next study, Drennan *et al.*[12] sought to determine if the effect of night lights was limited to women with long and irregular menstrual cycles, or whether there was also an effect on the ovulation of women with normal cycles. We did not know Dewan's empirical basis for originally recommending nights 14–17 of the menstrual cycle for bedside lights and for stating that such stimulation was ineffective a few days earlier.[10] Because many normal women might already have ovulated by days 13–17 of the menstrual cycle, we decided to test light effects somewhat earlier in the follicular phase on days 10–14 of their cycles.

Two groups of women were studied. The first group reported regular 28–32 day menstrual cycles. The second group reported long and irregular cycles averaging 41–45 days. As in our first study, women were selected whose physical examination and medical histories tended to rule our anorexia, androgenizing syndromes, or discrete psychiatric and endocrine disorders, however, the subjects with reported oligomenorrhea were not evaluated sufficiently to definitively diagnose functional hypothalamic oligomenorrhea. As in the previous study, subjects from each group were randomly assigned either to 100 watt beside lights or to dim red photographic safe light placebos.

Second San Diego replication: results

With the 100 watt bulbs, the women with long cycles again showed significant shortening from a baseline of 44.9 days to 33.2 days during the treatment cycle. Again the women treated with the dim red placebo showed no effect, so that the group responses were significantly different. Unlike the women in the first replication who had been treated from days 13–17 of their cycles and who reverted to their prior long cycles in the follow-up month, these second-replication women treated from days 10–14 had cycle durations of 35.4 days on follow-up, which was almost significantly shorter than baseline. This suggested some possibility of a carry-over effect from 100 watt treatments from days 10–14 of the cycle to the next cycle. The women with normal 28–32 day cycles showed no significant effects on cycle length either from 100 watt or from dim red lights. Endocrine studies of these women will be reported elsewhere.

Perhaps because these subjects were being paid to contribute blood specimens, in the second replication, drop-out rates were not as troublesome. Also, both subjects randomized to 100 watt and dim red treatments had spent 0.5 hour in bed reading, so this was balanced. Finally, positive expectations were found to actually be somewhat higher among the subjects who were randomized to placebo, so the likelihood of suggestion as confounder was remote. This replication added strong assurance that effects of 100 watt bulbs on long and irregular menstrual cycles are real biological effects.

Discussion

From the results of these two studies, we are quite confident that as Dewan had reported, 100 watt bedside lights can indeed shorten and regularize the menstrual cycles of women whose cycles were long and irregular. We agree with Dewan that these effects may have potential usefulness both for contraception[9,11] (by facilitating rhythm methods) and perhaps for treatments of infertility as well. If night lights influence reproductive endocrinology, they may have more extensive applications in endocrinology than we can currently envision. Unfortunately, elaborate clinical trials will be needed before we can develop more than speculations as to the mechanisms by which 100 watt lights achieve their effects.

We would question Dewan's theory that night lights are simulating effects of moonlight. Our placebo dim red lights, although only about 1.7 lux, are nevertheless of greater photopic brightness than the full moon. Red lights are much brighter than the moon in the red portion of the visible spectrum but perhaps dimmer in some blue and green regions, nevertheless, there is currently no basis for supposing that moonlight would have stronger effects than 1.7 lux red light. It would certainly be valuable to obtain dose/response data for night lights across the range of roughly 1–500 lux. Unfortunately, dose/response data curves will require laborious clinical studies.

Our data do not confirm Dewan's statements that days 14–17 of the menstrual cycle are the optimal times for bedside light treatment. We achieved comparable effects using days 13–17 and days 10–14, though both protocols included day 14, which Dewan had suggested might be the critical day. Note also that most women did not achieve a shortening of their cycles to 29–30 days, which would be expected if the effect simulated lunar synchronization.

The issue of brightness is complicated by uncertainty as to what illumination exposures bedside lights produce during sleep. If one places a photometer by the subject's head aimed toward a 100 watt bulb at 1 m, then the brightness is 235–250 lux, however, brightness is certainly much less if the photometer is turned away from the light, as the subject's head may be. Following Dewan's model, we have permitted our subjects to close their eyes and roll over, so that the illumination actually reaching the eyes must be extremely variable. Illumination reaching the cornea must be much less than 235 lux when subjects' eyes are closed, when they face away from the night lights, and when they may possibly obscure their eyes with pillows, limbs, or bedclothes. In future experiments, we hope to study women using Dr. Cole's lighted sleep mask,[7] which should be very helpful in defining and standardizing the illumination exposures of sleeping subjects. Until these studies can be done, the time/duration curves of illumination actually reaching our subjects' eyes can only be guessed, and indeed, it is surprising that the results seem so consistent when the illumination exposures were probably erratic.

We believe that light masks may prove the best method for administering night lights for several reasons besides standardization of light exposure. If they use light masks, women desiring to regulate their menstrual cycles need not disturb bed partners or roommates. Night lights must have real potential to disturb other occupants of the bedroom, if we admit that night lights are biologically active. In rustic and third-world settings, 100 watt bedside lights may be impractical, but light masks could probably be powered by low-cost solar batteries, since they require much less power than 100 watt bulbs.

Another variable is the timing of night light exposures. In our first two experiments, we had asked subjects to give themselves 30 minutes exposure to 235 lux before going to sleep, in order to assure the strongest effects possible. So far as we know, Dewan's subjects did not receive this preliminary exposure, but results were quite similar. Our current ongoing study exposes subjects to 235 lux only from 30 minutes *after* bedtime to 30 minutes *before* wake-up time, nevertheless, the first subject has already shown a typical shortening of the cycle. To systematically explore different durations of night light exposure given at varying times of night will require large-scale clinical experimentation, multiplying the complexity of intensity/response studies.

The mode of action of Dewan's effect is currently purely a matter of speculation. Neither Dewan's subjects nor ours have been well-characterized from the viewpoint of reproductive endocrinology. Having attempted to exclude volunteers with a variety of psychiatric and endocrine disorders, anorexia, and androgenizing syndromes, we believe that most of our subjects were experiencing forms of functional hypothalamic oligomenorrhea,[2] a syndrome resulting primarily from inadequate pulsatile secretion of gonadotropin releasing hormone, with resultant inadequate ovulation and decreased corpus-luteum production of progesterone. The mechanisms producing such hypothalamic failures are poorly characterized. In a series of patients with hypothalamic amenorrhea (a more extreme disorder, possibly with similar etiology), Berga *et al.*[3] found that nocturnal melatonin secretion was elevated. In patients with amenorrhea related to strenuous exercise, similar elevations of melatonin have been described.[17] Thus, it is tempting to speculate that our subjects may have been suffering from excessive nocturnal melatonin secretion, and that night lights blunt this excessive secretion, nevertheless, currently we have no empirical confirmation of this speculation.

Parry *et al.* have described a group of women with premenstrual depressions who appeared to have an early offset of nocturnal melatonin secretion, associated with decreased total area under the nocturnal melatonin curve.[23] Perhaps premenstrual depressions resemble the "low melatonin syndrome" described among major depressive patients by Beck-Friis and colleagues.[1] Although light is generally conceived as melatonin-suppressing, patients with major depressions do respond to bright light treatment.[14] Thus, although our subjects had no overt major depressions, the alternative hypothesis that night lights somehow combat a low-melatonin condition cannot presently be excluded.

Unfortunately, a barrier of disbelief delayed exploitation of Dewan's discovery for two decades. We hope that others will add to the replications from our laboratory, and that night light effects on the menstrual cycle will become an accepted reality. This apparently powerful—yet simple and seemingly safe—methodology should lead to exciting progress in our understanding of human reproductive endocrinology and probably to very useful clinical approaches.

References

1. Beck-Friis J., Kjellman B.F., Aperia B., Unden F., von Rosen D., Ljunggren J.-G. and Wetterberg L. (1985) Serum melatonin in relation to clinical variables in patients with major depressive disorder and a hypothesis of a low melatonin syndrome. *Acta Psychiatr. Scand.* **71**, 319–330.
2. Berga S.L. and Daniels T.L. (1991) Use of the laboratory in disorders of reproductive neuroendocrinology. *Journal of Clinical Immunoassay* **14**(1), 23–28.
3. Berga S.L., Mortola J.F. and Yen S.S.C. (1988) Amplification of nocturnal melatonin secretion in women with functional amenorrhea. *J. Clin. Endocrinol. Metab.* **66**, 242–244.
4. Bojkowski C.J., Aldhous M.E., English J., Franey C., Poulton A.L., Skene D.J. and Arendt J. (1987) Suppression of nocturnal plasma melatonin and 6-sulphatoxymelatonin by bright and dim light in man. *Horm. Metab. Res.* **19**, 437–440.
5. Brainard G.C., Lewy A.J., Menaker M., Fredrickson R.H., Miller L.S., Weleber R.G., Cassone V. and Hudson D. (1988) Dose-response relationship between light irradiance and the suppression of plasma melatonin in human volunteers. *Brain Res.* **454**, 212–218.
6. Byerley W.F., Risch S.C., Gillin J.C., Janowsky D.S., Parker D., Rossman L.G. and Kripke D.F. (1989) Biological effect of bright light. *Prog. Neuropsychopharm. and Biol. Psychiatry* **3**, 683–686.
7. Cole R.J. and Kripke D.F. (1991) Light exposure during sleep with a bright light mask: effects on human body temperature rhythm. *Sleep Res.* **20**, 450 (Abstract).
8. Czeisler C.A., Kronauer R.E., Allan J.S., Duffy J.F., Jewett M.E., Brown E.N. and Ronda J.M. (1989) Bright light induction of strong (type 0) resetting of the human circadian pacemaker. *Science* **244**, 1328–1332.
9. Dewan E.M (1967) On the possibility of a perfect rhythm method of birth control by periodic light stimulation. *Am. J. Obstet. Gynecol.* **99**, 1016–1018.
10. Dewan E.M. (1969) Rhythms. *Science and Technology* January, 20–28.
11. Dewan E.M., Menkin M.F. and Rock J. (1978) Effect of photic stimulation on the human menstrual cycle. *Photochem. Photobiol.* **27**, 581–585.
12. Drennan M.D., Kripke D.F. and Berga S.L. (1991) Night light shortens long and irregular, but not regular, menstrual cycles. *Society for Light Treatment and Biological Rhythms. Abstracts* **3**, 20 (Abstract).
13. Jewett M.E., Kronauer R.E. and Czeisler C.A. (1991) Light-induced suppression of endogenous circadian amplitude in humans [letter]. *Nature* **350**, 59–62.
14. Kripke D.F., Mullaney D.J., Klauber M.R., Risch S.C. and Gillin J.C. (1992) Controlled trial of bright light for nonseasonal major depressive disorders. *Biol. Psychiatry* **31**, 119–134.

15. Kronauer R.E. (1987) A model for the effect of light on the human "deep" circadian pacemaker. *Sleep Res.* **16**, 621.
16. Lacey L. (1975) *Lunaception: A Feminine Odyssey into Fertility and Contraception*, Coward, McCann & Geoghegan, New York.
17. Laughlin G.A., Loucks A.B. and Yen S.s. (1991) Marked augmentation of nocturnal melatonin secretion in amenorrheic athletes, but not in cycling athletes: unaltered by opioidergic or dopaminergic blockade. *J. Clin. Endocrinol. Metab.* **73**(6), 1321–1326.
18. Lewy A.J., Nurnberger J.I., Wehr T.A., Pack D., Becker L.E., Powell R. and Newsome D.A. (1985) Supersensitivity to light: possible trait marker for manic-depressive illness. *Am. J. Psychiatry* **142**, 725–728.
19. Lewy A.J., Wehr T.A., Goodwin F.K., Newsome D.A. and Markey S.P. (1980) Light suppresses melatonin secretion in humans. *Science* **210**, 1267–1269.
20. Lin M.C., Kripke D.F., Parry B.L. and Berga S.L. (1990) Night light alters menstrual cycles. *Psychiatry Res.* **33**, 135–138.
21. McIntyre I.M., Norman T.R., Burrows G.D. and Armstrong S.M. (1989) Human melatonin suppression by light is intensity dependent. *J. Pineal Res.* **6**, 149–156.
22. Nelson D.E. and Takahashi J.S. (1991) Comparison of visual sensitivity for suppression of pineal melatonin and circadian phase-shifting in the golden hamster. *Brain Res.* **554**, 272–277.
23. Parry B.L., Berga S.L., Laughlin G.A., Klauber M.R., Yen S.S.C., Kripke D.F. and Gillin J.C. (1990) Altered waveform of plasma nocturnal melatonin secretion in premenstrual depression. *Arch. Gen. Psychiatry* **47**, 1139–1146.
24. Sykes S.M., Robison W.G.Jr., Waxler M. and Kuwabara T. (1981) Damage to the monkey retina by broad-spectrum fluorescent light. *Investigative Ophthalmology & Visual Science* **20**(4), 425–434.
25. Terman J.S., Terman M., Schalger D., Rafferty B., Rosofsky M., Link M.J., Gallin P.F. and Quitkin F.M. (1990) Efficacy of brief, intense light exposure for treatment of winter depression. *Psychopharmacol. Bull.* **26**(1), 3–11.
26. Wetterberg L. (1978) Melatonin in humans: physiological and clinical studies. *J. Neural Transm.* **Suppl. 13**, 289–310.
27. Wever R.A., Polasek J. and Wildgruber C.M. (1983) Bright light affects human circadian rhythms. *Pflugers Arch.* **396**, 85–87.

22

Circannual Variation of some Endocrine and Neuroendocrine Functions in Humans

YVAN TOUITOU[1] and ERHARD HAUS[2]

[1]*Department of Biochemistry, Faculté de Médecine Pitié-Salpêtrière, 91, bd de l'Hôpital, 75634 Paris Cedex 13, France*

[2]*Department of Anatomic and Clinical Pathology, Ramsey Clinic, Saint Paul Ramsey Medical Center, 640 Jackson Street, Saint Paul, Minn, 55101–2595, USA*

Introduction

THE MAMMALIAN organism is characterized by a complex time structure of biologic rhythms which are present at all levels of organization ranging from population and group rhythm to individuals, organ systems, organs and tissue, cells and subcellular structures. These rhythms cover a large frequency range (infradian, circadian, ultradian) providing multiple and highly complex interactions. Most authors support the view that the natural apparently genetically fixed period of biologic rhythms will continue in the absence of environmental forces e.g. light/dark, magnetism, cosmic radiations . . . although certain endogenous or exogenous factors can exert a considerable influence upon them.[60]

Seasonal variations of various body functions may be induced environmentally and may represent the organism's response to variations in the environment such as changes in temperature, in the length of the daily light and dark spans (photoperiod), and in seasonal availability of food for, e.g. free-living animals. However, such seasonal rhythms persist in many functions even in a controlled environment of the laboratory, i.e. under constant conditions of temperature, lighting regimen and food uptake.[19] It appears likely that many of the apparent seasonal variations of body functions are endogenous circannual rhythms which may be synchronized (determined in their timing) by environmental factors but are not directly induced by them.

Circannual rhythms are defined as rhythms that persist for more than one cycle under conditions that provide no information about the duration of a year. The period of these rhythms may deviate under constant environmental conditions from 12 months (± 2 months as a rule) thus attesting to their endogenous nature.[5] In humans it is not possible, for obvious reasons, to provide a constant environmental condition over 2 or more years and it is therefore not possible to ascertain if such a rhythm is totally endogenous or not. However such changes persist in some variables studied under as far as feasible constant conditions of temperature, lighting regimen, and food uptake.[19,68] In addition, a circannual free-running from the calendar year has been shown for 17-ketosteroids[17] and for human blood pressure.[12,21] It is also interesting to note that a circannual rhythm established in one hemisphere was found to persist with the same timing several years after the subject changed from the southern to the northern hemisphere.[41] Therefore, it appears that some, if not many, human seasonal rhythms are endogenous and probably generated by a circannual oscillator. The term **seasonal variations** should be used whenever environmental factors like light, temperature, etc . . . seem to determine the changes observed. The term **circannual rhythm** should be reserved for variations with a period of about 1 year for which an endogenous rhythm component has been shown, irrespective of whether it may be modulated and/or synchronized by the seasonal changes in the environment. The question of potential Zeitgebers for circannual rhythms is of great interest and experimental work suggests that the same kind of cues (photoperiod, temperature, social cues . . .) in the environment act upon both circadian and circannual rhythms. The photoperiod is one of the major cues with ample evidence for the photoperiodic regulation of seasonal reproduction. Temperature is another major Zeitgeber, e.g. transfer of ground squirrels from warm to cold can phase shift the circannual rhythm of hibernation. Finally, social cues have also to be taken into account, especially in humans.

The physiological basis of circannual rhythmicity is poorly understood. Organisms frequently restrict energetically expensive activities to a specific time of the year, e.g. reduction of activity or animal migration when food availability is low. Reproduction, preparation for migration and other various energy demanding activities also coincide with abundant local food resources and environmental conditions that promote survival. Seasonally breeding animals frequently detect and respond to environmental cues that accurately signal well in advance, the arrival or departure of seasons favoring reproductive success. However, one of the most important factors is the day length (photoperiod). Seasonal changes in behavior are also observed in tropical animals despite relatively constant photoperiod and temperature conditions.

They could be due, e.g. to regular timing of the onset of rain in these latitudes.

The neural pathways involved in the generation of circannual rhythms are unkown. For instance the integrity of the suprachiasmatic nuclei (SCN) which is essential for the normal expression of a number of circadian rhythms is not required for the generation of circannual rhythms in ground squirrels.[72]

Methodologic considerations

Sampling

For a given physiologic function bioperiodicity can be expressed in several frequency domains whatever the species. When sampling is limited to once or twice daily at monthly intervals over the period of a year, collected data may reflect a change of the circadian phase rather than the existence of a circannual rhythmicity. A seasonal change of a morning or an evening value in blood could reflect a modification of one (or more) of the parameters characterizing a biological rhythm, e.g. the phase, the amplitude or the mesor, but without indicating which one is affected. Circannual rhythms can be shown in longitudinal studies of the same subjects by repeatedly sampling over one or several years or by the study of different subjects only once or a few times during different seasons.

The interactions of rhythms of different frequencies have also to be taken into account. Single sampling at a fixed clock hour, throughout the year may be misleading since circannual changes in circadian amplitude or acrophase would not be recognized which could lead to misinterpretation of the results.

Ethnic-geographic differences

Ethnic-geographic differences (climate, diet, social customs . . .) have been reported in the rhythm parameters of several frequencies, including circannual rhythms.

Melatonin

Melatonin is mainly secreted by the pineal gland, the secretion being larger at night both in diurnal and nocturnal animals.[33] The photoperiod (changes in day length in non-equatorial regions) controls seasonal cycles in many species, e.g. the reproductive cycles, behavior, coat growth and color, etc. . . In a large number of species the pineal gland plays a major role in the transmission of the information on photoperiod to the body. As the onset darkness is followed by an increase in the synthesis of melatonin,[33] the greatest activity of the gland can be expected to occur during the shortest days, i.e. in winter. This dogma, commonly accepted for animals in their natural state[48] has not been verified in healthy human subjects.[58] Whatever the length of natural light, longer in summer and shorter in winter, the circadian acrophases were located around 03.00 hours all year long (Figure 1). This stability suggests that the plasma level

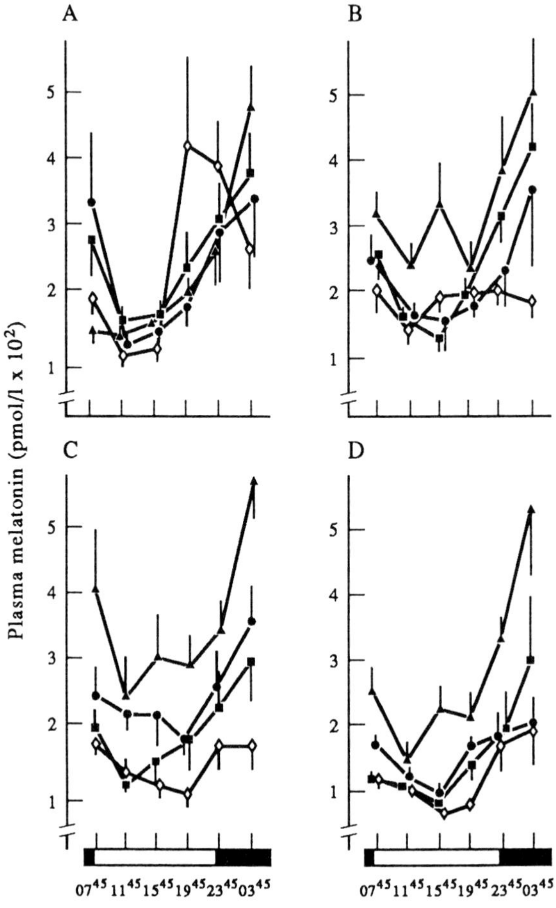

FIG. 1. Circadian patterns of plasma melatonin in young men (▲), elderly men (■), elderly women (●) and elderly demented subjects (◇) documented in January (A), March (B), June (C) and October (D). From Touitou *et al.*[58]

of melatonin is relatively independent of the onset of darkness outdoors.[58] Humans are not considered to be a photoperiodic species. However the control of indoor light (i.e. artificial photoperiod), as well as indoor temperature by humans particulary in urban situations, might cue the circadian (and seasonal) rhythm of melatonin although it has been shown that exposure to artifical light during the dark phase does not reduce plasma melatonin levels in human subjects[28,40,64,65] except when very bright artificial light is used.[37,38] In northern temperate zones and Antarctica a phase advance of the melatonin circadian rhythm occurs in summer compared to winter. A similar phase advance can be achieved when submitting human subjects to bright light (>2,500 lux) treatment during the Antarctica winter.[7] In seasonally breeding species the circadian melatonin rhythm may be the hormonal signal transducing environmental light information to the reproductive system.[48–50] However in nonseasonal breeders the physiologic role of this molecule remains unclear. The circadian rhythm of melatonin in human body fluids has been demonstrated by several groups with a decrease of melatonin secretion in the elderly (review in 59). Evaluation of seasonal differences in the circadian 24-hour mean of plasma melatonin concentration showed apparent peaks in January/March in the elderly and in June in young men both with significant peak-trough differences (Figure 2). These data extend and clarify those obtained by others[2,3,69] with one or two samplings per day at monthly intervals. Indeed in such designs a shift in the acrophase or a change in the circadian amplitude would also result in yearly fluctuating concentrations of the hormone documented at a single sampling time without overall modification of the circadian mean of melatonin secretion.

A study was performed by Tamarkin *et al.*[54] in 20 women with stage I

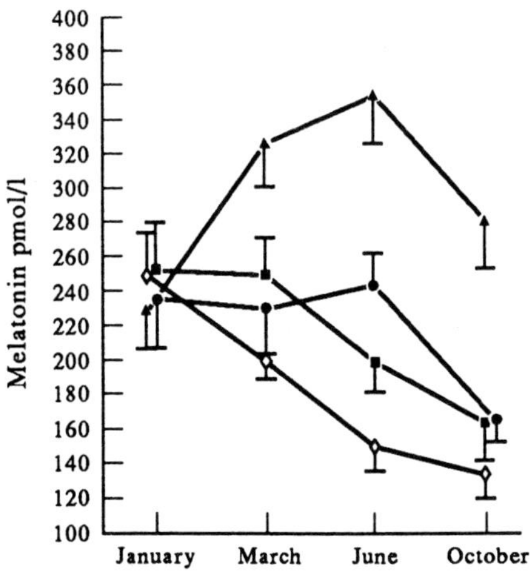

FIG. 2. Circannual patterns of 24 hour-mean levels of plasma melatonin in young men (▲), elderly men (■), elderly women (●) and elderly demented subjects (◇). From Touitou *et al.*[58]

or II breast cancer and eight normal control women. Samples were obtained every 3 hours 1 day before or 5 days after surgery. A section of the malignant breast tissue resected was analyzed for estrogen receptors (ER) The circadian melatonin rhythms of some patients from breast cancer showed differences with those of normal subjects. Although all subjects had a nocturnal peak of plasma melatonin around 02.00 it was much smaller in ER positive patients. The authors were able to put in evidence a significant inverse relation between ER concentration and peak melatonin concentration. Unfortunately no mention was made of the stage of the menstrual cycle at the time of sampling which could have introduced a bias since large variations of melatonin concentrations have been reported by Wetterberg *et al.*[69] at the time of menstruation. The question remains unanswered whether the decline in the nocturnal peak of melatonin represents a long-term pineal failure or a change occurring at the time of breast cancer development. To know whether the alteration of the melatonin rhythm precedes the presence of malignant tissue will require longitudinal and familial studies.

Nevertheless these data suggest that the lower the nocturnal peak of plasma melatonin the greater the probability that the tumor may be hormonally dependent. Thus, the modification of the melatonin rhythm may serve as a marker for increased risk of ER positive breast cancer development. Seasonal fluctuations of ER concentrations in human breast cancer have been demonstrated.[25,26] The possible relations between these variations and the seasonal variations of plasma melatonin should be investigated as they might be consistent with the seasonal incidence of breast cancer mortality[35]and seasonal occurrence of the disease.[9]

Prolactin

Prolactin (PRL), a polypeptide hormone secreted by the pituitary, is released episodically over the 24 hour span and the frequency and magnitude of the episodes are, in part, under the control of central nervous system rhythms (time of day, sleep-wake cycle). The circadian pattern is characterized by a minimum around noon and a nighttime increase with a maximum around midsleep (Figure 3). Whereas the part played by PRL in conditions such as lactation and amenorrhea is well known, its role in the genesis of mammary cancer in human beings is still controversial.[20,22,27,67]

A number of hormones are known to affect the secretion of prolactin in humans, e.g. administration of estrogens[14,71] or melatonin[31] which increase plasma prolactin levels in young adult and adult men and women. Although the secretion of both melatonin[59] and estrogens[66] is reduced in

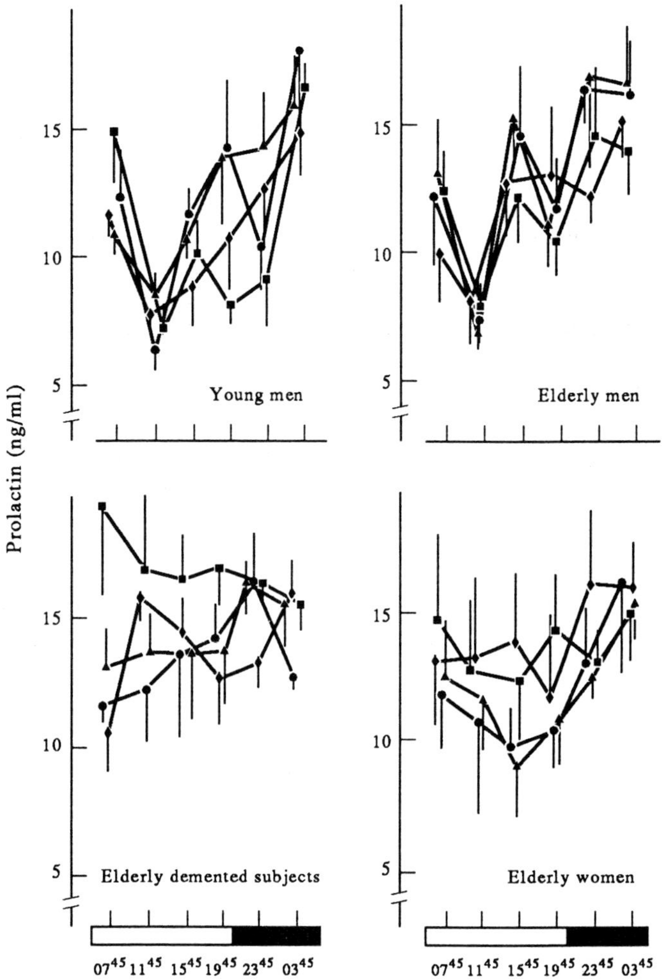

FIG. 3. Circadian patterns of plasma prolactin in young men (▲), elderly men (■), elderly women (●) and elderly demented subjects (◆). From Touitou *et al.*[57]

the elderly, no apparent effect could be seen on prolactin levels of elderly subjects when compared to young men.[59] A number of studies agree with the absence of detectable circannual rhythm of prolactin in young or elderly men (Figure 4)[11,15,47] whereas circannual rhythmicity was validated in healthy young and elderly women.[55]

Prolactin acts on the estrogen-primed breast to stimulate and maintain lactation. It is associated with the functioning of differentiated cells rather than with cell growth and thus would be expected to decrease breast cancer risk. The literature reports conflicting results on plasma prolactin

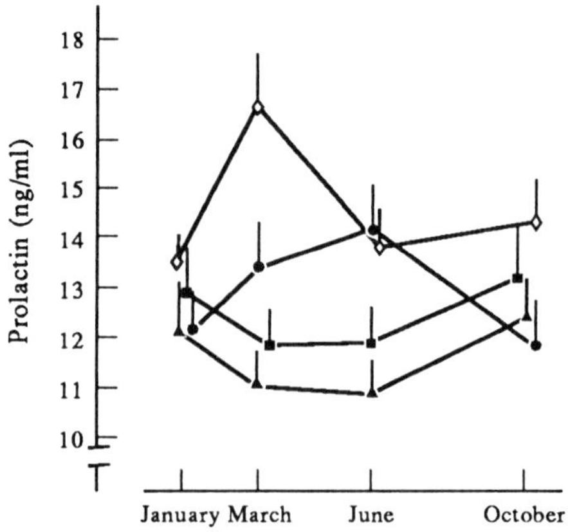

FIG. 4. Circannual patterns of 24 hour-mean level of plasma prolactin in young men (◆), elderly men (■), elderly women (●) and elderly demented subjects (◇). From Touitou *et al.*[57]

concentrations in human breast cancer.[6,10,13,14,16,24,34,51] This may be due to the rhythmic characteristics of prolactin secretion as well as to geographic location[52] and ethnic differences[20] although this latter point might in fact merely represent environmental and nutritional factors.[4,56]

Many of the controversies in the study of the potential role of prolactin in the etiology and/or pathogenesis of breast cancer are due to a lack of regard paid to the multifrequency time structure in prolactin secretion and serum concentrations. Studies at a single clock hour only or at different clock hours in different studies and/or limitation to a single, e.g. the circadian frequency, may often be the source of nonreproducible, controversial and even misleading results.

Haus *et al.*[20] studied the ethnic-geographic differences in prolactin serum concentrations between Japanese women in Kyushu, Japan and in American women in Minnesota, USA. The Japanese and half the Minnesota population were selected for absence of epidemiologically determined risk factors to develop breast cancer, while in the American women, half of the participants in the study were selected to present such risk factors (especially a history of first degree female relatives with breast cancer). This study extended over three age ranges, namely adolescent, young adult, and menopausal women. Sampling at 20 minute intervals was carried out over a 24-hour span four time in each woman, once during each of the four seasons. The rhythmometric analysis of the data indicated statistically highly significant differences between Japanese and American women with higher values of serum prolactin concentration in the Japanese only during the night hours in winter and spring. During the day time (when most investigators choose to sample), there was usually no difference in plasma prolactin concentrations between the two groups of women. These differences disappeared during summer and began to reappear in fall. The rhythmometric analysis revealed that this difference was due to a larger circadian amplitude and a circannual rhythm (or seasonal variation) in the Japanese women with acrophase during the second part of November, which was not statistically significant in the Americans. A repeat study during March confirmed the nocturnal differences in plasma prolactin concentrations during spring. Depending on the clock hour of sampling along the circadian and circannual scale, the Japanese had either higher, comparable or lower plasma prolactin concentrations than the Americans.

The double amplitude of the circannual variation (a measure of the extent of predictable change) amounted in the Japanese women to about 40% of the mesor.

The circadian rhythm characteristics between the low and high risk groups of Minnesota women showed a difference in the circadian amplitude of the two risk groups in Minnesota during fall only ($p=0.04$).

In menstrually cycling subjects, there was a difference in prolactin circadian mesor between Japanese and Americans at the end of the luteal phase. A menstrual cycle related rhythm in plasma prolactin concentrations was described only in the USA high risk group with an acrophase in the luteal stage.

An age trend was found with lower circadian prolactin mesor in postmenopausal women in both countries. An age related decrease of the circadian amplitude, however, was found in the Japanese only.

Since dietary and climatic conditions were not controlled, it remains to be clarified whether or not these facts reflect a true ethnic-geographic difference between American and Japanese women (the latter of which are epidemiologically classified as low-risk).

The observation of a higher circannual amplitude associated with a lower risk to develop breast cancer may correspond to the results of a study performed in Italy[55] involving women with a related yet different risk, namely with fibrocystic mastopathy. As compared with healthy controls, the subjects with fibrocystic mastopathy revealed a lack of circannual variation of plasma prolactin as group phenomenon. All subjects in this study were followed at the same geographic location with comparable diet and climate exposure. In both studies involving Italian, Japanese and American women, a negative correlation was found between breast cancer risk and amplitude of the plasma prolactin circannual variation.

It would be desirable to follow a larger group of women of different risk to develop breast cancer over their entire reproductive life using chronobiologic techniques and rhythmometric analyses. This has thus far not been feasible. In the meantime, chronobiologic methodology will have to be followed to obtain meaningful results. The accumulating data will have to be carefully evaluated and any conclusions to be drawn will have to keep the multifrequency time structure of prolactin secretion in mind.

Vitamin D, parathyroid hormone, calcitonin and electrolyte-related metabolism

Before vitamin D can exert its biologic effects, two hydroxylation reactions are required. It is first metabolized to 25-hydroxyvitamin D (25-OH-D) in the liver and subsequently in the kidney to 1α-25-dihydroxyvitamin D (1,25-$(OH)_2$D), which is the biologically active metabolite. 1,25-$(OH)_2$D plays a major role in the regulation of the mineral metabolism by stimulating the active intestinal absorption of calcium (Ca) and phosphorus (P) and regulating their flux into and out of bone. The renal synthesis and the circulating concentrations of 1,25-$(OH)_2$D are, in turn, controlled by the serum levels of Ca and inorganic phosphate (Pi) either directly or, in the case of Ca, through parathyroid hormone (PTH).

The serum concentration of 1,25-$(OH)_2$D does not undergo large variations along the 24-hour scale but is maintained within relatively narrow limits throughout the day[1,46] though some young men were found to exhibit a circadian rhythm with a small amplitude of about 10% or less.[18] Circulating levels of calcitonin are also circadian periodic in normal subjects, with a peak around midday.[23]

The vitamin D_3 metabolite concentration would be expected to vary according to climate and sunlight exposure. A seasonal variation (with higher levels during the summer) has been described for 25-OH-D,[36,53] for 24,25-$(OH)_2$D[32] and for 1,25-$(OH)_2$D[30] in young and elderly subjects though seasonal variations were not found by some investigators for this latter metabolite.[8,39]

In recent years, osteomalacia has been named as a possible contributory factor in femoral neck fractures, and lower values of vitamin D metabolites are often encountered in the elderly.[36] Meller *et al.*[44] have observed in geriatric patients with long bone fractures that the serum concentrations of vitamin D metabolites correlate positively while that of parathyroid hormone correlates negatively with day length and global solar radiation. The high prevalence of fractures in young and elderly people during the winter months can be related, at least in part, to these biologic variations.

Seasonal variations in serum concentrations of PTH were described in elderly subjects with higher levels in winter when concentrations of 25-OH-D and 24,25-$(OH)_2$D were lowest, which suggests a possible functional relationship between these hormones.[36,44]

Plasma total Ca and Mg undergo circadian and seasonal variations in the elderly with a small rhythm amplitude.[45,60,61] The circadian variations of both cations (Figure 5) have been thought to reflect predominantly the diurnal changes in plasma proteins.[29,42] However, this

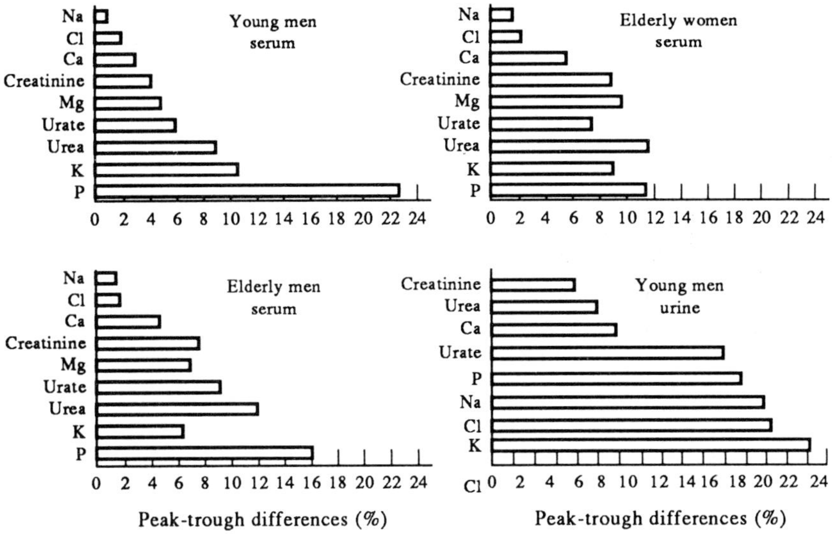

FIG. 5. Circadian variations of serum constituents. From Touitou *et al.*[63]

does not fit well with the large differences in the circadian amplitudes of the plasma total proteins as compared to the much smaller amplitudes of Ca and Mg as found by Touitou *et al.*[62,63] and the phase difference between plasma Ca and plasma proteins as reported by Nicolau *et al.*[45] Markowitz *et al.*[43] have described the temporal relationship between the circadian rhythms of serum PTH and Ca concentrations and have found, at least in healthy young men, that the changes in ionized calcium (Cai) concentrations precede inverse changes in PTH levels by 2 hours whereas changes in PTH precede similar directional alterations in Cai by about 4 hours.

The circadian variations in plasma inorganic phosphorus (Pi) are large with a peak located at night in both young and elderly subjects.[29,63] In both diurnal and nocturnal animals the peak values are observed at the end of the resting period.

References

1. Adams N.D., Gray R.W. and Lemann J. (1979) The effects of oral $CaCO_3$ loading and dietary calcium deprivation on plasma 1,25-dihydroxyvitamin D concentrations in healthy adults. *J. Clin. Endocrinol. Metab.* **48**, 1008.
2. Arendt J., Wirz-Justice A. and Bradtke J. (1978) Annual rhythm of serum melatonin in man. *Neurosci. Lett.* **7**, 327–330.
3. Arendt J., Wirz-Justice A., Bradtke J. and Kornemark M. (1979) Long term studies on immunoreactive human melatonin. *Ann. Clin. Biochem.* **16**, 307–312.
4. Armstrong B. and Doll R. (1975) Environmental factors and cancer incidence and mortality in different countries with special reference to dietary practices. *Int. J. Cancer* **15**, 617–631.
5. Aschoff J. (1960) Exogenous and endogenous componenets in circadian rhythms. *Cold Spring Harbor Symp. Quant. Biol., Biological Clocks* **25**, 11–28.

6. Boyns A.R., Cole E.N., Griffith K. *et al.* (1973) Plasma prolactin in breast cancer. *Lancet* **ii**, 99–100.
7. Broadway J., Folkard S. and Arendt J. (1987) Bright light phase shifts the human melatonin rhythm in Antarctica. *Neurosci. Lett.* **79**, 185–189.
8. Chesney R.W., Rosen J.F., Hamstra A.J., Smith O., Mahafrey K. and de Luca H.F. (1981) Absence of seasonal variation in serum concentration of 1,25 $(OH)_2$ vitamin D despite a rise in 25 (OH) vitamin D in summer. *J. Clin. Endocrinol. Metab.* **53**, 139–142.
9. Cohen P., Wax Y. and Modan B (1983) Seasonality in the occurrence of breast cancer. *Cancer Res.* 43, 892–896.
10. Cole E.N., England P.C., Sellwood R.A. *et al.* (1976) Prolactin in benign and malignant breast disease. *J. Endocrinol.* **69**, 49P.
11. Djursing H., Hagen C., Müller J. and Christiansen C. (1981) Short- and long-term fluctuations in plasma prolactin concentration in normal subjects. *Acta Endocrinol.* **97**, 1–6.
12. Engel R., Sothern R.B. and Halberg F. (1985) Circadian and infradian aspects of blood pressure in treated elderly mesor-hypertensive physician (Abstr). *Chronobiologia* **12**, 243.
13. Frants S., Ralphs D.N.L., Seagrott V. *et al.* (1974) Prolactin concentration in patients with breast cancer. *Br. Med. J.* **4**, 320–323.
14. Frantz A.G., Kleinberg D.L. and Noel G.L. (1972) Studies on prolactin in man. *Recent Progress in Hormone Research* **28**, 527–573.
15. Gala R.R., Van De Walle C., Hoffman W.H., Lawson D.M., Piefer D.R., Smith S.W. and Subramanina M.G. (1977) Lack of a circannual cycle of daytime serum prolactin in man and monkey. *Acta Endocrinol.* **86**, 257–262.
16. Halberg F., Cornelissen G., Sothern R.B., Wallach L.A., Halberg E., Ahlgren A., Kuzel M., Radke A., Barbosa J., Goetz F., Buckley J., Mandel J., Shuman L., Haus E., Lakatua D., Sackett L., Berg H., Kawasaki T., Ueno M., Uezono K., Matsuoka M., Omae T., Tarquini B., Cagnoni M., Garcia Sainz M., Perez V.E., Griffiths K., Wilson D., Donati L., Tatti P., Vasta M., Locatelli I., Camagna A., Lauro R., Tritsch G., Wetterberg L. and Wendt H.W. (1981) International geographic studies of oncological interest on chronobiologic variables. In *Neoplasms—Comparative Pathology of Growth in Animals, Plants, and Man* (ed. Kaiser H.E.), pp. 553–596. Williams and Wilkins, Baltimore.
17. Halberg F., Engeli M., Hamburger C. and Hillman D. (1965) Spectral resolution of low frequency, small amplitude rhythms in excreted ketosteroids: probably androgen induced circaseptan desynchronization. *Acta Endocrinol.* Suppl. 103.
18. Halloran B.P., Portale A.A., Castro M., Morris R.C. and Goldsmith R.S. (1985) Serum concentration of 1,25-dihydroxyvitamin D in the human:diurnal variation. *J. Clin. Endocrinol. Metab.* **60**, 1104–1110.
19. Haus E. and Halberg F. (1970) Circannual rhythm in level and timing of serum corticosterone in standardized inbreed mature C-mice. *Environ. Res.* **3**, 81–106.
20. Haus E., Lakatua D.J., Halberg F., Halberg E., Cornelissen G., Sackett-Lundeen L., Berg H.G., Kawasaki T., Ueno M., Uezono K., Matsuoka M. and Omae T. (1980) Chronobiological studies of plasma prolactin in women in Kyushu, Japan and Minnesota, USA. *J. Clin. Endocrinol. Metab.* **51**, 632–640.
21. Haus E., Sackett L.L., Haus M. Sr, Babb W.K. and Bixby E.K. (1981) Cardiovascular and temperature adaptation to phase shift by intercontinental flights—longitudinal observations. *Adv. Biosci.* **30**, 375–390.
22. Henderson B.C. and Pike M.C. (1981) Prolactin—An important hormone in breast neoplasia? In *Hormones and Breast Cancer* (eds. Pike M.C., Siiteri P.K. and Welsh C.W.), pp 115–127. Cold Spring Harbor Laboratory, New York.
23. Hillyard C.J., Cooke T.J.C., Coombes R.C., Evans I.M.A. and MacIntyre I. (1977) Normal plasma calcitonin: circadian variation and response to stimuli. *Clin. Endocrinol.* **6**, 291–298.
24. Hill P., Wynder E.L., Kumar H. *et al.* (1976) Prolactin levels in population at risk for breast cancer. *Cancer Res.* **36**, 4102–4105.
25. Hrushesky W., Teslow T., Halberg F. *et al.* (1979) Temporal components of predictable

variability along the 1-year scale in estrogen receptor concentration of primary human breast cancer, abstracted. *Proc. Am. Soc. Clin. Oncol.* **20**, 331.

26. Hughes A., Jacobson H.I., Wagner R.K. *et al.* (1976) Ovarian independent fluctuations of estradiol receptor levels in mammalian tissues. *Mol. Cell. Endocrinol.* **5**, 379–388.
27. Ingram D.M., Nottage E.M. and Roberts A.N. (1990) Prolactin and breast cancer risk. *Med. J. Australia* **153**, 469–473.
28. Jimmerson D.C., Lynch H.J., Post R.M., Wurtman R.J. and Bunney W.E. Jr (1977) Urinary melatonin rhythms during sleep deprivation in depressed patients and normals. *Life Sci.* **20**, 1501–1508.
29. Jubiz W., Canterbury J.M., Reiss E. and Taylor F.H. (1972) Circadian rhythms in parathyroid hormone concentration in human subjects : correlation with serum calcium, phosphate, albumin and growth hormone levels. *J. Clin. Invest.* **51**, 2040–2046.
30. Juttman J.R., Visser T.J., Buurman C., De Kam E. and Birkenhager J.C. (1981) Seasonal fluctuations in serum concentrations of vitamin D metabolites in normal subjects. *Br. Med. J.* **282**, 1349–1352.
31. Kamberi I.A., Mical R.S. and Porter J.C. (1971) Effect of melatonin and serotonin on the release of FSH and prolactin. *Endocrinology* **88**, 1288–1293.
32. Kano K., Yoshida H., Yata J. and Suda T. (1980) Age and seasonal variations in the serum levels of 25-OH vitamin D and 24,25 $(OH)_2$ vitamin D in normal humans. *Endocrinol. Jpn.* **27**, 215–221.
33. Klein D.C. (1979) Circadian rhythms in the pineal gland. In *Endocrine Rhythms* (ed. Krieger D.T.), pp. 203–223. Raven Press, New York.
34. Kwa H.G., Delong-Baker M., Engelsman E. *et al.* (1974) Plasma prolactin in human breast cancer. *Lancet* **ii**, 438–439.
35. Langlands A.O., Simpson H., Sothern R. *et al.* (1977) Different timing of circannual rhythms in mortality of women with breast cancer diagnosed before and after menopause. In *Proceedings of the 8th International Scientific Meeting of the International Epidemiological Association*, 17–23 Sept 1977. San Juan, Puerto Rico.
36. Lester E., Skinner R.K. and Wills M.R. (1977) Seasonal variation in serum 25-OH vitamin D in the elderly in Britain. *Lancet* **i**, 979–980.
37. Lewy A.J., Wehr T.A., Goodwin F.K., Newsome D.A. and Markey S.P. (1980) Light suppresses melatonin secretion in humans. *Science* **210**, 1267–1269.
38. Lewy A.J., Wehr T.A., Goodwin F.K., Newsome D.A. and Rosenthal N.E. (1981) Manic-depressive patients may be supersensitive to light. *Lancet* **i**, 383–384.
39. Lips P., Hackeng W.H.L., Jongen M.J.M., Van Ginkel F.C. and Netelenberg J.C. (1983) Seasonal variation in serum concentrations of parathyroid hormone in elderly people. *J. Clin. Endocrinol. Metab.* **57**, 204–206.
40. Lynch H.J., Wurtman R.J., Moskowitz M.A., Archer M.C. and Ho M.H. (1975) Daily rhythm in human urinary melatonin. *Science* **187**, 169–175.
41. Manton W.I. (1977) Sources of lead in blood. *Arch. Environ. Health.* **32**, 149–159.
42. Markowitz M., Rotkin L., Rosen J.F. (1981) Circadian rhythms of blood minerals in humans. *Science* **213**, 672–674.
43. Markowitz M.E., Arnaud S., Rosen J.F., Thorpy M. and Laximinarayan S. (1988) Temporal interrelationships between the circadian rhythms of serum parathyroid hormone and calcium concentrations. *J. Clin. Endocrinol. Metab.* **67**, 1068–1073.
44. Meller Y., Kestenbaum R.S., Gralinsky D. and Shany S. (1986) Seasonal variation in serum levels of vitamin D metabolites and parathormone in geriatric patients with fractures in southern Israel. *Israel J. Med. Sci.* **22**, 8–11.
45. Nicolau G.Y., Haus E., Lakatua D.J., Bogdan C., Petrescu E., Robu E., Sackett-Lundeen L. and Swoyer J. (1985) Chronobiologic observations of calcium and magnesium in the elderly. *Rev. Roum. Med. Endocrinol.* **23**, 39–53.
46. Prince R.L., Wark J.D., Omond S., Opie J.M., Eagle M.R. and Eisman J.A. (1983) A test of 1,25-dihydroxyvitamin D secretory capacity in normal subjects for application in metabolic bone diseases. *Clin. Endocrinol.* **18**, 127.
47. Reinberg A., Lagoguey M., Cesselin F., Touitou Y., Legrand J.C., Delasalle A., Antréassian J. and Lagoguey A. (1978) Circadian and circannual rhythms in plasma hormones and other variables of five healthy young human males. *Acta Endocrinol.* **88**, 417–427.

48. Reiter R.J. (1974) Circannual reproductive system in mammals related to photoperiod and pineal function: a review. *Chronobiologia* **1**, 365–395.
49. Reiter R.J. (1978) Interaction of photoperiod, pineal and seasonal reproduction as examplified in the hamster. In *The Pineal and Reproduction. Progress in Reproductive Biology* (ed. Reiter R.J.), Vol. 4, pp. 169–190. Karger, Basel.
50. Reiter R.J., Rollag M.D., Pankee E.S., Banks A.F. (1978) Melatonin: reproductive effects. *J. Neural Transm.* **Suppl. 13**, 209–223.
51. Rolandi E., Barreca T., Masturzo P. *et al.* (1974) Plasma prolactin in breast cancer. *Lancet* **ii**, 845–846.
52. Segi M. (1977) Graphic presentation of cancer incidence by site and by area and population. Segi Institute of Cancer Epidemiology, Nagoya, Japan.
53. Stamp T.C.B. and Round J.M. (1974) Seasonal changes in human plasma levels of 25-hydroxyvitamin D. *Nature* **247**, 563–565.
54. Tamarkin L., Danforth D., Lichter A., De Moss E., Cohen M., Chabner B. and Lippman M. (1982) Decreased nocturnal plasma melatonin peak in patients with estrogen receptor positive breast cancer. *Science* **216**, 1003–1005.
55. Tarquini B., Gheri R., Romano S., Costa A., Cagnoni M., Lee J.K. and Halberg F. (1979) Circadian mesor hyperprolactinemia in fibrocystic mastopathy. *Amer. J. Med.* **66**, 229–237.
56. Thomas D.B. (1980) Epidemiologic and related studies of breast cancer etiology. In *Reviews in Cancer Epidemiology* (ed. Lilienfeld A.M.), Vol. 1. Elsevier, New York, Amsterdam.
57. Touitou Y., Carayon A., Reinberg A., Bogdan A. and Beck H. (1983) Differences in the seasonal rhythmicity of plasma prolactin in elderly human subjects: detection in women but not in men. *J. Endocr.* **96**, 65–71.
58. Touitou Y., Fèvre M., Bogdan A., Reinberg A., De Prins J., Beck H. and Touitou C. (1984) Patterns of plasma melatonin with ageing and mental condition: stability of nyctohemeral rhythms and differences in seasonal variation. *Acta Endocrinol.* **106**, 145–151.
59. Touitou Y., Fèvre M., Lagoguey M., Carayon A., Bogdan A., Reinberg A., Beck H., Cesselin F. and Touitou C. (1981) Age and mental health-related circadian rhythm of plasma levels of melatonin, prolactin, luteinizing hormone and follicule-stimulating hormone in man. *J. Endocrinol.* **91**, 467–475.
60. Touitou Y. and Haus E. (eds.) (1992) *Biologic Rhythms in Clinical and Laboratory Medicine.* Springer, Berlin, New York, London, Paris.
61. Touitou Y., Touitou C., Bogdan A., Beck H. and Reinberg A. (1978) Serum magnesium circadian rhythm in human adults with respect to age, sex and mental status. *Clin. Chim. Acta* **87**, 35–41.
62. Touitou Y., Touitou C., Bogdan A., Reinberg A., Auzéby A., Beck H. and Guillet P. (1986) Differences between young and elderly subjects in seasonal and circadian variations of total plasma proteins and blood volume as reflected by hemoglobin, hematocrit and erythrocyte counts. *Clin. Chem.* **32**, 801–804.
63. Touitou Y., Touitou C., Bogdan A., Reinberg A., Motohashi Y., Auzéby A. and Beck H. (1989) Circadian and seasonal variations of electrolytes in aging humans. *Clin. Chim. Acta* **180**, 245–254.
64. Vaughan G.M., Bell R. and de la Peña A. (1979) Nocturnal plasma melatonin in humans: episodic pattern and influence of light. *Neurosci. Lett.* **14**, 81–84.
65. Vaughan G.M., Pelham R.W., Pang S.W., Loughlin L.L., Wilson K.M., Sandock K.L., Vaughan M.K., Koslow S.H. and Reiter R.J. (1976) Nocturnal elevation of plasma melatonin and urinary 5-hydroxyindoleacetic acid in young men: attempts at modification by brief changes in environmental lighting and sleep and by autonomic drugs. *J. Clin. Endocrinol. Metab.* **42**, 752–764.
66. Vermeulen A. (1976) The hormonal activity of the postmenopausal ovary. *J. Clin. Endocrinol. Metab.* **42**, 247–253.
67. Völker W., Leipert K.P., Gehrlings H., Stauch G., von zur Muhlen A. and Schneider J. (1986) Mastopathie und Mammacarcinom: Besteht ein typisches Hormon Profil? *Geburtsch u. Frauenheilk.* **46**, 284–289.

68, Wallace A.L.C. (1979) Variations in plasma thyroxin concentrations throughout one year in penned sheep on a uniform food intake. *Aust. J. Biol. Sci.* **32**, 371–374.
69. Wetterberg L., Arendt J., Paunier L., Sizonenko P.C., van Donselaar W. and Heyden T. (1976) Human melatonin changes during the menstrual cycle. *J. Clin. Endocrinol. Metab.* **42**, 185–188.
70. Wirz-Justice A. and Richter R. (1979) Seasonality in biochemical determinations: a source of variance and a clue to the temporal incidence of affective illness. *Psychiatry Res.* **31**, 53–60.
71. Yen S.S.C., Ehara Y. and Silver T.M. (1974) Augmentation of prolactin secretion by estrogen in hypogonadal women. *J. Clin. Invest.* **53**, 652–655.
72. Zucker I., Boshes M. and Dark J. (1983) Suprachiasmatic nuclei influence circannual and circadian rhythms of ground squirrels. *Am. J. Physiol.* **244** (*Regulatory Integrative Comp. Physiol.* 13), R472–R480.

23

Peptic Ulcer and Melatonin

F. I. KOMAROV, S. I. RAPOPORT, N. K. MALINOVSKAYA, M. V. NEVEROVA and V. A. BORILKO

Center of Gastroenterology, Department of Propedeutical Therapy, I.M. Setchenov Mockow Medical Academy, Pogodinskaya str.-5, Moscow 119435, Russia

Summary

The epiphysis extract—epithalamine—was used in treatment of duodenal ulcer. Epithalamine was administered i.m. in a dose of 10 mg a day. A rapid response to the treatment was obtained. Duodenal ulcer remission was noted at the 10th day of treatment in four patients and on the 20th day in five patients. This effect of epithalamine can be explained by its adaptogene activity and synchronization of human biological rhythms. The role of epithalamine in treating of desynchronosis is discussed.

Introduction

IN THIS report the first results of the work on this subject are presented.

The problem of human adaption to the environment and the changes in physiological rhythms at their disturbance attract the attention of all investigators. The most important feature of this condition for the clinicians is desynchronosis—divergence of the natural physiological rhythms of the organism between themselves, and with time. The degree of the desynchronosis can determine the severity of the disease and in some cases it can act as a separate disease: after an abrupt change in climate, during "sleep-awakening" disturbances, after changes in social rhythm sensors, and during chronic stress and emotional tension.[4]

The phase dyssynchrony hypothesis includes a central, integrative and controlling mechanism that provides a framework for the mode in which the pharmacological preparation works in rhythm disorders. In the present study we test the hypothesis that treatment with pineal gland extract forms a role in the entrainment of disturbed biological rhythms in patients with duodenal ulcer.

It is obvious, that the epiphysis is not a rudimentary formation, but a special neuroendocrinal gland, having an unusual evolution and combining morphological features of the neural and endocrinal tissues and producing a broad number of biological active substances and hormones.[3,6] These very features are the cause of the specialty of this neuroendocrinal gland and its broad participation in the vegetative functions and homeostasis system regulation, which realize its action through direct and indirect connections with hypothalamus, hypophysis, endocrinal glands, immune system, central and peripheral sympathic nervous system.

The study of the epiphysal hormonopoesis of serotonin, melatonin, and enzymes participating in their synthesis, demonstrated the clear circadian, ovarian, menstrual, and seasonal rhythms of this process.[2,13] In morphological studies on some species of primary vertebrates and mammals was mentioned, that neuroendocrinal cells of the epiphysis (pinealocytes) are the rudimentary photoreceptor cells.[11,12] This fact, unique evolutional epiphysis transformation from the photoreceptive organ to the neuroendocrinal gland and its extreme biochemical and functional rhythms are the basis of the special studies on the relation of this gland to the human and animal "biological watch" system. The fundamental studies on the gland innervation in epiphysis and hypophysis tissues and experimental works on the gland denervation detected, that epiphysis is one of the leaders of the "biological watch" system and it realizes the photoperiodical synchronization of the physiological functions.[5] This circumstance allowed us to use the extract of epiphysis in correction of desynchronosis.

Earlier in our works, fulfilled in our laboratory under the leadership of the academics F. I. Komarov and Prof. S. I. Rapoport, it was demonstrated, that any duodenal ulcer relapse is accompanied by the divergence of the organism physiological functions and the degree with which it occurs correlates with the severity of the disease. During the remission the degree of the desynchronosis decreased and it was moderate during the recovering.[7]

The role of substances, secreted by epiphysis, in the pathogenesis and clinic of peptic ulcer was not studied earlier. At the same time the periodicity of this disease, the seasonal character of its relapses, and clear rhythms in pain and dyspepsia indicate the possible role of this substance in the pathogenesis and formation of the clinical features of this disease. Taking into account a cyclical melatonin secretion during the day and its seasonal changes one can explain the periodicity of peptic ulcer.

All these facts allowed the study to be carried out, in which the acetic cattle extract of bovine epiphysis—epithalamine—was used as a medicine. The substratum was obtained in Leningrad Military Medicine Institute under the leadership of Prof. V. Ch. Havinsone. Chemically it is a polypeptide substance, consisting at least of five homogenous active peptides without indol substances.[6] According to the literature to date epithalamine was successfully used in oncology and gerontology (it was confirmed in clinical and experimental studies). In some papers its high effectiveness as an adaptogene is noted.[8]

Materials

Monotherapy of epithalamine was carried out in 10 patients with uncomplicated duodenal ulcer and without accompanying pathology (all men 20–45 years old); the patients had frequent relapses, equal biorhythms ("evening type"), hyperacidity, and duration of anamnesis over 2 years. Patient's examination and treatment was carried out in the hospital. Examination was done twice and included gastroscopy, blood examination, plasma secretory IgA concentration by the Manchini test, urine examination, examination of feces, ECG findings, sonography, chest R-ray, intragastric pH-metry with histamine stimulation. Special biorhythmological examinations were administrated before and after therapy: 3-hours body temperature and pulse measuring (at 8 a.m., 11 a.m., 2 p.m., 5 p.m., 8 p.m., and 11 p.m.); as psychophysiology was used, a "line-test" was carried out at the same time: the patient was ordered to draw on a sheet of paper the maximal number of a vertical lines 10 mm long. It is known, that its number correlates well with the neuropsychical condition of a patient and with the speed ($r = +0.85$) of a simple sensorimotor reaction. This test is used for the estimation of the circadian "sleep-awakening" rhythm.[10]

Daily 3-hour portions of urine before and after treatment were obtained (eight portions a day) for the measurement of melatonin metabolites excretion.

Epithalamine was injected at 10 a.m. at a dose of 10 mg i.m. (this scheme was recommended by the producers of epithalamine) after general and biorhythmological examination of a patient. The calculation was done using cosinor and correlation tests.

Results

The data, describing rapid healing of ulcer after the epithalamin administration, are presented in Table 1; the pain and dyspepsia relieved on the 3rd–5th day of treatment. Duodenal ulcer remission was noted at the 10th day of the treatment in four patients and on the 20th day in five patients. In one patient the ulcer was indolent. The healing of duodenal ulcer was accompanied by the remaining inflammation of duodenal mucosa. One month after healing there was no inflammation in the duodenum. The statistically unreliable increase of plasma secretory IgA was noted at the moment of ulcer relapse (0.86 ± 0.09 g/l before and 0.93 ± 0.09 g/l after treatment).

TABLE 1
Clinical characteristics of the patients treated by epythalamine

Code of patients	Before treatment								After treatment												
	Clinical characteristics					Endoscopical size of ulcer			Pain relief			Dyspepsy relief			Ulcer healing			Presence of erosion and bulbitis after treatment		Presence of ulcer a month after treatment	
	Pain			Dyspeptic syndrome																	
	Severe	Moderate	Absent	Yes	No	0.5 cm	0.5–1 cm	>1 cm	3 days	5 days	>5 days	3 days	5 days	>5 days	10 days	20 days	>20 days	Yes	No	Yes	No
1	+			+			+		+			+				+		+			+
2	+			+				+	+			+			+				+		+
3		+			+		+		+			+					+	+			+
4		+			+			+	+				+		+			+			+
5		+		+			+			+		+				+		+			+
6	+			+				+	+			+			+				+		+
7		+		+				+		+				+		+		+			+
8		+		+				+	+				+			+		+			+
9	+			+		+			+			+			+			+			+
10	+			+			+			+		+				+		+			+

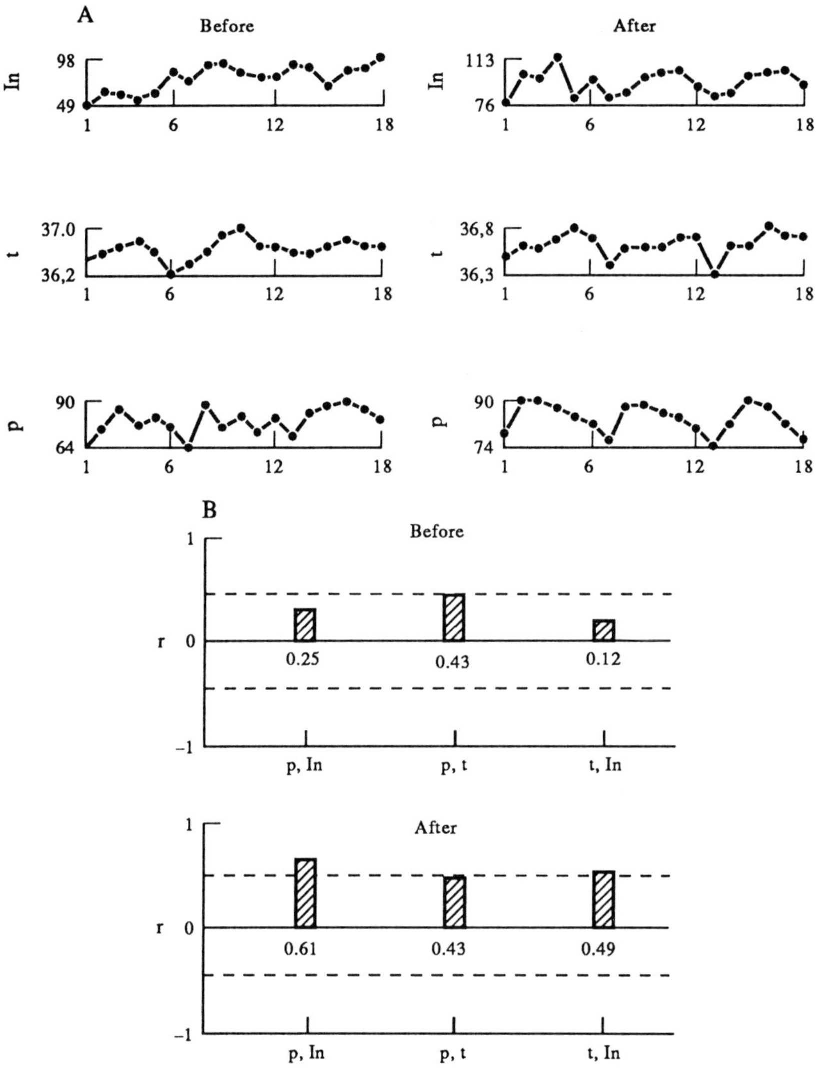

FIG. 1. Patient C. Duodenal ulcer. (a) Temperature, pulse and "line" test before and after treatment. (b) Calculation coefficient before and after treatment.

There was no changes in blood and urine analysis during epithalamin treatment, and no allergic reactions.

Discussion

The rapid healing indicates high effectiveness of epithalamine. The common duration of duodenal ulcer epithelisation is greater than 30 days.

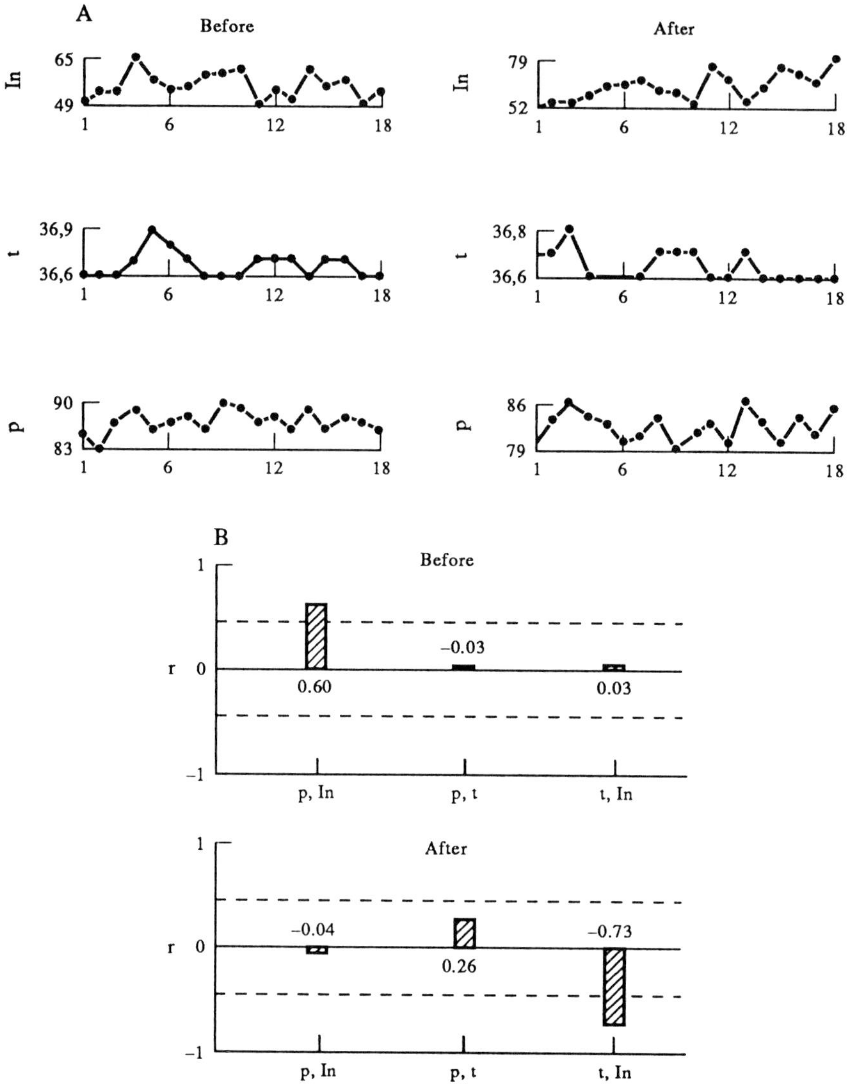

FIG. 2. Patient K. Duodenal ulcer. (a) Temperature, pulse and "line" test before and after treatment. (b) Calculation coefficient before and after treatment.

H_2-histamine receptor blocking drugs heal duodenal ulcer during 3 weeks treatment and epithalamine does it within the same period with very rapid reduction of pain and dyspepsia. The clinical effect of epithalamine can be explained by its adaptogen activity, hypothetically working through mechanisms which synchonize the human biological rhythms. It is well demonstrated on the following clinical examples: patient C (Figure 1) had intensive divergence of pulse, temperature and "line-test" rhythms and positive correlation analysis data after ulcer healing (after 10 days treatment). The opposite observation is demonstrated in patient K (Figure 2): the duodenal ulcer did not heal in 20 days, and after treatment all biorhythmological data worsened; negative correlation appeared between pulse and "line-test" and between temperature and "line-test", the circadian rhythm did not appear before, but after the treatment. It indicates the increasing degree of desynchronosis. We must note that some patients did not demonstrate a severe decrease in degree of desynchronosis after the ulcer healing. It is connected with the presence of duodenal erosions and bulbitis. Earlier in our works on the gastric mucosal protein synthesis rhythms, the absence of rhythm restoration to the normal level when inflammation in duodenal mucosa is present was demonstrated.[1] It means the absence of desynchronosis decreases.

The absence of statistically reliable changes in plasma secretory IgA level means that immunological mechanisms are not dominant in the epithalamine effect. The decrease in the patients' desynchronosis after the epithalamine treatment confirms the connection of its therapeutical effect with its synchronizing activity on the patients' biorhythms. We must note, that the method of automeasuring, used in our work, was to some extent imperfect due to patients' incompetence.

Undoubtedly the monitoring of physiological data is optimal.

In conclusion it must be noted that our results of the treatment of desynchronosis are very perspective for the future study and for the therapeutical administration of epithalamine as a medicine. These results are the first step of the work on this topic. In future we intend to enlarge our investigations and to study the dynamics of melatonin metabolites' excretion.

References

1. Brodski V.L. and Fateeva V.I. (1984). The presence of about hour rhythms of the protein synthesis in human gastric mucosal biopsies. *Bull. Exper. Biol.* **5**, 612–614.
2. Cardinali D.P. (1981). Melatonin. A mammalian pineale hormone. *Endocr. Rev.* **2**, 237–258.
3. Chasov E.I. and Isachenkov V.A. (1974). *Epiphysis: the Role in the Neuroendocrine Regulation.* Medicina, Moscow.
4. *Chronobiology and chronomedicine* (1989). Medicina, Moscow.
5. Dilman V.M. (1986) *The Large Biological Watch.* Znanie, Moscow.
6. Ebadi M. (1984) Regulation of the syntesis of melatonin and its significance to neuroendocrinology. In *The Pineal Gland* (ed. Reiter R.J.), pp. 1–38. Raven Press, New York.
7. Eremina L.V. (1988) *The Circadian Rhythms of Psychophysiological Functions in Patients with Peptic Ulcer.* Moscow.
8. Karpov R.S., Slepushkin V.D., Mordovkin V.F. *et al.* (1985) *The Using of Epiphysis Extracts in Clinic.* Tomski University, Tomsk.
9. Maestoni G.I., Conti A. and Pierpaoli W. (1987) The pineal and the circadian opiatergic, immunoregulatory role of melatonin. *Ann. N.Y. Acad. Sci.* **496**, 67–77.
10. Makarov V.I. (1986) Biorithmologycal test. *Nauka i jizn.* **1**, 79.

11. Reiter R.J. (1980) The pineal and its hormones in the control of reproduction in mammals. *Endocr.* **1**, 109–131.
12. Reiter R.J. (1981) The mammalian pineal gland: structure and function. *J. Anat.* **162**, 287–313.
13. Vaskovski B.V., Michaleva I.I., Ivanov V.T. *et al.* (1990) The excretion of the biological active peptides from bovin epiphysis extract. In *Vsesousni Symposium on the Peptide Chemistry*. Riga. pp. 16.

24

The Relationship between the Pineal Gland and Cancer: Seasonal Aspects

H. BARTSCH,[1] C. BARTSCH,[1] D. MECKE[2] and T. H. LIPPERT[1]

[1]*Division of Clinical Pharmacology, Department of Gynaecology and Obstetrics, University of Tübingen, Tübingen, Germany*
[2]*Institute of Physiological Chemistry, University of Tübingen, Tübingen, Germany*

DESPITE THE invention of electric light and central heating as well as constant food availability throughout the year modern man has, to a certain extent, maintained seasonal rhythms in many body functions[13] including reproduction.[23] It is therefore not surprising that also numerous diseases can show very strong seasonal rhythmicity in their occurrence, one of the most prominent examples being seasonal affective disorder (see elsewhere in this volume). There are also many reports on the seasonality of the occurrence of human cancer of which only some examples are given in the following chapter.

Seasonality in the occurrence of human cancer

Cohen *et al.*[12] reported more cases of breast cancer being diagnosed in spring and Ownby *et al.*[21] observed that breast tumors at the time of operation were bigger in spring. Increased appearance of breast cancer in spring and of prostate and testicular cancer in winter are described by Jacobson *et al.*[15] When drawing conclusions from such data one has to be very careful since the observed manifestations of the tumor, like e.g. first symptoms, size, appearance of metastases, depend on the type of tumor, its localization, and growth kinetics. Furthermore, the methods of detection vary widely for different types of cancer. Therefore, the rhythm parameters can differ but the occurrence of seasonality may still be due to the same biological factors.

Relationship between the pineal gland and cancer

The pineal gland is affected by growth of malignant tumors, particularly of breast and prostate, and the pineal hormone melatonin as well as other yet unidentified pineal substances are capable of inhibiting a number of experimental tumors *in vivo* and *in vitro* (for review see Ref. 8). The contribution to this book on the subject deals with the seasonal rhythmicity of both aspects of this relationship: in humans and experimental animals seasonal rhythms have been observed in melatonin production by the pineal gland and the unidentified tumor-inhibiting substance(s) could be found in human urine and in rat pineal glands mainly in summer.

Seasonal rhythmicity of serum melatonin is depleted in human cancer patients

Human patients suffering from breast and prostate cancer show a tumor-stage-dependent depression of circadian serum melatonin rhythms when compared to age-matched patients with benign tumors.[2,4,8] The rhythm parameters obtained by a circadian cosinor analysis[19] were subjected to a subsequent circannual analysis giving possible evidence for seasonal rhythmicity. The results are shown in Tables 1–4.

Among the female patients only the group with benign breast diseases showed significant seasonal rhythms of the circadian MESOR (rhythm adjusted mean of all time points over the 24-hour cycle analyzed) and

TABLE 1

Circannual cosinor analysis of the circadian MESOR of melatonin in patients with breast cancer and their controls

Hormone	Group	No.	Rhythm detect. ($p<$)	MESOR ± SEM	Amplitude mean range	% of MESOR	Acrophase mean range
Melatonin [pmol/ml]	A	28	0.005	0.146 ± 0.029	0.082 0.030–0.134	56	2/I 1/XII–2/II
	B	23	NS (0.90)	0.078 ± 0.030	0.009	12	1/XII
	A (19–41 years)	13	0.01	0.142 ± 0.031	0.107 0.031–0.183	75	1/XII 2/XII–2/II
	A (>41–72 years)	15	NS (0.25)	0.148 ± 0.053	0.069	47	1/I
	A(Pre-MP)	19	0.005	0.146 ± 0.025	0.078 0.024–0.132	53	2/I 1/XII–2/II
	A (PMP)	9	NS (0.50)	0.144 ± 0.086	0.094	65	2/I

Abbreviations: A: controls, patients with benign breast disease ($n=28$, average age: 43 ± 3 years), i.e. fibrocystic mastopathy ($n=14$), fibroadenoma ($n=10$) and other breast diseases ($n=4$); B: patients with primary unoperated breast cancer ($n=23$, average age: 58 ± 2 years). Pre-MP: pre-menopausal; PMP: post-menopausal. 1 = first, 2 = second half of a given month; months are indicated by Roman numbers.

TABLE 2

Circannual cosinor analysis of the circadian amplitude of melatonin in patients with breast cancer and their controls

Hormone	Group	No.	Rhythm detect. ($p<$)	MESOR ± SEM	Amplitude mean range	% of MESOR	Acrophase mean range
Melatonin [pmol/ml]	A	28	0.01	0.162 ± 0.035	0.087 0.022–0.152	54	2/I 1/XII-1/III
	B	23	NS(0.95)	0.081 ± 0.035	0.009	11	2/II
	A (19–41 years)	13	0.025	0.167 ± 0.038	0.119 0.021–0.217	71	1/II 2/XII–1/III
	A (>41–72 years)	15	NS (0.25)	0.153 ± 0.063	0.074	48	1/I
	A (Pre-MP)	19	0.025	0.168 ± 0.029	0.087 0.039–0.135	52	1/II 2/XII–1/III
	A (PMP)	9	NS (0.50)	0.148 ± 0.102	0.107	72	2/I

Abbreviations: see Table 1.

TABLE 3

Circannual cosinor analysis of the circadian MESOR of hormones in patients with benign and malignant prostate tumors

Hormone	Group	No.	Rhythm detect. ($p<$)	MESOR ± SEM	Amplitude mean range	% of MESOR	Acrophase mean range
Melatonin [pmol/ml]	BPH	13	0.01	0.065 ± 0.019	0.048 0.013–0.083	74	2/II 1/I-1/IV
	PC	9	NS(0.50)	0.033 ± 0.023	0.015	46	1/XI
Prolactin	BPH	12	NS(0.25)	226 ± 66	79	35	1/V
[μIU/ml]	PC	9	NS(0.975)	229 ± 74	12	5	2/IV
TSH	BPH	13	NS(0.75)	1.14 ± 0.65	0.40	35	2/III
[μIU/ml]	PC	8	NS(0.75)	0.33 ± 0.40	0.29	88	1/II
hGH	BPH	12	NS(0.75)	2.77 ± 1.00	0.62	22	1/IV
[ng/ml]	PC	9	NS(0.50)	2.55 ± 1.24	1.71	67	2/II
LH	BPH	13	NS(0.50)	16.10 ± 7.60	7.60	47	2/XI
[mIU/ml]	PC	9	NS(0.50)	13.90 ± 9.30	10.00	72	2/VIII
FSH	BPH	12	NS(0.50)	7.59 ± 4.02	3.09	40	2/XI
[mIU/ml]	PC	9	NS(0.50)	11.82 ± 9.78	11.19	95	1/IX
Thyroxine	BPH	13	NS(0.25)	6.97 ± 0.91	1.21	17	1/X
[μg/dl]	PC	9	NS(0.95)	7.89 ± 1.51	0.46	6	2/VII
Testosterone	BPH	13	NS(0.50)	20.50 ± 5.30	4.30	21	2/I
[pmol/ml]	PC	9	NS(0.50)	18.70 ± 6.70	9.90	53	1/VIII
Cortisol	BPH	12	NS(0.50)	293 ± 107	108	37	2/VI
[pmol/ml]	PC	8	NS(0.10)	263 ± 53	118	45	2/II
Serotonin	BPH	13	NS(0.90)	50 ± 17	8	16	1/VIII
[ng/ml]	PC	8	NS(0.75)	63 ± 36	22	35	2/IX

Abbreviations: BPH = patients with benign prostatic hyperplasia; PC = patients with prostate carcinoma. 1 = first, 2 = second half of a given month; months are indicated by Roman numbers.

TABLE 4

Circannual cosinor analysis of the circadian amplitude of hormones in patients with benign and malignant prostate tumors

Hormone	Group	No.	Rhythm detect. ($p<$)	MESOR ± SEM	Amplitude mean range	% of MESOR	Acrophase mean range
Melatonin [pmol/ml]	BPH	13	0.05	0.068 ± 0.024	0.047 0.005–0.089	69	1/II 2/XI-2/IV
	PC	9	NS (0.25)	0.025 ± 0.020	0.018	72	1/XI
Prolactin	BPH	12	NS (0.50)	72 ± 30	29	40	1/V
[μIU/ml]	PC	9	NS (0.75)	63 ± 28	17	27	2/VII
Cortisol	BPH	12	NS (0.25)	140 ± 31	40	29	2/V
[pmol/ml]	PC	8	NS (0.50)	131 ± 62	43	33	2/IV

Abbreviations: see Table 3.

amplitude (half of the rhythmic variability per 24 hours) with acrophases (peak time of the cosine function used to approximate the rhythm) in January while there was no indication of seasonal rhythmicity in patients with malignant breast tumors. When the group of patients with benign breast diseases was sub-divided between individuals below and above 40 years of age or between pre- and postmenopausal women seasonal rhythmicity was detectable only in the younger (<40 years) and the premenopausal women. Therefore, the lack of a seasonal melatonin rhythm in cancer patients can be attributed to age and not to the presence of a malignant as opposed to a benign tumor whereas the observed decrease in circannual MESORs and amplitudes is specific to the cancer patients and is independent of age (see Tables 1 and 2). Interestingly, seasonality in the detection of breast cancer as reported by Cohen *et al.*[12] was most pronounced in patients younger than 55 years.

Elderly men suffering from benign prostatic hyperplasia show significant seasonal rhythms of circadian melatonin MESOR and amplitude with acrophases in February but age-matched patients with prostate carcinoma lack such rhythmicity. A number of other hormones were also analyzed in these patients but none of them showed any significant seasonal rhythmicity in both groups (see Tables 3 and 4).

These data indicate that in humans not only the acrophase undergoes a seasonal shift[10,14] but that also MESOR and amplitude of the circadian melatonin rhythm show seasonal rhythmicity. In women, this rhythm gets lost after menopause while it is maintained in elderly men free of malignant tumors.

Seasonal rhythmicity of pineal melatonin production in female rats is disturbed by malignant but not benign tumors

Melatonin production by the pineal gland can be estimated by determination of its metabolic end-product, 6-sulphatoxymelatonin (aMT6s), in urine.[11] The nocturnal urinary excretion (23.00–07.00 hours) of 6-sulphatoxymelatonin was determined by radioimmunoassay[1] in 21 inbred female Fischer F344 rats over one year. Fourteen animals had received a single intragastral dose of the carcinogen 7,12-dimethylbenz[a]anthracene (DMBA, 10 mg/100 g body weight in 1 ml of peanut oil) on their 50th day of life to induce mammary tumors. Seven animals developed malignant mammary tumors (mostly adenocarcinomas) and seven animals showed benign mammary tumors (mostly fibroadenomas). In the control group which had received only vehicle nocturnal aMT6s showed clear changes over the year with higher values in summer than in winter despite constant environmental conditions (photoperiod, temperature, air humidity). The presence of seasonal rhythmicity was tested by the cosinor analysis and a highly significant rhythm ($p<0.005$) with an acrophase in mid-August was detected in controls (Table 5). Animals with benign tumors also showed a significant rhythm ($p<0.025$) with an acrophase in mid-July. Animals with malignant tumors, however, had no significant seasonal rhythm ($p<0.25$), showing a less marked depression of aMT6s in winter as compared to the controls (see Figure 1). At this time, the tumor development began which

TABLE 5
Results of the cosinor analysis of nocturnal 6-sulphatoxymelatonin excretion (ng/hour) in female Fischer rats

Group	*n*	Rhythm detect. ($p<$)	MESOR ± SEM	Amplitude mean range	% of MESOR	Acrophase* mean range
Controls	210	0.005	39.50 ± 1.52	3.95 1.70–6.20	10.0	15/VIII 7/VII-17/IX
Benign tumors	221	0.025	41.99 ± 1.65	3.33 0.87–5.79	7.9	15/VII 19/V-31/VIII
Malignant tumors	223	0.25	40.72 ± 1.47	1.84	4.5	4/VIII

* Day of month is given in Arabic, month in Roman numbers.

may be the cause for the observed elevated melatonin production. Using another tumor model, i.e. spontaneous carcinoma of the endometrium in female rats, a similar observation was made: a group of animals developing malignant tumors had no seasonal rhythm in its 6-sulphatoxymelatonin excretion while another group of rats which did not develop tumors due to hormonal treatment showed a highly significant rhythm with a peak in May (Deerberg and Bartsch, unpublished observations).

Unidentified pineal substances with tumor-inhibiting activity

It is known that the stimulatory effect of pinealectomy on the development and growth of experimental mammary tumors induced by DMBA cannot

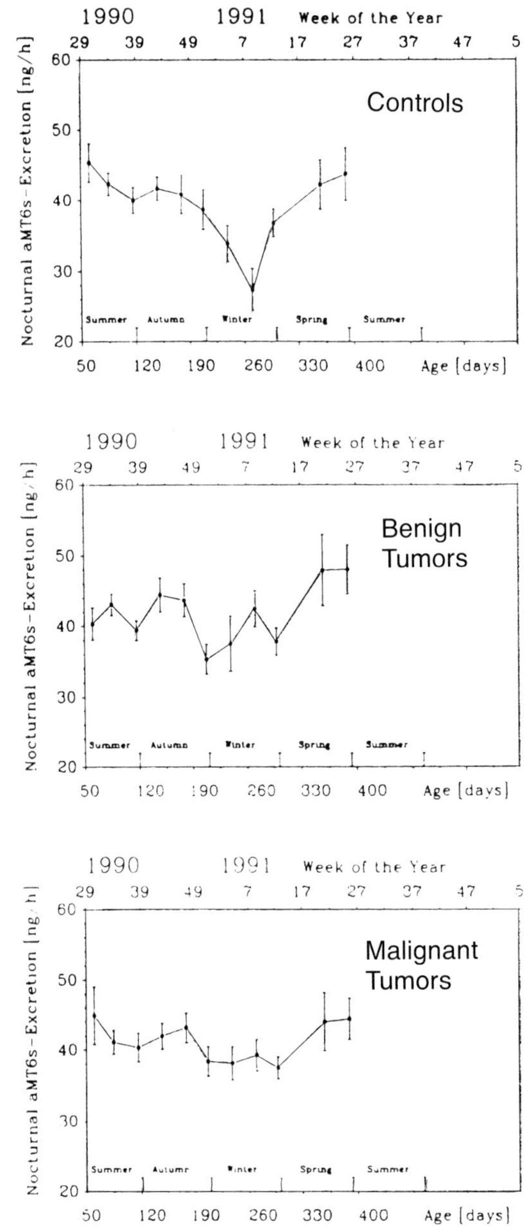

FIG. 1. Nocturnal (23.00–07.00 hours) 6-sulphatoxymelatonin excretion over a complete year of healthy female rats (controls, top), seven rats developing benign mammary tumors (centre), and seven rats with malignant mammary tumors (bottom). Each value represents the monthly average of 2–4 determinations per month. aMT6s = 6-sulphatoxymelatonin.

be completely reversed by the application of melatonin.[25] This observation leads to the assumption that the pineal gland may produce other substances which can inhibit tumor growth. Evidence has existed for quite some time for the presence of cytostatic factors in pineal extracts free of melatonin.[3] Such substances have been found in urine as well.[3]

Anti-tumor activity in human urine was detectable only in summer

In an attempt to purify tumor-inhibiting substance(s) with possible pineal origin from human urine a number of experiments were carried out using three different experimental tumor systems.[9] The aim was to find a suitable tumor system for the detection of the anti-tumor activity during the course of its purification. The experiments were carried out over 4 years. Strong tumor-inhibition as well as complete lack of effect was observed when rats and mice were treated with human urine administered in drinking water. Only after analyzing the data on a seasonal basis was a clear result obtained: the activity was present only in summer. The cosinor analysis revealed a significant rhythm ($p<0.005$) with an acrophase in August (see Figure 2).

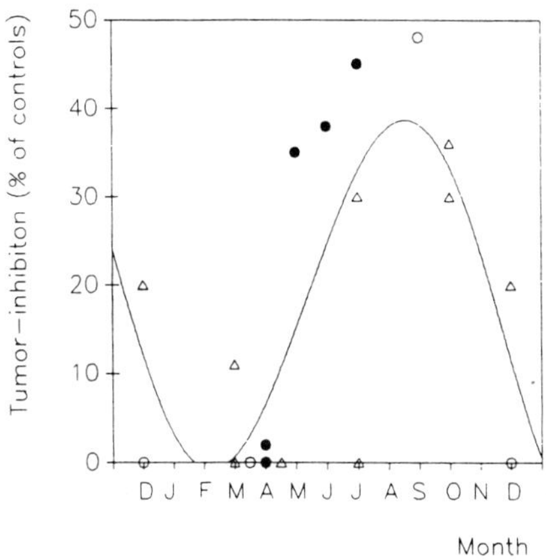

FIG. 2. Seasonal rhythm of tumor-inhibiting activity in human urine as assessed in three tumor models (open circles: adenocarcinoma of the breast in female Holtzman rats; filled circles: solid Ehrlich tumors in Swiss mice; open triangles: fibrosarcoma ascites in Swiss mice) [from 9].

Anti-tumor activity in the rat pineal gland peaks in summer

The tumor-inhibiting activity in individual rat pineal glands was analyzed over a complete year in two separate experiments.[6,7,9] The first one was

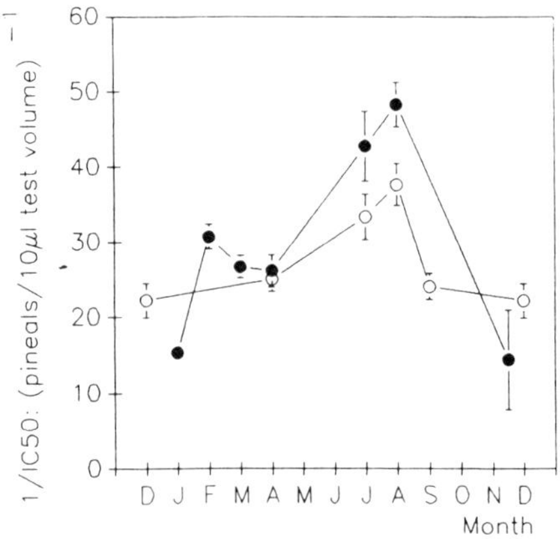

FIG. 3. Seasonal rhythm of anti-tumor activity in rat pineal glands as determined in two different experiments (1986/87: filled circles; 1989/90: open circles). Anti-tumor activity was defined as the reciprocal IC_{50}-value, i.e. the concentration necessary to obtain 50% inhibition of cell growth as compared to untreated controls, and expressed as (pineals/10 μl test volume)$^{-1}$ (from Ref. 9).

carried out in 1986/87 with 167 pineal glands from rats of both sexes, aged between 20 and 120 days, killed at different times of the day. Ethanol extracts were prepared and applied to human erythroleukemia cells K562. Anti-tumor activity was defined as the reciprocal IC_{50}-value (i.e. the concentration necessary to obtain 50% inhibition of cell growth as compared to untreated controls) and expressed as (pineals/10 μl test volume)$^{-1}$. A significant seasonal rhythm was observed with all animals ($p<0.0001$, acrophase in July, the absolute peak occurring in August, see Figure 3, filled circles) as well as with a sub-group consisting of only female rats, aged 100–120 days, killed at 9 a.m. ($p<0.001$, acrophase in May). The second experiment was carried out in 1989/90 with 52 270 day old male rats killed at 11 a.m. Again, a significant seasonal rhythm was obtained ($p<0.001$, acrophase in July, absolute peak in August, see Figure 3, open circles). The decreased peak values and lower amplitude and MESOR may be due to the advanced age of these animals as compared to the first experiment.

Conclusions

The data presented above suggest that pineal function in man as well as in laboratory animals shows seasonal rhythmicity. As far as human beings are concerned this observation is not surprising since it has been shown that although the modern way of life screens them from their natural surroundings to a great degree they have still maintained seasonality in many body functions. However, laboratory animals kept under constant environmental conditions of light, temperature and air humidity throughout the year for many generations are expected to be "standardized" also with respect to lack of seasonality. Nevertheless, there is an increasing number of reports about the maintenance of seasonal rhythms even under constant environmental conditions. In this contribution we will confine ourselves to those related to pineal activity: rat pineal content of immunoreactive arginine-vasotocin[22] and the related nonapeptides vasopressin and oxytocin[17] were found to have pronounced peaks in August; in an electron microscopic study over 2 years synaptic ribbons and spherules were found to display distinct seasonal rhythms in the rat pineal gland;[16] nocturnal plasma melatonin in rats was approximately twice as high in July as compared to March,[5] an observation which agrees with the elevated 6-sulphatoxymelatonin excretion in summer described above. Thus, it appears that pineal activity of laboratory animals is generally higher in summer than in winter, in its production of melatonin, its content of nonapeptides, and in the presence of yet unidentified tumor-inhibiting substance(s). In humans, the existence of anti-tumor activity in urine shows its peak also in summer although it still has to be verified whether this substance really is of pineal origin. Human melatonin, however, showed its maximum in winter—as expected due to its dependence on the length of the photoperiod.

The question however arises, what may be the cause of the observed seasonal rhythmicity in laboratory animals? It is of course possible and likely that the rhythm is of an endogenous nature. In this case, however, it would be expected to free-run if there is no external zeitgeber as observed with circadian rhythms. The described observations suggest a rather stable acrophase over various years, particularly in the case of the nonapeptides and the anti-tumor activity in the rat pineal gland but the number of experiments may not be enough to actually prove it. If the peaks are really as stable as they appear to be then an unknown exogenous zeitgeber must exist—even under so-called standardized laboratory conditions. It is easily conceivable that olfactory stimuli reaching the animals via the air-conditioning may be such factors. Another possibility is an influence of the geomagnetic field on the activity of the pineal gland. It is becoming more and more evident that the pineal is part of a neural system in the brain that perceives and responds to changes in the horizontal component of the geomagnetic field,[20,24,26] which shows distinct seasonal rhythms.[18] To prove the involvement of olfactory or geomagnetic factors as zeitgebers for seasonal rhythms experiments will have to be performed under complete shielding of such influences which require highly sophisticated experimental setups.

Irrespective of the factors acting as zeitgebers for the synchronization of endogenous rhythms under controlled laboratory conditions the fact remains that such rhythms cannot be easily extinguished, neither in humans nor in experimental animals. Thus the question arises: what may be the physiological significance for these highly stable rhythms? It has to be assumed that the organism requires regular seasonal "ups and downs" in diverse body functions in order to maintain health. This should be considered if we try to eliminate the natural "downs" for the sake of higher professional performance since they may have considerable importance as a reminescent of hibernation or as a resting period serving regeneration of the organism which may be as indispensible as is our daily sleep.

References

1. Aldous M.E. and Arendt J. (1988) Radioimmunoassay for 6-sulphatoxy- melatonin in urine using an iodinated tracer. *Ann. Clin. Biochem.* **25**, 298–303.
2. Bartsch C., Bartsch H., Flüchter S.-H., Attanasio A. and Gupta D. (1985) Evidence for modulation of melatonin secretion in men with benign and malignant tumors of the prostate: relationship with the pituitary hormones. *J. Pineal Res.* **2**, 121–132.
3. Bartsch H. and Bartsch C. (1988) Unidentified pineal substances with anti-tumor activity. In *The Pineal Gland and Cancer* (eds. Gupta D., Attanasio A. and Reiter R.J.), pp. 369–376. Brain Research Promotion, London, Tübingen.
4. Bartsch C., Bartsch H., Fuchs U., Lippert T.H., Bellmann O. and Gupta D. (1989) Stage-dependent depression of melatonin in patients with primary breast cancer: correlation with prolactin, TSH and steroid receptors. *Cancer* **64**, 426–433.
5. Bartsch H., Bartsch C. and Gupta D. (1990) Seasonal variations of endogenous defence mechanisms against cancer. In *Neuroendocrinology: New Frontiers* (eds. Gupta D., Wollmann H.A. and Ranke M.B.), pp. 333–339. Brain Research Promotion, London, Tübingen.
6. Bartsch H., Bartsch C. and Gupta D. (1990) Tumor-inhibiting activity in the rat pineal gland displays a circannual rhythm. *J. Pineal Res.* **9**, 171–178.
7. Bartsch H., Bartsch C. and Lippert T.H. (1991) Melatonin und chronobiologische Aspekte von Krebs. *Münch. med. Wschr.* **133**, 113–116.
8. Bartsch C., Bartsch H. and Lippert T.H. (1992) The pineal gland and cancer: facts, hypotheses and perspectives. *The Cancer J.* **5**, 194–199.
9. Bartsch H. and Bartsch C. (1993) Anti-tumor activity in pineal glands and in urine display similar circannual rhythmicity. In *Chronobiology and Chronomedicine* (eds. Gutenbrunner C., Hildebrandt G. and Moog R.), pp. 390–395. Peter Long, Frankfurt.
10. Bojkowski S.J. and Arendt J. (1988) Annual changes in 6-sulphatoxymelatonin excretion in man. *Acta Endocrinol. (Copenh.)* **117**, 470–476.
11. Brown G.M., Bar-Or A., Grossi D., Kashur S., Johannson E. and Yie SM. (1991) Urinary 6-sulphatoxymelatonin, an index of pineal function in the rat. *J. Pineal Res.* **10**, 141–147.
12. Cohen P., Wax Y. and Modan B. (1983) Seasonality in the occurrence of breast cancer. *Cancer Res.* **43**, 892–896.
13. Halberg F., Lagoguey M. and Reinberg A. (1983) Human circannual rhythms over a broad spectrum of physiological processes. *Inter. J. Chronobiol.* **8**, 225–268.
14. Illnerova H., Zvolsky P. and Vanecek J. (1985) The circadian rhythm in plasma melatonin concentration of the urbanized man: the effect of summer and winter time. *Brain Res.* **328**, 186–189.
15. Jacobson H., Janerich D.T., Nasca P., Langevin T., Steiner B. and Hrushesky W.J.M. (1983) Circannual rhythmicity in the incidence of endocrine malignancy: evidence for neurohumoral control of cancer development and growth. *Chronobiologia* **10**, 135.
16. Karasek M., Lewinski A. and Vollrath L. (1988) Precise annual changes in the numbers of "synaptic" ribbons and spherules in the rat pineal gland. *J. Biol. Rhythms* **3**, 41–48.
17. Liu B. and Burbach J.P.H. (1987) Detection and high performance liquid chromatography identification of the summer rises of vasopressin and oxytocin immunoreactivity in the rat pineal gland. *Endocrinology* **121**, 1716–1720.
18. Münchener Universitätsschriften, Fakultät für Geowissenschaften, Veröffentlichungen

des Geophysikalischen Observatoriums Fürstenfeldbruck der Ludwig Maximilians Universität München. Serie A, Nr. 22–33, Ergebnisse der Beobachtungen am Erdmagnetischen Observatorium Fürstenfeldbruck im Jahre 1980–1990, München 1981–1991.

19. Nelson W., Tong Y.L., Lee J.-K. and Halberg F. (1979) Methods for cosinor-rhythmometry. *Chronobiologia* **6**, 305–323.
20. Olcese J., Reuss S. and Vollrath L. (1988) Geomagnetic field detection in rodents. *Life Sci.* **42**, 605–613.
21. Ownby H.E., Frederick J., Mortensen R.F., Ownby D.R. and Russo J. (1986) Seasonal variation in tumor size at diagnosis and immunologic responses in human breast cancer. *Invasion Metastasis* **6**, 246–256.
22. Prechel M.M., Audhya T.K. and Simmons W.H. (1984) Pineal arginine vasotocin activity increases 200-fold during August in adult rats and hamsters. *J. Pineal Res.* **1**, 175–180.
23. Roenneberg T. and Aschoff J. (1990) Annual rhythm of human reproduction: I. Biology, sociology, or both? *J. Biol. Rhythms* **5**, 195–216.
24. Semm P., Schneider T. and Vollrath L. (1980) Effects of an earth-strength magnetic field on electrical activity of pineal cells. *Nature* **288**, 607–608.
25. Shah P.N., Mhatre M.C. and Kothari L. (1984) Effect of melatonin on mammary carcinogenesis in intact and pinealectomized rats in varying photoperiods. *Cancer Res.* **44**, 3403–3407.
26. Welker H.A., Semm P., Willig R.P., Commentz J.C., Wiltschko W. and Vollrath L. (1983) Effects of an artificial magnetic field on serotonin N-acetyltransferase activity and melatonin content of the rat pineal gland. *Exp. Brain Res.* **50**, 426–432.

SECTION V

Light Treatment of Disorders with Rhythmic Disturbances

25

Light Treatment of Depressive States: Swedish Experiences at Latitude 59° North

BENGT F. KJELLMAN, BJÖRN-ERIK THALÉN and LENNART WETTERBERG

Karolinska Institute, Department of Psychiatry, St. Göran's Hospital, Stockholm, Sweden

Abstract

To explore the possible mechanisms mediating the clinical effect of light treatment, the covariation between biochemical rhythm markers and the clinical outcome of light treatment was evaluated. Another aim of the study was to test for a biochemical predictor of the clinical outcome of light treatment. Results from 109 patients with major depression studied autumn 1988 to spring 1991 were included. On Day 1 blood sampling was done 20.00–08.00 hours measuring serum melatonin, cortisol, prolactin and TSH. On Day 3 clinical rating with the use of the Comprehensive Psychopathological Rating Scale (CPRS) and the Hamilton Depression Rating Scale (HAM-D) was performed. The patients were exposed to bright light (luminance: 350 candela/m^2) for 1 hour (22.00–23.00 hours) and blood sampling was done. The light treatment was given for 10 consecutive days in a room with indirect light of 350 candela/m^2 luminance. The patients were treated with light 06.00–08.00 hours (morning light) or 18.00–20.00 hours (evening light). Of the 109 depressive patients 94 completed the treatment procedures and of the patients with winter depression treated with morning light 75% were improved. When the serum melatonin level during the light test was lower at 23.00 hours than at 22.00 hours, clinical improvement in the patients with winter depression treated with morning light was 88%. The area under the profile time curve (AUC) of melatonin decreased significantly more after light treatment in the improved. The time center of gravity (TC) of melatonin was phase advanced by light treatment in the whole group of depressive patients; 06.00–08.00 hours light induced a phase advance and 18.00–20.00 hours light a phase delay of TC. The AUC of cortisol was increased by morning light and decreased by evening light in the whole group of depressive patients. In the improved patients a phase advance of cortisol TC was seen in opposition to a phase delay in the nonimproved patients. In the patients with winter depression a phase advance of TC was also seen in improved patients. Morning and evening light had differential effects on the AUC and TC of TSH in the whole group of depressive patients with a decrease of AUC and a phase advance of TC with morning light. The clinical effects of light therapy seem to be related to alteration of biological rhythm markers.

SUNLIGHT IS of great importance for living beings. One major effect is the entraining and synchronizing effect on circadian rhythms. Thus the nonvisual effects of light entering the eye influence a rhythm generating system in which the suprachiastic nucleii (the circadian clock mechanism) and the pineal gland with its main hormone melatonin form the major parts.

The first controlled study of light treatment in depression was performed by Kripke and coworkers in San Diego, USA treating male patients with nonseasonal depression with bright white light or red (placebo) light.[8] Lewy and coworkers in 1982 reported the successful treatment with bright light in a patient with seasonal depression.[11] Winter depression occurs during the winter with seasonal pattern according to Diagnostic and Statistical Manual of Mental Disorders (DSM-III-R).[1] Some of the patients with winter depression report atypical symptoms as increased appetite, weight gain and hypersomnia. Impairment of mood, cognitive function, and fatiguability are other main features.[15] Light treatment has been used during the last decade mainly in patients with Seasonal Affective Disorder (SAD) of the winter type (winter depression).[19]

Light treatment of depressive states hypothetically works through effects on the rhythm generating system. Lewy and coworkers had proposed that an advancement of the evening rise of melatonin during dim light indicates a phase advancement of the circadian rhythm.[12,13] They have published data relating this phase advancement to the clinical effect of light treatment in the morning.[12] Rosenthal and coworkers support other hypotheses claiming the rhythm changing effects of light to be of less importance and emphasize the probable effect of light treatment on the function of transmitter substances in the brain.[7,15,22]

This presentation will comprise results of light treatment from a 3 year period from autumn 1988 to the spring 1991. To explore the response to light treatment in different diagnostic categories we included patients meeting the DSM-III-R criteria of major depressive episode with either seasonal or nonseasonal patterns.

To explore the possible mechanisms mediating the clinical effects of light treatment we tested the hypothesis that there were biological rhythm marker changes covarying with the clinical outcome of light treatment. Another aim was to evaluate if there were biochemical markers that could help to predict the therapeutic outcome of light treatment.

Materials and methods

Altogether 109 patients with a major depressive episode were included in the study, 83 were female. The mean age was 47 years with a range of 18–74 years. The patients were referred by physicians or contacted us directly. Patients with nonseasonal depression and patients with seasonal pattern according to the DSM-III-R were included. The latter were of three different types: spring, fall, and winter depression. The definition of the

TABLE 1
DSM-III-R criteria for major affective disorder according to seasonal pattern at latitude 59°N

Seasonal pattern	Time of onset	Duration
Spring	After 1st February before 1st June	Longer than 2 weeks
Autumn	After 1st August before 1st November	Longer than 2 weeks Shorter than 3 months
Winter	After 1st September before 1st February	Longer than 3 months
Nonseasonal	Nonseasonal	Longer than 2 weeks

four categories is shown in Table 1. On latitude 59°N fall and spring depressions are recognized types of seasonal affective disorders.

The number of patients in each diagnostic group is shown in Table 2. The majority of the patients were of the winter depression type. The patients with spring and fall depression who had documented episodes in both seasons were diagnosed according to the season when they were treated.

TABLE 2
*Number of patients in different diagnostic groups (total $n = 109$)**

Seasonal type of depression	Female	Male	Total
Winter	47	17	64
Spring	14	2	16
Fall	5	3	8
Nonseasonal	17	4	21
Total	83	26	109

* 15 of them did not complete light treatment.

Light treatment and light test

The light treatment at St. Göran's Hospital was applied in a light treatment room with white colored ceiling, walls, and floor. The reflection of 24 fluorescent tubes in the ceiling gave indirect light with a luminance of 350 candela/m^2 (1,500 lux, 0.8 m above the floor). The patients were also clad in white clothes to minimize light absorption.

The light source was a full spectrum light with two different light temperatures, 4,000 Kelvin ($n = 29$) and 6,500 Kelvin ($n = 65$). The time of light treatment was either 06.00–08.00 hours (morning light) or 18.00–20.00 hours (evening light) for 10 consecutive days. The decision of the time of light treatment was not random, but to some degree influenced by the clinical symptoms, predominantly the sleep-wake rhythm. Late sleepers were treated with morning light and patients with early morning awakening with evening light.

The light test was performed between 22.00 hours and 23.00 hours using the light tubes of 6,500 Kelvin with a luminance of 350 candela/m^2 as the light source. The light test was performed in the light room and for the patients treated with 4,000 Kelvin in another room.

Design

The design of the study is shown in Table 3. On Day 1 a clinical interview rating according to DSM-III-R took place and the patients also performed a self-rating using the Beck's Depression Inventory (BDI).[4]

TABLE 3
Design (overview) of the investigation schedule

	Day 1	Day 3	Day 15	Day 17
Clinical ratings	BDI	CPRS HAM-D	BDI	CPRS HAM-D
Biochemical rhythm markers	20.00–08.00 hours B1 MT CORT TSH PRL	20.00–08.00 hours Light Test 1 MT CORT TSH PRL	20.00–08.00 hours B2 MT CORT TSH PRL	20.00–08.00 hours Light Test 2 MT CORT TSH PRL
Day 4–14 Light treatment 06.00–08.00 hours (M) or 18.00–20.00 hours (E) for 10 days				

For further information see text.

During the night at 20.00, 22.00, 22.30, 23.00, 24.00, 02.00, 04.00, 06.00 and 08.00 hours blood samples were drawn from an indwelling catheter measuring melatonin, cortisol, prolactin, and TSH (Basal 1 (B1)). The patients slept in darkness between 23.00 and 06.00 hours. Two days later clinical ratings were made by two doctors independently, using the Comprehensive Psychopathological Rating Scale (CPRS) items measuring depression and the first 18 items of the Hamilton Depression Rating Scale (HAM-D).[3,5] In the night the blood samples were taken at the same time points as on Day 1, but the patient underwent the light test between 22.00 and 23.00 hours. After 10 days of light treatment the same protocol was applied as before treatment. These results are named Basal 2 (B2) and Test 2, respectively.

Laboratory measurements

The serum samples of melatonin were analyzed by the method of Wetterberg and coworkers[23] (1978). The lower limit of detection was 0.01 nmol/l, intra-assay coefficient of variation (CV) 7.4% and the inter-assay CV for samples above 0.015 nmol/l was 4.8%. The serum samples for cortisol and prolactin were analyzed by commercial RIA kits (Farmos Diagnostica, Turku, Finland). Intra- and inter-assay CV were for cortisol 2.9% and 5.8%, and for prolactin 2.9% and 5.1%, respectively. TSH was analyzed with two commercial kits (Spectra TSH IRMA, Farmos, Turku, Finland and from 910301 Birilux h TSH, Behringewerke, Marburg, Germany). Sensitivity was 0.05 mU/l and 0.02 mU/l, respectively and inter-assay CV in the normal range (0.2—4.0 mU/l) was 5% and 3%, respectively.

Statistical methods

In this design different measures before treatment were compared to measures after treatment to evaluate the treatment outcome. Several measures describing the status of the patients and of their hormonal levels were defined. The nightly hormone results before and after treatment are called B1 and B2, respectively, and the nightly hormone curves where the light test are included at 22.00–23.00 hours before and after treatment are called Test 1 and Test 2, respectively.

A set of repeated measurement of serum hormone levels during one night from 8.00 p.m. to 8.00 a.m. is called the hormone profile. Single lacking data points in the profiles were substituted by linearly interpolated points (or extrapolated points at the ends).

Patients with two adjacent lacking data points were excluded from the material. For each hormone and each individual the area under the profile time curve (AUC) was calculated by means of the trapezoidal rule. Since the time points are not equidistant, AUC is a better estimate of the average hormone level than the arithmetic mean value.

As an estimate of the "mean appearance time" the first-order time moment, TC="time center of gravity", of the profile time curve was calculated, also using an trapezoidal rule (linear interpolation between time points).

A negative difference (TC B2–TC B1) is defined as a phase advance and a positive difference as a phase delay. When the differences between B1 and B2 in AUC and TC were normally distributed, a two sided *t*-test was used to test the difference between B1 and B2. When the difference was non-normally distributed a logarithmic transformation was performed to obtain a normal distribution or the nonparametric Wilcoxon signed rank test was used. A two-factor ANOVA was used to examine the significance of the clinical response (nonresponse vs. response) and light timing condition (morning vs. evening). P-values <0.05 were considered statistically significant.

Results

Some of the clinical features of the 64 patients with winter depression are shown in Table 4. The majority of the patients had been on sick-leave and had been treated for depressive states. Carbohydrate craving was reported by 60% but only 30% had increased appetite. Fatiguability was common (88%), but only 33% were hypersomnic and as much as 53% were hyposomnic.

TABLE 4
Patients with winter depression ($n=64$, $F=47$, $M=17$): selected clinical history, symptoms, and diagnosis

	Percent
Clinical history:	
Earlier hospitalization (depression)	36
Sick leave for depression	75
Consulted physician for depression	89
Earlier treatment for depression	83
Drug treatment for depression	79
Psychotherapy	38
Reported symptoms:	
Increased appetite	30
Decreased appetite	48
Carbohydrate craving	60
Increased body weight	38
Decreased body weight	31
Fatiguability	88
Hypersomnia	33
Hyposomnia	53
Diagnosis:	
Unipolar	73
Bipolar I	5
Bipolar II	22

The therapeutic outcome in the different diagnostic groups is shown in Table 5. Fifteen patients of the initial 109 did not complete the research protocol mostly due to technical problems. A patient was considered as improved (responder), if a 50% or more reduction of depression scores in HAM-D and/or CPRS ratings was found after 10 days of light treatment.

Of the patients with winter depression 37 (63%) improved and 22 (27%) were nonimproved. In the patients with nonseasonal depression 4 (21%) improved and 15 (79%) were nonimproved.

In the total group of patients and in the patients with winter depression

TABLE 5

Outcome of 10 days of light treatment in the morning 06.00–08.00 hours (M) or in the evening 18.00–20.00 hours (E) in 94 patients

Diagnosis		Female		Male		Total			
		I	NI	I	NI	I		NI	
Winter	M	21	7	9	3	30	(75%)	10	(25%)
	E	6	8	1	3	7	(39%)	11	(61%)
Spring + Fall	M	2	2	0	3	2	(29%)	5	(71%)
	E	3	6	1	0	4	(40%)	6	(60%)
Nonseasonal	M	1	6	1	1	2	(22%)	7	(78%)
	E	1	7	1	1	2	(20%)	8	(78%)
Total	M	24	15	10	7	34	(61%)	22	(39%)
	E	10	21	3	4	13	(34%)	25	(66%)

Improved (I) = at least 50% reduction of ratings in HAM-D and/or CPRS (responders). Nonimproved (NI) = others (nonresponders).

In the patients with winter depression, and in the total group, a statistically significant difference ($2p = 0.02$) between morning and evening light was found (Fischer exact test).

the morning light treatment was significantly better than light treatment in the evening ($p < 0.05$). No statistically difference in improvement rate was found between the two light sources (4,000 Kelvin vs. 6,500 Kelvin).

In a group of 33 patients with winter depression treated with morning light the results of the light test are given in Table 6. (Patients with zero levels of melatonin at 22.00, 22.30 and 23.00 hours at B1 were excluded in this calculation.)

TABLE 6

Light test as a predictor of treatment outcome in 33 patients with winter depression treated with light 06.00–08.00 hours

Melatonin (MT) changes	I	NI
MT levels higher 23.00 hours than 22.00 hours	3 (37%)	5 (63%)
MT levels lower 23.00 hours than 22.00 hours or unchanged	22 (88%)	3 (12%)

($2p = 0.0099$) (Fischer exact test).

Of the seven patients with zero melatonin levels 22.00–23.00 hours, five (71%) improved and two were rated as nonimproved.

Out of 8 patients with higher serum melatonin levels at 23.00 hours than 22.00 hours, 3 (37%) improved and 5 (63%) did not improve. Out of 25 patients with lower or the same levels of melatonin at 23.00 hours compared to 22.00 hours, 22 (88%) were improved and 3 (12%) were nonimproved. The difference was statistically significant ($p < 0.05$). Thus, when the melatonin level after the light test was lower or the same at 23.00 hours than at 22.00 hours, the chance of improvement with morning light in winter depression was increased from 37% to 88%.

Biological rhythm markers

Melatonin

The serum melatonin levels before (B1) and after (B2) 10 days of treatment with morning light and evening light in the whole group of depressive patients are shown in Figures 1A and 1B. Figures 2A and 2B show B1 and B2 of the improved and nonimproved patients. There was a significant decrease of AUC between B1 and B2 in the whole group of depressive patients (Table 7).

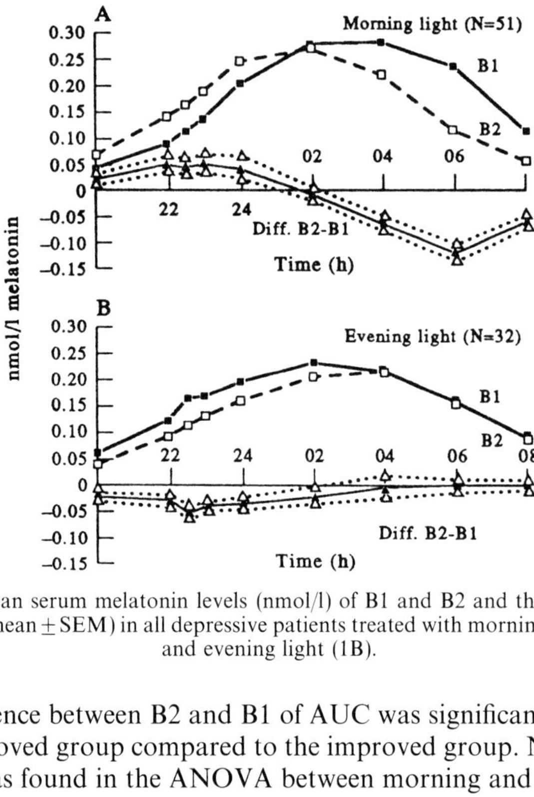

FIG. 1. Mean serum melatonin levels (nmol/l) of B1 and B2 and the difference (B2–B1) (mean ± SEM) in all depressive patients treated with morning light (1A) and evening light (1B).

The difference between B2 and B1 of AUC was significantly smaller in the nonimproved group compared to the improved group. No significant difference was found in the ANOVA between morning and evening light treatment (Table 8). The TC for the whole patient group was significantly earlier for B2 than for B1 (Table 7).

There was a significant difference in the ANOVA between the difference (B2–B1) in TC between morning light treatment (-0.99 ± 0.14 hours) and evening light treatment (0.405 ± 0.14 hours), but not between the improved and the nonimproved patients (Table 8).

The serum melatonin results for B1 and B2 in the patients with winter depression treated with morning light are shown in Figure 3. The mean ± SEM for log AUC of B1 was 0.30 ± 0.04 (log $h \times$ nmol/l) and for B2

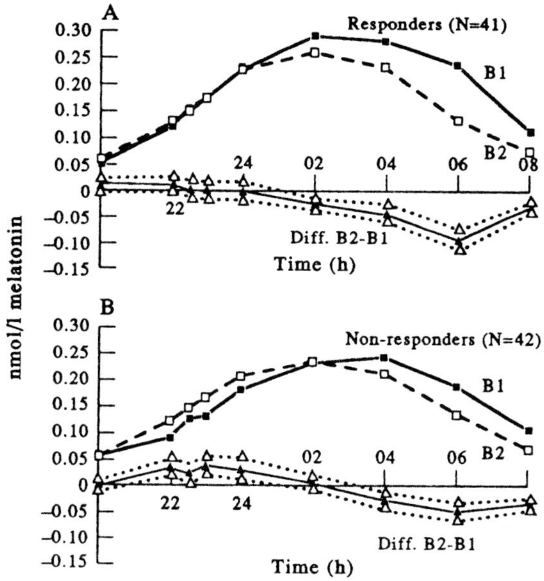

FIG. 2. Mean serum melatonin levels (nmol/l) of B1 and B2 and the difference (B2–B1) (mean ± SEM) in all depressive patients, improved patients (responders) (2A) and nonimproved (nonresponders) (2B).

0.24 ± 0.04 (log *h* × nmol/l), a statistically significant difference ($t = -2.88$, $2p = 0.006$).

The mean ± SEM of the TC was 2.73 ± 0.15 hours before light treatment (B1) and 1.74 ± 0.14 hours after (B2) ($t = -5.88$, $2p$ 0.0001). In the ANOVA no significant differences due to sex or improvement and nonimprovement were found in AUC and TC.

Cortisol

The serum cortisol levels of B1 and B2 in all depressive patients treated with morning and evening light are shown in Figures 4A and 4B, and the improved and nonimproved patients in Figures 5A and 5B.

In the ANOVA the difference of AUC increased with morning light (275 ± 103) and decreased with evening light treatment (−555 ± 112) (Table 8). The difference was significant. No significant difference in AUC was found between improved and nonimproved patients in the ANOVA, and no significant difference in AUC was found in B1 and B2 in the whole group (Table 7).

Table 7 also shows that there was no statistically significant change of the TC for the whole group between B1 and B2. In the ANOVA a significant difference was found between the improved and the non-

TABLE 7
AUC and TC of B1 and B2 and the difference B2–B1 for melatonin (MT), cortisol (CORT), prolactin (PRL), and TSH

	MT B1	MT B2	MT Diff.	CORT B1	CORT B2	CORT Diff.
AUC (mean ± SEM) ($h \times$ nmol/l)	2.21 ± 0.12	1.98 ± 0.12	−0.23 ± 0.07 $t = -3.21$ $2p = 0.0018$	2679 ± 95	2625 ± 77	54 ± 87 $t = -0.62$ $2p > 0.5$
TC (mean ± SEM) (h)	2.55 ± 0.1	2.11 ± 0.1	−0.44 $t = -3.55$ $2p = 0.0006$	3.59 ± 0.07	3.64 ± 0.06	0.04 $t = 0.50$ $2p > 0.5$

	PRL B1 (h × mg/l)	PRL B2	PRL Diff.	TSH B1 (h × mU/l)	TSH B2	TSH Diff.
AUC (mean ± SEM)	102.9 ± 4.8	116.1 ± 7.3	13.2 ± 5.57 $2p = 0.0097$	27.35 ± 2.78	25.21 ± 2.43	−2.14 ± 1.2 $2p = 0.23$
TC (mean ± SEM)	2.69 ± 0.06	2.59 ± 0.05	-0.10 ± 0.06 $t = -1.74$ $2p = 0.09$	1.97 ± 0.04	1.86 ± 0.04	−0.11 ± 0.04 $t = -2.88$ $2p = 0.005$

The TC is stated as hundredth of an hour instead of minutes.

TABLE 8
The statistically significant differencies of melatonin (MT), cortisol (CORT), prolactin (PRL), and TSH between nonresponders and responders and between morning and evening light in the ANOVA

	MT Diff.	CORT Diff.	TSH Diff.
AUC:	($h\times$nmol/l)	($h\times$nmol/l)	($h\times$mU/l)
Morning		275 ± 103	-3.89 ± 1.86
Evening		-555 ± 112	$+0.12\pm1.06$
		$p<0.0001$	$p=0.0360$
Nonresponders	-0.085 ± 0.108		
Responders	-0.374 ± 0.098		
	$p=0.04$		
TC:			
Morning	-0.99 ± 0.141		-0.245 ± 0.045
Evening	0.405 ± 0.144		0.082 ± 0.054
	$p<0.0001$		$p<0.0001$
Nonresponders		0.323 ± 0.114	
Responders		-0.225 ± 0.117	
		$p=0.0024$	

TC is stated as hundredth of an hour instead of minutes.

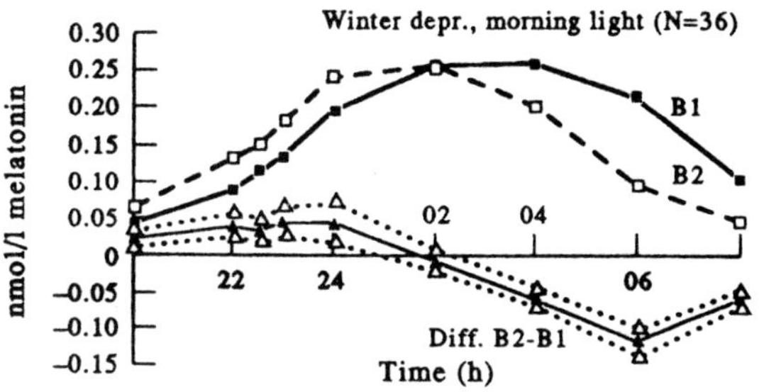

FIG. 3. Mean melatonin serum levels (nmol/l) of B1 and B2 and the difference (B2–B1) (mean$\pm$SEM) in the winter depressive patients treated with morning light.

improved patients in TC (improved -0.225 ± 0.12 hours, nonimproved 0.323 ± 0.11 hours) (Table 8).

The B1 and B2 levels of serum cortisol in the patients with winter depression treated with morning light are shown in Figure 6.

The mean$\pm$SEM of log AUC was 3.40 ± 0.02 (log $h\times$nmol/l) for B1 and 3.44 ± 0.02 (log $h\times$nmol/l) for B2. The increase of log AUC was statistically significant ($t=2.42$, $2p=0.02$). The mean$\pm$SEM of the TC was for B1 3.62 ± 0.10 hours and for B2 3.56 ± 0.09 hours. The difference was not statistically significant ($t=0.31$, $2p=0.76$).

In the ANOVA no significant influence due to sex or improvement and nonimprovement was found on AUC. A significant effect of improvement

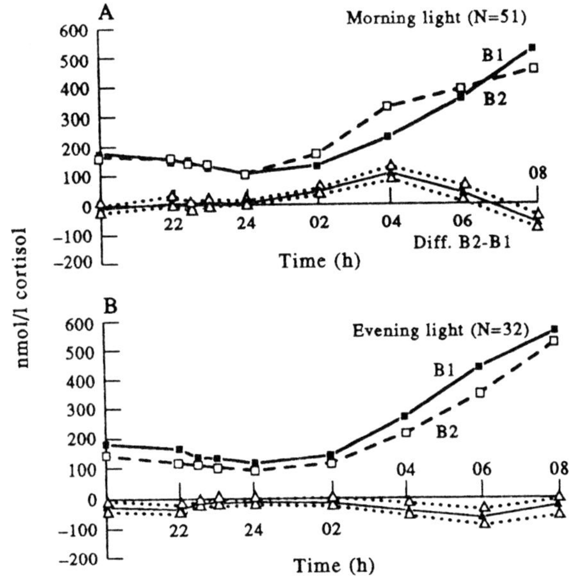

FIG. 4. Mean serum cortisol levels (nmol/l) of B1 and B2 and the difference (B2–B1) (mean ± SEM) in all depressive patients treated with morning light (4A) and evening light (4B)

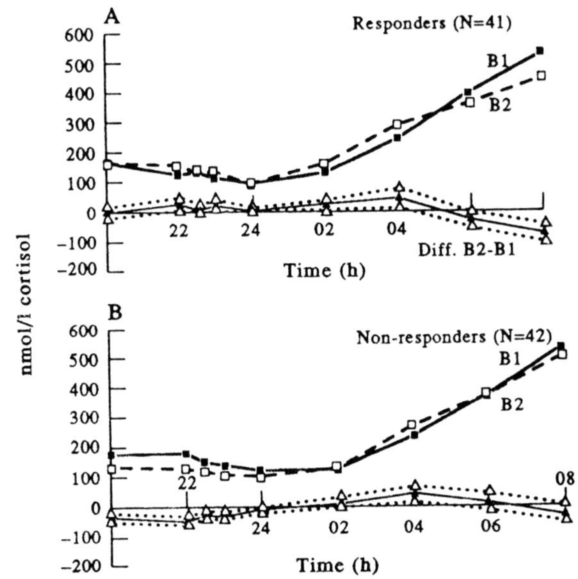

FIG. 5. Mean serum cortisol levels (nmol/l) of B1 and B2 and the difference (B2–B1) (mean ± SEM) in all depressive patients, improved (responders) (5A) and nonimproved (nonresponders) (5B).

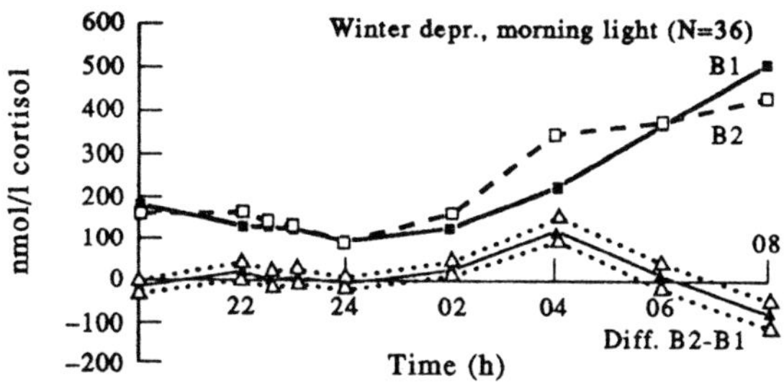

FIG. 6. Mean serum cortisol levels (nmol/l) of B1 and B2 and the difference (B2–B1) (mean ± SEM) in the winter depressive patients treated with morning light.

was found for TC ($2p=0.05$) with a phase advance of TC in the improved group but not in the nonimproved group.

Prolactin

Figures 7A and 7B show the serum prolactin levels of B1 and B2 in all depressive patients treated with morning or evening light. Figures 8A and 8B show B1 and B2 in the same group divided in improved and nonimproved patients. Table 7 shows a significantly larger AUC for B2 than for B1. There were no significant differences in the ANOVA (Table 8). Table 7 shows that the TC in the whole group for B1 was 2.69 ± 0.06 hours and for B2 2.59 ± 0.05 hours ($t=1.74$, $2p=0.09$). No significant differences were found in the ANOVA (Table 7).

The prolactin levels of winter depressive patients treated with morning light are shown in Figure 9. The mean ± SEM for log AUC was for B1 1.94 ± 0.03 (log $h \times \mu g/l$) and for B2 1.96 ± 0.04 (log $h \times \mu g/l$) ($t=0.6$, $2p=0.55$). The mean ± SEM for TC was for B1 2.77 ± 0.09 hours and for B2 2.53 ± 0.09 hours. The difference was statistically significant ($t=-2.78$, $2p=0.0084$). In the ANOVA no significant influences of sex and improvement or nonimprovement could be found.

TSH

The nightly serum TSH levels of the whole group of depressive patients divided into those treated with morning, and respectively evening light are shown in Figures 10A and 10B, while the TSH levels of the improved and nonimproved patients are displayed in Figures 11A and 11B. In the whole group there was no significant change of AUC between B1 and B2, while the TC was significantly phase advanced (Table 7).

Evening and morning light treatment had differential effects both on mean TSH, AUC, and TC in the whole group of depressive patients with a

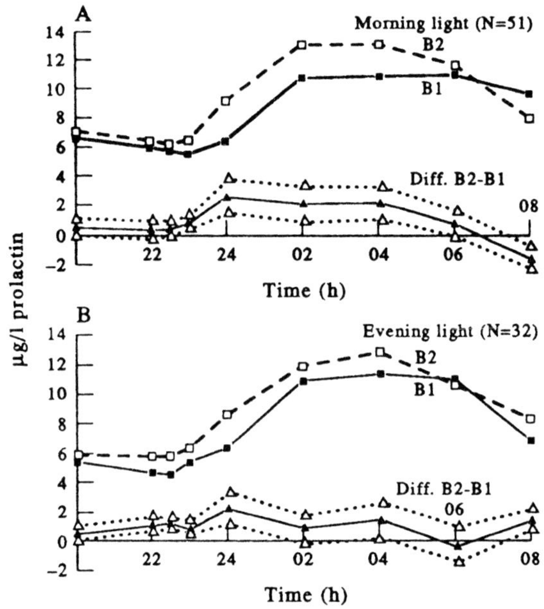

FIG. 7. Mean serum prolactin levels (μg/l) of B1 and B2 and the difference (B2–B1) (mean $\pm$ SEM) in all depressive patients treated with morning light (7A) and evening light (7B).

decrease of AUC and a phase advance of TC with morning light and the opposite with evening light treatment.

The B1 and B2 levels of TSH in the patients with winter depression treated with morning light are displayed in Figure 12. There was no significant difference in AUC between B2 and B1. But TC in B2 was significantly earlier than B1 (B1 1.94 $\pm$ 0.06, B2 1.74 $\pm$ 0.06, $t = -2.71$, $2p = 0.0102$). No significant effects of clinical outcome and sex were found.

Discussion

A comparison with other materials of patients with winter depression sampled in a similar manner as ours in Washington, US, Basel, Switzerland, and London, UK shows that hypersomnia was less common in our sample (33%) than in the other three (78–97%). Increased appetite was most common in the US and the UK samples (60–74%).[15,21,25] The percentage of the Swiss sample (45%) was intermediate and our sample had the lowest frequency (30%). The difference in frequency of carbohydrate craving was less pronounced. In our sample 60% compared to 77–82% in the other three. The lower percentage of atypical symptoms in our sample is in accordance with the findings of Sakamoto and coworkers in Japan reported in patients, like ours, not recruited via the media.[17]

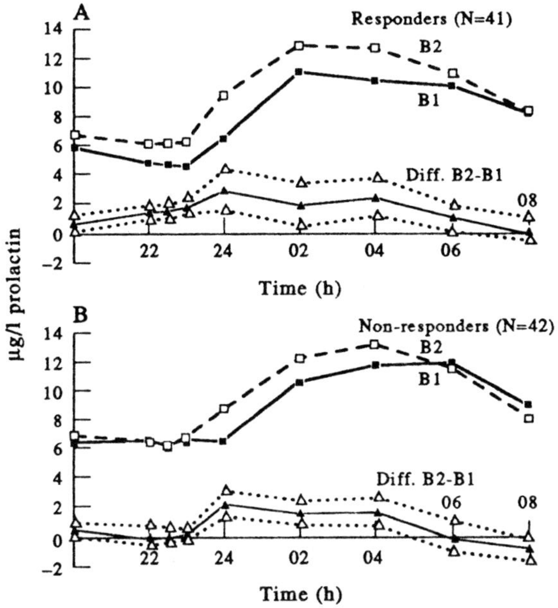

FIG. 8. Mean serum prolactin levels (μg/l) of B1 and B2 and the diffference (B2–B1) (mean ± SEM) in all depressive patients, improved (responders) (8A) and nonimproved (nonresponders) (8B).

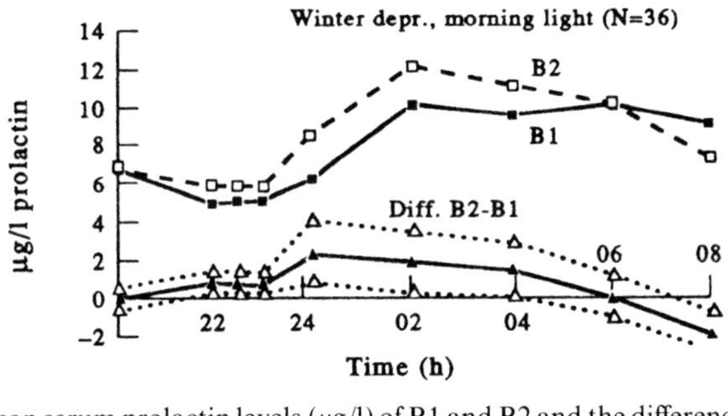

FIG. 9. Mean serum prolactin levels (μg/l) of B1 and B2 and the difference (B2–B1) (mean ± SEM) in the winter depressive patients treated with morning light.

These differences show that the symptom picture could vary in different parts of the world possibly due to latitude, genetic and cultural factors, and contradicts the creation of an international diagnose system based on symptom profiles.

The diagnosis and symptomalogy of winter depression are discussed in Chapter 30 of this volume by Johan Beck-Friis.

The therapeutic outcome in winter depression is in agreement with the findings of other groups. The differential effect of morning and evening

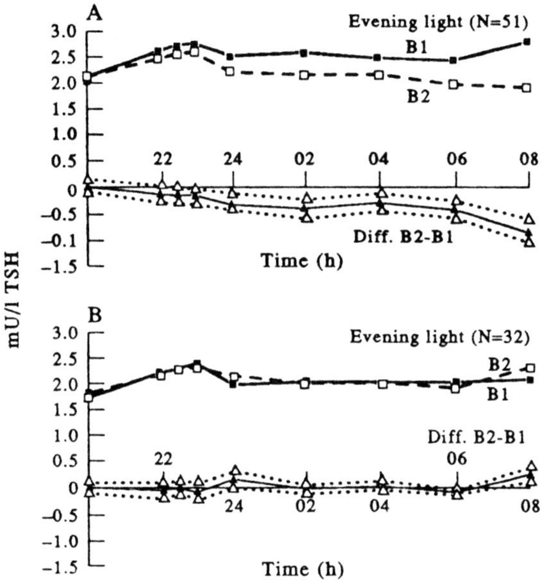

FIG. 10. Mean serum TSH levels (mU/l) of B1 and B2 and the difference (B2–B1) (mean ± SEM) in all depressive patients treated with morning light (10A) and evening light (10B).

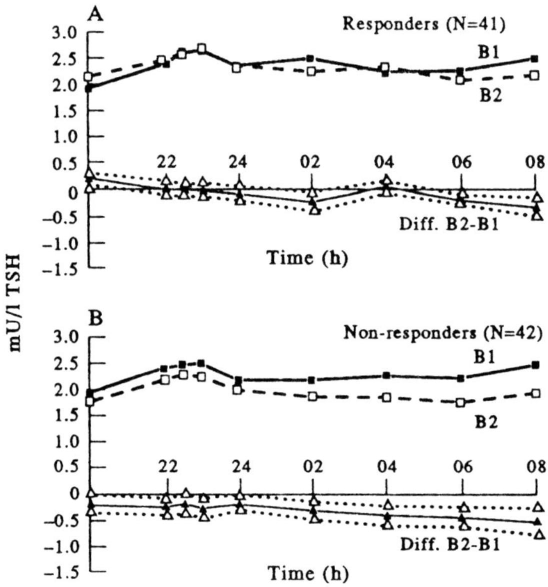

FIG. 11. Mean serum TSH levels (mU/l) of B1 and B2 and the difference (B2–B1) (mean ± SEM) in all depressive patients improved (responders) (11A) and nonimproved (nonresponders) (11B).

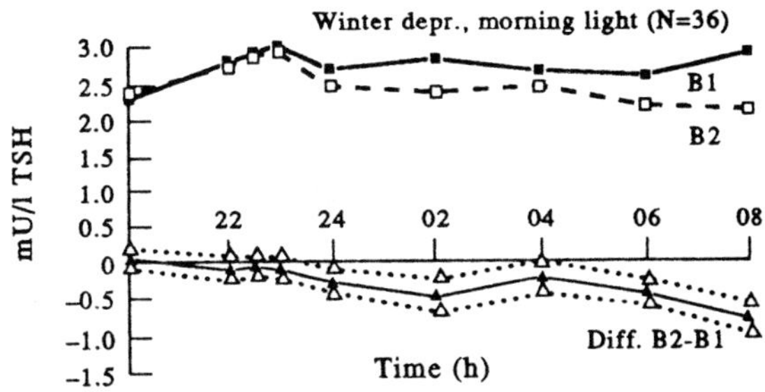

FIG. 12. Mean serum TSH levels (mU/l) of B1 and B2 and the difference (B2–B1) (mean ± SEM) in the winter depressive patients treated with morning light.

light in our sample of patients with winter depression is in agreement with the results of the meta analysis performed by Terman and coworkers[19] (1989). They reported on 322 patients from 14 research centers and found that the remission rates were 62% under morning light exposure compared to 28% under evening light. The difference was statistically significant in the patients with low severity on the baseline HAM-D ratings.

Most of these studies used a crossover design. Later additional support to the hypothesis that morning light is more effective than evening light has been published by Avery and coworkers and Sack and coworkers.[2,16] In a recent study Wirtz-Justice and coworkers found no statistical difference between morning (06.00–08.00 hours) and evening light (21.30–23.00 hours).[26]

The differential clinical effect of morning and evening light in winter depression and between winter and nonseasonal depression supports the hypothesis that light treatment has more than a placebo effect.

The clinical effects on nonseasonal patients were small, but in accordance with the results of Kripke and coworkers.[9,10]

Like in our earlier report the light test also in these larger samples had a predictive value.[20] The chances of improvement were reduced by 58% when the melatonin level was higher at 23.00 hours than 22.00 hours after the light test. Thus the nonimproved patients seem to be less sensitive to light induced suppression of melatonin levels at 22.00–23.00 hours.

In the whole group of depressive patients light treatment caused a lowered secretion of melatonin measured as AUC more in the improved patients than in the nonimproved patients. Evening light treatment caused a phase delay of the TC of melatonin while morning light treatment led to a phase advance of the TC. There was no significant difference in the phase advancing effect between the improved and the nonimproved patients.

Morning light treatment caused an increased AUC of cortisol and evening light a decreased AUC in the whole group of depressive patients.

In the improved patients the TC was phase advanced and in the nonimproved it was phase delayed.

In the patients with winter depression treated with morning light a significant decrease of AUC of melatonin and a significant phase advancement of the TC was seen, while no significant differences due to therapeutic outcome could be recorded. Regarding cortisol a significant increase of AUC was seen and the TC was significantly more phase advanced in the improved than in the nonimproved group.

Regarding prolactin an overall increase of AUC was seen after light treatment, but no differential effects were found with evening or morning light treatment or between improved and nonimproved patients.

The TSH rhythm was differentially affected by morning and evening light with a decrease of AUC, and a phase advancement of TC with morning light and the opposite with evening light. No significant differences were found between improved and nonimproved patients. In the interpretation of these findings the strong influence of the patients with winter depression in the results of the patients in the whole group should be taken into consideration.

The effects of light on melatonin have been studied by Terman and coworkers and Lewy and coworkers, both groups reporting phase advancing effects of morning light in patients with winter depression.[12,13,18] Lewy and coworkers also show these changes of melatonin phase to be of therapeutic importance, relating a phase advancement of the melatonin rise in the evening after treatment with a clinical effect.[13] Rao and coworkers on the other hand found no significant effects of morning light treatment on melatonin rhythms in nonseasonal depressive patients.[14] Winton and coworkers in a small patient group ($n=7$) found that the clinical effect of light treatment was dissociated from the effect on the nightly melatonin rhythm.[24]

Plasma cortisol and the dexamethasone suppression test had not been found to be abnormal. In a recent report, Joseph-Vanderpool and coworkers report of a delayed and blunted ACTH response to CRH stimulation in nine patients with winter depression, suggesting hypofunctional CRH neurons.[7] After 9 days of light treatment a significant increase of the ACTH and cortisol responses to CRH was observed. To our knowledge, this report is the first on cortisol, prolactin, and TSH rhythms in a larger group of patients with winter depression.

The increased prolactin levels in the patients after light treatment could be interpreted as effects of lowered dopaminergic and/or increased noradrenergic and/or serotonergic function.

Perhaps the most interesting result of this preliminary report on an ongoing study is the differential effect regarding the clinical outcome of light treatment on the changes of cortisol TC, both in the whole depressive group and in the group of patients with winter depression.

Thus a relation between therapeutic beneficial effects of light and a phase advancement of the cortisol rhythm seem possible, but the findings have to be evaluated further in a larger sample and in other patient groups.

Whether or not clinical data as diagnosis and reported symptoms could predict the clinical outcome of light treatment is still unsolved. We are examining this question in a larger sample of patients.

Acknowledgements

The authors acknowledge the grants from KI fonder, Svenska läkaresällskapet, Medicinska forskningsrådet (3371) och Gadeliusstiftelsen. We also thank Ylva Friberg, Klas Levin, Lars Mörkrid and Mariann Zettergren for excellent help.

References

1. APA. (1987) *Diagnostic and Statistic Manual of Mental Disorders* (DSM-III-R). American Psychiatric Association, Washington D.C.
2. Avery D.H., Khan A., Dager S.R., Cox G.B. and Dunner D.L. (1990) Bright light treatment of winter depression: Morning versus evening light. *Acta. Psychiatr. Scand.* **82**, 335–338.
3. Åsberg M., Montgomery S., Perris C., Schalling D. and Sedvall G. (1978) A Comprehensive Psychopathological Rating Scale (CPRS). *Acta. Psychiatr. Scand.* **271**, Suppl., 5–27.
4. Beck A.T., Ward C.H., Mendelson M., Mock J. and Erbaugh J. (1961) An inventory for measuring depression. *Arch. Gen. Psychiatry.* **4**, 53–63.
5. Hamilton M. (1967) Development of a rating scale for primary depressive illness. *Br. J. Clin. Psychol.* **6**, 278–296.
6. James S., Wehr T.A., Sack D. and Parry B. (1988) The dexamethasone suppression test in seasonal affective disorder. *Compr. Psychiatry.* **27**, 224–226.
7. Joseph-Vanderpool J.R., Rosenthal N.E., Chrousos G.P., Wehr T.A., Skwerer R., Kasper S. and Gold P.W. (1991) Abnormal pituitary-adrenal responses to corticotropin-releasing hormone in patients with seasonal affective disorder: Clinical and pathophysiological implications. *J. Clin. Endocrinol. Metab.* **72**(6), 1382–1387.
8. Kripke D.F. (1981) Photoperiodic mechanisms for depression and its treatment. In *Biological Psychiatry* (eds. Perris C., Struwe G. and Jansson B.), pp. 1249–1252. Elsevier-North Holland Biomedical Press, Amsterdam.
9. Kripke D.F., Mullaney D.J., Savides T.J. and Gillin J.C. (1989) Phototherapy for nonseasonal major depressive disorders. In *Seasonal Affective Disorders and Phototherapy* (eds. Rosenthal N.E. and Blehar M.C.), pp. 342–356. Guilford, New York.
10. Kripke D.F., Mullaney D.J., Klauber M.R., Risch S.G. and Gillin J.C. (1992) Controlled trial of bright light for nonseasonal major depressive disorders. *Biol. Psychiatry* **31**, 119–134.
11. Lewy A.J., Herbert K.A., Rosenthal N.E. and Wehr T.A. (1982) Bright artificial light treatment of a manic depressive patient with a seasonal mood cycle. *Am. J. Psychiat.* **139**, 1496–1498.
12. Lewy A.J., Sack R.L., Miller L.S. and Hoban T.M. (1987) Antidepressant and circadian phase-shifting effects of light. *Science* **235**, 352–354.
13. Lewy A.J., Sack R.L., Singer C.M. (1985) Bright light, melatonin and biological rhythms: Implications for the affective disorders. *Psychopharmacol. Bull.* **2**, 368–372.
14. Rao M.L., Müller-Oerlinghausen B., Mackert A., Strebel B., Stieglitz R.D. and Voltz H.P. (1992) Blood serotonin, serum melatonin and light therapy in healthy subjects and in patients with nonseasonal depression. *Acta Psychiatr. Scand.* **86**, 127–132.
15. Rosenthal N.E., Sack D.A., Gillin J.-C., Lewy A.J., Goodwin F.K., Davenport Y.,

Mueller P.S., Newsome D.A. and Wehr T.A. (1984) Seasonal affective disorder: A description of the syndrome and preliminary findings with light therapy. *Arch. Gen. Psychiatry.* **41**, 72–80.

16. Sack R.L., Lewy A.J., White D.M., Singer C.M., Fireman M.J. and Vandiver R. (1990) Morning vs. evening light treatment for winter depression. Evidence that the therapeutic effects of light are mediated by circadian phase shifts. *Arch. Gen. Psychiatry.* **47**, 343–351.
17. Sakamoto K., Kamo T., Nakadaira S., Tamura A. and Takahashi K. (1993) A nationwide survey of seasonal affective disorder at 53 outpatient university clinics in Japan. *Acta. Psychiatr. Scand.* **87**, 258–265.
18. Terman M., Terman J.S., Quitkin F.M., Cooper T.B., Lo E.S., Gorman J.M., Stewart J.N. and McGrath P.J. (1988) Response of melatonin cycle to phototherapy for seasonal affective disorder. *J. Neural. Transm.* **72**, 147–165.
19. Terman M., Terman J.S. and Quitkin F.M. (1989) Light therapy for seasonal afffective disorder: A review of efficacy. *Neuropsychopharmacology* **2**, 1–22.
20. Thalén B-E., Kjellman B. and Wetterberg L. (1993) Phototherapy and melatonin in relation to seasonal affective disorder and depression. In *Melatonin, Biosynthesis, Physiological Effects, and Clinical Application* (eds. Yu H.-S. and Reiter R.J.), pp. 495–511. CRC Press, Inc., Bota Raton.
21. Thompson C. and Isaacs G. (1988) Seasonal affective disorder—A British sample. Symptomatology in relation to mode of referral and diagnostic subtype. *J. Affective Disord.* **14**, 1–11.
22. Wehr T.A., Jacobsen F.M., Sack D.A., Arendt J., Tamarkin L. and Rosenthal N.E. (1986) Phototherapy of seasonal affective disorder. Time of day and suppression of melatonin are not critical for antidepressant effects. *Arch. Gen. Psychiatry.* **43**, 870–875.
23. Wetterberg L., Eriksson O., Friberg Y. and Vangbo B. (1978) A simplified radioimmunoassay for melatonin and its application to biological fluids: Preliminary observation on the half-life of plasma melatonin in man. *Clin. Chem. Acta.* **86**, 169–177.
24. Winton F., Corn T., Huson L.W., Franey C., Arendt J. and Checkley S. (1989) Effects of light treatment upon mood and melatonin in patients with seasonal affective disorders. *Psychol. Med.* **19**, 585–590.
25. Wirtz-Justice A., Bucheli C., Graw P., Kielholz P., Fish H.-U. and Woggon B. (1986) Light treatment of seasonal affective disorder in Switzerland. *Acta Psychiatr. Scand.* **74**, 193–204.
26. Wirtz-Justice A., Graw P., Kräuchi K., Gisin B., Jochum A., Arendt J., Fish H.-U., Buddeberg C. and Pöldinger W. (1993) Light therapy in seasonal affective disorder is independent of time of day of circadian phase. *Arch. Gen. Psychiatry* (in press).

26

A Comparison of Two Different Placebo-Controlled SAD Light Treatment Studies

C. I. EASTMAN,[1] M. A. YOUNG[2] and L. F. FOGG[2]

[1]*Biological Rhythms Research Laboratory, Psychology Department, Rush-Presbyterian-St. Luke's Medical Center, Chicago, IL, USA*
[2]*Psychiatry Department and Psychology Department, Rush-Presbyterian-St. Luke's Medical Center, Chicago, IL, USA*

Abstract

High intensity light is widely used as an antidepressant for the treatment of winter depression. There is no doubt that treatment with light can substantially reduce and even eliminate all symptoms. However, establishing that this effect is greater than what could be produced by a placebo-control treatment is hampered by the difficulty in finding an appropriate placebo for light treatment. This paper compares two light treatment studies with good placebo controls. The first study employed a crossover design between morning light and morning placebo. The second, which is still in progress, employs a parallel design with morning light, morning placebo and evening light groups. The first study did not show a statistically significant difference between light and placebo. Although the response rates in the second study are much greater than in the first study, there is still no significant difference between light and placebo. The response to light treatment is much greater than in the first study, but the response to placebo treatment is also much greater. Several possible reasons for the greater antidepressant effects in the second study are discussed, such as a larger placebo component in all three treatments due to a more optimistic staff. Although it would be premature to draw final conclusions about efficacy, we have shown that for light treatment, like antidepressant drug treatment, most of the antidepressant effect can be attributed to placebo effects.

Many researchers and clinicians assert that bright light treatment is an effective antidepressant for winter depression or seasonal affective disorder (SAD). However, even a placebo can be "effective" in that it can produce large reductions in depression.[6,18] The difficulty in finding an appropriate placebo control for light treatment makes conclusions about efficacy beyond placebo effects problematic.[3,4] A major goal of our studies is to establish the efficacy of high intensity light treatment for winter depression. In this paper we will report results from two studies which compare high intensity light treatment to a placebo control treatment. The first was a crossover design between light and placebo, conducted during the two consecutive winters from 1988 to 1990.[5] The second study, a parallel design, began the next winter and is currently in progress. Since we can only report the results from the first two winters of this study, conclusions must be tentative. These studies are repeated for more than one winter in order to obtain a larger sample of subjects.

We decided to test light treatment in the morning, because of the evidence that treatment at this time of day produces the greatest antidepressant effect.[21] Since the question of efficacy is our major focus, it was important to use an appropriate placebo control treatment for morning light treatment. We did not want to use dim light as a control, because we did not think our patients would believe it to be a plausible treatment (especially in a crossover design). The media in our city, and other northern cities in the USA, has focused so much attention on "bright" light as a treatment, that we think that our patients would expect bright light to work better than dim. Even in the first multi-patient study of bright light treatment,[14] patients predicted that the bright light would work better than the dim. If patients expect one treatment to work better, then the placebo effect may be greater for that treatment. Furthermore, several studies have not shown the expected differences between bright and dim light,[9,19,24] leading to suggestions that dim light might be an active treatment.[12,24] Our goal was to find a placebo control treatment that is inert, plausible to patients, and that produces similar expectations for success as light treatment. Some investigators have asserted that evening light is a good control treatment for morning light.[10,11] Although evening light may be a plausible treatment to patients, many researchers claim that evening light is an effective antidepressant,[15,25] i.e. that it is not inert.

Morning light treatment can produce profound changes in the sleep schedule. Patients usually have to wake up earlier in order to have time for the morning treatment. They may go to bed earlier and/or they may become partially sleep deprived. Morning treatment may also serve to regularize the sleep schedule, if patients are asked to awaken at the same time every day. We wanted a placebo control that would produce the same changes in sleep as morning light treatment. Light treatment also necessitates a regular period of inactivity. This may be an enjoyable, restful time for some people, or alternatively, an aversive bore or inconvenience. In any case, we wanted a placebo treatment that would also require the same period of inactivity.

There is a large scientific and popular literature about how negative air ions can improve mood, energy and affect serotonin transmission. There is even evidence that our exposure to negative ions varies seasonally. A negative ion generator is similar to a light box, in that it is an electrical device that produces a natural environmental factor which travels through the air to impinge upon the subject. We thought that negative ion treatment could be plausible to our patients, and could generate good expectations for improvement. However, in order to have a control treatment that was inert, some of the negative ion generators were altered so that they did not actually produce negative ions. This report only includes data from the patients who received "deactivated" generators, i.e. the placebo treatment.

Method—Study No. 1

Subjects

Subjects were diagnosed by the usual criteria for seasonal affective disorder,[14] but in addition we required the "atypical" symptoms of increased appetite or weight, and increased sleep. Most SAD patients have these "atypical" symptoms, and this gave us a more homogenous and more typical sample of SAD subjects. Subjects also had to score 21 or more on the Hamilton Depression Scale composed of the 17 original items[8] and 7 "atypical" items[13] (HAM 17 O + 7 A). All items were administered with the SIGH-SAD semi-structured interview.[22] All subjects were free of psychotropic medications for several months, and none had previously received bright light treatment.

Experimental protocol

The design was a 5-week counterbalanced crossover, with a baseline (B) week, 2 consecutive weeks of morning light (ML) treatment and 2 consecutive weeks of morning placebo (MP) treatment. Both treatments were 1 hour in duration beginning immediately upon waking in the morning.

The light box was 65 cm wide and 43.5 cm tall (Apollo Light Systems, Orem, Utah). It contained six horizontally mounted cool-white fluorescent lamps. Subjects sat at a table or desk with the light box directly in front of them at a distance of 12 inches (30.5 cm). Subjects were provided with 12 inch rulers. They usually read during the treatment time, which produced an illuminance of about 7,000 lux.

The negative ion generator was a commercially available air cleaner (28 × 13 × 18 cm). It was placed on the same table or desk as the light box, positioned 61 cm away from the subject, so that the negative ions were blown across the table in front of the subject. Some of the generators were deactivated by disconnecting a circuit within the generator. However, this procedure did not disable the indicator light or fan. Subjects were informed that they might receive an active or a deactivated generator, but that they would not be able to tell which type they had received.

Dr Lahmeyer directed the staff, at the Psychiatry Department at the University of Illinois Medical Center, that conducted the diagnoses and weekly interviews for depression ratings (SIGH-SAD). This staff was "blind" to treatment condition. After the weekly interviews the subjects walked a few blocks to the Biological Rhythms Research Laboratory at Rush-Presbyterian-St Luke's Medical Center where they received or exchanged their equipment and delivered their questionnaires and logs. This division of functions between the two medical centers helped to maintain the blind. Furthermore, patients were coached to use the word "equipment" rather than "lights" or "generator" in order to help maintain the blind. Subjects completed the Beck Depression Inventory[1] at the end of each week.

Before subjects agreed to participate, Dr Eastman, at Rush Medical Center, showed them a light box and negative ion generator and explained the rationale for these treatments. They were given a packet of information from the scientific and lay literature about negative ions and light treatment. They were told that some people responded better to ions than light, some responded better to lights than ions, some responded to both, and some to neither. The attitude that the research staff at both medical centers displayed to subjects with regard to both treatments was generally neutral and cautious rather than enthusiastic. However, both staffs were frequently coached to use the word "equipment" rather than "lights" or "generators" as much as possible, and to show equal interest in both treatments.

Together with Dr Eastman each subject was required to select a fixed wake-up time for the entire 5-week study, including weekends. Subjects were told that an early and regular wake-up time was part of the treatment. Subjects were required to go to bed at least 7 hours before their scheduled wake-up time. They were permitted to go to bed earlier, but not later than this time. Naps were permitted as long as they started at least 2 hours after the scheduled wake-up time. During the study, compliance was encouraged and monitored by periodic morning phone calls to the subjects and by reviewing the daily sleep logs once or twice a week.

Subjects were informed of their order group assignment at the beginning of the baseline week. At the end of the baseline week subjects completed an Expectation Questionnaire containing four 100 mm visual analog scales, one for each of the upcoming four treatment weeks. Thus, all ratings were performed before treatment began. The 100 mm lines were anchored with "will feel much better" and "will feel much worse." The treatment planned for each of the 4 weeks, "lights" or "ions," was written onto each individual's Expectation Questionnaire.

Data analysis

Statistical analyses were performed on the composite Hamilton Depression Scale (HAM 17 O + 7 A), the original 17-item Hamilton scale and the Beck scale. Since these scales are intercorrelated measures of depression, we conducted a multivariate analysis of variance (MANOVA) using a repeated measures design.[23] The between subject factor was order group (light first, light last) and the within-subjects factor was treatment (baseline, end of light treatment, end of placebo treatment). In all statistical tests we used a significance level of 0.05, two-tailed, except where noted.

Method—Study No. 2

The methods are the same as for the first study with the following changes.

Experimental protocol

Treatment durations were increased from 1 to 1.5 hours. We are using a parallel design with three treatment groups: morning light (1.5 hours upon awakening), morning placebo (1.5 hours upon awakening), and evening light (1.5 hours before bed, ending 1 hour before the latest permitted bedtime).

The same model light box is used, but it is placed 15 inches (38 cm) in front of the patient, or 3 inches (7.6 cm) further away than in the first study. This reduces the average illumination while reading to about 6,000 lux, but gives the patients a larger work area in front of the lights. Patients are provided with 15 inch rulers.

New negative ion generators were built for this study. They are enclosed in shiny black cylinders, 32 cm tall, diameter 15 cm. Three small lights on the front change rapidly between red and green. Since the ion generator makes a soft hissing sound, a white noise generator was added to all the units. This makes it impossible to tell from listening whether the generator is active or not. The ion generators used in the previous study contained a fan which masked the noise. Two generators are set up on a desk or table in front of the patient, 15 inches (38 cm) apart, with 15 inches between the patient and each generator. Patients are provided with rulers.

There is a different staff who performs diagnoses and depression ratings, directed by Dr Young in the Psychiatry Department at Rush Medical Center. After the weekly SIGH-SAD ratings patients only have to walk about one block to go to the Biological Rhythms Research Lab. However, there is the same division of functions between the two groups which helps to maintain the blind. In the meeting with potential patients, Dr Eastman acts more enthusiastic and confident about the success of the treatments than in the previous study. She coaches the staff in both departments to be optimistic in general, as well as to be equally enthusiastic about light and ion treatment. We try to refer to subjects as patients, rather than subjects, and to emphasize that they are receiving treatment, rather than that this is a research study.

Together with Dr Eastman each patient is required to select a wake-up time that is earlier than usual, so that they will have time in their schedule for treatment if they are assigned to a morning treatment group. They are required to go to bed between 7 and 8 hours before their scheduled wake-up time. Naps are permitted within a 6-hour window in the middle of their waking period. Thus, naps are not permitted near the time of morning or evening treatment. Patients are told that an early and regular sleep schedule is part of the treatment. Compliance is encouraged and monitored by requiring that patients call a time-stamp answering machine three times each morning, and by reviewing the daily sleep logs.

As in the first study, patients are told their group assignment at the beginning of the baseline week. At the end of the baseline week they complete an expectation questionnaire which asks them to rate, on a 7-point scale, how they expect to feel at the end of the 4-week treatment, from "will feel as good as summer" (1) to "no change" (7).

Data analysis

Although we are theoretically opposed to data presentation and publication before a study is completed, preliminary analyses were performed for the purpose of comparing these two studies for this paper. A repeated measure ANOVA was performed on the HAM 17 O+7 A scores. The between-subjects factor was treatment group (morning light, evening light, morning placebo), and the within-subjects factor was time (treatment weeks 1, 2, 3 and 4). All scores were input as difference from baseline scores.

Results—Study No. 1

Depression ratings

Figure 1 shows that the Hamilton depression scores declined over the course of the study in both order groups (light first and light last). The Beck scale produced a similar pattern of results. The MANOVA followed by *post-hoc* tests indicated that depression was significantly reduced with both morning light and morning placebo treatments, but that there was no significant difference between them in their antidepressant effects.

A responder to treatment was defined as a subject whose Ham 17 O+7 A score was reduced to at least 50% of baseline. By this criterion, 18.8% (6/32) of the subjects responded to morning light but not morning placebo,

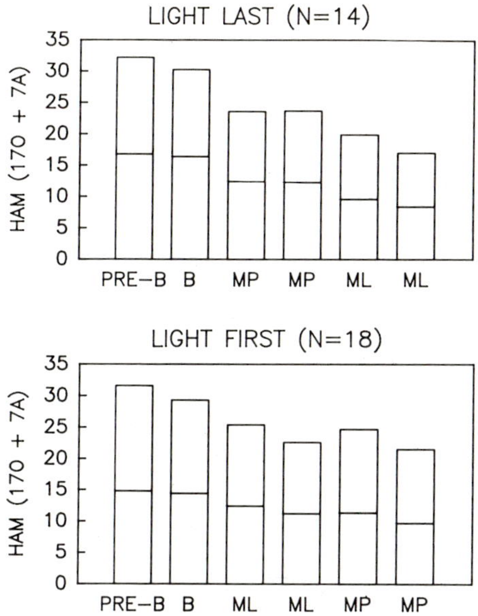

FIG. 1. Average Hamilton depression scores over the course of study No. 1. The bottom of each bar represents the original 17-item scale, the top represents the 7-item atypical scale.

6.3% (2/32) responded to morning placebo but not morning light, 15.6% (5/32) responded to both treatments, and 59.3% (19/32) responded to neither. Thus, 34.4% (11/32) were morning light responders and 21.9% (7/32) were morning placebo responders. Using a 1-tailed exact probability test,[2] this difference in proportions was not significant.

Expectations

Subjects had good expectations for both treatments, but they were slightly better for light treatment than for ion treatment. The mean ± SD expectation scores were 18.8 ± 10.9 mm for the end of light treatment and 25.8 ± 13.4 mm for the end of deactivated ion generator treatment ($p < 0.01$, by a paired t-test).

Sleep

Sleep parameters were averaged for the week before the study (pre-baseline) and each of the 5 weeks of the protocol. The only significant changes occurred between the pre-baseline and baseline weeks. Wake-up time advanced, sleep onset did not change significantly, and nap time was small, so that total sleep time was reduced. Sleep parameters remained stable during the 5-week protocol.

The changes from pre-baseline to baseline were: Wake-up time advanced 1.5 hours from 7.8 ± 1.4 hours after midnight to 6.3 ± 0.9 ($p < 0.001$, paired t-test). Total sleep time (including naps) decreased 1.4 hours, from 8.9 ± 1.1 hours to 7.5 ± 0.7 hours ($p < 0.001$, paired t-test). Thus, the experimental protocol caused an advance in wake-up time and some sleep loss, but these changes occurred prior to both light and placebo treatment. Sleep parameters did not differ between the light and placebo weeks. Thus, in terms of sleep, the placebo treatment was a good control for light treatment.

Results—Study No. 2

Depression ratings

Figure 2 shows that the Hamilton depression scores declined over the course of the study in all three groups (morning light, evening light and morning placebo). The Beck scale produced a similar pattern of results. The ANOVA showed a significant time effect. The treatment group factor was not significant, nor was the treatment group by time interaction. Thus, patients improved over time regardless of treatment group. The results so far do not reveal any significant differences between light and placebo treatments.

A responder to treatment was defined the same way as for experiment No. 1, as a subject whose Hamilton score was reduced by at least 50%. By this criterion, 84.6% (11/13) of the subjects responded to morning light, 75.0% (9/12) responded to evening light, and 66.7% (8/12) responded to morning placebo. This difference was not significant by a Pearson's Chi-square test.

Expectations

Subjects had good expectations for all treatments, and there were no significant differences in expectations between the three groups. The mean $\pm$ SD expectation scores were 3.3 ± 1.5 for morning light, 3.3 ± 1.1 for evening light, and 3.8 ± 1.3 for morning placebo.

Sleep

As in study No. 1, the only significant change in sleep occurred between the pre-baseline and baseline weeks. Wake-up time advanced 1.7 hours, from 7.5 to 5.8, for the morning light group ($p < 0.001$, paired t-test), 0.9 hours, from 7.3 to 6.4, for the evening light group ($p < 0.001$), and 1.8 hours, from 8.1 to 6.3, for the morning placebo group ($p < 0.001$). Total sleep time (including naps) decreased 1.2 hours, from 8.2 to 7.0 hours, for the morning light group ($p < 0.001$, paired t-test), 0.6 hours, from 7.7 to 7.1

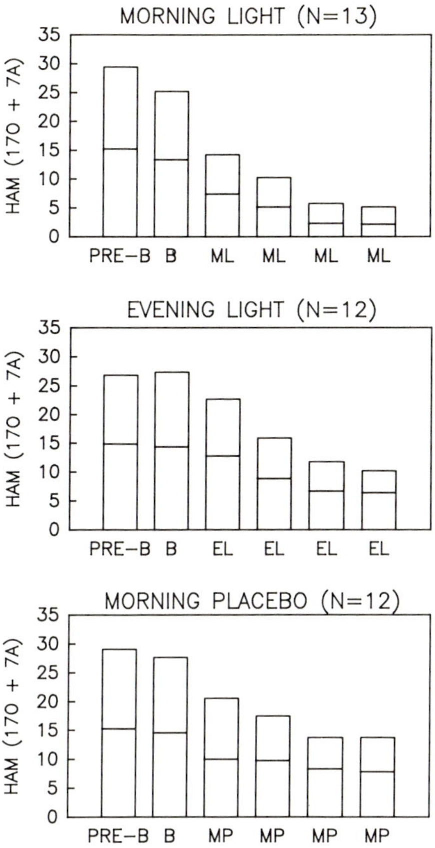

FIG. 2. Average Hamilton depression scores for the data collected so far in study No. 2.

hours, for the evening light group ($p<0.01$), and 1.1 hours, from 8.1 to 7.0 hours, for the morning placebo group ($p<0.001$). As for study No. 1, these sleep changes occurred prior to all treatments, and sleep parameters during the 5-week protocol did not differ among the treatments. However, the changes from pre-baseline to baseline were more similar between the morning light and morning placebo groups. Thus, the placebo group was a better control for the morning light group than for the evening light group regarding potential antidepressant effects caused by sleep changes.

Discussion

The first study did not demonstrate efficacy for morning light treatment; morning light did not produce a significantly greater antidepressant effect than the placebo treatment. This result is even more striking given the fact that patients expected the light treatment to work slightly better than the placebo treatment. Expectations are thought to account for at least part of the placebo effect. If there had been a larger difference between light treatment and placebo, it might have been explained by this difference in expectations.

For the next study, several design changes were made in an attempt to demonstrate a superiority for light treatment over placebo. The placebo literature shows that the physician's attitude toward a treatment, his or her belief, enthusiasm, conviction, commitment, optimism, etc. may be necessary for the success of that treatment. Not only are positive attitudes necessary to enhance the placebo effect, which is an important component of all treatments, but positive attitudes may work synergistically with a treatment and be necessary to release the therapeutic effect of a treatment.[4,7,16] Since the attitude of the research staff was neutral and cautious in the first study, we changed it to be more enthusiastic and confident in the second study.

Another change in the study design was to increase the duration of light treatment from 1 to 1.5 hours, in case the 1 hour treatment did not provide a large enough "dose." We also changed the length of light treatment from 2 to 4 weeks, in case a significant difference between light and placebo could only be detected after several weeks, as in antidepressant drug studies.[17]

Two other design changes were made in an attempt to make subsequent interpretation of results more straightforward. We changed to a parallel groups design, to avoid the potential complications of order and carryover effects. We switched to a more impressive negative ion generator, and gave each patients two generators instead of one, in order to generate equal expectations for light and ion treatment. Another design change was the addition of an evening light treatment group, to include a test of evening light against placebo.

So far, the results of the second study do not show a statistically significant difference in antidepressant effect between light and placebo. We do not know whether the difference will become significant after more subjects are added to the sample. We can, at this point, estimate the magnitude of the benefit of light treatment over placebo, and the proportion of light treatment effect that can be attributed to placebo effects, if we assume that the total antidepressant effect of light treatment is a simple addition of the placebo effect and an active light effect. If we use Hamilton (17 O + 7 A) difference scores (the difference between baseline and the last week of treatment), then morning light treatment reduced the average score 20.0 points, and morning placebo reduced it 13.9 points. Therefore, the benefit of using morning light over placebo was 6.1 points. Furthermore, 70% (13.9/20) of the reduction in score produced by morning light can be attributed to placebo effects. If we base estimates on the response rates previously calculated, then the benefit of morning light treatment is 17.9% (84.6–66.7). In other words, we can assume that 17.9% of the patients who responded to morning light would not have responded to placebo. Also, 79% (66.7/84.6) of the patients who responded to morning light treatment would have responded to the placebo treatment.

Although these differences between light and placebo may seem small, they are actually similar to the differences between many commonly used antidepressant drugs and placebo pills.[4,6] Studies that show statistically significant benefits for antidepressant drugs over placebo often include hundreds of patients.[17] It is more difficult to obtain large sample sizes for SAD light treatment studies, because data can only be collected during a few months of the year, and because patients are less likely to agree to the more demanding treatment procedures and protocols. Those who believe that the benefit of antidepressant drugs over placebo is clinically important, probably feel the same way about the benefit of light treatment. Others may consider the active drug and light effects to be relatively unimportant compared to the placebo effects.

It is most interesting to compare the results of the two studies. All three treatment groups in the second study improved more than the groups in the first study. This is not due to the fact that the treatments were increased from 2 to 4 weeks, because the differences between the studies were already evident during the first two treatment weeks. Nor can this be ascribed to different styles of Hamilton ratings between the two research teams, because the Beck self-ratings showed the same differences between the studies.

One possibility is that the patients in the second study did better because their wake-up times were earlier. The median scheduled wake-up time was 6:30 a.m. in the first study and 6:00 a.m. in the second study. However, there was no correlation between clinical improvement and wake-up time in either study, nor was greater clinical improvement associated with any particular range of wake-up times.

Another possible reason for the improved antidepressant response in the second study could be the increase in treatment durations from 1 to 1.5 hours. This increased investment of time on the part of the patients could increase the placebo effect of all three treatments. It is also possible that the increased light exposure could have a greater physiological effect, accounting for part of the improved morning light response. The change from one to two negative ion generators, and to a more impressive generator, could be responsible for part of the improved placebo response. Another factor which probably increased the placebo effect in all three treatments was the more optimistic attitude of the research staff.

This discussion illustrates that the magnitude of the antidepressant response to light depends on many factors besides intensity, duration, and other parameters that might relate to the physiological dose. Similarly, antidepressant response rates to placebo pills vary widely among studies—from 0 to 91% according to one review.[6] Therefore, concepts such as "a placebo response rate for SAD" or "a response rate to 1 hour morning light at 6 a.m." are erroneous. It is usually meaningless to compare response rates between studies. For example, a new light delivery device, such as a visor, cannot be considered effective just because it produces a response rate similar to a more conventional light box, or even because it produces a high response rate. Each study must have its own control treatment. Reviews of response rates from many studies, such as the study by Terman *et al.*[21] can be of heuristic value. However, as noted by Terman,[20] this type of review cannot be used to draw final conclusions or take the place of well-controlled studies.

It may take many years to establish whether light treatment is more effective than a placebo, and this work is important in order to help understand the mechanisms and etiology of winter depression. Meanwhile, there is no doubt that treatment with light can greatly reduce the suffering of those with SAD. Given the low risk and minimal side effects of this treatment, at least one of us (CIE) believes that it is ethical to prescribe its use. As long as the research, clinical and medical professionals continue to be enthusiastic about light treatment, and patients do not read papers like this one, we can continue to improve the lives of patients with winter depression.

Acknowledgements

We thank Eva Cherin, Marilyn Perlman and Kay Peterson for diagnoses and SIGH-SAD ratings, and Linda Gallo, Suk-fong Tang, Michael Mahoney and Charles Splete for technical assistance. Supported by National Institutes of Mental Health grant MH42768.

References

1. Beck A.T., Ward C.H., Mendelson M., Mock J. and Erbaugh J. (1961) An inventory for measuring depression. *Arch. Gen. Psychiatry* **4**, 53–63.
2. Bishop Y.M.M., Fienberg S.E. and Holland P.W. (1975) *Discrete Multivariate Analysis*. The MIT Press, Cambridge, MA.
3. Brown W.A. (1990) Is light treatment a placebo? *Psychopharmacol. Bull.* **26**, 527–530.
4. Eastman C.I. (1990) What the placebo literature can tell us about light therapy for SAD. *Psychopharmacol. Bull.* **26**, 495–504.
5. Eastman C.I., Lahmeyer H.W., Watell L.G., Good G.D. and Young M.A. (1992) A placebo-controlled trial of light treatment for winter depression. *J. Affect. Disord.* **26**, 211–222.
6. Greenberg R.P. and Fisher S. (1989) Examining antidepressant effectiveness: Findings, ambiguities, and some vexing puzzles. In *The Limits of Biological Treatments for Psychological Distress* (eds. Fisher S. and Greenberg R.P.), pp. 1–37. Lawrence Erlbaum Associates, Hillsdale, NJ.
7. Grunbaum A. (1986) The placebo concept in medicine and psychiatry. *Psychol. Med.* **16**, 19–38.

8. Hamilton M. (1960) A rating scale for depression. *J. Neurol. Neurosurg. Psychiatry* **23**, 56–62.
9. Issacs G., Stainer D.S., Sensky T.E., Moor S. and Thompson C. (1988) Phototherapy and its mechanisms of action in seasonal affective disorder. *J. Affect. Disord.* **14**, 13–19.
10. Lewy A.J., Sack R.L., Miller L.S. and Hoban T.M. (1987) Antidepressant and circadian phase-shifting effects of light. *Science* **235**, 352–354.
11. Lewy A.J., Sack R.L., Singer C.M., White D.M. and Hoban T.M. (1989) Winter depression: the phase angle between sleep and other circadian rhythms may be critical. In *Seasonal Affective Disorder* (eds. Thompson C. and Silverstone T.), pp. 205–221. CNS Clinical Neuroscience, London.
12. Moul D.E., Hellekson C.J., Oren D.A., Frank A., Brainard G., Murray M.G., Wehr T.A. and Rosenthal N.E. (1990) Treating SAD with a light visor: A multicenter study. *Soc. Light Treatment Biol. Rhythms Abst.* **2**, 15.
13. Rosenthal N.E. and Heffernan M.M. (1986) Bulimia, carbohydrate craving, and depression: A central connection? In *Nutrition and the Brain* (eds. Wurtman R.J. and Wurtman J.J.), Vol. 7, pp. 139–166. Raven Press, New York.
14. Rosenthal N.E., Sack D.A., Gillin J.C., Lewy A.J., Goodwin F.K., Davenport Y., Mueller P.S., Newsome D.A. and Wehr T.A. (1984) Seasonal Affective Disorder: A description of the syndrome and preliminary findings with light therapy. *Arch. Gen. Psychiatry* **41**, 72–80.
15. Rosenthal N.E., Sack D.A., Skwerer R.G., Jacobsen F.M. and Wehr T.A. (1988) Phototherapy for seasonal affective disorder. *J. Biol. Rhythms* **3**, 101–120.
16. Shapiro A.K. (1971) Placebo effects in medicine, psychotherapy, and psychoanalysis. In *Handbook of Psychotherapy and Behavior Change: An Empirical Analysis* (eds. Bergin A.E. and Garfield S.L.), pp. 439–473. John Wiley & Sons, Inc., New York.
17. Stark P. and Hardison C.D. (1985) A review of multicenter controlled studies of fluoxetine vs. imipramine and placebo in outpatients with major depressive disorder. *J. Clin. Psychiatry* **46**, 53–58.
18. Stewart J. (1990) Placebos in evaluating light therapy for seasonal affective disorder. *Psychopharmacol. Bull.* **26**, 525–526.
19. Teicher M.H., Glod C.A., Oren D.A., Luetke C., Schwartz P., Brown C. and Rosenthal N.E. (1992) The phototherapy light visor: there is more to it than meets the eye. *Soc. Light Treatment Biol. Rhythms Abst.* **4**, 20.
20. Terman M. (1989) Editorial: Developing the case for efficacy. *Soc. Light Treatment Biol. Rhythms Newsletter* **1**, 4.
21. Terman M., Terman J.S., Quitkin F.M., McGrath P.J., Stewart J.W. and Rafferty B. (1989) Light therapy for seasonal affective disorder: A review of efficacy. *Neuropsychopharmacology* **2**, 1–22.
22. Williams J.B.W., Link M.J., Rosenthal N.E. and Terman M. (1988) *Structured Interview Guide for the Hamilton Depression Rating Scale—Seasonal Affective Disorder Version* (SIGH-SAD). NY State Psychiatric Institute, New York.
23. Winer B.J. (1971) *Statistical Principles in Experimental Design.* McGraw-Hill, New York.
24. Wirz-Justice A., Bucheli C., Graw P., Kielholz P., Fisch H.U. and Woggon B. (1986) Light treatment of seasonal affective disorder in Switzerland. *Acta Psychiatr. Scand.* **74**, 193–204.
25. Wirz-Justice A., Graw P., Krauchi K., Arendt J., Aldhous M., Gisin B., Jochum A. and Poldinger W. (1989) Most SAD patients are phase delayed in winter but respond equally to morning or evening light. *Soc. Light Treatment Biol. Rhythms Abst.* **1**, 19.

27

Effect of Light on Unmasked Circadian Rhythms in Winter Depression

ANNA WIRZ-JUSTICE, PETER GRAW, KURT KRÄUCHI,
HANS-JOACHIM HAUG, GEORG LEONHARDT and DANIEL P. BRUNNER

Psychiatric University Clinic, CH-4025 Basel, Switzerland

Abstract

Nine women with winter depression carried out a "constant routine" protocol of 40 hours sustained wakefulness to measure endogenous circadian rhythms (rectal temperature, mood, alertness, performance). Bright light given in the middle of the day reduced depressive symptoms and improved mood, as did the sleep deprivation of the constant routine protocol itself. Mood and alertness showed a bimodal rhythm under unmasked conditions, with both a nocturnal and afternoon trough. Light did not modify the endogenous rhythm of alertness but did improve a simple performance task at all circadian phases both in winter and in summer.

Introduction

Seasonal Affective Disorder (SAD), winter pattern, is a syndrome of recurrent depressive phases in autumn and winter often associated with increased appetite, carbohydrate craving, weight gain, and hypersomnia, which improves spontaneously in spring and summer.[2,24] Although a consensus has been reached within the field that bright light is the therapy of choice for SAD,[25] there is little consensus as to mechanisms underlying the rapid antidepressant effect.[6]

Two major hypotheses for the pathophysiology of SAD involve the circadian system: that SAD arises as a consequence of abnormally phase-delayed circadian rhythms[22] or of diminished circadian amplitude.[14] Light is therefore antidepressant because of its phase-advancing properties (when given in the morning) or its ability to enhance amplitude (when given in the daytime).[21] Testing these circadian hypotheses requires more stringent experimental methods than usual in clinical trials, since under nychthemeral conditions circadian rhythms are masked by sleep, motor activity, meals, temperature and variable light exposure.[11,13] Thus measurement of phase and amplitude, for example of core body temperature in SAD, have yielded equivocal results. Only one study has yet investigated endogenous circadian rhythms using a 27-hour "constant routine" unmasking protocol. SAD patients were found to have phase-delayed circadian rhythms in core body temperature, melatonin, TSH and cortisol, which were advanced by therapeutic morning light.[3,4]

In an ongoing study, we are investigating the effect of light on amplitude and phase of circadian rhythms in SAD in more detail. Pilot studies were set up in autumn 1989 and regular investigation began in autumn 1990. Women diagnosed as SAD were studied in winter, at the beginning of their depressive phase, and in summer when euthymic, as were age-matched controls. When possible, four separate 40-hour constant routine (CR) protocols were carried out: in winter and in summer, before and after 5 days treatment with 6,000 lux light from 10–14 hours, a time predicted to augment circadian amplitude.[21] As of summer 1992, nine SAD patients have been studied in winter; six SAD patients and four controls have completed the seasonally paired protocol.

There is as yet insufficient data to test the main hypothesis of whether circadian rhythms in SAD patients are phase delayed and/or diminished in amplitude, or whether midday light treatment can modify phase or amplitude of the core body temperature circadian rhythm. However, some of the effects of light appear not to be related to winter depression *per se*. We present three parameters—mood, alertness, and performance—in order to illustrate the effects of light treatment on unmasked circadian rhythms in SAD. This is a first, preliminary presentation of our CR data, and is intended as a descriptive rather than analytic or exhaustive summary.

Experimental procedure

SAD patients (27–66 years) fulfilled the criteria for winter depression.[24] They were studied in the follicular phase of the menstrual cycle, if applicable. The constant routine protocol as developed by Czeisler (details in[13]) was carried out in a room controlled for temperature (22°C), humidity (67%) and light (<80 lux). Subjects remained awake in semi-recumbent posture for 40 hours, with isocaloric meals every 2 hours. Rectal temperature, peripheral skin temperature (10 positions) and heart rate were measured on line. Self-rating of mood, alertness, tension and hunger were made at half-hourly intervals with linear nonnumeric 100-mm bipolar Visual Analogue Scales (VAS),[1] together with collection of saliva for assay of melatonin[16] and cortisol. Performance was assessed at 2-hour intervals with a shortened form of a validated letter-cancellation test for concentration/attention (d2), which measures the speed and precision of discriminatory perception of similar visual stimuli.[7]

Results

Light administered in the middle of the day was an effective treatment for SAD patients in winter ($N=9$). The observer rating of depression (Hamilton Rating Scale[18]) decreased from 14.0 ± 5.5 [SD] to 2.9 ± 1.4, the Atypical Items score[23] from 7.7 ± 3.4 to 1.1 ± 1.3, depression self-ratings[26] from 16.0 ± 6.2 to 8.8 ± 3.3. A paradox of the CR protocol is that in order to unmask circadian rhythms, subjects must undergo a complete sleep deprivation. Sleep deprivation itself can be an antidepressant intervention, and indeed, all SAD patients improved on day 2 of the protocol, as can be seen in the mood self-ratings. This "unwanted" effect in terms of changed clinical state may thus have introduced a new source of masking.

Mood

Our SAD patient collective has previously been characterized by diurnal variation of mood during depression that persists, albeit dampened, when successfully treated with light, and is present also in summer (measured with VAS in the morning upon awakening and in the evening before retiring.[17,19] The CR provides an opportunity to follow the endogenous rhythm of mood with great precision.

The time course of half-hourly mood ratings is presented in Figure 1. Depressed SAD patients in winter began the CR with low morning mood which improved in the afternoon and declined again at night. The sleep deprivation response could be seen after 6 a.m. Around midday, mood was higher on day 2 than day 1, but was followed by an afternoon slump, with mood improvement again in the evening. Thus, under the unmasking conditions of the CR, diurnal variation of mood reveals itself to be a complex phenomenon. Not only is there a circadian component of low mood, but sleep deprivation also "unmasks" a second 12-hour component.

After light treatment a bimodal rhythm of mood was also found. The timing and extent of the afternoon dip was identical on days 1 and 2. SAD patients during depression (day 1) appeared to have an inverse diurnal rhythm of mood. On day 2 after response to sleep deprivation, the afternoon dip was identical in timing to that after light treatment. This indicates that sleep deprivation induced a normal, non-depressed diurnal mood pattern. The nocturnal trough of mood lasted from *ca.* 9 p.m. to 6 a.m. before and from *ca.* 1 a.m. to 10 a.m. after light treatment.

When the smaller group of SAD patients studied both winter and summer were compared ($N=6$, data not shown), mood was increased by light treatment in winter to the same level as found in both CRs in summer

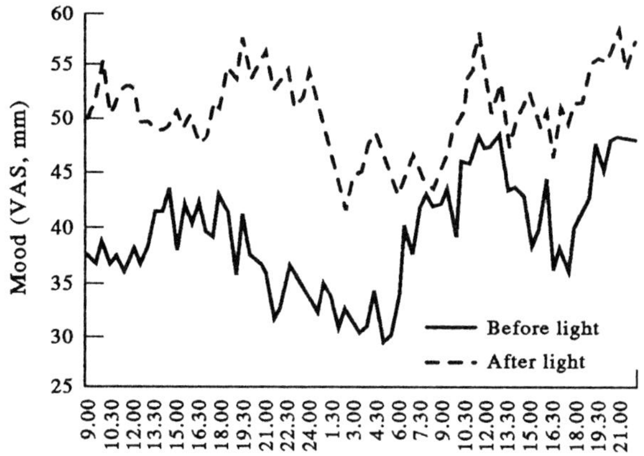

FIG. 1. Endogenous circadian rhythm of half-hourly self-ratings of mood (VAS: 0 mm = extremely depressed mood; 100 mm = extremely good mood) in SAD patients in winter ($N=9$). 2-way ANOVA for repeated measures using 2-hour means revealed a significant effect of light ($F_{(1,8)}=5.6$, $p=0.05$), a significant time course ($F_{(17,136)}=3.2$, $p<0.001$), and interaction n.s.

(three-way ANOVA [season, before and after light, time course] interaction $F_{(1,5)}=9.0$, $p=0.03$). Light treatment in summer had no significant mood augmenting effect. VAS mood ratings were averaged across the CR ($N=74$ scores/person) and then across individuals ($N=6$). This mean mood score (in mm $\pm$ SD) illustrated the mood-normalizing effect of light treatment in winter: winter before light: 38 ± 7; after light: 54 ± 14; summer before light: 53 ± 18; after light: 53 ± 21.

Alertness

The circadian rhythm of alertness in depressed SAD patients in winter also showed a bimodal pattern: a slight midday dip in alertness was present on day 1 and accentuated on day 2, with a clear circadian trough from *ca.* 8 p.m. to 6 a.m. In contrast to mood, which improved after sleep deprivation, alertness declined (values on day 2 were lower than the equivalent values on day 1). Light treatment had no effect on any aspect of the alertness rhythm (Figure 2).

When the smaller group of SAD patients studied both winter and summer were compared ($N=6$, data not shown), the circadian rhythm of alertness in all four CRs was similar. Thus surprisingly, light treatment did not modify the endogenous circadian rhythm of alertness in winter nor in summer, nor was there a difference between winter or summer. VAS alertness ratings were averaged across the CR ($N=74$ scores/person) and then across individuals ($N=6$). Mean alertness scores (in mm $\pm$ SD) were in winter before light: 33 ± 15; after light: 32 ± 7; summer before light: 38 ± 7; after light: 35 ± 9.

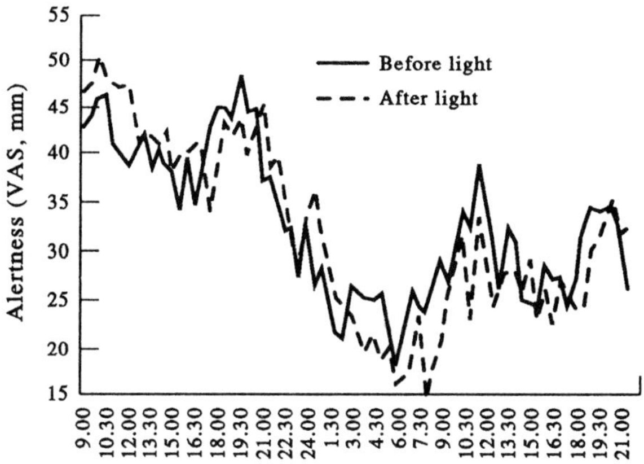

FIG. 2. Endogenous circadian rhythm of half-hourly self-ratings of alertness (VAS: 0 mm = extremely tired; 100 mm = extremely alert) in SAD patients in winter ($N=9$). ANOVA for repeated measures using 2-hour means revealed a significant time course ($F_{(17,136)}=9.2$, $p<0.001$), but no significant effect of light or interaction term.

Performance

A learning effect (decline in number of errors) was found in the first four tests of the first CR. The number of errors thereafter remained constant. The quantitative performance (total number of letters cancelled minus errors) followed a significant circadian rhythm. The rhythm of performance in SAD patients in winter was low in amplitude, with a wide trough from *ca.* 2 a.m.–8 a.m. On the afternoon following sleep deprivation, performance rose to similar values as on day 1, unlike alertness, which declined. After light treatment, performance was markedly increased at all times of day, as was the amplitude of the rhythm. The wave form also changed, with only a narrow trough of low performance around 6 a.m. Following sleep deprivation, performance rose later, after 4 p.m., to similar values as on day 1. The 2-hourly interval of testing did not reveal any mid-afternoon dip in performance (Figure 3).

When the smaller group of SAD patients studied both winter and summer were compared ($N=5$, data not shown), there was no seasonal difference in performance levels. However the duration of the nocturnal trough in summer was shorter than that in winter (interaction term $F_{(18,72)}=2.1, p<0.01$). Light treatment increased performance to the same extent in winter and summer. The duration of the trough after light was decreased compared with before light. Performance was averaged across the CR ($N=19$ tests/person) and then across individuals ($N=5$). This mean score ($\pm$SD) illustrates the performance-enhancing effect of light treatment independent of season: winter before light: 176 ± 38; after light: 200 ± 42; summer before light: 168 ± 28; after light: 196 ± 31.

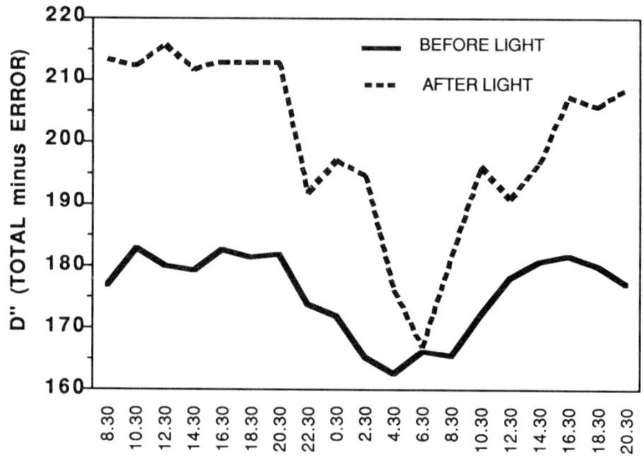

FIG. 3. Endogenous circadian rhythm of performance in a simple letter cancellation test (d2) in SAD patients in winter ($N=9$). ANOVA for repeated measures revealed a significant effect of light ($F_{(1,8)}=61.5$, $p<0.001$), a significant time course ($F_{(18,144)}=6.6$, $p<0.001$), and a significant interaction ($F_{(18,144)}=1.7$, $p<0.05$).

Temperature

The rectal temperature data have not yet been subjected to the appropriate wave-form fitting and statistic analyses. A group mean of the unfitted raw data (not shown) showed minimum temperature to be around 5 a.m. Analysis by two-way ANOVA (using 2-hour mean values as for the VAS) revealed a highly significant time course ($F_{(17,136)}=33.9$, $p<0.0001$) but no effect of midday light treatment (interaction NS).

Conclusions

These preliminary data exemplify the complex interactions between clinical state, season, and the biological effects of light on unmasked circadian rhythms in SAD patients. In the nine SAD patients studied in winter, midday light exposition for 5 days effectively reduced depressive symptoms. However, the circadian rhythms of core body temperature, alertness, and salivary melatonin (data not shown, collaborative study with J. Arendt and J. English, University of Surrey) were not significantly modified by light. Whether the timing of the temperature minimum around 5 a.m. is phase delayed, or the amplitude diminished, requires future comparison with age-matched control women. However the clinical improvement that occurred was not associated with any apparent phase-shift or augmentation of amplitude of three major circadian rhythms driven by the endogenous pacemaker.

Mood was clearly modified by sleep deprivation, light and season. Our data support other unpublished observations (T. A. Wehr, D. Avery) that

SAD patients, as other sub-groups of depressed patients, respond to sleep deprivation and relapse after recovery sleep. The mood-elevating effect of sleep deprivation was, however, less profound and less long-lasting than after light.

Diurnal variation of mood during depression could be clearly followed in the constant routine protocol and revealed that the endogenous mood rhythm was bimodal. The extent of this diurnal variation was enhanced during the day following sleep deprivation, but reduced after light treatment and in summer. This replicates our previous finding of a reduced morning-evening mood difference after light treatment and in summer.[19]

The timing of the afternoon dip in mood during euthymia was parallel to the afternoon dip in alertness. This latter was small but increased after sleep deprivation. There was no significant afternoon dip in alertness in summer. No afternoon dip in alertness before or after the sleep deprivation of the constant routine has been found in healthy young men.[8,15]

An important finding was that light could increase performance independent of clinical state or season. Light and summer also modified the wave form of the circadian rhythm of performance, reducing the duration of the trough of low performance. Previous studies have shown that light can improve performance in both healthy controls,[5,9,10,12,] Grünberger *et al.*, unpublished data) and severely depressed non-seasonal patients, independent of clinical state.[20] We now show that this augmentation occurs at all circadian phases, is not associated with a phase shift, and is independent of season. It is of particular interest that an increase of performance throughout the entire 24 hours could be attained by increasing light intensity only during the daytime. A practical consequence of this would be daylight augmentation of light intensity to improve nocturnal performance without necessarily inducing phase shifts.

It is difficult to make comparisons of these findings with previous studies in SAD patients, since most have administered light in the morning and/or evening, and not over midday, as well as having been carried out under the usual masked conditions. Whether SAD is a circadian rhythm disorder, as hypothesized, is still unclear, until completion of the study and comparison with age-matched controls. However, these preliminary results suggest that the antidepressant effect of midday light is not necessarily mediated by a phase-advance nor by amplitude enhancement.

Acknowledgements

Our studies were designed and initiated in collaboration with J. Anderson (supported by an NSF exchange fellowship SNF #83NI-028030). We thank J. Duffy and C.A. Czeisler, also from the Laboratory for Circadian and Sleep Disorders Medicine, Harvard Medical School, for advice in setting up the CR protocol; G. Moll, C. Hetsch, P. van der Velde,

A. Sarrafzadeh, J. Schulman, T. Brühl, E. Weber and E. Holsboer-Trachsler for collaborative support; the students who worked the shifts and the subjects who volunteered. This study was supported by SNF Grant #32-28741.90, the Horten Foundation, and the Roche Research Foundation (DB).

References

1. Aitken R.C.B. (1969) Measurement of feelings using visual analogue scales. *Proc. Royal Soc. Med.* **62**, 17–21.
2. American Psychiatric Association (1987) *Diagnostic and Statistic Manual of Mental Disorders* (*DSM-III-R*). Washington DC.
3. Avery D.H., Khan A., Dager S.R., Cohen S., Cox G.B. and Dunner D.L. (1989) Phase-typing seasonal affective disorder using a constant routine. *SLTBR Abstracts* **1**, 14.
4. Avery D., Dahl K., Savage M., Brenglemann G., Larson L., Vitiello M. and Prinz P. (1990) The temperature rhythm is phase-delayed in winter depression. *Biol. Psychiatry* **27**, 99A.
5. Badia P., Culpepper J., Myers B., Boecker M. and Harsh J. (1990) Psychophysiological and behavioral effects of bright and dim light. *Sleep Res.* **19**, 387.
6. Blehar M. C. and Lewy A. J. (1990) Seasonal mood disorders: consensus and controversy. *Psychopharmacol. Bull.* **26**, 465–530.
7. Brickenkamp R. (1981) Test d2: Aufmerksamkeits-Belastungs Test. Verlag für Psychologie Dr. C. J. Hogrefe, Göttingen.
8. Brunner D.P., Wirz-Justice A., Kräuchi K., Graw P., Haug H.-J. and Leonhardt G. (1992) Is there an endogenous mid-afternoon dip of subjective alertness? *J. Sleep Res.* **1** (Suppl. 1), 33.
9. Campbell S. and Dawson D. (1990) Enhancement of night-time alertness and performance with bright ambient light. *Physiol. Behav.* **48**, 317–320.
10. Clodoré M., Foret J., Benôit O., Touitou Y., Aguirre A., Bouard G. and Touitou C. (1990) Psychophysiological effects of early morning bright light exposure in young adults. *Psychoneuroendocrinology* **15**, 193–205.
11. Czeisler C.A., Brown E.N., Ronda J.M., Kronauer R.E., Richardson G.S. and Freitag W.O. (1985) A clinical method to assess the endogenous circadian phase (ECP) of the deep circadian oscillator in man. *Sleep Res.* **14**, 295.
12. Czeisler C.A., Johnson M.P., Duffy J.F., Brown E.N., Ronda J.M. and Kronauer R.E. (1990) Exposure to bright light and darkness to treat physiologic maladaptation to night work. *NEJM*. **322**, 1253–1259.
13. Czeisler C.A., Kronauer R.E., Allan J.S., Duffy J.F., Jewett M.E., Brown E.N. and Ronda J.M. (1989) Bright light induction of strong (type 0) resetting of the human circadian pacemaker. *Science* **244**, 1328–1333.
14. Czeisler C.A., Kronauer R.E., Mooney J.J., Anderson J.L. and Allan J.S. (1987) Biologic rhythm disorders, depression, and phototherapy: a new hypothesis. *Psy. Clin. N. Amer.* **10**, 687–709.
15. Dijk D.-J., Duffy J.F. and Czeisler C.A. (1992) Circadian and sleep/wake dependent aspects of subjective alertness and cognitive performance. *J. Sleep Res.* **1**, 112–117.
16. English J., Middleton B.A., Arendt J. and Wirz-Justice A. (1993) Rapid direct measurement of melatonin in saliva using an iodinated tracer and solid phase second antibody. *Ann. Clin. Biochem.* **30**, 415–416.
17. Graw P., Kräuchi K., Wirz-Justice A. and Pöldinger W. (1991) Diurnal variation of symptoms in seasonal affective disorder. *Psychiatry Res.* **37**, 105–111.
18. Hamilton M. (1967) Development of a rating scale for primary depressive illness. *Br. J. Soc. Clin. Psychol.* **6**, 278–296.
19. Haug H.-J., Graw P., Haug M. and Wirz-Justice A. (1991) Course of diurnal variation of mood in patients with "seasonal affective disorder" (SAD). Abstract, *VII Annual Meeting of the European Society for Chronobiology*, Marburg.
20. Holsboer-Trachsler E., Stohler R., Müller J., Gerhard U. and Hobi V. (1991) Cognitive

and psychomotor performance during sleep deprivation and light therapy. *Biol. Psychiatry.* **29**, 194S.

21. Jewett M.E., Kronauer R.E. and Czeisler C.A. (1991) Light-induced suppression of endogenous circadian amplitude in humans. *Nature* **350**, 59–62.
22. Lewy A., Sack R., Singer C. and White D. (1987) The phase shift hypothesis for bright light's therapeutic mechanism of action: theoretical considerations and experimental evidence. *Psychopharmacol. Bull.* **23**, 349–353.
23. Rosenthal N.E. and Heffernan M.M. (1986) Bulimia, carbohydrate craving and depression: a central connection? In *Nutrition and the Brain* (eds. Wurtman R.J. and Wurtman J.J.), Vol. 7, pp. 139–166. Raven Press, New York.
24. Rosenthal N.E., Sack D.A., Gillin J.C., Lewy A.J., Goodwin F.K., Davenport Y., Mueller P.S., Newsome D.A. and Wehr T.A. (1984) Seasonal affective disorder: a description of the syndrome and preliminary findings with light therapy. *Arch. Gen. Psychiatry.* **41**, 72–80.
25. Society of Light Treatment and Biological Rhythms. (1990) Consensus statement on the efficacy of light treatment for SAD. *LTBR Bulletin* **3**, 5–9.
26. Zerssen D.v. and Koeller D.M. (1976) Paranoid-Depressivitäts-skala. Beltz Testgesellschaft, Weinheim.

28

Light Treatment versus Drug Treatment in Winter Depression

O. LINGJÆRDE, T. REICHBORN-KJENNERUD, A. HAGGAG, E. M. BERG, K. NARUD and I. GÄRTNER

Gaustad Hospital, Oslo, Norway

NUMEROUS STUDIES seem to prove beyond reasonable doubt that light is an effective treatment for winter depression. This has been demonstrated also in *controlled* studies, but not in truly *double blind* studies. In fact, since it is literally impossible to study the effect of light treatment under double blind conditions, it is still uncertain how much of the observed improvement after light treatment must be ascribed to the patient's expectation of effect, and how much to a true therapeutic, physiological effect of light—if indeed these two aspects can be clearly separated!

It is easier to perform a true double blind treatment study with drugs, although marked side effects of a drug sometimes can make the blindness illusory in at least some patients. Therefore, besides being of interest in their own right since drugs are an alternative to light treatment, drug treatment studies in winter depression can possibly contribute to an elucidation of placebo responsiveness in persons suffering from Seasonal Affective Disorder. It may therefore be of interest at this meeting to present some relevant findings from a study we carried out in Oslo last winter, comparing light treatment, the selective MAO-A inhibitor moclobemide (Aurorix "Roche"), and placebo. First, however, we will give a short survey of other drug studies in winter depression.

In general, there have been surprisingly few such studies, and most of them have been uncontrolled.

As to conventional tricyclic antidepressants (TCA), we are not aware of any systematic study in winter depression, but anecdotal reports seem to indicate that such drugs often have an unsatisfactory effect, or give too many side effects.

Since TCA seems to be less effective in winter depression than in endogenous depression—which is in itself an interesting observation—several newer and unconventional drugs have been tried. The beta-adrenergic blocker atenolol was found to be without effect in a small cross-over study by Rosenthal and coworkers.[9] But other drugs like L-tryptophan,[6] fenfluramine,[8] alprazolam,[10] tranylcypromine,[2] and bupropion[3] have all been reported to have more or less effect, although most often in uncontrolled studies. In a small, likewise uncontrolled study, Lingjærde and Haggag[4] observed a rapid effect of moclobemide in five women with winter depression. This prompted us to perform the following study from November 1991 to April 1992.[5] Details of the study will be published elsewhere.

A total of 79 outpatients with winter-SAD or subsyndromal winter-SAD, recruited mainly by newspaper advertising, were included. Our plan was to allocate the patients randomly to light treatment and coded drugs. However, there was a surprisingly high number of patients who would only accept light treatment. We chose to place these patients non-randomly in the light group, instead of excluding them from the study. This, besides the fact that the comparison between drug and light could not be made double blind, means that comparison between the results in the light treatment group and the drug groups has to be made with due reservations. However, the comparison of the effects of moclobemide (400 mg daily) and placebo is unbiased for the first 3 weeks (the more so since moclobemide gave no more side effects than did placebo). For ethical reasons, nonresponders after 3 weeks on coded drugs (less than 20% reduction in total score on the extended Montgomery-Åsberg Depression Rating Scale, MADRS[7]) were changed to open moclobemide. Therefore, in the following we will ony present results from the first 3 weeks after the treatment started.

For light treatment, we used the only type of light equipment that was easily obtainable and in some use in Norway at that time, with fluorescent white light (Truelite) giving a light intensity of 1,500 lux at eye level. This is less than usually recommended, but we considered it of great interest, both practically and theoretically, to see whether this light intensity—which, as far as we know, has never been tested before as treatment for SAD—is also effective in winter depression. Light treatment was administered for 2 hours in the morning—but not earlier than from 7 a.m.—for 6 consecutive days.

Our main instrument for assessment of therapeutic outcome was the MADRS, extended with another nine symptoms often seen in SAD, most of them taken from the Comprehensive Psychopathological Rating Scale (CPRS[1]) (Increased sleep, Increased appetite, Carbohydrate craving, Reduced sexual feelings, Hypochondria, Getting easily tired, Reduced speech, Motor retardation and Motor agitation).

Assessments with extended MADRS, Clinical Global Impression, and a side effect checklist were performed at baseline, and after 1, 3, 6, 10 and possibly 14 and 18 weeks.

Five patients dropped out after less than 1 week of treatment, leaving 74 patients for analysis. Some of their main characteristics are shown in Table 1.

Table 2 shows the extended MADRS scores in the three treatment groups during the first 3 weeks of the study.

TABLE 1
Main characteristics of treatment groups

	Light	Placebo	Moclobemide
N	31	18	16
Male/female	4/27	5/13	4/12
Age, mean	42.9	43.2	43.0
Age, range	24–63	20–62	20–65
Major depression, present episode	18	13	13
Major depression, worst episode	28	15	12
Age 1st WD, mean	23.2	26.1	23.1
Age 1st WD, range	5–50	12–45	11–38
No. WD, mean	16.9	16.5	15.9
No. WD, range	4–59	4–45	3–30

TABLE 2
Mean total score on extended MADRS at baseline and after 1 and 3 weeks. SD in parenthesis

		Light	Moclobemide	Placebo
Baseline:	*N*	31	16	18
	Ext. MADRS	31.5 (10.1)	38.1 (9.3)	32.3 (7.9)
One week:	*N*	31	16	18
	Ext. MADRS	16.7 (11.8)	31.3 (13.2)	26.8 (7.7)
Three weeks:	*N*	27	16	16
	Ext. MADRS	12.9 (11.5)	23.9 (13.7)	21.4 (11.3)

The main findings to be read from this table are the following: (1) Light treatment is reasonably effective, with marked improvement after 1 week (immediately after the end of treatment), and—what is perhaps more noteworthy—the improvement is even more marked 2 weeks afterwards. (In fact, most patients maintained the improvement for the rest of the season.) (2) Improvement in the drug groups is inferior to the improvement on light, and with little difference between moclobemide and placebo. In fact, to our disappointment there were about 50% non-responders in both groups after 3 weeks. However, after having changed from coded tablets to open moclobemide, most of these patients also showed a rapid improvement (not shown).

Since about half of the patients on placebo showed marked improvement after 3 weeks, we tried to identify possible predictors of placebo response, and we found one factor explaining a surprisingly high percentage of the response variance, namely the patients' age.

TABLE 3
Improvement after 3 weeks (per cent reduction of total extended MADRS score) in relation to age. r = linear regression coefficient

	Light	Moclobemide	Placebo
N	26	16	15
r	−0.05	−0.48	−0.76
P	0.80	0.06	0.001

As seen from Table 3, regression of per cent improvement after 3 weeks on age shows a highly significant negative correlation, with a linear regression coefficient as high as −0.76 and $p = 0.001$, explaining more than half of the variance. There is a non-significant tendency in the same direction in the moclobemide group, but not in the light treatment group.

The negative correlation between per cent improvement and age might be due to the older patients being more severely ill, but this was not the case. Neither was there a significant correlation between improvement and total illness duration, or number of winter depression periods.

Due to the particularly marked placebo effect in younger patients, we compared the effect of moclobemide and placebo separately for patients below and above the median age of 45 years. No difference was found below 45 years, whereas the difference was barely significant ($p < 0.05$, one-sided test) in patients above 45 years.

Since to our knowledge such a negative correlation between age and placebo improvement has never been found in non-seasonal depression, the question arises whether this is characteristic of patients with SAD. Is it possible that the pathophysiology of SAD, including whatever mechanisms are mediating the placebo response, is different for young and somewhat older SAD patients?

As to placebo response on light treatment, the lack of age effect would seem to argue against the improvement being a pure placebo effect—at least in patients of middle or higher age. At any rate, our findings should prompt anyone doing treatment studies in SAD to look for possible effects of age.

References

1. Åsberg M., Perris C., Schalling D. and Sedvall G. (1978) The CPRS—Development and application of a psychiatric rating scale. *Acta Psychiatr. Scand.* **Suppl. 271**.

2. Dilsaver S.C., Del Medico V.J., Quadri A. and Jaeckle R.S. (1990) Pharmacological responsiveness of winter depression. *Psychopharmacol. Bull.* **26**, 303–309.
3. Dilsaver S.C., Quamar A.B. and Del Medico V.J. (1992) The efficacy of bupropion in winter depression: results of an open trial. *J. Clin. Psychiatry* **53**, 252–255.
4. Lingjærde O. and Haggag A. (1992) Moclobemide in winter depression: some results from an open trial. *Nord. J. Psychatry* **46**, 201–203.
5. Lingjærde O., Haggag A., Reichborn-Kjennerud T., Berg E.M., Narud K. and Gärtner I. (1992) Treatment of winter depression: a comparative trial with bright light, moclobemide, and placebo. *Clin. Neuropharmacol.* **15**, Suppl. 1, Part B, 183B.
6. McGrath R.E., Buckwald B. and Resnick E.V. (1990) The effect of *l*-tryptophan on seasonal affective disorder. *J. Clin. Psychiatry* **51**, 162–163.
7. Montgomery S.A. and Åsberg M. (1979) A new depression rating scale designed to be sensitive to change. *Brit. J. Psychiatry* **134**, 382–389.
8. O'Rourke D., Wurtman J.J., Wurtman R.J., Chebli R. and Gleason R. (1989) Treatment of seasonal depression with *d*-fenfluramine. *J. Clin. Psychiatry* **50**, 343–347.
9. Rosenthal N.E., Jacobsen F.M., Sack D.A., Arendt J., James S.P., Parry B.L. and Wehr T.A. (1988) Atenolol in seasonal affective disorder: A test of the melatonin hypothesis. *Amer. J. Pschiatry* **145**, 52–56.
10. Teicher M.H. and Glod C.A. (1990) Seasonal affective disorder: rapid resolution by low-dose alprazolam. *Psychopharmacol. Bull.* **26**, 197–202.

29

Light Therapy of Premenstrual Depression

BARBARA L. PARRY

Department of Psychiatry, UCSD, 0804, La Jolla, CA 92093, USA

Abstract

The purpose of this study was to test the efficacy of bright light treatment in patients with premenstrual depression. In a 3 month crossover design, we administered bright (>2,500 lux) white morning, bright white evening, and placebo dim (<10 lux) red evening light, daily for 1 week of the premenstrual phase to 19 patients with late luteal phase dysphoric disorder (LLPDD) and to 11 normal controls. In contrast to our previous pilot study in this area, all light treatments significantly reduced depressive ratings from baseline. None of the light treatments differed significantly from each other. Thus, neither morning or evening bright light were shown to have incremental benefits over placebo in this study of patients with premenstrual mood disorders.

BRIGHT LIGHT has been used to treat seasonal affective disorder,[1] some patients with major depressive disorders,[2] seasonal premenstrual syndrome,[3] and most recently nonseasonal premenstrual syndrome.[4] In our pilot study,[4] we reported greater effectiveness of evening, compared with morning bright light in the treatment of six patients with prospectively documented premenstrual depression. To extend our preliminary investigation, in this study we report on effects of light treatment in an additional 19 patients with late luteal phase dysphoric disorder and in 11 normal controls. In our initial pilot study, morning bright light treatment, hypothesized to exacerbate symptoms by phase-advancing already pathologically advanced circadian rhythms in patients with premenstrual depression, was intended to serve as an internal control. In that study, depression ratings, although not statistically different from ratings obtained at baseline (before treatment), did decrease after morning light treatment. Thus, morning bright light did not serve as a sufficient control condition. In the present study, we added an additional placebo control condition: dim red evening light. In this study, we report the effects on mood of morning bright white, evening bright white, and evening dim red light administered in a crossover design to patients with LLPDD depression and to normal control subjects.

Subjects and methods

The patients were referred by local professionals or were recruited by advertisement. Screening procedures, inclusions and exclusion criteria have been described previously (4). Subject characteristics are described in Table 1.

TABLE 1
Reported mood, behavior or physical symptoms for PMS ($N = 19$) and normal controls ($N = 11$)

Mood, behavior or physical change	% Moderate-severe PMS	NC
Appetite	66.7	0
Anxious or jittery	66.7	0
Increase in accidents	27.8	0
Thoughts of hurting self	16.7	0
Difficult to enjoy self	64.7	0
Irritable	83.3	0
Difficulty concentrating/making decisions	68.4	9.1
Feel sad/cry	77.8	0
Feel guilty	38.9	0
Decrease in efficiency	55.6	0
Crave specific foods	72.2	9.1
Excessive sleeping	27.8	0
Decrease in sexual drive	38.9	0
Loss of energy or fatigue	77.8	0
Seasonal change in energy level	21.1	0
Seasonal change in mood	26.3	0
Experienced postpartum depression	57.1	0

$p < 0.05$ for all variables listed.

Subjects selected for the study underwent, in random order, a crossover trial of bright (>2,500 lux) white morning light (6.30 a.m.–8.30 a.m.), bright white evening light (7.00 p.m.–9.00 p.m.), or placebo dim (<10 lux) red evening light treatment. Each treatment was administered for 7 consecutive days during the luteal phase of a separate menstrual cycle (7–10 days before the onset of menses). We administered the dim red light placebo in the evening to contrast with the evening bright light, which our preliminary study had indicated was more effective than morning light in alleviating PMS symptoms.

Menstrual cycle phase was determined by using a colorimetric urinary immunoassay (Ovustick Company, Irvine, CA) for the midcycle luteinizing hormone surge. Patients were asked to sit 3 feet from the light source. The bright light was a portable illumination box from Apollo Light which produced 2,500 lux at 3 feet by cool white fluorescent light bulbs. The dim light was one fluorescent bulb behind a red filter encased in the same size box producing 10 lux at 3 feet. Subjects were asked to keep their eyes open and to gaze every few minutes at the light. During each month of study, patients were asked to maintain their normal sleep patterns, to take no naps, and to document their sleep times on daily sleep logs. They also were asked to wear dark sunglasses or goggles to block out light when outdoors between 6:30 a.m. and 8:30 a.m. when they were being treated with evening light, or between 7:00 p.m. and 9:00 p.m. when they were being treated with morning light, in order to balance the amount of light exposure on both treatment conditions. Subjects also completed daily light logs, expectation forms (100 mm visual analog scales from "much better than usual" to "much worse than usual" for each light treatment condition administered at the beginning of the study), and Horne-Ostberg ratings[10] for morningness and eveningness.

At the end of 7 days of light treatment in the luteal phase, mood over the past several days was assessed by two raters, who were blind to the treatment condition, using the Hamilton depression scale, using an addendum to the Hamilton scale that assessed atypical depressive symptoms such as fatigue, social withdrawal, carbohydrate craving, weight gain, and hypersomnia (maximum of 23 points), and using a mania rating scale.[11] Criteria of Terman *et al.*[12] were used to determine responders to light treatment: a 50% reduction in scores on the 21-item Hamilton Scale to a score of less than 8. In addition, after each treatment, subjects completed the Beck Inventory for Depression, and they were asked to continue filling out the daily ratings forms throughout the light study.

Subjects had been off psychoactive medication for at least 2 months before the initiation of the study. All gave written informed consent after the procedures had been fully explained.

Follow up study

After completion of the study, PMS subjects who wished to continue with the light treatments were invited to do so. Menstrual phase continued to be monitored by urinary LH assay, and that light treatment (bright white a.m., bright white p.m. or dim red p.m.), which was found to be most effective in the initial study, was administered for a subsequent 6–18 month period in the premenstrual phase. Each month Hamilton ratings were done following a week of the specified light treatment. Patients continued to complete daily mood ratings, sleep and light logs and Beck inventories for depression during each light treatment.

Analyses

Baseline measures for each rating scale (Hamilton, Beck, atypical, mania, or daily ratings) were determined by taking the mean of the ratings during the premenstrual weeks of the 2–3 month evaluation phase before treatment. There was no spontaneous improvement in symptoms over a 2-month interval before the treatment protocol was begun. For each of the rating scales normality was tested by the Shapiro-Wilk approach. There were no statistically significant effects by order of treatment; hence, treatment order was not taken into account in the statistical model. The nonparametric tests used were those of Friedman and Wilcoxon. LLPDD and normal control subjects were analysed separately. Since this was a small-sample exploratory study, no adjustment for multiple comparisons was used. The 5% level was used for statistical significance.

Results

Clinical rating scales (see Figures 1–4)

Hamilton, Beck, and atypical ratings all showed statistically significant ($p<0.005$) effects of condition (baseline, a.m. bright white, p.m. bright white, p.m. dim red light) in LLPDD subjects. Baseline ratings (before treatment) were significantly ($p<0.01$, Wilcoxin) greater than those obtained after treatment with each light treatment (see Figure 1) but none of the light treatments differed significantly from each other.

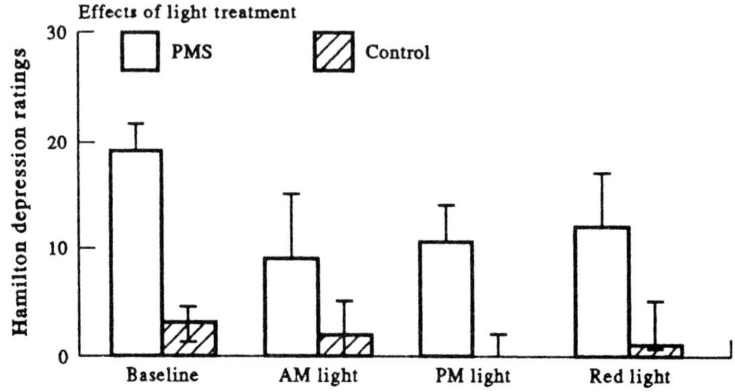

FIG. 1. Median, first and third quartiles of Hamilton depression ratings in PMS ($N=19$) and normal control ($n=11$) subjects, by baseline and treatment values. Each of the light treatments (bright white a.m., bright white p.m., dim red p.m.) were significantly ($p<0.01$) different from baseline, but none of the light treatments differed significantly from each other.

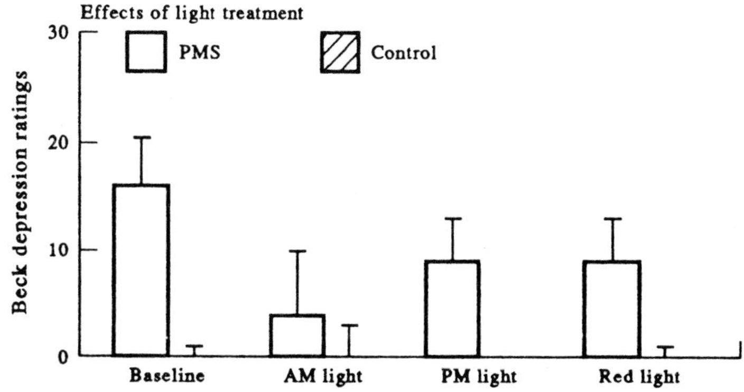

FIG. 2. Median, first and third quartiles of Beck depression ratings in PMS ($N=19$) and normal control ($n=11$) subjects, by baseline and treatment values. Each of the light treatments (bright white a.m., bright white p.m., dim red p.m.) were significantly ($p<0.01$) different from baseline, but none of the light treatments differed significantly from each other.

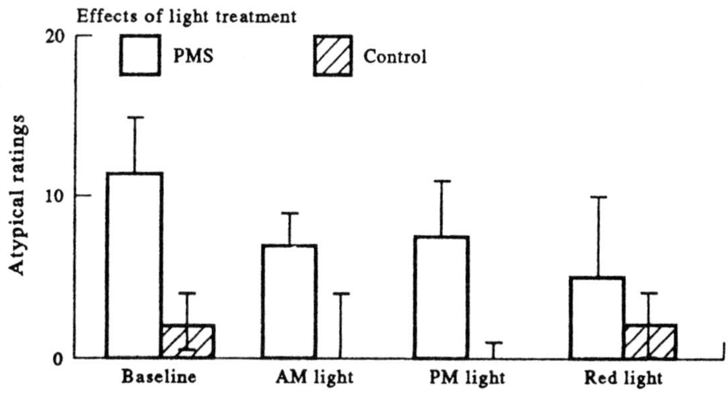

FIG. 3. Median, first and third quartiles of atypical depressive ratings in PMS ($n=19$) and normal control ($n=11$) subjects, by baseline and treatment values. Each of the light treatments (bright white a.m., bright white p.m., dim red p.m.) were significantly ($p<0.01$) different from baseline, but none of the light treatments differed significantly from each other.

Using the criteria of Terman *et al.*,[12] in which a response to light therapy is determined by a 50% reduction in Hamilton ratings to a value less than 8, 8 (39%) of patients responded to a.m. bright light, 4 (22%) responded to p.m. bright light and 5 (28%) responded to p.m. dim light.

In normal controls, Hamilton, Beck and atypical ratings showed no statistically significant effects of light treatment.

Mania ratings showed significant ($p<0.02$, Friedmans) effects of condition for normal controls (see Figure 4 results of Wilcoxin tests), but

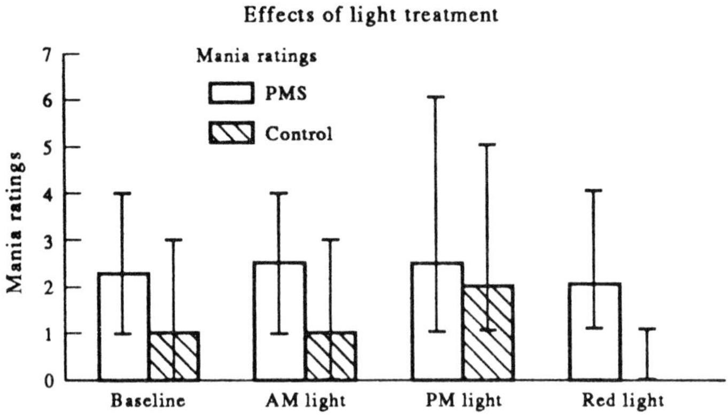

FIG. 4. Median, first and third quartiles of mania ratings in PMS ($n=19$) and normal control ($n=11$) subjects, by baseline and treatment values. In normal controls, there was a significant difference in mania ratings of comparing baseline vs. dim red p.m. light ($p<0.05$) and in bright white p.m. vs. dim red p.m. light ($p<0.05$).

not for LLPDD patients. None of the elevation of mania score ratings were considered clinically important.

Daily ratings

Only daily ratings after 4 days of light treatment were used in the analysis, because lights generally take 3–4 days to exert their therapeutic effects.[1] The daily visual analog ratings of depression ($p<0.04$), irritability ($p<0.001$), and physical symptoms ($p<0.002$) showed significant effects of light treatment (Friedman's test). Improvement in each of these symptoms was observed between baseline vs. a.m., baseline vs. p.m., and baseline vs. red light treatments ($p<0.05$, Wilcoxin). Anxiety, appetite, fatigue, and withdrawal symptoms did not show significant effects of light treatment.

Subject expectations

For LLPDD patients, the only statistically significant correlations between expectation of improvement with a.m., p.m., or red light and clinical response by observed ratings were seen in Beck ratings for bright white evening light ($p=0.015$) and in Hamilton ratings for dim red evening light ($p<0.03$).

Horne-Ostberg ratings[11]

There were no statistically significant correlations of morning—eveningness with response to a.m., p.m., or red light treatment based on Hamilton, Beck, atypical, or mania ratings.

Sleep logs

In PMS patients, there was a statistically significant difference (Friedman's test, $p < 0.02$) in amount of sleep reported by condition (baseline premenstrual months before treatment vs. treatment months after either a.m., p.m., or red light). By Wilcoxon tests, there was a statistically significant ($p < 0.01$) reduction of sleep reported during a.m. compared with red light treatment in PMS patients. There were no important differences in sleep duration during different conditions in normal control subjects.

Side effects checklist

Only PMS patients reported significant side effects from the light treatments. The most frequently reported side effect of a.m. light was increased activity (agitation) reported in 6 of 10 patients (3 reported mild symptoms, 3 moderate). Eye strain was the most frequently reported symptom from p.m. light reported in 5 of 10 patients (3 reported mild, 1 moderate, and 1 marked symptoms). Drowsiness was the most frequently reported symptom under red light, reported in 7 of 10 patients (1 mild, 5 moderate, 1 marked).

Follow-up study (see Figure 5)

No patient who participated in the follow-up study did the best on red light treatment during the initial study and, therefore, red light treatment was not used in the follow-up study. Four patients (two who chose to continue on a.m. light and two who chose to continue on p.m. light) completed at least 12 months and three patients 18 months of follow-up bright light treatment. Mean ratings for Hamilton ($p < 0.008$), Beck ($p < 0.005$), and atypical scales ($p < 0.04$) continued to show statistically significant reductions in depressive symptoms compared with baseline at 18 months after bright light treatment. Mania ratings did not differ during the months of light treatment. There were no statistically significant differences between a.m. vs. p.m. light treatment on any of the rating scales (Hamilton, Beck, atypical, mania). Furthermore, there were no statistically significant differences in Hamilton, Beck, atypical, or mania ratings when month 1–3 vs. month 4–6 were compared or when months 1–6 were compared with months 7–12, or months 13–18 in the follow-up study.

Daily ratings

The daily ratings, which after follow-up bright light treatment showed statistically significant improvement compared with baseline ratings,

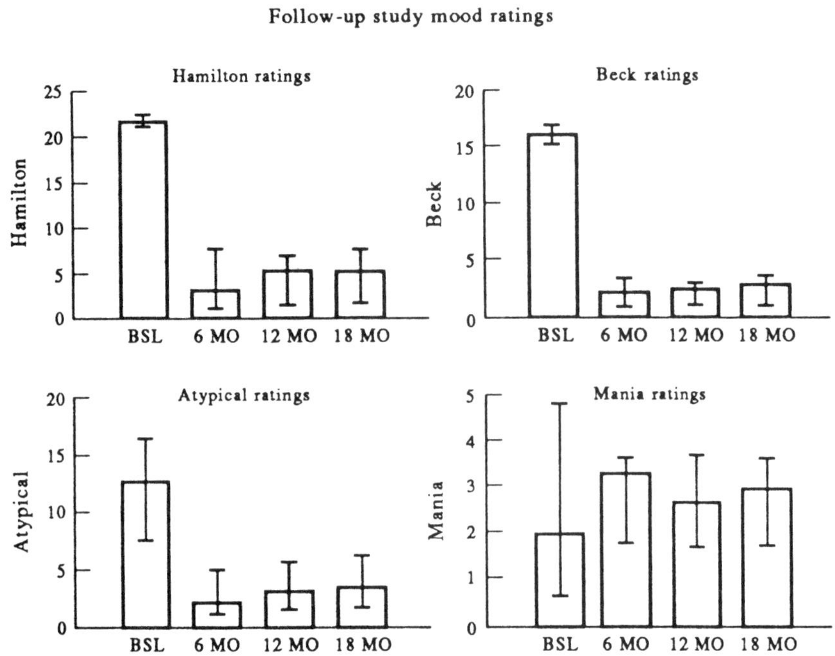

FIG. 5. Results of an 18 month follow-up study of light treatment in PMS subjects comparing baseline (before treatment) premenstrual Hamilton Beck, atypical and mania ratings with the same premenstrual ratings obtained after bright white morning, or bright white evening light treatment. Compared with baseline ratings, mean Hamilton, Beck and atypical ratings were significantly ($p<0.05$) reduced during follow up months 1–6, 7–12 and 13–18. Ratings obtained during month 1–6, vs. months 7–12 vs. months 13–18 did not differ from each other. Mania ratings did not significantly change during the months of light treatment compared with baseline.

included withdrawal ($p<0.02$), irritability ($p<0.04$) and physical discomfort ($p<0.04$).

Discussion

In summary, light treatments had no significant effects on mood ratings in normal control subjects. In LLPDD, depression ratings fell significantly from baseline after bright white morning, bright white evening, and dim red evening light treatments. The lack of discernable order of treatment effects is consistent with the observation that the three treatments were nearly equally effective. Because the active treatments did not show incremental benefits over placebo treatments, we cannot claim efficacy for light treatment in LLPDD based on the present study.

Thus, a placebo effect of light therapy cannot be ruled out, as Eastman[13] described in her discussion of the placebo response of phototherapy in seasonal affective disorder. Further controlled follow-up trials are indicated in a larger number of subjects. Such a trial should include a completely untreated, i.e. negative control group, especially to evaluate any long term effects. It is possible that we have not yet developed an appropriate placebo control and that the different light interventions may be working through different mechanisms, i.e. through altering phase or amplitude of circadian rhythms or serving as a dark pulse.

Acknowledgements

This work was supported in part by NIMH grant #R29MH42831, CRC grant #M01RR00827, NIMH CRC grant #MH30914–14. D.F. Kripke, M.D., served as a consultant.

References

1. Rosenthal N.E., Sack D.A., Carpenter C.J., Parry B.L., Mendelson W.B. and Wehr T.A. (1985) Antidepressant effects of light in seasonal affective disorder. *Am. J. Psychiatry* **142**, 163–170.
2. Kripke D.F., Klauber M.R., Risch S.C. and Gillin J.C. (1992) Controlled trial of bright light for nonseasonal major depressive disorders. *Biol. Psychiatry* **31**, 119–134.
3. Parry B.L., Rosenthal N.E., Tamarkin L. and Wehr T.A. (1987) Treatment of a patient with seasonal premenstrual syndrome. *Am. J. Psychiatry* **144**, 762–766.
4. Parry B.L., Berga S.L., Mostofi N., Sependa P.A., Kripke D.F. and Gillin J.C. (1989) Morning versus evening bright light treatment of late luteal phase dysphoric disorder. *Am. J. Psychiatry* **146**, 1215–1217.
5. Roy-Byrne P.P., Rubinow D.R., Hoban C.M., Parry B.L., Rosenthal N.E., Nurnberger J.I. and Byrnes S. (1986) Premenstrual Changes: A comparison of five populations. *Psychiatry Research* **17**, 77–85.
6. Spitzer R.L., Williams J.B.W., Gibbon M. and First M.B. (1990) *Structured Clinical Interview for DSM-III-R Patient Edition* (SCID-P, Version 1.0). American Psychiatric Press, Washington DC, 1990.
7. Hamilton M. (1967) Development of a rating scale for primary depressive illness. *Br. J. Soc. Clin. Psychol.* **6**, 278–296.
8. Beck A.T., Ward C.H., Mendelson M., Mock J.E. and Erbauch J. (1961) Inventory for measuring depression. *Arc. Gen. Psychiatry* **4**, 561–569.
9. American Psychiatric Association, Committee on Nomenclature and Statistics. *Diagnostic and Statistical Manual of Mental Disorders*. Revised 4th Ed., Washington DC, American Psychiatric Association, 1987.
10. Horne J.A. and Ostberg O. (1976) A self-assessment questionnaire to determine moringness-eveningness in human circadian rhythms. *International Journal of Chronobiology* **4**, 97–110.
11. Rosenthal N.E. and Heffernan M.M. (1986) Bulimia, carbohydrate craving and depression: A central connection? In *Nutrition and the Brain* (eds. Wurtman R.J. and Wurtman J.J), Vol. 7, pp. 139–166. Raven Press, New York.
12. Terman M., Terman J.S., Quitkin F.M., McGrath D.J., Stewart J.W. and Rafferty B. (1989) Light therapy for seasonal affective disorder. *Neuropsychopharmacology* **2**, 1–22.
13. Eastman C.I. (1990) What the placebo literature can tell us about phototherapy for SAD. *Psychopharmacol. Bull.* **26**(4), 495–504.

30

Seasonality and Seasonal Affective Disorders: Diagnostic Implications for the Outcome of Phototherapy

JOHAN BECK-FRIIS

Karolinska Institute, Department of Psychiatry, St Görans Hospital, S-11281 Stockholm

SEASONS ARE the natural consequences in nature due to the earth's rotation around the sun. Photoperiod and temperature are examples of basic elements forming the seasons. By seasonality we refer to a seasonal rhythm, which may impact any physiological or behavioral variable. Seasonal rhythms are circarhythms like circadian rhythms. We talk about circannual rhythms because seasons vary from year to year. There is also a direct correlation between circadian and seasonal/circannual rhythms, insofar as the circadian rhythm relates to the year according to the photoperiod. This means that seasons do vary from latitude to latitude.

These aspects of seasonality also have important implications for the diagnosis of seasonal affective disorders (SAD) and its subdromal form i.e. subdromal-seasonal affective disorder (S-SAD). In this paper the diagnosis of SAD, S-SAD as well as mood disorders in general will be discussed. The art of diagnosing always holds at least two aspects: reliability and validity. An ideal system of classification would be one with highest possible reliability as well as validity. This is the aim we as researchers may all aspire to.

However, in a commentary Fabrega[7] discussed social and cultural factors influencing systems of psychiatric classification and diagnosis. He concluded that the process of developing psychiatric knowledge and diagnosing instruments must be recognized as interminable, since psychiatric disorders are constituted by a dialectical interplay of biology, history, society and culture.

Aspects of reliability

The reliability of the SAD diagnosis is one example of this complexity referring to Fabrega's conclusion. As seasons vary from year to year due to the weather, from latitude to latitude due to the photoperiod and temperature and as cultural aspects of seasons vary between societies (e.g. Christmas celebration varies between societies from the Northern and Southern hemispheres according to different seasons) problems in reaching high reliability are evident. Thus, reliability primarly has to be established on the different latitudes.

In the ongoing work on developing the DSM-IV the efforts to improve the reliability have been focused on the duration aspects of the SAD diagnosis, leaving the problems of symptomatology besides. There is an agreement that the DSM-III-R criterium[2] concerning a 60 day period of symptom on/offset is too narrow and therefore is under evaluation.[3] In this way a broadening of the diagnosis is proposed, probably resulting in a higher reliability.

However, regarding the diagnoses of SAD (an episode of Bipolar or Bipolar II Disorder or Major Depressive Disorder, Recurrent), the "depressive atypical" symptoms[24] such as hypersomnia, weight gain and carbohydrate craving have been shown to be of importance in the symptomatology of SAD but do vary acccording to the seasonal type: seen in the winter-type but not in the spring-type or summer-type. Furthermore, also these "atypical" winter symptoms seem to vary with latitudes. Kjellman[15] has reported that the frequency of "atypical" symptoms at the latitude of Sweden differ from the corresponding symptoms at the latitude of, e.g. Washington DC. This seems also to be true for seasonal symptoms in healthy subjects. In their review Lacoste and Wirz-Justice showed a seasonal variation in several variables related to, e.g. climatologic factors.[17]

Latitude differences are also obvious for meteorological conditions. An example is the fact that in springtime at the latitude of Stockholm (59°N) the air is less cloudy than in the autumn. This implies that there is a disconformity in the springtime between radiancy and photoperiod at this latitude, which differs from other seasons and might differ from conditions at other latitudes.

Also differences in psychological aspects are obvious in this respect. Anniversaries[22] have been proposed to trigger some of the mood reactions, which might correspond to episodes of SAD although they by definition should be separated. Precipitating factors for an anniversary reaction have mainly been described as memories of losses by different causes, with reappearance in almost a predictable/compulsive manner. Such a memory may act disturbingly by its own but might also be charged with social factors as, e.g. seasonal unemployment, resulting in a mood dysfunction appearing in a seasonal way. The question whether such an episode should be indicated as an episode of SAD or not is relevant in this discussion.

These aspects are examples of factors being relevant to consider in discussing the reliability of the SAD-diagnosis and in evaluating the relationship between reliability and the validity of SAD.

Aspects of validity

The validity of the diagnosis of depression and SAD seems to be even more intriguing.

It is implicit that seasonality primarily refers to health and not to disease. But as we all know, diseases may also be closely linked to seasons. The question is, what is the difference between health and disease according to the rhythmic function? Or is there no difference? Regarding SAD: is there a difference in rhythmicity between normal seasonal mood swings, S-SAD and SAD?

I assume seasonal diseases, like SAD and peptic ulcers, to be stress reactions to different seasonal factors. They have to be caused either by the patient's maladaptive response to elements or functions of the season itself, like for example high or low temperature, short or long photoperiod or even psychological/social factors as, e.g. unemployment or it could be caused by the patient's difficulty in adapting to the changes from one season to another. To understand whether it may be the elements of the season or the seasonal change which are most likely to be the pathogenic factor, studying the quality of the disease symptoms seems to be relevant.

We may ask: which is the core feature of the major depression? In almost all diagnostic manuals including the DSM-III-R and the DSM-IV the depressed mood of distinct quality is the core characteristic in depressions, and the elevated mood corresponding in manic states. However, van Praag and coworkers[4,23] have discussed anxiety and aggression to be the core characteristic in some subgroups of depressions, the mood lowering being merely a derivative phenomenon from aggression, in which they hypothesize serotonin disturbances to play an essential pathogenic role. They point to something crucial, namely the role of aggression in the pathogenesis of mood disorders, an hypothesis made by Abraham[1] and discussed by many authors (e.g. Ref. 8). In this context it is of interest that the modern serotonin reuptake inhibiting agents seem to be effective not only in mood disorders but also in e.g. panic disorders, social phobia, Premenstrual Syndrome[27] and Late Luteal Phase Dysphoric Disorder,[26] suggesting serotonin to be involved in different clinical states involving anxiety and irritability as important features.

Before discussing the aggressive drive in depression in more detail I first would like to contribute to the discussion of the distinct quality of the depressed mood. The important question is: what characterizes the distinct quality of depressed mood? Available diagnostic systems including the DSM-III-R and DSM-IV seem to lack the answer. However, the question is relevant and has to be addressed.

The delimitation between mourning and depression

Clinically, two states seem often to be mistaken for a proper depressive state: mourning and the asthenic reaction. Freud[9] discussed the important similarity and distinction between mourning and melancholia and made the classical formulation: "in mourning it is the world outside which has been poor and empty, in melancholia it is the ego itself." (Melancholia is here used in the historical sense, corresponding to major depressive episode in the modern DSM sense). What has been lost within the ego making it poor and empty in melancholic patients has by many authors[13] been translated into **the loss of the self-regard** and has by them been described as the main feature in all kinds of depressions. The self-regard is defined as a concept involving the libidinal as well as the aggressive components of the self-preservation drive and including concepts such as self-esteem, self-confidence, self-evaluation and self-respect. The struggle for maintaining the self-confidence is central in personalities apt for depressive reactions; the loss of self-regard being the line of demarcation for a depressive episode proper.

Self-regard is closely connected to the concept of narcissism. A normal and healthy development of a person's narcissism is the ground for a healthy self-regard in adult life. A loss of the self-regard could be initiated in disposed individuals for different reasons.[12] It could be a loss of an important and introjected object or it could be a loss of an ideal. A loss of self-regard could also be initiated by the loss of energy, giving rise in disposed persons to an impairment of the self-preservation defence mechanisms.

Freud together with Abraham[1] also discussed the importance of the inward direction of aggressions in melancholic patients. They described the inward direction of the aggressive drive directed to a beloved but dependency evoking object, internalized into the self. What is important about this aggressive drive towards introjected objects is its oral character, a finding which may be of interest in understanding the reported relation between mood and food in winter-depressive SAD patients.[16]

With these characteristics of the self-regard including the direction of aggressive drive the border between mourning and depression should be distinct.

The delimitation between the asthenic reaction and depression

This demarcation is in clinical practice a work of the same importance and frequency, but less well described or documented. The importance of this task is exemplified by the fact that a genuine asthenic reaction, contrary to a major depressive episode, does not respond to, e.g. tricyclic antidepressants. The asthenic reaction, which seems to be one of the most common syndromes in psychiatry, is a syndrome distinct from the major depression. In the DSM-III-R it is not specifically described but seems closest to be related to the diagnosis of Dysthymia or an Organic Mood Syndrome.

According to the Swedish textbook in psychiatry by Ottosson[19] the reaction is characterized by its main symptoms, which are **fatigue and irritability**. The loss of energy seems to be the core in this syndrome and could be triggered by various wearing stress factors: (1) longstanding psychic stress; (2) physical stress after surgery, infections etc; deficiences of minerals, vitamins etc.; or secondary to endocrine diseases, cancer etc.

In many ways this state resembles a depressive one and like the state of dysthymia the asthenic reaction may transit into a major depressive episode. But what from a clinical point of view is of importance is the fact that in its genuine form the asthenic reaction has no depressive quality, insofar as it does not present a depressed mood of distinct quality. Thus, a major depressive state is separated from both the state of mourning and that of asthenia by the regulation of the self-regard and the direction of the aggressive drive. This is illustrated in Table 1, for comparison together with the equivalences for mania.

TABLE 1
Differentiation between depression and some related affective states according to the maintenance of self-regard and the main direction of the aggressive drive

	Self-regard	Main direction of aggressions
Mourning	Intact	Neutral
Depression	Impaired or lost	Inward
Asthenia	Intact	Outward
Mania	Elevated	Outward

The delimitation between SAD and S-SAD

The diagnosis of S-SAD has been proposed to be characterized by the same symptoms as in the SAD-diagnosis, excluding the symptom of depressed mood.[14] At the same time it was reported that the subsyndromal states of winter-SAD equally benefit from phototherapy.[14] Therefore, it seems likely that in the SAD-patient the core symptom sensitive to light treatment might not be the depressed mood but something else. Terman[25] has reported a study about Depression Scale Predictors in SAD patients according to the balance of atypical and typical symptoms. The "atypical" symptoms are the increased eating, the carbohydrate craving, weight gain, hypersomnia, pronounced afternoon slump and the fatigue. He found the atypical balance to be the strongest predictor for responsiveness to light-therapy in SAD patients. He also found **fatigue** to be the most universal symptom. In the Stockholm study,[15] fatigue was seen in 88% of the patients.

In many ways although not overall these "atypical" symptoms resemble those of the genuine asthenic reaction described by Ottosson. They are seen both in the SAD and in the S-SAD patients, suggesting the possibility of a complex of neurovegetative symptoms being the core symptoms of SAD and S-SAD. In the same context, it is no doubt that the depression of winter-type (true SAD winter-type) fulfills the criteria for a major depressive episode including the above mentioned criterion of a loss of self-regard and the inward direction of the aggressive drive, seen in the SAD but not in the S-SAD winter type.

I therefore suggest the S-SAD of winter-type to correspond to a classical form of asthenic reaction, which is very similar to "the seasonal anergy syndrome", reported by Lacoste and Wirz-Justice[17] and the SAD to correspond to a major depressive winter episode, which has been superimposed in patients especially disposed. Such a disposition could be the narcissistic vulnerability for a loss of the self-regard.

Discussion

The aim to reach a high degree of validity in diagnosing mental dysfunctions will always be of significance. Modern diagnostic systems as, e.g. DSM, which use symptomatology as the main diagnostic instrument, however seem to obtain diagnoses with high reliability at the expense of a lower validity. Any trial to rise the extent of validity without losing the sharpness of reliability then must be important. The definition about the loss of self-regard as the core symptom in major depression and the turning of the aggressive drive has here been proposed for improvement of the validity of the depression diagnosis by adding the classical psycho-analytical concept of depression to the modern way of diagnosing.

In this context a valid diagnosis of SAD must contribute to a proper discrimination between SAD and S-SAD. This discrimination focuses both on what separates the conditions but also on what they share in common. The "atypical" symptoms of fatigue etc. seem to be what link the conditions of winter-type and therefore is here proposed to be the main target for the effect by phototherapy. If this holds true the classification as affective disorder may be misleading. According to the winter-type therefore a more relevant nomenclature is here suggested by the use of "seasonal anergy syndrome", as described by Wark, Wirz-Justice and Graw (see Ref. 17). The effect by phototherapy on the affective component in the syndrome—i.e. the depressed mood of distinct quality—then should be interpretated as secondary to the primary effect on the energy component. This is in line with the considerations by Mueller and Davies,[18] who proposed the classification as "seasonal energy syndrome". They discussed the symptoms of anhedonia and low energy level in the fall–winter and hyperphoria, racing thoughts, and agitation in the spring–summer period as the hallmarks of this syndrome. However, they did not discuss in detail the discrimination between the energy-anhedonia symptoms and the true affective complaints; i.e. the delimination between the symptoms of SAD and S-SAD respectively nor the distinct quality of the depressed mood.

This differentation between SAS and S-SAD might be relevant also to the diagnosis of Premenstrual Syndrome (PMS), corresponding to a differentation between PMS and Late Luteal Phase Dysphoric Disorder[3] with reference to a distinct quality of depressed mood. The effect by phototherapy on these conditions described by Parry *et al.*[20] in analogue might then be interpretated as secondary to the effect on an energy/neuro-vegetative component.

Furthermore, the validation of the distinct quality of depressed mood could be of interest in understanding the mechanisms of placebo in phototherapy.

This possible placebo effect in the bright light treatment of SAD reported by Eastman at this symposium[6] might be menacing to those who regard the placebo effect in any way as contradictory to a biologic effect, seeing the placebo effect as an unspecific psychological background effect.

The issue of placebo as a non-specific, active or intentional phenomenon has been discussed by e.g. Grünbaum.[10,11] In placebo effects unconscious thoughts, emotions and memories as well as magical and unrealistic expectations, claims, yearnings etc. may be involved. The role of faith in therapy and treatment has been discussed[21] as well as the role of expectancy in behavior change.[5] It seems likely that these factors might be of a specific nature. Concerning the placebo effect in phototherapy these factors might be specific to the patients' experiences of light and dark as well as colors, temperature, social contact etc.

The psychological experiences of light and dark *per se* certainly then are factors of importance to study in this context. From the large anthropological literature we also can learn about the importance of sun-worship in old cultures, as e.g. in Babylonia and old Egypt. Also in the north as among the sames/laps worship of the sun and "returning of the sun" has been of central importance. The sames regard themselves as "the sons and daughters of the sun".

Human beings experiences of seasons also involve the process of mourning. At higher latitudes seasons are more distinct and definite and therefore also remind people more clearly about the passage of time. This implies in a certain way the necessity of being able to mourn. You have to mourn the loss of the old season before you may enter the new one. Maybe the unmistakeable sound of tender sadness and sweet melancholy, which is so typical in Nordic folklore music, reflects this need of mourning.

However, the close similarity and the distinct delimitation between mourning and a depressive state again should be kept in mind. The validation of the distinct quality of depressed mood by the use of the concept "loss of self-regard" could be one step forward in this ongoing work of differential diagnosing and its implications for SAD and the outcome of phototherapy.

Conclusion

I discuss the importance of a valid diagnosis of depression by focusing on the distinct quality of the depressed mood and by proposing a definition of this quality. This definition I propose by introducing the concept of self-regard and I stress the loss of the manifestation of self-regard as the core symptom of any depression. This core symptom is suggested to demarcate SAD from its subdromal form S-SAD.

In the further work of reaching reliable and valid diagnoses of SAD and S-SAD the use of psychological models seem to be of value and therefore are suggested to be added among the scientific tools. The use of "the loss of self-regard" in diagnosing a major depressive disorder might be one possible tool in this continued work.

References

1. Abraham K. (1911) Notes on the psycho-analytical investigation and treatment of manic-depressive insanity and allied conditions. In (Abraham) *Selected Papers*. pp. 137–156.
2. American Psychiatric Association (1987) *Diagnostic and Statistical Manual of Mental Disorders*. 3rd ed., rev. American Psychiatric Association, Washington, DC.
3. American Psychiatric Association (1993) *DSM-IV Draft Criteria*, 1 March 1993, American Psychiatric Assocation, Washington, DC.
4. Apter A., van Praag H.M., Plutchik R., Sevy S., Korn M. and Brown S.L. (1990) Interrelationships among anxiety, aggression, impulsivity, and mood: a serotonin-ergically linked cluster? *Psychiatry Res.* **32** (2), 191–199.
5. Bootzin R.C. (1985) The role of expectancy in behavior change. In *Placebo. Theory, Research, and Mechanisms* (eds. White L., Tursky B. and Schwartz G.E.), pp. 196–210. The Guilford Press, New York.
6. Eastman C. See this volume.
7. Fabrega H. Jr. (1992) Commentary: Diagnosis interminable: toward a culturally sensitive DSM-IV. *The Journal of Nervous and Mental Disease* **180**, 1, 5–7.

8. Farmer R. (1987) Hostility and deliberate self-poisoning: the role of depression. *Br. J. Psychiatry* **150**, 609–614.
9. Freud S. (1917) Mourning and melancholia. In *The Standard Edition of the Complete Psychological Works of Sigmund Freud* (ed. Strachey J.), Vol. XIV, pp. 239–260. The Hogarth Press, London.
10. Grünbaum A. (1985) Explication and implications of the placebo concept. In *Placebo. Theory, Research, and Mechanisms* (eds. White L., Tursky B. and Schwartz G.E.), pp. 9–36. The Guilford Press, New York.
11. Grünbaum A. (1986) The placebo concept in medicine and psychiatry. *Psychological Medicine* **16**, 19–38.
12. Grunberger B. (1979) *Narcissism. Psychoanalytic Essays.* International Universities Press, Inc. New York.
13. Jackson S.W. (1986) Chapter Nine: Melancholia and depression in the twentieth century. In (Jackson) *Melancholia and Depression—From Hippocratic Times to Modern Times*, pp. 188–246. Yale University Press, New Haven and London.
14. Kasper S., Rogers S.L.B., Yancey A., Schulz P.M., Skwerer R.G. and Rosenthal N.E. (1989) Phototherapy in individuals with and without subsyndromal seasonal affective disorder. *Arch. Gen. Psychiatry* **46**, 837–844.
15. Kjellman B.K. See this volume.
16. Kräuchi K., Wirz-Justice A. and Graw P. (1990) The relationship of affective state to dietary preference: winter depression and light therapy as a model. *J. Aff. Disorders* **20**, 43–53.
17. Lacoste V. and Wirz-Justice A. (1989) Seasonal variation in normal subjects: an update of variables current in depression research. In *Seasonal Affective Disorders and Phototherapy* (eds. Rosenthal N.E. and Blehar M.C.), pp. 167–229. The Guilford Press, New York.
18. Mueller P.S. and Davies R.K. (1986) Seasonal affective disorders: Seasonal energy syndrome? Letter to Ed. (with reply by Rosenthal N.E.), *Arch. Gen. Psychiatry* **43**, 188–189.
19. Ottosson J.O. (1983) *Psykiatri.* Almqvist & Wiksell, Stockholm.
20. Parry B. See this volume.
21. Plotkin W.B. (1985) A psychological approach to placebo: the role of faith in therapy and treatment. In *Placebo. Theory, Research, and Mechanisms* (eds. White L., Tursky B. and Schwartz G.E.), pp. 237–254. The Guilford Press, New York.
22. Pollock G.H. (1970) Anniversary reactions, trauma and mourning. *Psychoanal. Q.* **39**, 347–371.
23. van Praag H.M., Kahn R.S., Asnis G.M., Wetzler S., Brown S.L., Bleich A. and Korn M.L. (1987) Review. Denosologization of biological psychiatry or the specificity of 5-HT disturbances in psychiatric disorders. *J. Aff. Disorders* **13**, 1–8.
24. Rosenthal N.E., Sack D.A., Gillin J.C., Lewy A.J., Goodwin F.K., Davenport Y., Mueller P.S., Newsome D.A., and Wehr T.A. (1984) Seasonal affective disorder. A description of the syndrome and preliminary findings with light therapy. *Arch. Gen. Psychiatry* **41**, 72–80.
25. Terman M. See this volume.
26. Stone A.B., Pearlstein T.B. and Brown W.A. (1991) Fluoxetine in the treatment of late luteal phase dysphoric disorder. *J. Clin. Psychiatry* **527**, 290–293.
27. Wood S.H., Mortola J.F., Chan Y.F., Moossazadeh F. and Yen S.S.C. (1992) Treatment of premenstrual syndrome with fluoxetine: a double-blind, placebo-controlled, crossover study. *Obstetrics and Gynecology* **80**, 3, 1, 339–344.

31

Problems and Prospects for use of Bright Light as a Therapeutic Intervention

MICHAEL TERMAN

Department of Psychiatry, Columbia University and New York State Psychiatric Institute, New York, USA

THIS WENNER-GREN symposium has brought together specialists in light presentation, light reception, and light therapeutics, and has served in delineating certain ambiguities in current research and pointing towards solutions. At issue are definition of the dosing dimensions of light presentation (above and beyond listings of obvious physical parameters), and whether light therapy promotes antidepressant response (specifically, in winter seasonal affective disorder, SAD) exceeding placebo rates. If the latter issue were resolved in the negative, then the dosing issue either becomes moot or degenerates into a question of how lighting manipulations affect the rate of placebo response (Eastman *et al.*, this volume).

Stimulus specification

Radiometric measures of light intensity (e.g. irradiance in $\mu w/cm^2$) specify the stimulus in units of physical power without accounting for wavelength-specific variations in retinal sensitivity. A physiological transform of the proximal stimulus would more closely describe the central nervous system-activating event. Photometric measures of light intensity make such a transform for a "standard photopic observer", though the result for any individual may be inaccurate. As Wibom (this volume) and others (for review, see Ref. 7; for proposed standards, see Ref. 18) explain, the stimulus can be specified in terms of the properties of the illuminating source (e.g. luminance in cd/m^2) or of the amount of light reaching the target surface (e.g. illuminance in lux, as has been conventionally reported in the light therapy literature). Wibom argues for the luminance measure because, for a uniform diffuse light source, the visual sensation of brightness is invariant with the observer's spatial position.

On the other hand, the observer can actively influence proximal stimulus intensity by head and body movements, and direction of gaze relative to the light source. Some researchers (e.g. Kjellman *et al.*, this volume) have opted for whole-room illumination systems, in which case the observer's momentary position introduces less variability in retinal stimulation. Others have opted for head-mounted "visor" systems, in which the light source moves with the subject, obviating the effect of head movements.[20,25] Under standard light therapy, the patient sits before a box of lamps covered with a diffusion screen.[22] Although the patient is not ambulatory, even small changes in posture, head position and gaze can lead to large variations in proximal stimulus intensity (see, e.g. Ref. 3).

It has been generally assumed—but never demonstrated—that consistency of bright light exposure enhances the treatment effect. Treatment protocols have often instructed patients to look directly into the light at frequent intervals, to maximize stimulation. In studies in my laboratory, patients have been instructed *not* to look into the light, but to concentrate on the illuminated table surface (e.g. Ref. 22), resulting in indirect illumination of the central visual field with some direct illumination in the periphery. Despite the potential problem of variable illuminance, clinical improvement in light box studies is not perforce inferior to that in whole-room and visor studies, though direct comparisons have yet to be made. If such clinical improvement were mainly a placebo response, the issue of illuminance variability might be moot. On the other hand, stimulus variability can serve to *potentiate* retinal responses.[1]

In therapeutic situations where illuminance varies dynamically, it can be approximately estimated by affixing a photosensor to the observer's head and time-sampling the received signal. This method does not account for variations in gaze at a given head position, eye and eyelid movements, pupil size, and retinal adaptation state. But it is sensitive to the major behavioral "gate", head position, and the measure serves to demonstrate that proximal variation in the stimulus does not necessarily reduce treatment effectiveness. On the other hand, it has yet to be determined whether partial responders or non-responders to light therapy tend to receive less stimulation during treatment sessions than do responders.

We attached an Actillume™ (Ambulatory Monitoring, Inc., Ardsley, New York) photosensor on a cap, centered on the forehead, and sampled maximum illuminance within successive 1-minute intervals during 10,000 lux treatment sessions at home. Subjects had previously shown remission of winter depressive symptoms under similar treatment conditions (for criteria, see Ref. 28), and had relapsed upon withdrawal from the lights. Figure 1a shows illustrative results across days for a depressed patient treated with evening light in 45-minute sessions beginning at 22.00 hours. Minute-to-minute variability in light exposure within sessions was large, as was variability across sessions. In a retrospective review of these data, the patient was unaware that his average head position varied substantially from day to day. (Illuminance of 10,000 lux would be achieved with the head facing the light box screen at about 30 cm distance.) On some days, 10,000 lux was never achieved even briefly; on other days, there was substantial time in which illuminance far exceeded 10,000 lux. Despite the daily and local variations, the depression remitted during the treatment period.

Figure 1b reduces these data into a frequency histogram, collapsing across treatment sessions. Modal illuminance was between 6,000–8,000 lux. There was an orderly decrease to either side of the mode, with a strong skew toward high intensities. The largest fall-off in exposure occurred between 8,000–10,000 lux and 10,000–12,000 lux. In this sense, the nominal value of 10,000 lux is a valid descriptor of the dose. Average illuminance, however, was $8{,}905 \pm 4{,}823$ lux (mean $\pm$ SD), which would more accurately describe the "effective" dose. Similar measurements have been performed with four SAD patients, and the shape (skewness, kurtosis) and mean of the frequency distributions differ individually. In three of the cases, however, modal illuminance was at 8,000–10,000 lux, and in all cases, the major fall-off in illuminance occurred just above 10,000 lux.

My associates and I[27] and others[33] have hypothesized that therapeutic dose is determined by an interaction of light intensity and exposure duration. Patients have shown similar positive response to (nominal) 2,500 lux exposure at 2 hours, and 10,000 lux at 30 minutes.[22] The putative tradeoff relation between intensity and time ($I \times t = c$) has not yet been systematically tested within subjects (but see Ref. 27, Figure 1), and the shape of the $I \times t$ function as well as its limits are unknown. If integrated momentary illuminance, in conjunction with exposure duration, comprised a basic metric, we might want to specify light therapy dose as "lux · min". Under this assumption, a dosimeter recently has been developed (Photodose™, Photon Therapeutics, Mission Viejo, California) for on-line monitoring of integrated light exposure, using a head-mounted photosensor. The lights can be automatically extinguished when a specified dose, in lux · min, has been attained. The patient can be alerted when momentary light levels fall above or below a specified intensity range, guiding adjustments of head position. However, limit parameters based on therapeutic and safety considerations have yet to be determined. The dosimeter can also serve as a compliance monitor. Whether the integrated lux · min metric improves upon simpler formulations of therapeutic dose is now a feasible question for research.

The placebo sword over our heads

As has been the case with antidepressant drugs, placebo response to light therapy for SAD and other disorders (e.g. Parry, this volume) may be

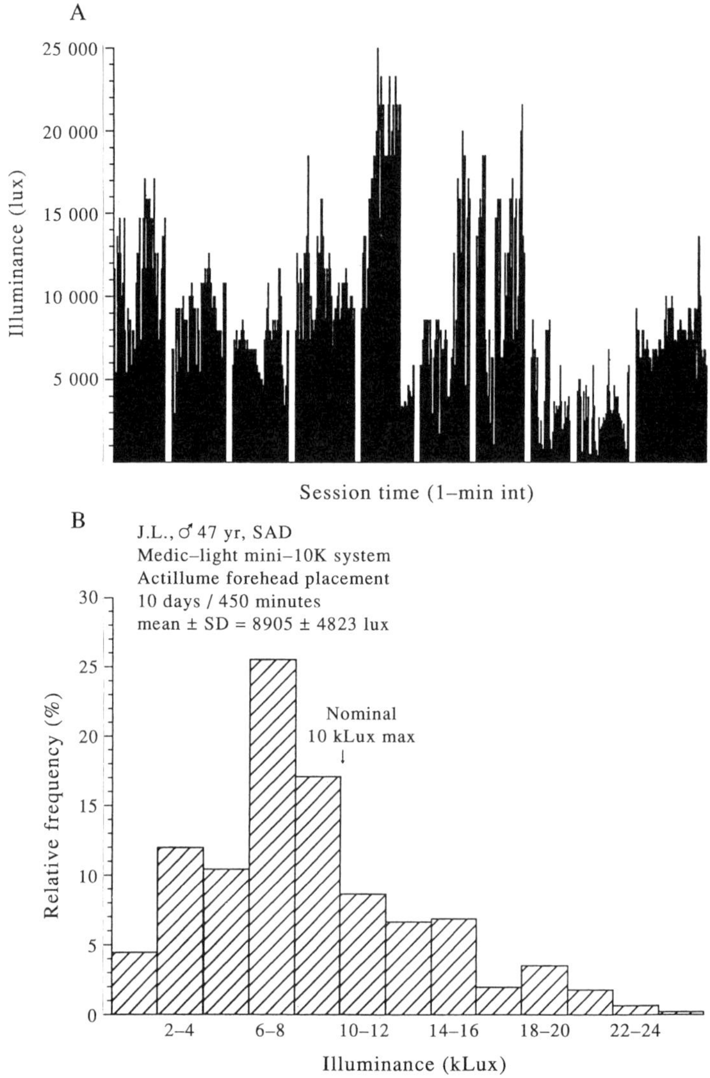

FIG. 1. (a) Momentary illuminance measured at the head of a patient during 10 successive daily light therapy sessions. Each session is separated by a space. Within-session patterns were measured as the maximum illuminance (in lux) recorded in successive 1-minute intervals. (b) Frequency histogram of illuminance received across sessions.

substantial.[4] The problem becomes pivotal in experiments that show no advantage of light beyond the placebo control. Inconsistent results, including several major trials with apparent light-placebo equivalence (e.g. Refs 6, 20), will keep this problem alive for years to come. As in psychopharmacology research, the apparent efficacy of placebo controls has varied widely. However, if there were at least a few compelling demonstrations of light-specific effectiveness, it would quell the nagging doubt that we have been chasing an illusory goal. There are indeed some promising signs.

Intensity and duration

The earliest experiments used, as controls, reduced light intensity or exposure duration, and a cross-center pooled data analysis clearly showed an advantage of higher intensity, longer duration exposure.[26] However, some of the underlying experiments could be faulted for revealing the ineffective treatment to patients, and reducing their expectations for the "inactive" controls, leading to a questionable conclusion of specific efficacy. Other experiments showed no significant difference between the putative placebo and active conditions, but the clinical response to both was poor (e.g. Ref. 8).

Time of day

A second type of control has been manipulation of time of day of treatment, stemming from the hypothesis that morning light is specifically antidepressant, and evening light is inactive or merely "activating".[11] Evening light would be ineffective because of its failure to elicit a circadian phase advance, and such light could exacerbate the depression if it elicited a phase delay. Phase advances and delays are measured in clock time, although in one elaboration of the hypothesis they should be measured against the sleep interval as a phase-angle difference.[13] Sleep phase, however, never has been directly measured in these experiments, though subjects are instructed to restrict sleep between 10 p.m. and 6 a.m. (which is a very early wake-up time for a depressed over-sleeper). The general hypothesis, relating differential phase shifts to clinical efficacy, has been sustained in several individual experiments (e.g. Ref. 12). The claim for a morning-evening light clinical differential—without actually measuring circadian phase—was strengthened by a cross-center analysis.[26] Because patients in these early studies did not expect the time of day of treatment to determine efficacy, evening light could be said to have succeeded as a placebo control.

However, when period 1 of a morning-evening crossover design was separately analyzed (e.g. Ref. 28), or parallel groups were studied without crossover (e.g. Ref. 34), the effect of time of day of treatment vanished. (For an unresolved exception to this rule, in which morning light was superior to evening light in parallel groups, see Kjellman *et al.*, this volume. In this study, subjects were not randomly assigned to morning or evening treatment groups, but rather scheduled according to convenience with respect to their habitual sleep schedule.)

Nondifferential morning-evening results raise a problem both for the phase-shift hypothesis and the claim of light-specific efficacy. It must be emphasized that for an experiment to refute the phase-shift hypothesis it is insufficient to show that morning and evening light are similar in their clinical effect. Rather, one must directly show that light-induced phase shifts are not correlated with clinical effect.

The clinical advantage of morning over evening light appears to be a byproduct of the crossover design.[28] The phenomenon becomes obvious under the "balanced", expanded crossover design[9] in which the two treatment times are alternated or are repeated within groups (Figure 2). Improvement scores do not differ for morning and evening light when (a) either treatment is given first, (b) either treatment follows itself, or (c) morning treatment follows evening treatment. However, *clinical improvement is reduced when evening treatment follows morning treatment*. Morning light appears to "prevent" evening light effect.[17] This is also true for cross-center pooled data analysis of morning-evening/evening-morning crossover studies using 2,500 lux light in 2-hour sessions (Figure 3). Thus, time of day is an important factor, though its influence is buried within a sequential dependency.

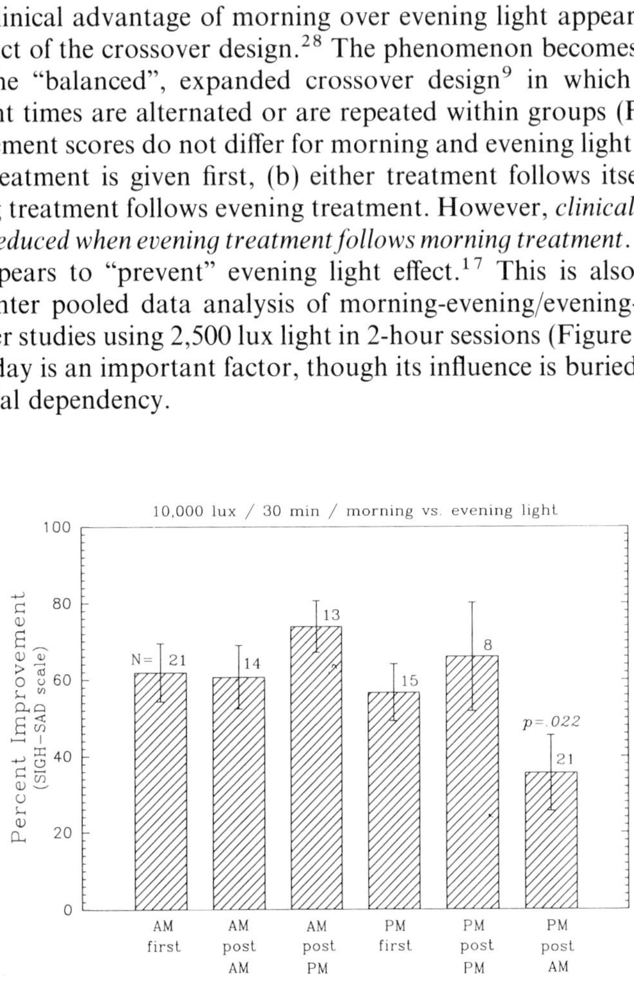

FIG. 2. Mean percent reduction ($\pm$SEM) of SIGH-SAD depression scale scores relative to baseline, in six cells of a "balanced" crossover design. Efficacy of evening light is reduced in relation to the other conditions only when it follows morning light in crossover. Statistical significance is illustrated in an independent *t*-test for groups receiving evening light as treatment 2, preceded either by morning light or evening light as treatment 1.

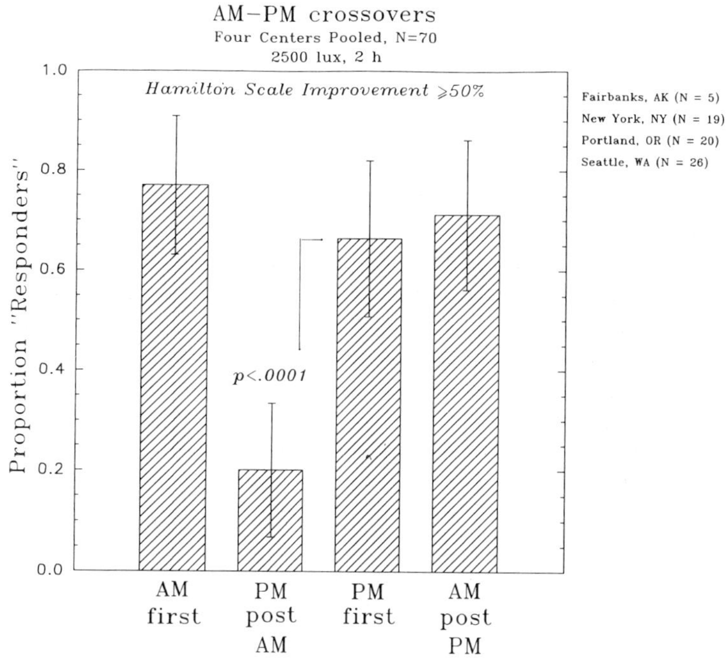

FIG. 3. Pooled cross-center analysis of morning-evening crossover studies. The outcome variable is the proportion of patients (±95% confidence interval) showing post-treatment reductions of at least 50% in Hamilton depression scores, relative to baseline. Efficacy of evening light is reduced in relation to the other conditions only when it follows morning light in crossover. Statistical significance is illustrated in a hypothesis test of proportions for independent groups receiving evening light as treatment 1, or evening light as treatment 2 following morning light as treatment 1.

The crossover experiments support a phase-shift hypothesis, though altered from Lewy and Sack's original formulation:[11] *light therapy is effective at any time of day as long as it does not induce a circadian phase delay*. Circadian phase advances to morning light do not mediate the antidepressant effect. Rather, phase delays to evening light weaken the treatment effect, or render it ineffective or even depressogenic. When evening light is given as initial treatment, it may not cause a phase delay, because it falls on or near the dead zone of the phase responsive curve.[16] Morning light, however, elicits a phase advance, making the patient vulnerable to phase delays when evening light follows morning light in a crossover.

This altered hypothesis is a current focus of research by my group, using the dim light meletonin onset (DLMO)[10] as the marker for circadian phase. However, it must be admitted that the experiment has already been done, if not analyzed in these terms. In their recent expansion on earlier findings, Sack and colleagues continue to report the relative superiority of morning light—accompanied by circadian phase advances—based on data pooled across the morning-evening crossover.[21] The DLMO was first assessed at "pre-baseline", prior to experimental interventions. Before administering light, subjects adjusted their sleep interval to fall between 10 p.m. and 6 a.m. ("baseline"). For the treatment periods, subjects were given morning or evening light treatment in a counterbalanced crossover.

In a reanalysis of Lewy and Sack's data for 24 patients, J. S. Terman, L. Amira and I teased apart HAM-D and DLMO effects for the two periods of crossover (Figure 4), with the following results and conclusions.

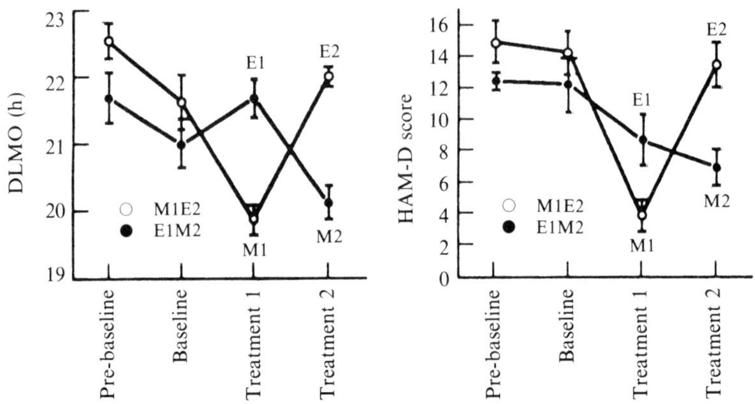

FIG. 4. Crossover analysis of dim light melatonin onset (DMLO) phase and Hamilton (HAM-D) score (mean ± SEM) for two groups of subjects who received 2,500 lux morning light treatment (6–8 a.m.) before evening light treatment (7–9 p.m. or 8–10 p.m.), and vice versa. Starting with baseline, sleep was permitted only between 10 p.m. and 6 a.m. Data were pooled from Ref. 12 ($N=16$) and Ref. 21 ($N=8$).

1. Pre-baseline to baseline change

A 2-way analysis of variance for the DLMO in pre-baseline and baseline conditions indicated a significant effect of time ($F=91.0$, $p=0.0002$), with an overall mean phase advance from pre-baseline to baseline of 0.75 hour. There was no significant group difference (morning or evening light first) or group × time interaction. A corresponding analysis of HAM-D scores showed a significant group difference ($F=13.7$, $p=0.034$), with morning-first subjects slightly more depressed than evening-first subjects at pre-baseline (14.8 vs. 12.4, $p=0.015$ by 2-tailed t-test; multiple-test correction, $p=0.03$), but not after sleep adjustment at baseline. Pointedly, HAM-D scores did not improve in parallel to DLMO phase advances from pre-baseline to baseline. Thus, the sleep adjustment served to advance a circadian rhythm, perhaps by exposing over-sleepers to dim ambient room light earlier in the morning. Sleep might even have acted as a zeitgeber in its own right (cf., Ref. 5). However, this phase advance was not therapeutic. If phase shifts of the DLMO and sleep interval were equal, there would be no net change in phase-angle difference. To know for sure, the sleep interval would have to be measured.

2. Changes in clinical state and circadian phase, with light treatment during morning-evening crossover

DLMO and HAM-D data were then compared for the two periods of the crossover. Because the hypothesis predicts that failure of response to light is caused by a circadian phase delay, DLMO data were analyzed as phase *shifts* from baseline to treatment period 1, or treatment period 1-to-2, rather than absolute values of the DLMO. HAM-D data were analyzed as absolute values. The DLMO showed significant effects of order of treatment ($F=10.5$, $p=0.01$), time of day of treatment ($F=69.0$, $p=0.0001$), and the interaction of order × time of day ($F=6.3$, $p=0.033$). The DLMO showed a large mean phase advance to morning light whether in treatment period 1 or 2 (1.77 hours vs 1.57 hours, *NS*). There was a small phase delay to evening light in treatment 1, but a large delay in treatment 2, following morning light (-0.69 hour vs. -2.12 hour, $p=0.0001$; multiple-test correction, $p=0.0002$). HAM-D scores showed significant effects of time of day of treatment ($F=16.9$, $p=0.003$) and the interaction of order of treatment × time of day ($F=5.6$, $p=0.042$), but not order of treatment *per se* ($F=0.577$, *NS*). The group receiving evening followed by morning light (E1M2) showed no significant difference in mean HAM-D score (8.64 vs. 6.86, *NS*); by contrast, the group receiving morning followed by evening light (M1E2) showed a strong disadvantage of evening light (3.8 vs. 13.4, $P=0.0001$; multiple-test correction, $p=0.0003$). *Thus, within-group evidence for a morning-evening light differential stems only from subjects receiving the M1E2 sequence.* This group is also the only one to show a major DLMO phase delay to evening light.

3. Sources of period 1 change

Nonetheless, subjects receiving evening light as treatment 1 appeared to show higher HAM-D post-treatment scores (poorer clinical response) than those receiving morning light, in an independent-groups comparison (8.64 vs. 3.8, $p=0.02$; multiple-test correction, $p=0.06$). Even if this trend were found to be reliable it would be important to recognize that the response to evening light was still an improvement relative to baseline (12.1 vs. 8.64, $p=0.003$). The DLMO phase shift to baseline sleep adjustments may have been responsible for the morning-evening trend differences of treatment 1. The HAM-D score is correlated with the DLMO phase shift (period 1, $r = -0.425$; period 2, $r = -0.464$; both $p < 0.05$). Since the baseline sleep adjustment produced a mean DLMO phase advance (0.75 hours), one would expect DLMO phase delays to evening light to exceed those which might be measured without preparatory sleep adjustment. My own studies, which have not imposed a 6 a.m. wake-up time, have shown no statistical difference in clinical response to morning and evening light in period 1.

A corollary of the original phase-shift hypothesis still seems viable: as fall and winter set in, SAD patients show drift toward later circadian phase, perhaps tracking the delay in outdoor dawn illumination (cf. Illnerova *et al.*, this volume). Following this spontaneous, seasonal depressogenic phase delay, artificial evening light loses its phase-delaying capacity because the delaying limb of the phase response curve has also drifted toward a later nighttime hour. Under this delayed condition, artificial morning and evening light are both antidepressant. Morning light, however, exerts a phase-advancing effect that reverses the naturally-occurring, seasonally delayed phase. Following morning light treatment, phase delays to evening light are enabled, and evening light loses its antidepressant quality. Given a situation under which evening light fails therapeutically, the crossover experiment provides a type of placebo control—albeit indirect—supporting the hypothesis that light, appropriately delivered, is a specific antidepressant.

Dummy treatment with a "behavioral placebo"

Eastman *et al.* (this volume) have used a deactivated negative ion generator as a placebo control. The patient sits either at a light box or an ion generator for equivalent treatment durations, thereby equating the daily behavioral "investment" in the procedure, which might affect expectations and placebo rate. In contrast, the placebo pill—in studies showing poorer response than to light (e.g. Lingjaerde *et al.*, this volume)—requires less behavioral investment in the therapy, and thus can be faulted for diminishing expectations. Eastman's study, which is still in progress, has thus far failed to find a differential benefit of light therapy over placebo in a 4-week treatment protocol. However, as early as the first week of treatment, morning light showed a substantial size of effect relative to placebo ($d = 0.9$; Cohen, 1988). As the weeks progressed, patients tended to improve regardless of treatment mode. It seems quite possible that larger sample size will eventually reveal a statistically significant overall advantage of light (whether for morning, or both morning and evening), with a size of effect similar to that of many large-sample placebo-controlled antidepressant drug trials.

Perhaps "brief treatment probes" (e.g. 7–10 days) would be likely to reveal such differences more expeditiously than longer duration treatment. On the one hand, light therapy studies have been criticized for reliance on brief treatment.[26] On the other hand, given the episodic nature of SAD, spontaneous remissions predictably increase as a function of treatment duration. In one sample of 28 SAD patients, spontaneous remissions, reported retrospectively, began shortly after the winter solstice, in January (4%). Twenty-two per cent of patients reported remission in February, 59% in March, and 13% in April (Ref. 27, Table 1). Thus, a placebo-controlled study showing continuous group improvement over the winter months—such as Eastman's—might be expected to show improvement in both experimental and control groups.

The end-of-season confound could be eliminated by withdrawing patients from successful treatment and retaining only those showing relapse for group comparisons. Fairly few patients might be retained, however, after a full month of treatment which terminates in February or March. A further complication is the probability of relapse even earlier in the season. Some centers have routinely observed relapse within a week or two of withdrawal (e.g. Bethesda,[19] New York City[22]) while others (e.g. Basel,[32] Groningen,[14] Oslo [Lingjaerde *et al.*, this volume], Rochester[35]) have observed little or no relapse. Indeed, the Groningen group has proposed a model in which 5 days of light therapy serves to "prevent" recurrence of symptoms across the entire winter season.[14,15] In an attempted replication of their paradigm in New York, however, only about 20% of patients resisted relapse across 8 weeks of withdrawal.[23] Whether the propensity for relapse is a function of particular treatment parameters, subjects' expectations, population differences, or other factors, is a question of increasing urgency.

Selective predictors of clinical response as prima facie evidence against a placebo explanation

If there are aspects of symptomatology that correlate with treatment outcome in the absence of differential expectations, a placebo explanation would be unlikely even in the absence of placebo control. We examined the two component depression scales of the SIGH-SAD[31]—the 21 items of the HAM-D scale and eight supplementary items characteristic of atypical depression (ATYP)—in a large set of patients while depressed, before they received bright light therapy.[30] For inclusion, the SIGH-SAD total score was at least 20, with at least 10 points on the HAM-D and 5 points on the ATYP subscales. Seventy-one patients were judged to be "responders" by the following strict criteria: SIGH-SAD score reduced by at least 50% from baseline, HAM-D and ATYP scores both 7 or below,[28] and Clinical

Global Impressions—Improvement (CGI-I) scores of 1 ("very much improved") or 2 ("much improved"). Fifteen patients were judged to be non-responders by the following criteria: SIGH-SAD scores increased, unchanged, or reduced by less than 25%, meeting original baseline inclusion criteria, with CGI-I score of 4 (unchanged) or higher (worsened). Additionally, although the analysis was restricted to initial treatment response, patients were not classified as non-responders if they responded even partially in *any* subsequent treatment period.

Figure 5 presents a scattergram of HAM-D vs. ATYP scores at baseline, with responders distinguished from non-responders. Positive response was found for every case with a HAM-D baseline score less than 15, regardless of ATYP score. Furthermore, there were no responders with atypical balance scores [(ATYP/SIGH-SAD) × 100] less than 29%. The higher the atypical balance at baseline, the greater the percent improvement in the post-treatment SIGH-SAD score ($r = 0.26$, $p < 0.05$). Expectation ratings, however, were not correlated with treatment outcome for either group. Thus, details of the symptom profile—above and beyond the global diagnosis of major depressive disorder with seasonal pattern—may prove useful in predicting treatment outcome. The definition of SAD might now be refined, as a light-responsive syndrome characterized by the presence of atypical symptoms (see also, Beck-Friis, this volume).

We examined the frequency and severity of all symptoms on the SIGH-SAD, and the derived scale scores, in an ordered array of effect size (d) between responders and non-responders at baseline (Figure 6). Positive effect size can be considered predictive of response to light therapy, with the mean item score of responders exceeding that of non-responders. In contrast, negative effect size can be considered predictive of non-response.

The strongest positive predictor is the atypical balance score while the strongest negative predictor is overall HAM-D severity—both show more than 1 SD unit difference between responders and non-responders, which is a very large size of effect. By comparison, total symptom severity, as measured by the overall SIGH-SAD score, is a weak negative predictor ($d = -0.42$). So is depressed mood itself ($d = -0.24$)—which, by definition, is seen in all patients at baseline, responders and non-responders alike. *The strong positive predictors of light response ($d > 0.5$) are all atypical symptoms*, with hypersomnia, afternoon or evening slump, and reverse diurnal variation ("pre-sleep worse") being most prominent. The appetite increase and weight gain symptoms follow, with decreasing effect size. The severity of fatigability at baseline (a symptom seen in 98% of patients), does not predict response to light therapy ($d = 0.03$) even though responders to light treatment always show increased energy. Social withdrawal—which was added to the supplementary ATYP scale even though it is not a neurovegetative symptom *per se*—falls away from the atypical cluster and is a weakly negative predictor ($d = -0.39$); this item

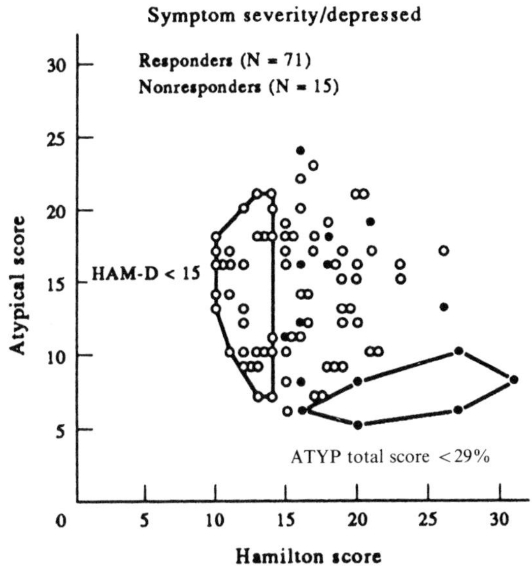

FIG. 5. Correlation of HAM-D and supplementary atypical (ATYP) scale scores at baseline, for 86 patients who met entry criteria for our light treatment studies: combined scale total (SIGH-SAD score) of at least 20, HAM-D score of at least 10, and ATYP score of at least 5. Open circles denote patients who met remission criteria following light therapy. Closed circles are for non-responders. Patients who showed intermediate response were omitted from this analysis. Two distinct clusters of data are drawn, for subgroups of patients who showed: (a) HAM-D scores less than 15, in which case the response rate was 100%; and (b) atypical balance scores less than 29%, in which case the response rate was 0%.

might well be excised from the ATYP scale. The strongest negative predictors of response to light treatment form a distinctly "melancholic" group, including retardation, suicidality, typical diurnal variation ("post awakening worse"), somatic anxiety and late insomnia ("early awakening"). Depersonalization/derealization—which stands apart from the core depressive symptoms of the HAM-D scale—also falls squarely in the middle of the negative predictor group ($d = -0.76$).

With such a distinction between positive and negative predictors at baseline, the clinician has a data-based tool to begin answering the inevitable question from patients, "How likely is it that light therapy will work for me?", and perhaps to counsel alternate treatment modes when the atypical balance is very low. We could not make such assertions if light were "merely" a placebo, unless there were an obscure mechanism linking clinical outcome to high expectations given an atypical profile, but no such correlation given a melancholic profile. The data show no such differences, and expectation ratings were not significantly correlated with SIGH-SAD, HAM-D, ATYP or CGI severity scores at baseline.

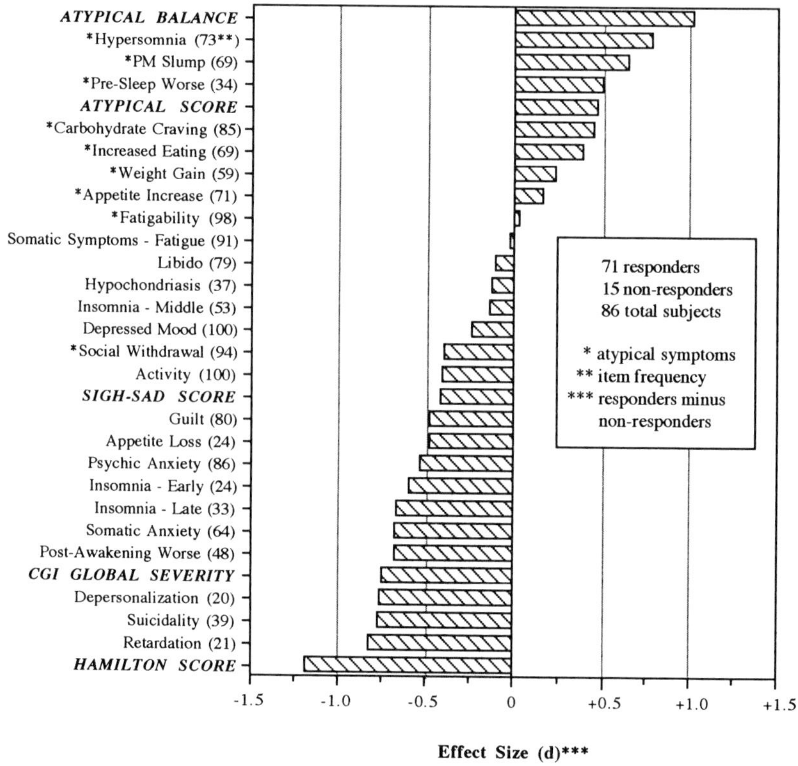

FIG. 6. Responders and non-responders to light treatment are compared in terms of effect sizes of scale and item-within-scale scores at baseline. Results are rank-ordered from the greatest positive effect size to the greatest negative effect size. The frequency of occurrence of each symptom is shown in parenthesis, with symptoms showing less than 20% frequency omitted from the analysis. Symptom frequency is not correlated with effect size—if nearly all patients show a symptom (e.g. fatigability), it cannot be a predictor of treatment response. However, these same symptoms show major improvement under successful light therapy, with little or no change among the non-responders.

Acknowledgements

This work was sponsored in part by Grant MH43219 from the US National Institute of Mental Health. I thank Jiuan Su Terman, Ph.D., for collaboration in all its aspects; Leora Amira, Ph.D., and Donald C. Ross, Ph.D., for contributions to data analysis; Charmane Eastman, Ph.D., Alfred J. Lewy, Ph.D., M.D., and Robert L. Sack, M.D., for sharing data; and William Gruen, M.S., of Ambulatory Monitoring, Inc., for loan of the Actillume™ apparatus. (The author is co-inventor of the 10,000 lux light therapy apparatus, with rights assigned to the Research Foundation for Mental Hygiene, Inc., New York Psychiatric Institute Division.)

References

1. Brown J.L. (1965) Flicker and intermittent stimulation. In: *Vision and Visual Perception* (ed. C.H. Graham), pp. 251–320. Wiley, New York.
2. Cohen J. (1988) *Statistical Power Analysis for the Behavioral Science*. Erlbaum, Hillsdale, NJ.
3. Dawson D. and Cambell S.S. (1990) Bright light treatment: Are we keeping our subjects in the dark? *Sleep* **13**, 267–271.
4. Eastman C.I. (1990) What the placebo literature can tell us about phototherapy for SAD. *Psychopharm. Bull.* **26**, 495–504.
5. Eastman C.I. (1990) Entrainment of circadian rhythms with 26-h bright light and sleep-wake schedules. *Am. J. Physiol.* **259**, R1189–R1197.
6. Eastman C.I., Lahmeyer H.W., Watell L.G., Good G.D. and Young M.A. (1992) A placebo-controlled trial of light treatment for winter depression. *J. Affect. Disord.* **26**, 211–222.
7. Hughes P.C., Bickford E.W. and Goldstein I.S. (1987) Optical radiation. *Pineal Res. Rev.* **5**, 1–67.
8. Isaacs C., Stainer D.S., Sensky T.E., Moor S. and Thompson C. (1988) Phototherapy and its mechanisms of action in seasonal affective disorder. *J. Affect. Disord.* **14**, 13–19.
9. Laska E., Meisner M. and Kushner H.B. (1983) Optimal crossover designs in the presence of carryover effects. *Biometrics* **39**, 1087–1091.
10. Lewy A.J., Sak R.L. and Singer C.M. (1985) Immediate and delayed effects of bright light on human melatonin production: Shifting "dawn" and "dusk" shifts the dim light melatonin onset (DMLO). *Ann. N.Y. Acad. Sci.* **453**, 253–259.
11. Lewy A.J. and Sack R.L. (1986) Minireview: light therapy and psychiatry. *Proc. Soc. Exp. Biol. Med.* **183**, 11–18.
12. Lewy A.J., Sack R.L., Miller L.S and Hoban T.M. (1987) Antidepressant and circadian phase-shifting effects of light. *Science* **235**, 352–354.
13. Lewy A.J., Sack R.L., Singer C.M., White D.M. and Hoban T.M. (1988) Winter depression and the phase-shift hypothesis for bright light's therapeutic effects: History, theory, and experimental evidence. *J. Biol. Rhythms* **3**, 121–134.
14. Meesters Y., Lambers P.A., Jansen J.H.C., Bouhuys A.L., Beersma D.G.M. and van den Hoofdakker R.H. (1991) Can winter depression be prevented by light treatment? *J. Aff. Disorders* **23**, 75–79.
15. Meesters Y., Jansen J.H.C., Beersma D.G.M., Bouhuys A.L. and van den Hoofdakker R.H. (1992) Early light treatment and the prevention of winter depression. *Soc. Light Treatment Biol. Rhythms Abst.* **4**, 28.
16. Minors D., Waterhouse J.M. and Wirz-Justice A. (1991) A human phase-response curve to light. *Neurosci. Lett.* **133**, 36–40.
17. Rafferty B., Terman M., Terman J.S. and Remé C.E. (1990) Does morning light prevent evening light effect? A statistical model for morning/evening crossover studies. *Soc. Light Treatment Biol. Rhythms Abst.* **2**, 18.
18. Remé C.E., Menozzi M. and Krueger H. (1991) Standards for specifying parameters of bright light stimulation for therapeutic applications: A proposal. *Light Treatment Biol. Rhythms* **3**, 40–43.
19. Rosenthal N.E., Sack D.A., Carpenter C.J., Parry B.L., Mendelson W.B. and Wehr T.A. (1985) Antidepressant effects of light in seasonal affective disorder. *Amer. J. Psychiatry* **142**, 163–170.
20. Rosenthal N.E., Moul D.E., Hellekson C.J., Oren D.A., Frank A., Brainard G.C., Murray M.G. and Wehr T.A. (1993) A multicenter study of the light visor for seasonal affective disorder: No difference in efficacy found between two different intensities. *Neuropsychopharm.* **8**, 151–160.
21. Sack R.L., Lewy A.J., White D.M., Singer C.M., Fireman M.J. and Vandiver R. (1990) Morning vs evening light treatment for winter depression: Evidence that the therapeutic effects of light are mediated by circadian phase shifts. *Arch. Gen. Psychiatry* **47**, 343–351.
22. Terman J.S., Terman M., Schlager D., Rafferty B., Rosofsky M., Link M.J., Gallin P.F. and Quitkin F.M. (1990). Efficacy of brief, intense light exposure for treatment of winter depression. *Psychopharm. Bull.* **26**, 3–11.

23. Terman J.S. and Terman M. (1993) Relapse rate following light treatment of winter depression. *Soc. Light Treatment Biol. Rhythms Abst.* **5**, 48.
24. Terman M. (1988) On the question of mechanism in phototherapy for seasonal affective disorder: Considerations of clinical efficacy and epidemiology. *J. Biol. Rhythms* **3**, 155–172.
25. Terman M. (1991) Clinical efficacy of the light visor, and its broader implications. *Light Treatment Biol. Rhythms* **3**, 37–40.
26. Terman M., Terman J.S., Quitkin F.M., McGrath P.J., Stewart J.W. and Rafferty B. (1989) Light therapy for seasonal affective disorder: A review of efficacy. *Neuropsychopharm.* **2**, 1–22.
27. Terman M., Terman J.S., Quitkin F.M., Stewart J.W., McGrath P.J., Nunes E.V., Wager S.G. and Tricamo E. (1989). Dosing dimensions of light therapy: Duration and time of day. In *Seasonal Affective Disorder* (eds. Silverstone T. and Thompson C.), pp. 187–204. Clinical Neuroscience Publishers, London.
28. Terman M., Terman J.S. and Rafferty B. (1990) Experimental design and measures of success in the treatment of winter depression by bright light. *Psychopharm. Bull.* **26**, 505–510.
29. Terman M., Williams J.B.W. and Terman J.S. (1991) Light therapy for winter depression: A clinician's guide. In *Innovations in Clinical Practice*, Vol. 10 (ed. P.A. Keller and S.R. Heyman), pp. 179–221. Professional Resource Exchange, Sarasota FL.
30. Terman M., Terman J.S. and Amira L. (1992) Light-refractory and light-responsive SAD patients: depression scale predictors. *Soc. Light Treatment Biol. Rhythms Abst.* **4**, 18.
31. Williams J.B.W., Link M.J., Rosenthal N.E., Amira L. and Terman M. (1992) *Structured Interview Guide for the Hamilton Depression Rating Scale—Seasonal Affective Disorder Version* (*SIGH-SAD*). Rev. ed. New York State Psychiatric Institute, New York.
32. Wirz-Justice A., Bucheli C., Graw P., Kielholz P., Fisch H.-U. and Woggon B. (1986) Light treatment of seasonsal affective disorder in Switzerland. *Acta Psychiatrica Scand.* **74**, 193–204.
33. Wirz-Justice A., Schmid A.C., Graw P., Kräuchi K., Kielholz P., Pöldinger W., Fisch H.-U. and Buddeberg C. (1987) Dose relationships of morning bright white light in seasonal affective disorder (SAD). *Experientia* **43**, 574–576.
34. Wirz-Justice A., Graw P., Kräuchi K., Gisin B., Jochum A., Arendt J., Fisch H.-U., Buddeberg C. and Pöldinger W. (1993) Light therapy in seasonal affective disorder is independent of time of day or circadian phase. *Arch. Gen. Psychiatry*, in press.
35. Yerevanian B.I., Anderson J.L., Grota L.J. and Bray M. (1986) Effects of bright incandescent light on seasonal and nonseasonal major depressive disorder. *Psychiatr. Res.* **18**, 355–364.

32

Melatonin: Historical Aspects*

AARON B. LERNER

Department of Dermatology, Yale University School of Medicine, New Haven, Connecticut 06510, USA

ONLY A handful of people were involved in the isolation of melatonin from the pineal gland. Dr Yoshiyata Takahashi and I initiated the work at Yale in 1955, and it was completed by Dr James Case and me in 1959. Assays of the pineal extracts for activity were carried out by Dr Takahashi, Dr Mori or Dr M. Ruth Wright. Biologic studies were done with Dr Wataru Mori and Jack Barchas. Fresh frozen bovine pineal glands were supplied to us by the thousands through Dr Joseph Fisher of the Armour Laboratories in Kankakee, Illinois. Assistance with the chemistry came from Dr Richard V. Heinzelman of the Upjohn Company in Kalamazoo, Michigan, and the determination of the structure of the methyl-ester of 5-methoxyindole acetic acid was done with Dr Klaus Biemann at the Massachusetts Institute of Technology.

In our research on normal and malignant pigment cells, we were studying the action of hormonal and neural factors on the dispersion and aggregation of melanosomes in frog melanocytes. It was 1955 and we had just come to Yale from the University of Oregon Medical School. At Oregon we had isolated a melanocyte stimulating hormone (MSH)—the dispersing factor—from the pituitary gland. Even though the MSH work was going to be continued at Yale, we wanted to know whether there were any neural aggregating factors in addition to adrenaline, noradrenaline, and acetylcholine. Dr Takahashi found a single reference to a paper by McCord and Allen in *The Journal of Experimental Zoology*, in 1917, in which the authors described the lightening—that is, aggregating action—of extracts from the pineal gland. In the pituitary work, there were numerous papers going back to 1914. But for the pineal gland there was only a single paper. Could the agent in the pineal gland be adrenaline,

* This paper is modified in parts from an essay written by the author for the European Pineal Study Group Newsletter, number 4, September 1980.

noradrenaline, or acetylcholine? It was easy for us to show that the chemical properties of the pineal factor did not match any of those three compounds. Was it a known factor or something not yet described? Dr Takahashi and I decided to make a determined effort to find out.

To test for the aggregation factor, which we later called melatonin, Dr Takahashi modified the assay that Dr. Kazuo Shizume and I had developed for MSH. In that *in vitro* assay, we measured by reflectance the darkening produced by MSH on isolated pieces of frog skin. The new assay for melatonin turned out to be more tricky because we had to first use MSH to darken the skin in order to later detect the presence of a lightening substance. Takahashi performed the bioassays, and I did all the extractions. We did not know that it would take 4 years of hard work and 250,000 pineal glands to finish the project.

At the end of 2 years, we had a reliable assay, and we knew many of the properties of melatonin. We could follow the activity through silicic acid column chromatography, and we were able to measure the agent's fluorescence spectrum. At that time, Dr Takahashi had to return to Japan. He was afraid that the people in Japan would forget about him and that he would not be able to find a job. The assays were taken over by Drs Wataru Mori and M. Ruth Wright. For help with the isolation, Dr James D. Case, whom I had known as a student at the University of Oregon Medical School, came to learn clinical dermatology and, at the same time, carry out biochemical investigations. He brought with him tremendous ability and a great capacity for work. His untimely death a year after he was made head of the section of dermatology at the University of Washington Medical School in Seattle was an enormous loss from which none of us will ever recover.

In about a year's time, Jim Case and I were able to show that melatonin was an indole derivative and that on electrophoresis it had no charge. We were also able to isolate, in approximately 10 times greater quantity than melatonin, a substance that had the same fluorescence and ultraviolet light absorption spectra and also gave the same color tests as melatonin. Yet, the compound was inactive biologically and behaved as an organic acid on electrophoresis. Since we could never get as much as 1 mg of melatonin to work with, it seemed that we should hold off working on its structure and concentrate on the biologically inactive acid.

I would take the 5 a.m. train from New Haven to Boston, carrying samples to Dr Klaus Biemann at M.I.T. Dr Biemann was a then relatively unknown young man doing research on the structure of organic compounds by using mass spectrometry to determine molecular weights of fragments obtained from subjecting the compounds to an electric discharge. Now Dr Biemann is world famous for his research. The biologically inactive organic acid turned out to be non-volatile. For this reason, we made the methyl-ester. With only a milligram or so of the methyl-ester, we got an accurate molecular weight determination of the entire molecule as well as the fragments. It was obvious that the structure of the molecule had to be 5-methoxy-indole acetic acid. There was no alternative. This compound was then made by the people at Upjohn, and they confirmed the structure that we had derived from mass spectrometry.

We still were unable to obtain enough pure melatonin to carry out experiments on structure. We could not even get an elementary analysis for carbon, hydrogen, nitrogen, and oxygen. The material was not volatile enough to use in the mass spectrometer. We realized that well over a million pineal glands would have to be collected and worked up in order to isolate as much as 1 mg of melatonin. At this point, because of our inability to get any sizeable quantities of melatonin, the work became very discouraging. Jim Case and I decided that if we could not find a good lead within a month's time, it would be best to drop the isolation project, go on to something else, and wait until one of the pharmaceutical houses would carry out the extractions of the millions of glands required to solve the problem. About a week before our deadline, which represented the time I was to go with my family to Cape Cod, I was discussing the metabolism of serotonin with another associate, Dr Joseph McGuire. We were reviewing the work done at the Cleveland Clinic which showed that the major product of serotonin metabolism was *N*-acetylserotonin. As the discussions progressed, I interrupted Joe McGuire and said, "I know what the structure of melatonin is. It undoubtedly is the methoxy derivative of *N*-acetylserotonin." Within an hour Jim Case and I looked over the numerous indole derivatives we had collected for reference and found that we had 5-ethoxytryptamine. We acetylated the compound and gave it to Ruth Wright to assay the next morning. A couple of hours later, Ruth came in and said, "This is it!" Never had she tested anything as potent as the new compound. We know we were on the right track and now had to make the methoxy instead of the ethoxy derivative. Jim Case did this a week later when I was at Cape Cod and everything checked. Even though we were not able to determine the structure of melatonin by a direct attack, perserverance, and a lucky guess paid off. The problem was solved.

Before we found the structure of melatonin, everyone said that biologic amines, such as adrenaline, noradrenaline, histamine, all the antihistamines and many of the tranquilizers, had to have a nitrogen carrying a positive charge. When acetylated, the charge was gone and biologic activity lost. But in pigmentation, removing the charge on the nitrogen greatly enhances the potency of the substance. Melatonin was 100,000 times more potent than noradrenaline or acetylcholine in lightening frog skin. Also, melatonin and 5-methoxy-indole acetic acid turned out to be the first methoxy-indoles isolated from mammalian tissue.

An important reason for the success of the melatonin project was the good fortune I had to work with people who were unusually capable. Drs Takahashi, Mori, Case, Barchas, and Wright were exceptional individuals. There were no technicians other than our excellent dishwasher, Mrs Kate Lehmann.

Biographical sketches

YOSHIYATA TAKAHASHI, M.D. After he was graduated from the University of Tokyo Medical School, Dr Takahashi did a residency in internal medicine before coming to the United States. We worked together for 1 year at Oregon on the mechanism of pigment granule movement and 2 years at Yale on the melatonin project. He became Chairman of the Department of Internal Medicine at Gifu University and then President of the hospital in that city. He was the leading hepatologist in Japan.

WATARU MORI, M.D. Dr Mori was graduated from the University of Tokyo Medical School and took a residency in pathology. We worked together at Yale on the biology of melatonin and the assay for almost 3 years. He returned to Japan as a member of the faculty in the Department of Pathology. He later became Chairman of that Department and then Dean of the Medical School. He recently completed a 4-year term as President of Tokyo University. He is a foreign associate of the Institute of Medicine of the National Academy of Sciences.

JAMES CASE, M.D. James Case received his medical degree from the University of Oregon Medical School. As a medical student he was familiar with our research on pigment cells. He came to Yale to do a residency in dermatology in our Department and to take part in the isolation of melatonin. Jim Case was a brilliant student. He left Yale to become head of the Dermatology Division at the University of Washington in Seattle. His death at an early age was a blow to all of us.

M. RUTH WRIGHT, M.D. During the Second World War, Dr Wright was a medical student at the University of Manchester in England. After 2 years, she was sent to Vanderbilt Medical School to complete her medical training. She and Dr Mori did the assays for melatonin after Dr Takahashi left.

JACK BARCHAS, M.D. As a medical student at Yale during the time that the melatonin project was going on, Dr Barchas did his work on a thesis with me. He was interested in neurochemistry and carried out research on the presence of melatonin in peripheral nerves. Dr Barchas became Co-chairman of the Department of Psychiatry at Stanford Medical School where he studied the process of sleep. He went on to become Dean of Neurosciences at UCLA and now is Professor and Chairman of the Department of Psychiatry at Cornell Medical School.

A reunion with Aaron Lerner in 1969 of previous fellows and their wives. Standing (left to right): Mrs Mori; Dr Nobu Hattori, Professor of Oncology; Dr Wataru Mori, Professor of Pathology; Dr Yoshiyata Takahashi, Professor of Medicine; Mrs Takahashi; Dr Kazuo Shizume, Professor of Medicine. Seated (left to right): Dr Taro Kawamura, Professor of Dermatology; Dr Marguerite Lerner; Dr Aaron Lerner; Mrs Kawamura; and Mrs Shizume.

KATE LEHMANN came from a well-to-do family in Germany. She was told that women did not need an education because they would have support from their husbands. She was an exceptionally able and alert person who would have received professional training under other circumstances. Her dedication to keeping our glassware uncontaminated so that the assays could be carried out was a critical and difficult task.

Afterword

The Wenner-Gren Center Foundation for Scientific Research

By DAVID OTTOSON

A VISITOR to Stockholm coming from the airport and entering the city from the North cannot escape from seeing a high building of modern architecture surrounded by a lower semicircular apartment house. This is the Wenner-Gren Center. It was created in 1955 through a donation by a wealthy Swedish industrialist Axel Wenner-Gren. He was a self-made man who started from nothing to become one of the richest men in Sweden and in Europe at this time. He started by selling vacuum-cleaners and was so successful in doing this that he after a few years established the now well-known company Elektrolux. He expanded his activities to a number of countries and by skilled investments soon created a world-wide industrial empire. At the same time, being a smart business man he was also an idealist and something of a dreamer. One of his dreams was to create an international science center where scientists could come together to exchange ideas and feel at home. He desired this center to also provide accommodation for their families so as not to be only a meeting point but also a place where scientists could stay for longer periods of time. He dreamed of such centers being established in various countries but he only saw his dream realized only in Sweden with the creation of the Wenner-Gren Center.

The center consists of three buildings named the Pylon, the Helicon and the Tetragon. The Pylon, 24 storeys high, provides space for administrative departments of many of Sweden's research organizations, the Helicon living quarters and departments for visiting scientists and their families and the Tetragon, an auditorium for conferences.

Within the framework of its scientific activities the Wenner-Gren Center Foundation is annually organizing 6–8 international and national conferences covering a variety of themes, but mainly in the field of neuroscience. The proceedings of these conferences are published in the Wenner-Gren International Series. The present volume is the 63rd in this series. In addition to the symposium programme the Foundation has developed an extensive fellowship programme for foreign scientists doing research in Sweden and for Swedish scientists going abroad for research work.

In 1992, 82 foreign scientists from 30 countries received fellowships for work in Sweden and 26 Swedish scientists for work abroad. The Foundation is also providing travel fellowships for young scientists to attend conferences abroad.

The Foundation has through these various programmes established itself as an important body for support of science in Sweden and the Wenner-Gren Center has been recognized as a truly international science center, thereby fulfilling the dream of its creator, Axel Wenner-Gren.

SUBJECT INDEX